FOOD SCIENCE

Food Science Texts Series

Series Editor
Dennis R. Heldman, University of Missouri

Editorial Board

Norman N. Potter and Joseph H. Hotchkiss *Food Science, 5th Edition*

Forthcoming Series Titles
Cameron Hackney, Merle D. Pierson and George J. Banwart,
 Basic Food Microbiology, 3rd Edition (1996)

Dennis R. Heldman and Richard W. Hartel, *Principles of Food
 Processing* (1996)

Hildegarde Heyman and Harry T. Lawless, *Sensory Evaluation of Food* (1997)

Other Chapman & Hall Texts Currently Available
James M. Jay, *Modern Food Microbiology, 4th Edition*
Ceirwyn S. James, *Analytical Chemistry of Foods*
Norman G. Marriot, *Principles of Food Sanitation, 3rd Edition*
A. M. Pearson and T. A. Gillett, *Processed Meats, 3rd Edition*
Syed S. Rizvi and Gauri S. Mittal, *Experimental Methods in Food Engineering*
Louis J. Ronsivalli and Ernest R. Viera, *Elementary Food Science, 3rd Edition*
Romeo T. Toledo, *Fundamentals of Food Process Engineering, 2nd Edition*

FOOD SCIENCE
FIFTH EDITION

NORMAN N. POTTER
JOSEPH H. HOTCHKISS

CHAPMAN & HALL

THOMSON PUBLISHING

New York • Albany • Bonn • Boston • Cincinnati • Detroit • London • Madrid • Melbourne
Mexico City • Pacific Grove • Paris • San Francisco • Singapore • Tokyo • Toronto • Washington

Cover design: Andrea Meyer, emDASH inc.

Copyright © 1995
Chapman & Hall

Printed in the United States of America

For more information, contact:

Chapman & Hall
115 Fifth Avenue
New York, NY 10003

Thomas Nelson Australia
102 Dodds Street
South Melbourne, 3205
Victoria, Australia

Nelson Canada
1120 Birchmount Road
Scarborough, Ontario
Canada, M1K 5G4

International Thomson Editores
Campos Eliseos 385, Piso 7
Col. Polanco
11560 Mexico D.F. Mexico

Chapman & Hall
2-6 Boundary Row
London SE1 8HN
England

Chapman & Hall GmbH
Postfach 100 263
D-69442 Weinheim
Germany

International Thomson Publishing Asia
221 Henderson Road #05-10
Henderson Building
Singapore 0315

International Thomson Publishing-Japan
Hirakawacho-cho Kyowa Building, 3F
1-2-1 Hirakawacho-cho
Chiyoda-ku, 102 Tokyo
Japan

    2 3 4 5 6 7 8 9 10 XXX 01 00 99 97 96 95

Library of Congress Cataloging-in-Publication Data

Potter, Norman N.
        Food science. —5th ed. / Norman N. Potter and Joseph H. Hotchkiss.
            p.   cm.
        Rev. ed of : Food Science / Norman N. Potter, 4th ed. 1986.
        Includes bibliographical references and index.
        ISBN 0-412-06451-0 (cloth : alk.paper)
        1. Food industry and trade.   I. Hotchkiss, Joseph H.   II. Title.
    TP370.P58   1995
    664—dc20   95-16000
                                                                CIP

British Library Cataloguing in Publication Data available

Please send your order for this or any Chapman & Hall book to **Chapman & Hall, 29 West 35th Street, New York, NY 10001, Attn: Customer Service Department.** You may also call our Order Department at 1-212-244-3336 or fax your purchase order to 1-800-248-4724.

For a complete listing of Chapman & Hall's titles, send your requests to **Chapman & Hall, Dept. BC,**

To our dear families whose support and encouragement make all things seem possible

Contents

Preface

It has been nearly 30 years since the first edition of *Food Science* was published. It and the subsequent three editions have enjoyed worldwide use as introductory texts for curriculums in food science and technology. This favorable response has encouraged us to adhere to the same basic format and objectives of the previous editions. Our goal is to provide readers with an introductory foundation in food science and technology upon which more advanced and specialized knowledge can be built. We are also aware that the book is widely used as a basic reference outside the academic environment. The fifth edition has been substantially updated and expanded where new information exists or was needed. The fifth edition continues to be aimed primarily at those with little or no previous instruction in food science and technology. The text introduces and surveys the broad and complex interrelationships among food ingredients, processing, packaging, distribution, and storage and explores how these factors influence food quality and safety. Foods are complex mixtures of mostly biochemicals and the number of methods available to convert raw agricultural commodities into edible foods is almost endless. It was not our intent to be comprehensive but rather to address the need for insight and appreciation of the basic components of foods and the processes most commonly used in food technology. We also hope to provide insight into the scope of food science for people considering food science as a career. As with previous editions, this one should continue to serve as a reference for professionals in food-related fields that service, regulate, or otherwise interface with food science and technology.

Food science and technology, like many other science-based disciplines, has advanced rapidly since the fourth edition was published in 1986. Although many of the basic unit operations have changed little, new knowledge and concerns about biotechnology and foods, food safety, environmental issues, packaging technologies, government regulation, globalization of foods, nutrition, and others, as well as new processing technologies such as ohmic heating and supercritical fluid extraction, have emerged. Many of the changes and additions to the fifth edition of *Food Science* reflect these and other developments which increasingly influence all involved in food processing as well as government agencies around the world. However, true change can only be measured against the broad principles and conventional food production practices of proven value. Therefore, most basic principles and practices continue to be described at an appropriate introductory level in the fifth edition.

We would like to acknowledge our colleagues at Cornell University and elsewhere who provided much of the insights and materials for this edition. We are indebted to Mrs. Terry Fowler for her technical assistance with the production of this text.

Joseph H. Hotchkiss
Norman N. Potter
Ithaca, New York

FOOD SCIENCE

1

Introduction: Food Science as a Discipline

Food Science can be defined as the application of the basic sciences and engineering to study the fundamental physical, chemical, and biochemical nature of foods and the principles of food processing. Food technology is the use of the information generated by food science in the selection, preservation, processing, packaging, and distribution, as it affects the consumption of safe, nutritious and wholesome food. As such, food science is a broad discipline which contains within it many specializations such as in food microbiology, food engineering, and food chemistry. Because food interacts directly with people, some food scientists are also interested in the psychology of food choice. These individuals work with the sensory properties of foods. Food engineers deal with the conversion of raw agricultural products such as wheat into more finished food products such as flour or baked goods. Food processing contains many of the same elements as chemical and mechanical engineering. Virtually all foods are derived from living cells. Thus, foods are for the most part composed of "edible biochemicals," and so biochemists often work with foods to understand how processing or storage might chemically affect foods and their biochemistry. Likewise, nutritionists are involved in food manufacture to ensure that foods maintain their expected nutritional content. Other food scientists work for the government in order to ensure that the foods we buy are safe, wholesome, and honestly represented.

At one time, the majority of scientists, technologists, and production personnel in the food field did not receive formal training in food science as it is recognized today. This was because very few universities offered a curriculum leading to a degree in food science. Many of these institutions had departments that were organized along commodity lines such as meats or dairy products. The food industry, government, and academic institutions continue to employ many persons who received their original technical training in dairy science, meat science, cereal chemistry, pomology, vegetable crops, and horticulture. Many others were trained as specialists in the basic sciences and applied fields of chemistry, physics, microbiology, statistics, and engineering. Such training has had the advantages generally associated with specialization. It also has resulted in certain limitations, especially for commodity-oriented individuals in segments of the food industry undergoing rapid technological change. Hence, the more general discipline of food science was established. Now, more than 40 universities in the United States and many more around the world offer degrees in food science.

PREPARATION FOR A CAREER IN FOOD SCIENCE

Industry and academic specialists have often differed about the definition of the term *food scientist,* and what should constitute appropriate formal training. Similarly, the major schools offering a degree in food science have not always agreed on the requirements for such a degree. The Education Committee of the Institute of Food Technologists (IFT) adopted a set of minimum standards for a university undergraduate curriculum in food science. These standards are followed by most universities which offer degrees in food science and reflect the scientific nature of food science. The most recent (1992) recommended minimum standards include both basic science courses and core food science and technology courses for the B.S. degree. The standards are based on a 120-semester-hour or 180-quarter-hour requirement for graduation. Courses should carry three to five semester hours or four to eight quarter hours of credit.

The core of food science and technology courses, representing a minimum of 24 semester hours or 36 quarter hours, includes the following, most of which include both lecture and laboratory components:

- *Food Chemistry* covers the basic composition, structure, and properties of foods and the chemistry of changes occurring during processing and utilization. Prerequisites should be courses in general chemistry, organic chemistry, and biochemistry.
- *Food Analysis* deals with the principles, methods, and techniques necessary for quantitative physical and chemical analyses of food products and ingredients. The analyses should be related to the standards and regulations for food processing. Prerequisites include courses in chemistry and one course in food chemistry.
- *Food Microbiology* is the study of the microbial ecology related to foods, the effect of environment on food spoilage and food manufacture, the physical, chemical, and biological destruction of microorganisms in foods, the microbiological examination of food stuffs, and public health and sanitation microbiology. One course in general microbiology is the prerequisite.
- *Food Processing* covers general characteristics of raw food materials; principles of food preservation, processing factors which influence quality, packaging, water and waste management, and good manufacturing practices and sanitation procedures.
- *Food Engineering* involves study of engineering concepts and unit operations used in food processing. Engineering principles should include material and energy balances, thermodynamics, fluid flow, and heat and mass transfer. Prerequisites should be one course in physics and two in calculus.

A senior-level "capstone" course that incorporates and unifies the principles of food chemistry, food microbiology, food engineering, food processing, nutrition, sensory analysis, and statistics should be taught after the other food science courses. The specific orientation of this course, that is, whether it's product development or product processing is left to the discretion of the university.

These courses are considered minimal. Additional required and optional courses should be integrated into the curriculum. Courses in computer science, food law and regulation, sensory analysis, toxicology, biotechnology, food physical chemistry, advanced food engineering, quality management, waste management, advanced food processing, and so on are important components of a food science program.

In addition to the core courses in food science and technology, other typical requirements for a food science degree include the following:

- Two courses in general chemistry followed by one each in organic chemistry and biochemistry.
- One course in general biology and one course in general microbiology which has both lecture and laboratory.
- One course dealing with the elements of nutrition.
- Two courses in calculus.
- One course in statistics.
- One course in general physics.
- A minimum of two courses which emphasize speaking and writing skills.
- Courses in the humanities and social sciences. This requirement is usually established by the college or university. In the absence of such requirements, about four courses may be selected from history, economics, government, literature, sociology, philosophy, psychology, or fine arts.

The above minimum requirements provide sound undergraduate training in the field of food science. The terms food scientist and food technologist are both commonly used and have caused some confusion. It has been suggested in the past that the term *food technologist* be used to describe those with a B.S. degree and the term *food scientist* be reserved primarily for those with an M.S. or Ph.D. degree as well as research competence. This distinction, however, is not definitive and both terms continue to be used widely and interchangeably.

ACTIVITIES OF FOOD SCIENTISTS

The educational requirements for a food science degree still fall short of an adequate description of food science. Some suggest that food science covers all aspects of food material production, handling, processing, distribution, marketing, and final consumption. Others would limit food science to the properties of food materials and their relation to processing and wholesomeness. The later view imposes serious limitations if it fails to recognize that the properties of food materials can be greatly influenced by such factors of raw material production as amount of rainfall, type of soil, degree of soil fertilization, genetic characteristics, methods of harvest or slaughter, and so on. At the other end, cultural and religious dictates and psychological acceptance factors determine the end use of a product.

Psychology and sociology prove important in an affluent society where there is choice, as well as in other areas where customs and taboos sometimes are responsible for malnutrition although there may be no shortage of essential nutrients. Since definitions can be misleading, the activities of today's food scientists can be illustrated by way of examples.

It has been estimated that as many as 2 billion people do not have enough to eat and that perhaps as many as 40,000 die every day from diseases related to inadequate diets, including the lack of sufficient food, protein, and/or specific nutrients. Many food scientists are engaged in developing palatable, nutritious, low-cost foods. Inadequate nutrition in extreme cases can produce in children an advanced state of protein deficiency known as kwashiorkor, or the more widespread protein–calorie malnutrition leading to marasmus. Dried milk can supply the needed calories and protein but is relatively expensive and is not readily digested by all. Fish "flour" prepared from fish of species not commonly eaten can be a cheaper source of protein. Incaparina, a cereal formulation containing about 28% protein, is prepared from a mixture of maize, sorghum, and cottonseed flour. Incaparina and similar products were developed to utilize low-cost crops grown in Central and South America. Miltone was developed from ingredients—peanut protein, hydrolyzed starch syrup, and cow or buffalo milk—that are readily available in India. As food losses during storage and processing can be enormous, food scientists are involved in adapting and developing preservation methods appropriate and affordable to various regions of the world.

Food scientists have developed thousands of food products including those used in the space shuttle program (Fig. 1.1). The first astronauts added a small quantity of water to dehydrated foods in a special pouch, kneaded the container, and consumed the food through a tube. They had to deal with space and weight limitations, little refrigeration and cooking equipment, special dietary requirements dictated by stress and physical inactivity, and weightlessness. There was concern that crumbs or liquid might get loose in the spacecraft and become a hazard. Currently, food scientists are developing systems which "recycle" foods for space voyages into deeper space. If astronauts are to be in space for extended periods without resupply, foods will have to be grown and processed in space. The problems inherent in such systems present unique challenges to the food scientist.

Perhaps the largest single activity of food scientists working in industrial organizations is the improvement of existing and development of new food products (Fig. 1.2). In the United States in 1993, there were over 12,000 new products introduced if one considers all products "new" even if they are simply a standard product with only a slight change. Consumers like to have new products available. Industrial food scientists

Figure 1.1. An astronaut consuming food aboard the space shuttle. *Courtesy of the Institute of Food Technologists.*

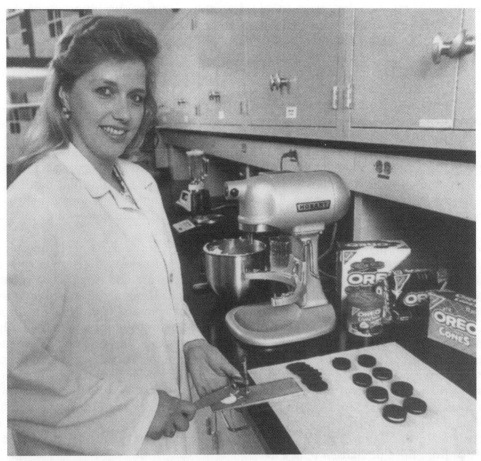

Figure 1.2. A food scientist works "at the bench" to optimize the formulation of a cookie product. *Courtesy of the Institute of Food Technologists.*

must find creative ways to meet this consumer demand for new and different products. Successful product development requires a blend of science and creativity.

Food scientists today are often involved in altering the nutrient content of foods, particularly reducing the caloric content or adding vitamins or minerals. Reducing the caloric content is accomplished in several ways, such as replacing caloric food components with low or non-nutritive components. The caloric content of soft drinks is reduced by replacing the nutritive sugar sweeteners (e.g., sucrose) with aspartame or saccharin. Aspartame goes by the trade name Nutrasweet. Aspartame contains the same number of calories as sugar but is 200 times sweeter, so much less is used for the same degree of sweetening, thus reducing the caloric content. In other cases, food scientists reduce the caloric content of fat containing foods by replacing the fat with substances which have similar properties but are not metabolized in the same way as fat. For example, low-fat ice cream can be made by removing the normal milk fat and adding specially treated proteins. These proteins are made into very small particles which give ice cream the smooth texture associated with the fat. Protein has four calories per gram, whereas fat has nine. Thus, the net effect is a decrease in the caloric content of the ice cream.

Food scientists also find ways to add desirable vitamins and minerals to foods. Breakfast cereals are good examples of such foods. Most cereals have some added nutrients and some have a whole day's supply of several nutrients. These vitamins and minerals must be added in such a way as to be evenly dispersed in the product and be stable. They must not adversely affect the flavor or appearance of the food. This requires considerable care.

Food processing technology is applied in the design and operation of ships which process fish at sea. These facilities include automatic separators for small and large fish, mechanized fish-cooling tanks, automatic oil extractors, ice-making equipment, complete canning factories, equipment for preparing fish fillets and cakes, and equipment for dehydrating fish and preparing dried fish meal. This factory approach prevents spoilage and minimizes protein and fat losses that otherwise would limit how long a fishing vessel could remain at sea. These factories can remain at sea for 2 months or more and range great distances from their home base. The Japanese and Russians have been most active in this kind of development.

An important application of food science technology is the controlled-atmosphere (CA) storage of fruits and vegetables. Fruits such as apples, after they are harvested, still have living respiring systems. They continue to mature and ripen. They require oxygen from the air for this continued respiration, which ultimately results in softening and breakdown. If the air is depleted of much of its oxygen and is enriched in carbon dioxide, respiration is slowed. For some fruits, the best storage atmosphere is one that contains about 3% oxygen and about 2–5% carbon dioxide, the rest being nitrogen. Such atmospheres are produced commercially by automatic controls which sample the atmosphere continuously and readjust it. Refrigerated warehouses using CA storage permit year-round sale of apples which previously was not possible due to storage deterioration. Low-oxygen CA storage is also currently used to preserve lettuce quality during refrigerated truck transport and to reduce spoilage of strawberries during air shipment. In the latter case, air in the storage compartment is displaced by carbon dioxide from the sublimation of dry ice.

Food science includes quick freezing of delicate foods with liquid nitrogen, liquid or solid carbon dioxide, or other low-temperature (cryogenic) liquids. When fruits and vegetables are frozen, ice crystals form within and between the cells which make up the pulpy tissue. If freezing is slow, large ice crystals form which can rupture the cell walls. When such a product is thawed, the pulp becomes soft and mushy and liquid drains from the tissue. The tomato is particularly susceptible. Under very rapid freezing conditions, as may be obtained with liquid nitrogen at $-196°C$, minute crystals form and the cellular structure is frozen before it can be ruptured; when thawed, the product retains its original appearance and texture much better than does material subjected to slow freezing. Nevertheless, even with such rapid freezing, only selected types of tomatoes can withstand freezing and thawing satisfactorily. However, less fragile frozen plant and animal foods of increasing variety owe their current commercial excellence to cryogenic freezing.

One of the most important goals of the food scientist is to make food as safe as possible. The judicious application of food processing, storage, and preservation methods helps prevent outbreaks of food poisoning. Food poisoning is defined as the occurrence of disease or illness resulting from the consumption of food. Food-borne diseases are caused by either pathogenic (i.e., disease causing) bacteria, viruses, parasites, or chemical contaminants. The incidence of such food-borne illnesses in the United States is higher than many think. According to the Center for Infectious Diseases, between

1983 and 1987 there were 91,678 confirmed cases. This probably only represents a small portion of the actual cases because of the strict criteria for classifying cases and the underreporting of cases where only one or two individuals were affected. Approximately 92% of these cases were due to pathogenic bacteria. However, processed food was implicated in only a tiny fraction of these cases.

The largest causes of outbreaks are improper food preparation, handling, and storage, most occurring in homes, institutions, or in restaurants. For example, in 1993, hamburgers containing undercooked ground beef were served in a fast-food restaurant resulting in several deaths. The causative bacteria was a type of *Escherichia coli* bacteria known as O157:H7 which has been associated with raw beef and other products. The number of cases of salmonellosis is also on the increase (Fig 1.3). The major cause of this increase is thought to be due to serving undercooked eggs and poultry contaminated with *Salmonella enteritidis*. Salmonella was found to be the cause of one of the largest food-borne disease outbreaks recorded. Approximately 16,000 people became ill from consuming contaminated milk. This single episode caused a large increase in the number of reported cases (Fig 1.3). Similar increases in botulism can occur due to large outbreaks (Fig 1.4).

As pointed out earlier, the majority of food poisoning cases are due to mishandling and not to errors in processing. The processed food industry has a outstanding record of preventing such mishaps when it is considered that billions of cans, jars, and pouches of food materials are consumed annually. Occasionally, however, this excellent record has been broken by a limited outbreak in which persons succumb to toxic food. This may occur when canned foods are not heated sufficiently to destroy the spores of the anaerobic bacterium *Clostridium botulinum* or susceptible products are not stored properly. For example, in the United States in 1989, three cases of botulism were due to consumption of garlic-in-oil. The product was not sufficiently acidified nor

Figure 1.3. Cases of salmonellosis in the United States reported to the Centers for Disease Control. Source: *Morbidity and Mortality Weekly Report,* 42:50. *1994.*

BOTULISM (foodborne) — by year, United States, 1975–1993

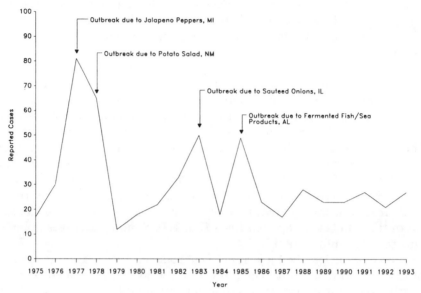

Figure 1.4. Cases of food-borne botulism in the United States reported to the Centers for Disease Control. Source: *Morbidity and Mortality Weekly Report, 42:24. 1994.*

refrigerated to prevent toxin formation. Also in 1989 in Great Britain, 27 people were diagnosed with botulism with one death from consumption of yogurt which contained a hazelnut preserve. Botulinum toxin had apparently developed in the hazelnut preserve before being added to the yogurt. Fish has been implicated in a number of botulism cases, most often because of a lack of knowledge by food handlers of the hazards associated with seafoods. An outbreak of botulism in Kapchunka is an example of how such misunderstanding can be lethal. Kapchunka is a whitefish which is soaked in a salt brine and air-dried without being eviscerated. The fish was then packaged and held at room temperature. *C. botulinum* in the intestinal tract of the fish grew and produced toxin, giving botulism to several individuals, one of whom died. Food scientists carefully study each outbreak in the hope of preventing future outbreaks.

When foods are heated to destroy pathogens and spoilage organisms, other changes in food components can affect color, texture, flavor, and nutrient values; thus, food scientists must optimize heat processes for specific products to be effective but not excessive. Sometimes pathogens enter food through faulty containers; this was the cause of major recalls of canned salmon contaminated with *C. botulinum*.

Food science researchers have developed methods for tenderizing beef. For example, the application of tenderizing enzyme–salt mixtures to the surface of meat cuts is common household practice. However, commercial tenderization has gone further. Proteolytic enzymes may be injected into the animal shortly before it is slaughtered so that the pumping action of the heart circulates the tenderizer throughout the tissues. When the animal is slaughtered and the meat cuts are prepared, they are more tender than they otherwise would be. Another approach to tenderization is the application of electric current to the carcass after slaughter.

Food scientists are studying the modification of beef muscle composition and properties through special feeding practices. The Japanese have produced beef of exceptionally high quality by the inclusion of beer in the ration, together with controlled exercise. In the United States and other countries, until recently, it was common practice to use the hormone diethylstilbestrol (DES) as a feed adjunct or as an animal implant to stimulate animal growth and reduce feed costs. This compound also slightly increases the moisture, protein, and ash, and decreases the fat levels of beef muscle and lamb. However, under certain conditions DES can cause cancer in mice and in humans, and in 1979, the U.S. Food and Drug Administration prohibited further use of DES in meat production.

Food scientists are also studying the production of milk by cows fed minimum synthetic diets, and the organoleptic and functional properties of this milk. It has long been known that many of the nutritional requirements of cows are met by microbial synthesis of complex compounds in the cow's rumen from simpler materials. Finnish scientists have shown that cows can maintain high levels of milk production on minimum rations containing purified carbohydrate and no protein, the nitrogen source being supplied by inexpensive urea and ammonium salts. Milk from cows thus fed is quite normal in gross composition, amino acid constitution of its protein, flavor properties, and functional characteristics. These findings are of particular significance since they provide a means for converting low-value cellulosic materials such as forest products and low-cost nitrogen compounds into valuable animal protein of a kind that is highly acceptable to humans.

Food scientists are working on the production of flavors by specific enzyme systems acting on basic raw material substrates. Cooked meat flavors have thus been produced from fats, and fruity flavors from carbohydrates.

Food scientists are beginning to use new techniques and products emerging from the fields of genetic engineering and biotechnology. Advances in recombinant DNA technology and related methods are providing improved microbial strains and new enzymes to increase yields and cut costs in the fermentation industries. For example, enzymes are required to coagulate the milk proteins for use in the manufacture of cheese. These enzymes have been isolated form natural sources. However, this limits availability and consistency. The genes for these enzymes have been cloned into bacteria which then biosynthesize the enzyme in a fermentation vessel. The enzyme is purified and used in cheese manufacture.

In addition to countless potential applications in the manufacture of such fermented foods as cheese, bread, wine, beer, sauerkraut, and sausages, the biotechnology is being applied in the production of vitamins, amino acids, flavors, colors, and other food ingredients. Improved cultures and enzymes also are being used by food scientists to convert cellulose and starch to a variety of sweeteners and to convert other substrates, when supplemented with nitrogen, into edible protein.

Food scientists always must be concerned with availability and cost of raw materials. Bakers' yeast is commonly grown on molasses; however, the price of molasses, which is a by-product of sugar manufacture, has increased in recent years because much sugar cane has been diverted to the production of alcohol for fuel use. Although other sources of carbohydrates have been used or modified for growing yeast, their costs fluctuate, they lack certain trace minerals and vitamins present in molasses and needed by yeast, and they tend to produce yeast with altered properties.

Food scientists are also studying the removal of ions from liquid foods through the combined use of selective membranes and electric current, a process known as electrodialysis. Special membranes can be made that allow passage of cations but restrict movement of anions. Others permit movement of anions but hold back cations.

Such membranes can be assembled into compartmented "stacks" which are connected to the anode and cathode of an electric circuit. Ions in liquids passing through the compartments tend to migrate to the poles of opposite charge, provided they are not rejected by a specific selective membrane. By proper choice of membranes and stack construction, certain anions, cations, or combinations of both may be removed from food liquids. In this fashion, tart fruit juices can be deacidified. Other membrane separation techniques (e.g., ultrafiltration and reverse osmosis) can separate proteins, sugars, and salts from liquid mixtures and can thus concentrate and change ratios of food constituents. These techniques are being used in new cheese-making processes and to fractionate valuable ingredients from whey, grain steep liquors, and other food liquids.

Today, many food production plants depend on computers in their daily operation. The heart of such automatic plants—making baked goods, frankfurters, margarine, ice cream, and dozens of other products—is a computer command center, such as that shown in Fig. 1.5. All formulas are calculated in advance and the metering of ingredients to mixers, ovens, freezers, and other equipment is controlled by microprocessors. Based on daily changes in the cost of formula ingredients, ratios of ingredients as well as operating conditions can be altered quickly by keyboard reprogramming. However, behind this automation is the skill of the formulator, food scientist, and the quality control laboratory.

Food scientists have become increasingly concerned with the safety of foods, as they may be affected by pathogenic microorganisms, toxic chemicals such as pesticides, and other environmental contaminants such as bits of glass or metal. In an industrialized society, concentrations of chemicals entering the environment must be monitored and

Figure 1.5. Food scientist at computerized controls in a large food processing plant.

limited, but since their complete elimination is impossible, foods may be expected to contain traces of "impurities," as do the air we breathe and the water we drink. Reports of unexpected chemicals in foods increase as the sensitivities of analytical instruments exceed nanogram-detection levels. This makes it imperative to know more about the toxicology of substances that may find their way into food, yet be harmless at low levels. In addition, more must be learned about the concentration of specific chemicals in the higher levels of the food chain. Thus, food scientists are analyzing meat, milk, and eggs from livestock fed crops grown near industrial sites. One interesting study involves potentially harmful levels of heavy metals in meat and eggs from poultry raised on feed containing ground mussels which, in turn, were grown in tanks of seawater on phytoplankton.

Food scientists are currently investigating the fate of pesticides during food processing. Their objective is to know if processes can be developed which eliminate or reduce pesticide residues and how this may affect food safety. They are also investigating, in conjunction with toxicologists, both naturally occurring as well as synthetic toxicants in foods. How these toxicants enter foods and how they might be eliminated are important research topics.

Food scientists are involved in establishing international food standards to promote and facilitate world trade and at the same time to assure the wholesomeness and value of foods purchased between nations. Standards generally cover ingredient composition, microbiological purity, and subjective quality factors that often are not agreed upon universally. Wherever possible, standards also must not be discriminatory against one nation or another. This sometimes poses highly perplexing problems. Many countries would agree that Cheddar cheese to be called by this name should be made from cow's milk and contain a moisture content of not more than 39%. In India, however, much of the milk comes from the buffalo. Further, a Cheddar-type cheese made from buffalo milk is of poor texture unless somewhat more moisture is retained. Thus, an international Cheddar cheese standard upholding cow's milk as the raw material source, and a maximum of 39% moisture, could be to the disadvantage of India were such a product produced for export. Similar problems are currently being studied for a wide variety of food items by international committees.

Food scientists work in conjunction with nutritionists to develop standards for the optimal nutritional content of the diet and to determine how food processing and storage affects nutrients. One important aspect of this is to investigate how food formulation affects the bioavailability of nutrients. For example, ascorbic acid (vitamin C) can increase the bioavailability of iron in the diet. Other food scientists study how storage affects nutritional content of foods. For example, storing milk in clear containers under display case lights can reduce the vitamin content of this important food.

As these examples show, food science is involved in many technical and scientific aspects of food. The activities of food scientists are involved in diverse areas with the common theme of food. Almost limitless other examples might have been chosen: food preservation by irradiation; freeze concentration to remove water without loss of volatiles; the use of chemical additives to enhance the physical, chemical, and nutritional properties of foods; mechanical deboning to increase the yields of flesh from red meat, poultry, and fish; quick cooking methods using infrared, dielectric, or microwave energy; optimization of processes to maximize nutrient retention and minimize energy expenditures; and the development of information meaningful to the public and essential to the creation of relevant, coherent food law.

All of these areas, and a great many more referred to in subsequent chapters, provide daily problems for food scientists. Together they help convey a better understanding of the term food science than can any simple definition.

References

Altschul, A.M. 1993. Low-Calorie Food Handbook. Marcel Dekker, New York.

Bauernfeind, J.C. and Lachance, P.A. 1991. Nutrient Additions to Food: Nutritional, Technological and Regulatory Aspects. Food & Nutrition Press, Trumbull, CT.

Catsberg, C.M.E. and Kempen-Van Dommelin, G.J.M. 1989. Food Handbook. E. Halsted Press, New York.

Charalambous, G. 1992. Food Science and Human Nutrition. Elsevier, Amsterdam.

Hall, R.L. 1992. Global Challenges for Food Science Students. Food Technol. *46*(9), 92, 94, 96.

Harlander, S. 1989. Introduction to Biotechnology. Food Technol. *43*(7), 44, 46, 48.

Hood, L.F. 1988. The Role of Food Science and Technology in the Food and Agriculture System. Food Technol. *42*(9), 130–132, 134.

Hui, Y.H. 1992. Encyclopedia of Food Science and Technology. Wiley, New York.

Institute of Food Technologists. 1992. IFT Undergraduate Curriculum Minimum Standards for Degrees in Food Science 1992 Revision. Institute of Food Technologists, Chicago.

Levine, A.S. 1990. Food Systems: The Relationship Between Health and Food Science/Technology. E.H.P. Environ. Health Perspect. *86*, 233–238.

Morton, I.D. and Lenges, J. 1992. Education and Training in Food Science. A Changing Scene. E. Horwood, New York.

O'Brien, J. 1991. An Overview of Online Information Resources for Food Research. Trends Food Sci. Technol. *2*(11), 301–304.

Ockerman, H.W. 1991. Food Science Sourcebook. 2nd ed. Chapman & Hall, London, New York.

Patlak, M. 1991. Looking Ahead to the Promises of Food Science in the Future. News Rep. *61*(2), 13–15.

Rayner, L. 1990. Dictionary of Foods and Food Processes. Food Science Publishers, Kenley, Surrey, England.

Ronsivalli, L.J. and Vieira, E.R., 1992. Elementary Food Science. 3rd ed. Chapman & Hall, London, New York.

Shewfelt, R.L. and Prussia, S.E. 1993. Postharvest Handling: A Systems Approach. Academic Press, San Diego, CA.

2

Characteristics of the Food Industry

Regardless of the criteria used to measure it, the food industry is very large, and if food production, manufacturing, marketing, restaurants and institutions are combined, it is the largest industrial enterprise in the United States. An appreciation of the size, components, interrelationships, and responsiveness to change of the food industry enlarges food scientists' understanding of the environment in which they operate.

The total U.S. food-producing system—including the agricultural sector, food processing and marketing functions, and supporting industrial activities—generates about 20% of the gross national product and draws on close to one-fourth of the work force. This is greater than the combined efforts employed in the steel, automobile, chemical manufacturing, communications, public utilities, mining, and several other industries.

The food-producing industry grows, processes, transports, and distributes our foodstuffs. Approximately, 3 million people work on farms, in orchards, on ranches, in fishing, and other areas and are directly involved in the production of the raw food materials that are subsequently processed into foods. As the chain of activities proceeds, people are engaged in such functions as produce buying, cattle feeding, dairy plant and grain elevator operations, and warehouse management. Food processing (manufacturing) operations convert raw agricultural commodities into canned, frozen, dehydrated, fermented, formulated, and otherwise modified forms of food. According to the *Survey of Current Business*, approximately 1.7 million people were directly employed in food manufacturing alone in 1991 (not counting related businesses). These people earned approximately $44 billion. Transportation of foodstuffs by rail, truck, water, and air, and the associated warehousing, employs an additional 2 million people. Wholesale distribution firms that keep retail outlets stocked are estimated to involve about 700,000 people. Retail distribution, including private stores, chain stores, and supermarkets account for an additional 2 million employees. Restaurants, drive-ins, cafeterias in hospitals, plants, and schools, vending machines, airline feeding, and other foodservice operations utilize about 5 million more people. Technically trained personnel serving the food industry in state, federal, and industrial positions of research, development, and quality control number over 25,000. Several million more persons from scores of occupations contribute directly or indirectly to the food production process.

The amount of food produced in the United States is enormous. For example, the production of cheese in 1991 was 6.1 billion pounds; wheat 74 million metric tons; poultry 24 billion pounds, meat 39 billion pounds, and eggs 188 million cases. The food industry is also very large when considered on the basis of sales. In 1980, consumers

Table 2.1. Percent of Total Personal
Consumption Expenditures Spent for Food
in Selected Countries in 1988

Country	Percent of Total
Phillipines	52
China	48
Korea	36
Greece	35
Portugal	33
Mexico	32
Former USSR	28
Israel	27
Japan	19
France	17
United Kingdom	14
Canada	12
United States	10

SOURCE: World Food Expenditures, National Food Review *12*(4) 26–29

spent $264 billion on food; in 1991, the figure was over $492 billion and rising. Interestingly, this is less than 12% of total disposable income. According to figures from the United States Department of Agriculture (USDA), this gives the United States the cheapest food supply in the world (Table 2.1). Nowhere else can people feed themselves as adequately on less than one-eighth of their disposable income. The cost of food when compared to disposable income has actually fallen from over 17% in 1960. The low cost of food was not always the case and certainly is not representative of less developed parts of the world where the majority of one's labor may be required to produce or purchase adequate food. In China, for example, it is estimated that more than 48% of disposable income must go for food.

The abundance of food in the United States is not due to more people, animals, and land on today's farms, but is the result of increased efficiency in agricultural operations resulting from application of advances in science and technology. Since 1920, cropland has not significantly increased, but production per hectare has gone up sharply; total numbers of livestock have changed very little, but production per animal has steadily risen. This increased production results from greater uses of fertilizers, agricultural chemicals, animal and plant genetic improvements, and farm mechanization. Each year more food for more people is being produced by fewer farmers. Thus, in 1940 one farmer supplied the food for about 12 people, in 1960 about 28 people, and presently one farmer can produce the food for approximately 80 people.

This has had both negative and positive consequences over the years. During the 1950s and 1960s, U.S. farmers produced large food surpluses, which lowered prices. This was alleviated by subsidizing farmers to limit their planting and production of certain commodities, by providing food aid to less developed areas of the world, and by developing processing methods to convert surplus foods into different forms to create new markets. Examples of the latter were conversion of surplus wheat into bulgur (a form of boiled wheat that can be dehydrated and consumed as a cereal) or peeled wheat, which acquires the appearance and some of the eating qualities of rice. Further examples were the conversion of perishable commodities such as sweet potatoes, apples, and milk to more stable dehydrated forms, either for export or for use in other manufactured foods. But surpluses have largely disappeared in recent years due to an increasing demand for food worldwide, and U.S. grain is being sought to feed the peoples of less developed countries as well as to fatten the livestock of the more affluent.

The food supply in the United States is marked not only by its quantity but also by its quality, variety, and convenience. Figure 2.1 shows how food was sold in large cities in the early 1900s. Fresh produce underwent heavy spoilage; variety was lacking; what variety there was, was seasonal, and the consequences of unsanitary handling were all too common. In contrast, today's large "superstores" stock as many as 50,000 individual food and nonfood items, and the battle is for shelf space and preferred position. The use of scanners and computers has helped the retail industry manage these large stores. This variety is largely the work of the food processor who may manufacture a basic food in 20 or more forms to entice the food purchaser. The many flavors of yogurt is one example.

COMPONENTS OF THE FOOD INDUSTRY

The food industry may be divided into segments, or components, in various ways. One of the simplest is a functional division into the four major segments of raw material

Figure 2.1. Big city marketing of food in the early 1890s. *Courtesy of U.S. Department of Agriculture.*

production, manufacture, distribution, and marketing. Raw material production encompasses the technologies of farming, orchard management, fishing, and so on, including the selection of plant and animal varieties, cultivation and growth, harvest and slaughter, and the storage and handling of the raw materials. Manufacturing converts the raw agricultural products into more refined or finished foods. Manufacturing includes the numerous unit operations and processes that many consider to be the core of food technology. Distribution is involved in product form, weight and bulk, storage requirements and storage stability, and product attributes conducive to product sales. Marketing is the selling of foods in commerce and involves wholesale, retail, institutions, and restaurants. This overall division is artificial and the segments flow into one another. The food industry is so geared that there is a highly planned organization and rhythm to the functions of the segments. In a well-developed food industry, this involves planning and scheduling of all phases to eliminate or at least minimize both shortages and surpluses among farmer, manufacturer, and distributor. Thus, it is common for large companies to own and manage farms or plantations, processing and distribution facilities, and even the outlets for sale of their manufactured products to ensure smooth operations and high profits. In recent years, for example, many food manufactures have opened national restaurant chains.

If the industry is divided according to function, then it is sometimes helpful to know the relative value of the different functions. However, this is not simple to determine because there are great differences between products, as analyses of the percentage of the consumer's dollar spent for production, processing, transportation, and selling of different foods has shown. Currently, in the case of beef, the greatest cost is in farm production and the smallest cost is in processing and packaging. In contrast, for canned tomatoes, the biggest cost is in processing and packaging, and farm production represents one of the smaller costs.

A more common way of dividing the food industry is along major product lines. Table 2.2 gives the per capita dollar spent in 1990 on major food categories as well as the per capita consumption of these foods. Thus, of the consumer dollar spent for all foods to be consumed at home in 1990, about $0.27 went for meat, poultry, fish, and egg products, about $0.16 for fresh and processed fruits and vegetables, about $0.12 for dairy products, $0.15 for cereal products, and $0.30 for other products such as edible oils. These values do not necessarily reflect the tonnage or per capita consumption of each of the food categories, as the cost of a unit of each food differs greatly. Of all dollars spent for food, about 42% were spent on food consumed away from home.

Food is most often consumed in a different form than that in which it is produced. For example, less than half of the 148 billion pounds of milk produced in the United States in 1991 was consumed as fluid milk. Approximately 17% went into the manufacture of butter, over 31% into cheese, about 9% into ice cream and other frozen desserts, and so on. As the yield is only about 4 kg of butter per 100 kg of milk, and about 10 kg of cheese per 100 kg of milk, this amounted to some 0.5 million tons of butter, and approximately 2.3 million tons of cheese. Other examples of commodity conversions include the transformation of cereal grains into breakfast cereals, soybeans into edible oils, and cereal starches into sugar syrups. One way to define the food industry is to say that it converts or changes raw agricultural commodities into more finished foods.

Americans have increased their consumption of fish and shellfish over the last decade but still consume less than in many other countries. Consumption in 1990 was approximately 15 pounds per person. This is 27% higher than it was in 1970–1974. The largest increases have come in fresh and frozen products, whereas canned fish

Table 2.2. Per Capita Food Spending and Selected Consumption Figures

	Dollars/Person		Consumption (Pounds per Capita)	
	1986	1990	1980–1984	1990
Food expenditures	1326	1652	—	—
Food at home	767	956	—	—
Cereals and bakery products	106	142	148	185
cereals and cereal products	36	50	—	—
bakery products	70	92	—	—
Meats, poultry, fish	216	257	182	191
beef	73	84	73	64
pork	45	51	48	46
other meats	30	38	—	—
poultry	33	42	45	64
fish and seafood	25	32	—	—
eggs	12	12	34	30
Dairy products	97	113	559	568
fresh milk and cream	47	54	—	—
other dairy products	49	60	—	—
Fruit and vegetables	123	157	—	—
fresh fruit	39	49	87	95
fresh vegetables	35	45	76	93
processed fruit	28	36	—	—
processed vegetables	21	27	—	—
Other food at home	213	287		
sugar and other sweets	28	36	148	—
fats and oils	20	26	64	62
miscellaneous foods	91	129	—	—
nonalcoholic beverages	74	82	—	—
Foods away from home	560	697	—	—
Alcoholic beverages	104	113	—	—

SOURCE: ERS-USDA 1991. Food and Nutrient Consumption. FoodReview *14*(3) 2–18.

rose only 11%. Lesser amounts are consumed in salted and smoked form. In 1991 the total domestic catch of fishery products for human consumption was 9.5 billion pounds of which 6.5 billion was for the fresh or frozen market, with 0.6 billion being canned. During the same period, the United States imported an additional 3 billion pounds.

Since everything about the food industry is big, it is to be expected that the number of people employed in food processing and the number of food processing plants are also large. The USDA estimated that there were 380,000 processing, wholesale, and retail firms in the United States in 1990. The estimated number of food processing plants with more than 20 employees in the United States is more than 20,000. The trend is for this number to become smaller and the remaining plants to become larger. This has concerned some who point out monopolistic tendencies in specific branches of the food industry. But a counterinfluence is a growing trend toward diversification within large food companies, including involvement in nonfood ventures. Small food processing establishments, with fewer than 20 employees, and retail outlets may number as many as 500,000. According to the USDA, in 1988 there were over 3 million people employed in the food processing and marketing and related industries. This number has declined somewhat over the recent years and is not expected to increase in the near future. By comparison, there were only 3 million people directly employed in farming and related activities in 1989.

Currently in the United States, nearly one-half of the dollars spent on food goes for food eaten away from home. The trend of increased spending for food eaten away from home has leveled off in recent years, but the general trend has been upward for the last few decades. Actually, the quantity of food consumed outside of the home is less than one-third that eaten at home, the dollar ratio representing higher prices for food away from home. This food away from home is consumed in restaurants, industrial and school cafeterias, hospitals, airplanes, and through vending machines and other outlets. In 1992 the USDA estimated that there were over 730,000 foodservice outlets in the United States. This included restaurants, fast-food outlets, schools, institutions, and cafeterias.

These types of foodservice have their own specific requirements for ease of preparation and serving speed, providing new challenges and opportunities for convenience-food manufacturers, suppliers of packaging materials, makers of quick cooking and reconstituting equipment, catering establishments, and an ever-increasing variety of carryout and fast-food franchise chains. Many companies have broadened their activities in response to these changes. Grocery stores have added restaurants and food processing facilities while food manufacturers have acquired restaurant chains and some have become direct marketers of their products.

ALLIED INDUSTRIES

Many companies which may not directly sell food are nonetheless deeply involved in the food industry. These are the companies that produce the nonfood components that are essential to the marketing of food. A good example is the packaging industry. For example, steel manufacturers make materials for the billions of cans used for food each year. They have worked in depth on the corrosive effects and interactions of different foods with the metals used in the manufacture of cans. They have supported extensive research on improved types of cans where the gauge of the metal may be reduced, thus lightweighting the cans and reducing costs. The same is true of the leading aluminum companies in the development of aluminum cans, aluminum dishes, and foil for food use. Other examples of companies which supply ingredients to the food industry include food ingredients such as color or flavor suppliers (Fig 2.2). The study of can closure, can closure machines, and heat transfer into cans for sterilization has kept food scientists and engineers busy. Companies do extensive research and development in food packaging in glass, paper, and plastic. Tinted glass that screens out ultraviolet rays and thus protects light-sensitive foods and plastic films that provide maximum moisture and oxygen barriers and resistance to heating and freezing are studied by scientists in these companies. Recent advancements in polymers have led to the development of many new types of plastic packages for foods such as those which will withstand the temperatures generated in the microwave oven or the pressures generated during retorting foods in plastic cans. These new technologies have reduced costs and provided many new products and more convenient foods.

Chemical manufacturers are important in the food industry because they supply many of the acidulants, preservatives, enzymes, stabilizers, and other chemicals used

Figure 2.2. A flavor chemist compounding flavors for a food. *Food Processing Magazine,* 55(5)15. *1994.*

in foods. All of these must be functional and fully satisfy specifications of safety set down by the Food & Drug Administration (FDA) or other regulatory agencies that have responsibility for food safey.

Food machinery and equipment manufacturers often are the prime innovators of new food processing methods and systems. They have developed pasteurizers and evaporators, microwave ovens and infrared cookers, freeze-drying systems and liquid-nitrogen freezers, and instrumentation and computer controls.

All of these, and many, many more companies in allied industries work directly with food.

In recent years, especially, all industries have become more accountable to government, consumers, their employees, and each other. This has taken many forms, such as providing the public with more information, assuming greater responsibility for the quality of the environment and the safety of products, conserving resources, and meeting increasingly rigorous government regulations. In the case of the food industry, this trend has led to greater dependence on outside consultants, testing laboratories, and legal expertise. Thus, more of these people are becoming involved in the food production process.

INTERNATIONAL ACTIVITIES

Food has become a global commodity. Foods are traded and shipped worldwide (Fig. 2.3). It is not unusual to find dozens of types of fine foods from around the world in a modern grocery store. This might include cheeses from Europe, lamb from New Zealand, fresh grapes from Chile, snowpeas from Guatemala, apples from Argentina, beef from Australia, and mangoes from South America. Many U.S. food companies have set up subsidiaries in other countries and many fast-food outlets such as McDonald's have opened stores around the world. The largest McDonald's is reported to be in Moscow.

Agricultural imports (food and other products) to the United States amounted to about $22 billion in 1991 which amounted to approximately 10% of all imports. These have included coffee, tea, cocoa, spices, and other products not grown in this country, as well as sugar, fish, and other products to supplement domestic production. Food exports were about $37 billion in 1991 and have grown due largely to increased world demand for cereal grains and soybeans, making the United States the world's largest food exporter. Most major American food companies have vigorous international divisions with manufacturing facilities in many parts of the world. Among those with extensive overseas operations are Kraft–General Foods, CPC International, H.J. Heinz, Borden, Campbell Soup, Nabisco Brands, Coca-Cola, Pepsico, Beatrice Companies,

Figure 2.3. Large cargo container ship transporting goods internationally. Source: *Handbook of Package Engineering*, J. P. Hanlon, Technomic Publishing Co. Lancaster, PA *1992*.

Ralston Purina, and General Mills. There is a recent trend to decrease the trade tariffs on many items including food. This can be expected to increase the international trade in food items. It is common for a U.S. grocery store to stock food items from around the world.

When these companies go into food production in foreign countries they do not simply build a plant and resume operations as in the United States. Experience has shown that often they must modify well-known products to suit local tastes. Even the most popular soft drink formulations may vary in certain parts of the world. Another problem facing companies entering operations in a new country is related to available food ingredients. In some countries, the food producer may not import certain essential or important ingredients but must utilize local ingredients such as wheat or cocoa. These may differ substantially from equivalent ingredients used in the United States and thus require extensive reformulation and process changes to achieve acceptable quality. This is further complicated by local food laws, which often prohibit the use of specific food acidifiers, preservatives, or food colors that are permitted in the United States.

RESPONSIVENESS TO CHANGE

It is often said that the food industry is a stable industry, resistant to the effects of recessions. In the sense that per capita consumption of total food is remarkably constant, and for many decades has remained at around 658 kg (1450 pounds) per year, this is true. However, the kind of foods consumed are continually changing and this contributes to great competition within the industry and makes it highly dynamic.

People choose the foods they eat in response to many influences. The food industry responds to these choices or more often tries to anticipate these changes so that they can have the desirable products. For example, changes in food use reflect demographic shifts such as the increasing numbers of older Americans, working women, and single-member households. The food products available in the marketplace also reflect the supply of ingredients and nonfood components, which are subject to extremes of weather, political barriers, and changing world demand. Availability and costs of energy influence all phases of food production. Advances in the areas of nutrition, health, and food safety often change eating habits whether the benefits that may result are real or just perceived.

According to the U.S. Department of Agriculture, between 1968 and 1988 U.S. residents increased their daily consumption of kilocalories from 3300 to 3600. However, the kinds of foods making up this consumption has changed substantially. For example, fresh and frozen fruits consumption has increased 25% over a the last 20 years. Attitudes with respect to fat, cholesterol, and fiber contents of foods have changed. During the same period, red meat consumption has declined by 16%. Government regulation of food additives, food composition standards, and labeling also influence product offerings. Technical innovations—from ingredient modifications to new processing and packaging methods to microwave oven and other cooking advances—alter our food supply.

The above determine the directions of new product development, marketing, and advertising. New products in great variety are characteristic of the modern food industry. In 1991 there were over 12,000 new food products introduced into commerce in the United States (Table 2.3). Currently, there are over 50,000 items representing

Food Science

Table 2.3. Numbers of New Food Products Introduced Between 1988 and 1993

Category	1988	1989	1990	1991	1992	1993
Baby food	55	53	31	95	53	7
Bakery products	968	1,155	1,239	1,631	1,508	1,420
Baking ingredients	212	233	307	335	346	383
Beverages	936	913	1,143	1,367	1,538	1,845
Breakfast cereal	97	118	123	108	122	99
Candy/gum/snacks	1,310	1,355	1,486	1,885	2,068	2,042
Condiments	1,608	1,701	2,028	2,787	2,555	3,148
Dairy	854	1,348	1,327	1,111	1,320	1,099
Desserts	39	69	43	124	93	158
Entrees	613	694	753	808	698	631
Fruits and vegetables	262	214	325	356	276	407
Pet food	100	126	130	202	179	276
Processed meat	548	509	663	798	785	454
Side dishes	402	489	538	530	560	680
Soup	179	215	159	265	211	248
Total, food	**8,183**	**9,192**	**10,301**	**12,398**	**12,312**	**12,897**

SOURCE: Gorman's New Product News, *29*(1) 1994.

different products, brand names, and sizes in large U.S. food stores, including as many as 3000 pet food items. Many of these products have short life spans.

INTERRELATED OPERATIONS

As has been stated, the production of specific foods in a highly advanced and organized food industry is a systematic and rhythmic process. The food manufacturer does not simply decide to produce 5000 tons of margarine and then casually do so. If they did, they might find themselves, at one end, unable to procure the necessary vegetable oils at a competitive price and, at the other, without a ready and adequate outlet for the product. These factors alone could make them unsuccessful in the highly competitive food field, where often fractions of a cent per kilogram or per package make the difference between economic success or failure. Throughout all production, manufacturing, and distribution operations, these fractions of cents per unit of food product are carefully controlled along with the quality aspects of the product. Since the food industry is a low-markup, high-volume industry, and numbers like several hundred thousand units per day—such as cartons of milk or loaves of bread—are common for a single plant, losses of fractions of a cent per unit anywhere along the chain from farmer to consumer can mean losses to the food producer of hundreds of thousands of dollars per year.

References

Caswell, J.A. and Preston, W.P. 1992. How new products find their place in the marketing system. *In* The Yearbook of agriculture. U.S. Department of Agriculture, Washington, DC, pp. 9–14.

Chou, M. 1992. Trends in food consumption during 1970–1990. Cereal Foods World *37*(4), 331–333.

Connor, J.M. and Barkema, A.D. 1992. Changing food marketing systems. *In* The Yearbook of Agriculture. U.S. Department of Agriculture, Washington, DC, pp. 15–21.

Donald, J.R. 1992. U.S. agricultural outlook. *In* Agriculture Outlook. U.S. Department of Agriculture, Washington, DC, pp. 84–106.

Donald, J.R. 1993. U.S. agricultural outlook. *In* Outlook. U.S. Department of Agriculture, Washington, DC, pp. 9–35.

Gady, R. 1993. Long term outlook and opportunities in food manufacturing. *In* Outlook. U.S. Department of Agriculture, Washington, DC, pp. 671–678.

Rankin, M.D. and Kill, R.C. 1993. Food Industries Manual. 23rd ed., Chapman & Hall, London, New York.

Ritson, C., Gofton, L. and McKenzie, J. 1986. The Food Consumer. Wiley, Chichester.

Russo, D.M. 1992. The year 2000: A food industry forecast. Agribusiness *8*(6), 493–506.

Senauer, B., Asp, E. and Kinsey, J. 1991. Food Trends and the Changing Consumer. Eagan Press, St. Paul, MN.

Stauber, K.N. 1992. Agricultural outlook '94. *In* Agriculture Outlook. U.S. Department of Agriculture, Washington, DC, pp. 32–36.

Uri, N.D. 1992. Industry structure and economic performance in the food manufacturing industries. J. Int. Food Agribus. Mark *4*(1), 95–123.

USDA. Economic Research Service. 1991. Food Review. USDA ERS Commodity Economics Division, U.S. GPO, Rockville, MD. *14*(1).

3

Constituents of Foods: Properties and Significance

A knowledge of the constituents of foods and their properties is central to food science. The advanced student of food science, grounded in the basic disciplines of organic chemistry, physical chemistry, and biochemistry, can visualize the properties and reactions between food constituents on a molecular basis. The beginning student is not yet so equipped. This chapter, therefore, will be more concerned with some of the general properties of important food constituents, and how these underlie practices of food science and technology.

Foods are made up mostly of biochemicals (i.e., edible biochemicals) which are mainly derived from living sources such as plants and animals. There are three main groups of constituents in foods: carbohydrates, proteins, and fats, and derivatives of these. In addition, there are inorganic and mineral components, and a diverse group of organic substances present in comparatively small proportions that include such substances as vitamins, enzymes, emulsifiers, acids, oxidants, antioxidants, pigments, and flavors. There is also the ever-present and very important constituent, water. These components are arranged in different foods to give the foods their structure, texture, flavor, color, and nutritive value. In some instances, foods also contain substances that can be toxic if consumed in large amounts. The general composition of a food as well as the way in which the components are organized give a food its individual characteristics. For example, whole milk and fresh apples have about the same water content, but one is a solid and the other a fluid because of the way the components are arranged.

The above constituents occur in foods naturally. Sometimes we are not satisfied with the structure, texture, flavor, color, nutritive value, or keeping quality of foods, and so we add other materials to foods to improve one or more properties. These may be natural or synthetic. For example, we may add natural or synthetic fruit flavors to beverages.

CARBOHYDRATES

Carbohydrates (from "hydrates of carbon") are organic compounds with the basic structure $C_x(H_2O)_y$. Among the most important types of carbohydrates in foods are the sugars, dextrins, starches, celluloses, hemicelluloses, pectins, and certain gums. Chemically, carbohydrates contain only the elements carbon, hydrogen, and oxygen.

Simple carbohydrates are called sugars. One of the simplest carbohydrates is the six-carbon sugar glucose. Glucose and other simple sugars form ring structures of the following form:

α-D-glucose   α-D-mannose   α-D-galactose

These simple sugars each contain 6 carbon atoms, 12 hydrogen atoms, and 6 oxygen atoms $[C_x(H_2O)_y$ where $x=6$; $y=6$]. They differ in the positions of oxygen and hydrogen around the ring. These differences in the arrangement of the elements result in differences in the solubility, sweetness, rates of fermentation by microorganisms, and other properties of these sugars.

Two glucose units may be linked together with the splitting out of a molecule of water. The result is the formation of a molecule of a disaccharide, in this case maltose:

Common disaccharides formed in similar fashion are sucrose (e.g., cane or beet sugar) made from glucose and fructose (a five-membered ring), maltose or malt sugar from two molecules of glucose, and lactose or milk sugar from glucose and galactose. These disaccharides also differ from one another in solubility, sweetness, susceptibility to fermentation, and other properties.

A larger number of glucose units may be linked together in polymer fashion to form polysaccharides (i.e., "many sugars"). One such polysaccharide is amylose, an important component of plant starches (Fig. 3.1). A chain of glucose units linked together in a slightly different way forms cellulose.

Thus, the simple sugars are the building blocks of the more complex polysaccharides, the disaccharides and trisaccharides, the dextrins, which are intermediate in chain length, on up to the starches, celluloses, and hemicelluloses; molecules of these latter substances may contain several hundred or more simple sugar units. Chemical derivatives of the simple sugars linked together in long chains likewise yield the pectins and carbohydrate gums.

The disaccharides, dextrins, starches, celluloses, hemicelluloses, pectins, and carbohydrate gums are composed of simple sugars, or their derivatives. Therefore, they can be broken down or hydrolyzed into smaller units, including their simple sugars. Such breakdown in the case of amylose, a straight chain fraction of starch, or amylopectin, a branched chain fraction (Fig. 3.1), yields dextrins of varying intermediate chain length, the disaccharide maltose, and the monosaccharide glucose. This breakdown or digestion can be accomplished with acid or by specific enzymes, which are biological catalysts. Microorganisms, germinating grain, and animals including humans possess various such enzymes.

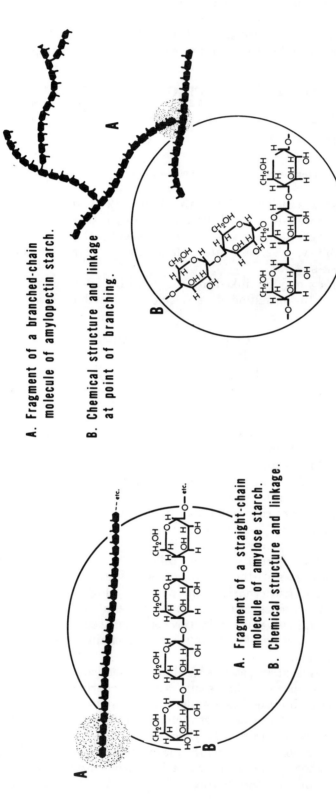

A. Fragment of a branched-chain molecule of amylopectin starch.

B. Chemical structure and linkage at point of branching.

A. Fragment of a straight-chain molecule of amylose starch.

B. Chemical structure and linkage.

Figure 3.1. Straight chain amylose and branched chain amylopectin fractions of starch. *Courtesy of Northern Regional Research Laboratory.*

The chemically reactive groups of sugars are the hydroxyl groups (—OH) around the ring structure, and when the ring is opened, the

$$-C\overset{\displaystyle O}{\underset{\displaystyle H}{\Big\backslash}} \text{ (aldehyde group) and the } -C\overset{\displaystyle O}{\Big\backslash} \text{ (ketone group).}$$

Sugars that possess free aldehyde or ketone groups are known as reducing sugars. All monosaccharides are reducing sugars. When two or more monosaccharides are linked together through their aldehyde or ketone groups so that these reducing groups are not free, the sugar is nonreducing. The disaccharide maltose is a reducing sugar; the disaccharide sucrose is a nonreducing sugar. Reducing sugars particularly can react with other food constituents, such as the amino acids of proteins, to form compounds that affect the color, flavor, and other properties of foods. In like fashion, the reactive groups of long-chain sugar polymers can combine in a cross-linking fashion. In this case the long chains can align and form fibers, films, and three-dimensional gellike networks. This is the basis for the production of edible films from starch as a unique coating and packaging material.

Carbohydrates play a major role in biological systems and in foods. They are produced by photosynthesis in green plants and are nature's way of storing energy from sunlight. They may serve as structural components as in the case of cellulose, be stored as energy reserves as in the case of starch in plants and liver glycogen in animals, and function as essential components of nucleic acids as in the case of ribose, and as components of vitamins such as the ribose of riboflavin. Carbohydrates can be oxidized to furnish energy. Glucose in the blood is a ready source of energy for animals. Fermentation of carbohydrates by yeast and other microorganisms can yield carbon dioxide, alcohol, organic acids, and a host of other compounds.

Some Properties of Sugars

Such sugars as glucose, fructose, maltose, sucrose, and lactose all share the following characteristics in varying degrees: (1) They are usually used for their sweetness; (2) they are soluble in water and readily form syrups; (3) they form crystals when water is evaporated from their solutions (this is the way sucrose is recovered from sugar cane juice); (4) they supply energy; (5) they are readily fermented by microorganisms; (6) they prevent the growth of microorganisms in high concentration, so they may be used as a preservative; (7) they darken in color or caramelize on heating; (8) some of them combine with proteins to give dark colors, known as the browning reaction; and (9) they give body and mouth feel to solutions in addition to sweetness.

A very important advance in sugar technology has been the development of enzymatic processes for the conversion of glucose to its isomer, fructose. Fructose is sweeter than glucose or sucrose. This has made possible the production of sugar syrups with the sweetness and certain other properties of sucrose starting from starch. Commonly, corn starch is hydrolyzed to provide the glucose, which is then isomerized. The United States produces enormous quantities of corn and with this technology has become less dependent on imported sucrose, the availability and price of which can fluctuate greatly.

Some Properties of Starches

The starches important in foods are primarily of plant origin and exhibit the following properties: (1) They are not sweet; (2) they are not readily soluble in cold water; (3) they form pastes and gels in hot water; (4) they provide a reserve energy source in plants and supply energy in nutrition; (5) they occur in seeds and tubers as characteristic starch granules (Fig. 3.2). When a suspension of starch granules in water is heated, the granules swell due to water uptake and gelatinize; this increases the viscosity of the suspension and, finally, a paste is formed which, on cooling, can form a gel. Because of their viscosity, starch pastes are used to thicken foods, and starch gels, which can be modified by sugar or acid, are used in puddings. Both pastes and gels can revert or retrograde back to the insoluble form on freezing or ageing, causing changes in food texture. Partial breakdown of starches yields dextrins, which are intermediate in chain length between starches and sugars and exhibit other properties intermediate between these two classes of compounds.

In recent years much has been learned about modifying the properties of natural starches by physical and chemical means. This has greatly increased the range of uses for starch as a food ingredient, especially with respect to controlling the texture of food systems and permitting the manufacture of food items that require minimum heating to achieve desired viscosity. This technology has been used to make such products as instant puddings which do not require cooking.

Modification techniques include reduction of a starch's viscosity by chemically or enzymatically breaking the molecules at the glucosidic linkages or by oxidation of some of the hydroxyl groups. The swelling properties of starch heated in water also

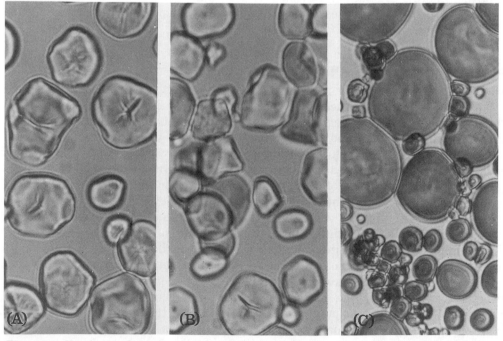

Figure 3.2. Ungelatinized starch granules. (A) Sorghum; (B) corn; (C) wheat. *Courtesy of Northern Regional Research Laboratory.*

can be slowed down by cross-linking reagents that react with hydroxyl groups on adjacent starch molecules to form chemical bridges between linear chains. The viscosity of such cross-linked starch also is less likely to break down in acid foods and at high temperatures as in cooking and canning. Starch further may be modified by reacting its hydroxyl groups with a range of reagents that form ester, ether, acetal, and other derivatives. A major effect of this type of modification is to interfere with the tendency of linear molecules to associate or retrograde to the insoluble form on freezing and ageing. Starch granules also may be precooked to produce a starch that will swell in cold water.

Some Properties of Celluloses and Hemicelluloses

Celluloses and hemicelluloses, which are abundant in the plant kingdom and act primarily as supporting structures in plant tissues, are relatively resistant to breakdown. They are insoluble in cold and hot water and are not digested by man, so do not yield energy. They are important, however, as dietary fiber. Long cellulose chains may be held together in bundles forming fibers, as in cotton and flax (Fig. 3.3); such structures make celery "stringy" and are often ruptured by the growth of ice crystals when vegetables such as lettuce are frozen. The fiber in food that produces necessary dietary roughage is largely cellulose, and the hard parts of coffee beans and nut shells contain cellulose and hemicellulose. These materials can be broken down to glucose units by certain enzymes and microorganisms. For example, cellulose from plants and from waste paper can be enzymatically converted to glucose, supplemented with nitrogen, and used for the growth of yeast and other microorganisms as an animal feed supplement or as a source of protein for humans.

Some Properties of Pectins and Carbohydrate Gums

Pectins and carbohydrate gums—sugar derivatives usually present in plants in lesser amounts than other carbohydrates—exhibit the following characteristics: (1) Like starches and celluloses, pectins are made up of chains of repeating units (but the units are sugar acids rather than simple sugars); (2) pectins are common in fruits and vegetables and are gumlike (they are found in and between cell walls and help hold

Figure 3.3. Electron micrograph (15,000×) showing cellulose fibers from plant cell walls. *Courtesy of R.D. Preston.*

the plant cells together); (3) pectins are soluble in water, especially hot water; (4) pectins in colloidal solution contribute viscosity to tomato paste and stabilize the fine particles in orange juice from settling out; (5) pectins in solution form gels when sugar and acid are added and this is the basis of jelly manufacture. Other carbohydrate gums from plants include gum arabic, gum karaya, and gum tragacanth (seaweeds yield the gums agar-agar, carageenan, and algin). In addition to their natural occurrence, pectins and gums are added to foods as thickeners and stabilizers.

PROTEINS

Proteins are made by linking individual amino acids together in long chains. Amino acids are made up principally of carbon, hydrogen, oxygen, and nitrogen. Some amino acids also contain other elements such as sulfur.

Proteins are essential to all life. In animals they help form supporting and protective structures such as cartilage, skin, nails, hair, and muscle. They are major constituents of enzymes, antibodies, many hormones, and body fluids such as blood, milk, and egg white.

Typical amino acids have the following chemical formulas:

$$\begin{array}{c} CH_3 \\ \diagdown \\ CH_3 \diagup \end{array} CHCH_2 \underset{\underset{NH_2}{|}}{CHCOOH}$$

leucine

$$CH_2 \underset{\underset{NH_2}{|}}{CH_2} CH_2 CH_2 \underset{\underset{NH_2}{|}}{CHCOOH}$$

lysine

$$\begin{array}{c} CH_3 CH_2 \\ \diagdown \\ CH_3 \diagup \end{array} CH \underset{\underset{NH_2}{|}}{CHCOOH}$$

isoleucine

$$\begin{array}{c} CH_3 \\ \diagdown \\ CH_3 \diagup \end{array} CH \underset{\underset{NH_2}{|}}{CHCOOH}$$

valine

Amino acids have the —NH_2 or amino group, and the —COOH or carboxyl group attached to the same carbon atom. These groups are chemically active and can combine with acids, bases, and a wide range of other reagents. The amino and carboxyl groups themselves are basic and acidic, respectively; the amino group of one amino acid readily combines with the carboxyl group of another. The result is the elimination of a molecule of water and formation of a peptide bond, which has the following chemical representation:

$$H_2N - \underset{\underset{H}{|}}{\overset{\overset{H}{|}}{C}} - \overset{\overset{O}{\|}}{C} - \underset{\underset{H}{|}}{N} - \underset{\underset{}{|}}{\overset{\overset{}{|}}{C}} - COOH$$

In this case, where two amino acids have reacted, a dipeptide is formed, with the peptide bond at the center. The remaining free amino and carboxyl groups at the ends can react in like fashion with other amino acids forming polypeptides. These and other reactive groups on the chains of different amino acids can enter into a wide range of reactions with many other food constituents. There are 20 different major amino acids and a few minor ones that make up human tissues, blood proteins, hormones, and enzymes. Eight of these are designated essential amino acids since they cannot be synthesized by humans in adequate amounts to sustain growth and health and must be supplied by the diet. The remaining amino acids also are necessary for health but can be synthesized by humans from other amino acids and nitrogenous compounds and so are designated as nonessential. The essential amino acids are leucine, isoleucine, lysine, methionine, phenylalanine, threonine, tryptophan, and valine. To this list of eight is added histidine to meet the demands of growth during childhood. The nonessential amino acids are alanine, arginine, aspartic acid, cysteine, cystine, glutamic acid, glycine, hydroxyproline, proline, serine, and tyrosine. The list of essential amino acids differs somewhat for other animal species.

The complexity of amino acid polymerization to form protein chains is illustrated in Fig. 3.4 for the protein of human hemoglobin. There is enormous opportunity for variation among proteins. This variation arises from combinations of different amino acids, from differences in the sequence of amino acids within a chain, and from differences in the shapes the chains assume. That is whether they are straight, coiled, or folded. These differences are largely responsible for the differences in the taste and texture of chicken muscle, beef muscle, and milk curd.

Protein chains can be oriented parallel to one another like the strands of a rope as in wool, hair, and the fibrous tissue of chicken breast. Or they can be randomly tangled like a tangled bunch of string. Thus, proteins taken from different foods such as egg, milk, and meat may have a very similar chemical analysis as to C, H, O, and N, and even with respect to their particular amino acids, yet contribute remarkably different structures to the foods containing them.

Further, the complex and subtle configuration of a protein can be readily changed, not only by chemical agents but by physical means. A given protein in solution can be converted to a gel or precipitate. This happens to egg white when it is coagulated by heat. Or the process can be reversed: a precipitate transformed to a gel or solution as in the case of dissolving animal hoofs with acid or alkali to make glue.

When the organized molecular or spatial configuration of a protein is disorganized, we say the protein is denatured. This can be done with heat, chemicals, excessive stirring of protein solutions, and acid or alkali. When egg white is heated, it becomes a solid rather than a liquid because it is irreversibly denatured.

These changes in food proteins are easily recognized in practice. When meat is heated, the protein chains shrink and so steak shrinks on cooking. When milk is coagulated by acid and heat, protein precipitates, forming cheese curd. If the heat or acid is excessive, the precipitated curd shrinks and becomes tough and rubbery.

Protein solutions can form films and this is why egg white can be whipped. The films hold entrapped air, but if you overwhip you denature the protein, the films break, and the foam collapses.

Like carbohydrate polymers, proteins can be broken down to yield intermediates of various sizes and properties. This can be accomplished with acids, alkalis, and enzymes. The products of protein degradation in order of decreasing size and complexity are protein, proteoses, peptones, polypeptides, peptides, amino acids, NH_3, and elemental nitrogen. In addition, highly odorous compounds, such as mercaptans, skatole, putrescine, and H_2S, can form during decomposition.

Controlled cheese ripening involves a desirable degree of protein breakdown. Putrefaction of meat is the result of excessive protein breakdown accompanying other changes. The deliberate and unavoidable changes in proteins during food processing

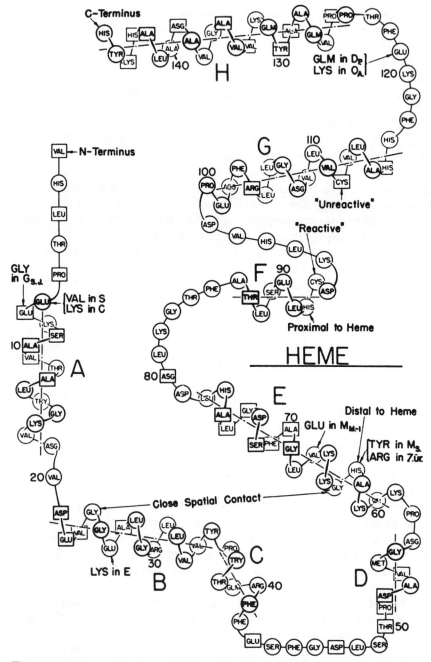

Figure 3.4. Diagram of amino acid chain of the human hemoglobin protein. *Courtesy of W.A. Schroeder.*

and handling are among the most interesting aspects of food science. Today, animal, vegetable, and microbial proteins are being extracted, modified, and incorporated into numerous manufactured food products. In addition to their nutritional value, they are selected for specific functional attributes including dispersibility, solubility, water sorption, viscosity, cohesion, elasticity, emulsifying effects, foamability, foam stability, and fiber formation. Additional properties of proteins are discussed in Chapter 4 dealing with nutrition, and in subsequent chapters.

FATS AND OILS

Fats differ from carbohydrates and proteins in that they are not polymers of repeating molecular units. They do not form long molecular chains, and they do not contribute structural strength to plant and animal tissues. Fats are smooth, greasy substances that are insoluble in water.

Fat is mainly a fuel source for the animal or plant in which it is found, or for the animal that eats it. It contains about 2¼ times the calories found in an equal dry weight of protein or carbohydrate. For this reason, reduction in the caloric content of foods is often accomplished by replacing fat with protein or carbohydrate. Fat always has other substances associated with it in natural foods, such as the fat-soluble vitamins A, D, E, and K; the sterols, cholesterol in animal fats and ergosterol in vegetable fats; and certain natural lipid emulsifiers designated phospholipids because of the presence of phosphoric acid in their molecules.

A typical fat molecule consists of glycerol combined with three fatty acids. Glycerol and butyric acid, a common fatty acid found in butter, have the following chemical formulas:

$$\begin{array}{l} H_2C-OH \\ | \\ HC-OH \qquad\qquad HOOC-CH_2-CH_2-CH_3 \\ | \qquad\qquad\qquad\qquad\quad \textbf{butyric acid} \\ H_2C-OH \\ \textbf{glycerol} \end{array}$$

Glycerol has three reactive hydroxyl groups, and fatty acids have one reactive carboxyl group. Therefore, three fatty acid molecules can combine with each glycerol molecule, eliminating three molecules of water. Such fats are called triglycerides.

There are about 20 different common fatty acids that are connected to glycerol in natural fats. These fatty acids differ in length and in the number of hydrogen atoms they contain. Formic acid ($HCOOH$), acetic acid (CH_3COOH), and propionic acid (CH_3CH_2COOH) are the shortest of the fatty acids. Stearic acid ($C_{17}H_{35}COOH$) is one of the longer common fatty acids. Some of the opportunities for variations in natural fats can be seen from the formula for a typical triglyceride:

$$\begin{array}{l} H_2C-O-C\overset{\displaystyle O}{\diagup}(CH_2)_{10}-CH_3 \\ | \qquad\qquad\quad \overset{\displaystyle O}{\diagup} \\ HC-O-C\overset{}{\diagup}(CH_2)_{16}-CH_3 \\ | \qquad\qquad\quad \overset{\displaystyle O}{\diagup} \\ H_2C-O-C\overset{}{\diagup}(CH_2)_7-CH=CH-(CH_2)_7-CH_3 \end{array}$$

In this case, the fatty acids reacting with glycerol from top to bottom are lauric acid, stearic acid, and oleic acid, with carbon chain lengths of 12, 18, and 18, respectively. Stearic and oleic acids, although of similar length, differ with respect to the number of hydrogen atoms in their chains. Stearic acid is said to be saturated with respect to hydrogen. Oleic acid with two fewer hydrogen atoms is said to be unsaturated. Another 18-carbon unsaturated fatty acid with four fewer hydrogen atoms and two points of unsaturation is linoleic acid. This unsaturated fatty acid is a dietary essential for health. The degree of unsaturation also affects the physical properties, such as melting temperature, of fats.

Fat molecules can differ with respect to the lengths of their fatty acids, the degree of unsaturation of their fatty acids, the position of specific fatty acids with respect to the three carbon atoms of glycerol, orientation in the chains of unsaturated fatty acids to produce spatial variations within these chains, and in still other ways.

Fat molecules need not have all three hydroxyl groups of glycerol reacted with fatty acids as in a triglyceride. When two are reacted, the molecule is known as a diglyceride; when glycerol combines with only one fatty acid molecule, the resulting fat is a monoglyceride. Diglycerides and monoglycerides have special emulsifying properties.

Natural fats are not made up of one type of fat molecule but are mixtures of many types, which may vary in any of the ways previously described. This complexity of fat chemistry today is well understood to the point where fats of very special properties are custom-produced and blended for specific food uses.

The chemical variations in fats lead to widely different functional, nutritional, and keeping-quality properties. The melting points of different fats are an example of this functional variation. The longer fatty acids yield harder fats, and the shorter fatty acids contribute to softer fats. Unsaturation of the fatty acids also contributes to softer fats. An oil is simply a fat that is liquid at room temperature. This is the basis of making solid fats from liquid oils. Hydrogen is added to saturate highly unsaturated fatty acids, a process known as hydrogenation. More will be said about changes in fat consistency in the chapter on fats and oils (Chapter 16).

Some additional properties of fats important in food technology are the following:

- They gradually soften on heating, that is, they do not have a sharp melting point. Since fats can be heated substantially above the boiling point of water, they can brown the surfaces of foods.
- When heated further, they first begin to smoke, then they flash, and then burn. The temperatures at which these occur are known as the smoke point, the flash point, and the fire point, respectively. This is important in commercial frying operations.
- Fats may become rancid when they react with oxygen or when the fatty acids are liberated from glycerol by enzymes.
- Fats form emulsions with water and air. Fat globules may be suspended in a large amount of water as in milk or cream, or water droplets may be suspended in a large amount of fat as in butter. Air may be trapped as an emulsion in fat as in butter-cream icing or in whipped butter.
- Fat is a lubricant in foods; that is, butter makes the swallowing of bread easier.
- Fat has shortening power; that is, it interlaces between protein and starch structures and makes them tear apart easily and short rather than allow them to stretch long. In this way, fat tenderizes meat as well as baked goods.
- Fats contribute characteristic flavors to foods and in small amounts produce a feeling of satiety or loss of hunger.

ADDITIONAL FOOD CONSTITUENTS

Whereas carbohydrates, proteins, and fats often are referred to as the major food constituents due to their presence in substantial amounts, there are other groups of substances which play in important role, out of proportion to their relatively small concentration in foods.

Natural Emulsifiers

Materials that keep fat globules dispersed in water or water droplets dispersed in fat are emulsifiers. Without emulsifiers, mayonnaise would separate into water and oil layers. The mayonnaise emulsion is stabilized by the presence of egg yolk, but the active ingredients in egg yolk stabilizing the emulsion are phospholipids, the best known of which is lecithin. There are many lecithins differing in their fatty acid contents. Chemically, a typical lecithin would have the following formula:

$$\begin{array}{l} H_2C-OOC-C_{17}H_{33} \\ | \\ HC-OOC-C_{17}H_{31} \\ | \qquad\qquad O \\ | \qquad\qquad \| \\ H_2C-O-P-O-CH_2CH_2N(CH_3)_3 \\ \qquad\qquad | \qquad\qquad\qquad + \\ \qquad\qquad O\ - \end{array}$$

Lecithins are structurally like fats but contain phosphoric acid. Most important, they have an electrically charged or polar end (the + and − at the bottom) and a noncharged or nonpolar end at the top. The polar end of this and similar molecules is water-loving or hydrophilic and easily dissolved in water. The uncharged or nonpolar end is fat-loving or hydrophobic and easily dissolved in fat or oil. The result in a water–oil mixture is that the emulsifier dissolves part of itself in water and the other part in oil. If the oil is shaken in an excess of water, the oil will form small droplets. Then the nonpolar ends of lecithin molecules orient themselves within the fat droplets and the polar ends stick out from the surface of the droplets into the water phase. This has the effect of surrounding the oil droplets with an electrically charged surface. Such droplets repel one another rather than having a tendency to coalesce and separate as an oil layer. The emulsion is thus stabilized. Such phenomena are common in foods containing oil and water. Lecithin and other phospholipid emulsifiers are present in animal and plant tissues and in egg, milk, and blood. Without them we could not have stable mayonnaise, margarine, or salad dressings. The monoglycerides and diglycerides mentioned earlier are also highly effective emulsifiers, as are certain proteins.

Emulsifiers belong to a broader group of chemicals known as surface active agents, designated as such because they exert their effects largely at surfaces. Today, a large number of natural and synthetic emulsifiers and emulsifier blends suitable for food use are available. Selection is based largely on the type of food system to be emulsified. With water and oil, one can have oil-in-water or water-in-oil emulsions. In an oil-in-water emulsion, water is the continuous phase and oil is the dispersed or discontinuous phase; mayonnaise is an example of this type of emulsion. In a water-in-oil emulsion, oil is the continuous phase and water is the dispersed phase; margarine is an example. Generally the phase present in greater amount becomes the continuous phase of the food system. In choosing an effective emulsifier for a manufactured food, oil-in-water emulsions are best stabilized with emulsifiers that have a high degree of water solubility (along with some oil solubility), whereas water-in-oil emulsions are best prepared with emulsifiers having considerable oil solubility and lesser water solubility.

Analogs and New Ingredients

In response to the desire to reduce the caloric content or improve the flavor of many foods, considerable effort has been directed at developing analogs of fat, sugar, and other food components. These analogs have the common objective of mimicking the functional properties such as flavor, mouthfeel, texture, and appearance of the indigenous components while at the same time reducing the caloric content of the food. Often these analogs are used to replace high-calorie sweeteners such as sugar or to replace fat. Replacing fat is especially desirable because as pointed out in Chapter 4, on an equal-weight basis, fat contains more than twice the calories of other food components.

The use of fat replacers in ice cream is a good example of the use of such analogs. Fat contributes smoothness, creaminess, and flavor to ice cream. A process has recently been developed where egg or milk proteins are formed into very small hard spheres. When suspended at the proper concentration in a liquid, these protein spheres also have a creamy or smooth texture. This is similar to chocolate in which the finely ground cocoa produces smoothness. These protein spheres can then replace some or all of the fat in ice cream or other products. Because protein on an equal-weight basis has fewer calories than fat, the net result is a reduction in caloric content of the ice cream.

Other analogs or substitutes have been developed. Fat substitutes which can be used as a cooking oil, yet are not absorbed by humans have been developed. This means that fried foods such as potato chips could have their caloric content substantially reduced. However, government approval of these substitutes is still pending. Aspartame, which is made up of amino acids, is an example of a sugar substitute. On an equal-weight basis, aspartame has about the same caloric content as sucrose but is 200 times sweeter. Therefore, less is used and the caloric content of the food is reduced.

Several challenges remain in the use of analogs. For example, analogs may mimic some important functions but often do not behave in exactly the same way as the food component they are substituting. Fat replacers may have the mouthfeel of fat but do not carry the fat-soluble flavors or vitamins of the natural fat. Sometimes, sugar substitutes can leave undesirable after tastes. Some fat or sugar substitutes are not heat stable and decompose when heated.

Organic Acids

Fruits contain natural acids, such as citric acid of oranges and lemons, malic acid of apples, and tartaric acid of grapes. These acids give the fruits tartness and slow down bacterial spoilage.

We deliberately ferment some foods with desirable bacteria to produce acids and thus give the food flavor and keeping quality. Examples are fermentation of cabbage to produce lactic acid and yield sauerkraut, and fermentation of apple juice to produce first alcohol and then acetic acid to obtain vinegar. In the manufacture of cheese, a bacterial starter culture is added to milk to produce lactic acid. This aids in curd formation and in the subsequent preservation of the curd against undesirable bacterial spoilage.

Besides imparting flavor and aiding in food preservation, organic acids have a wide range of textural effects in food systems due to their reactions with proteins, starches, pectins, gums, and other food constituents. The rubbery or crumbly condition of Cheddar cheese depends largely on acid concentration and pH, as does the stretchability of bread dough, the firmness of puddings, the viscosity of sugar syrups, the spreadability of jellies and jams, and the mouthfeel of certain beverages. Organic acids also influence the colors of foods, as many plant and animal pigments are natural pH indicators. Acids are also important inhibitors of bacterial spoilage in foods, particularly of bacteria which can cause human disease. For example, under anaerobic conditions and slightly above a pH of 4.6, *Clostridium botulinum* can grow and produce lethal toxin. *C. botulinum* does not grow in foods high enough in organic acids to have a pH of 4.6 or less, so it is not a hazard.

Oxidants and Antioxidants

Many food constituents are adversely affected by oxygen in the air. This is so of fats, oils, and oily flavor compounds which may become rancid on excessive exposure to air. Carotene, which yields vitamin A, and ascorbic acid, which is vitamin C, also are diminished in vitamin activity by oxygen. Oxygen is an oxidant; it causes oxidation of these materials. Oxygen is always present in and around foods, although it may be minimized by nitrogen or vacuum packaging.

Certain metals such as copper and iron are strong promoters or catalysts of oxidation. This is one of the reasons why copper and iron have largely been replaced in food processing equipment by stainless steel. Many natural foods, however, contain traces of copper and iron, but they also contain antioxidants.

An antioxidant, as the term implies, tends to prevent oxidation. Natural antioxidants present in foods include lecithin (which also is an emulsifier), vitamins C and E, and certain sulfur-containing amino acids. However, the most effective antioxidants are synthetic chemicals approved by the Food and Drug Administration for addition to foods.

Enzymes

Enzymes are biological catalysts that promote a wide variety of biochemical reactions. Amylase found in saliva promotes digestion or breakdown of starch in the mouth. Pepsin found in gastric juice promotes digestion of protein. Lipase found in liver promotes breakdown of fats. There are thousands of different enzymes found in bacteria, yeasts, molds, plants, and animals. Even after a plant is harvested or an animal is killed, most of the enzymes continue to promote specific chemical reactions, and most foods contain a great number of active enzymes. Enzymes are large protein molecules which, like other catalysts, need to be present in only minute amounts to be effective.

Enzymes function by lowering the activation energies of specific substrates. They do this by temporarily combining with the substrate to form an enzyme–substrate complex that is less stable than the substrate alone. This overcomes the resistance to reaction. The substrate thus excited plunges to a still lower energy level by forming new products of reaction. In the course of reaction, the enzyme is released unchanged. The release of the enzyme so that it can continue to act explains why enzymes are effective in such trace amounts.

The reactions catalyzed by a few enzymes of microbial origin are indicated in Table 3.1. Some properties of enzymes important to the food scientist are the following: (1) In living fruits and vegetables, enzymes control the reactions associated with ripening; (2) after harvest, unless destroyed by heat, chemicals, or some other means, enzymes continue the ripening process, in many cases to the point of spoilage—such as soft melons and overripe bananas; (3) because enzymes enter into a vast number of biochemical reactions in foods, they may be responsible for changes in flavor, color, texture, and nutritional properties; (4) the heating processes in food manufacturing are designed not only to destroy microorganisms but also to inactivate food enzymes and thereby extend the storage stability of foods; (5) when microorganisms are added to foods for fermentation purposes, the important agents are the enzymes the microorganisms produce; and (6) enzymes also can be extracted from biological materials and purified to a high degree. Such commercial enzyme preparations may be added to foods to break down starch, tenderize meat, clarify wines, coagulate milk protein, and produce many other desirable changes. Some of these additional changes are indicated in Table 3.2.

Sometimes we wish to limit the degree of activity of an added enzyme but cannot readily inactivate the enzyme without adversely affecting the food. One way to accomplish this is to immobilize the enzyme by attaching it to the surface of a membrane or another inert object in contact with the food being processed. In this way reaction time can be regulated without the enzyme becoming part of the food. Such immobilized

Table 3.1. Examples of Extracellular Hydrolytic Enzymes

Enzyme	Substrate	Catabolic Products
Esterases		
lipases	Glycerides (fats)	Glycerol + fatty acids
phosphatases		
lecithinase	Lecithin	Choline + H_3PO_4 + fat
Carbohdrases		
fructosidases	Sucrose	Fructose + glucose
α-glucosidases	Maltose	Glucose
(maltase)		
β-glucosidases	Cellobiose	Glucose
β-galactosidases	Lactose	Galactose + glucose
(lactase)		
amylase	Starch	Maltose
cellulase	Cellulose	Cellobiose
cytase	—	Simple sugars
Nitrogen-carrying compounds		
proteinases	Proteins	Polypeptides
polypeptidases	Proteins	Amino acids
desamidases		
urease	Urea	CO_2 + NH_3
asparaginase	Asparagine	Aspartic acid + NH_3
deaminases	Amino acids	NH_3 + organic acids

Courtesy of H. H. Weiser.

Table 3.2. Some Commercial Enzymes and Their Application

Type	Typical Use
Carbohydrases	Production of invert sugar in confectionary industry; production of corn syrups from starch; conversion of cereal starches into fermentable sugars in malting, brewing, distillery, baking industry; clarification of beverages and syrups containing fruit starches.
Proteases	Chill-proofing of beers and related products; tenderizing meat; production of animal and plant protein hydrolyzates.
Pectinases	Clarification of fruit juices; removal of excess pectins from juices such as apple juice before concentration; increase of yield of juice from grapes and other products; clarification of wines; dewatering of fruit and vegetable wastes before drying.
Glucose oxidase —Catalase	Removal of glucose from egg white before drying; removal of molecular oxygen dissolved or present at the surface of products wrapped or sealed in hermetic containers.
Glucosidases	Liberation of essential oils from precursors such as those present in bitter almonds; destruction of naturally occurring bitter principles such as those occurring in olives and the bitter principle glycosides in cucurbitaceae (cucumber and related family).
Flavor enzymes (flavorases)	Restoration and enrichment of flavor by the addition of enzymes capable of converting organic sulfur compounds into the particular volatile sulfur compounds responsible for flavor in garlic and onions, e.g., conversion of alliin of garlic into garlic oil by alliianase; conversion of sulfur-containing flavor precursors of cabbage and related species (watercress, mustard, radish) by enzyme preparations from related rich natural sources of enzymes; addition of enzyme preparations from mustard seeds to rehydrated blanched dehydrated cabbage to restore flavor; production of natural banana flavor in sterilized banana puree and dehydrated bananas by naturally occurring banana flavor enzyme; improvement in flavor of canned foods by an enzyme preparation from fresh corn.
Lipases	Improvement in whipping quality of dried egg white and flavor production in cheese and chocolate.
Cellulase	Mashing of grain and brewing, clarification and extraction of fruit juices, tenderization of vegetables.

Courtesy of M. A. Joslyn

enzymes are presently being used to hydrolyze the lactose of milk into glucose and galactose, to isomerize the glucose from corn starch into fructose, and in many other industrial food processes.

Pigments and Colors

Foods may acquire their color from any of several sources. One major source is natural plant and animal pigments. For example, chlorophyll imparts green color to lettuce and peas, carotene gives the orange color to carrots and corn, lycopene contributes to the red of tomatoes and watermelons, anthocyanins contribute purple to grapes and blueberries, and oxymyoglobin gives the red color to meats.

These natural pigments are highly susceptible to chemical change—as in fruit ripening and meat ageing. They also are sensitive to chemical and physical effects during food processing. Excessive heat alters virtually all natural food pigments. Chopping and grinding also generally change food colors because many of the plant and animal pigments are organized in tissue cells and pigment bodies, such as the chloroplasts which contain green chlorophyll. When these cells are broken, the pigments leach out and are partially destroyed on contact with air.

Not all food color comes from true plant and animal pigments. A second source of color comes from the action of heat on sugars. This is referred to as caramelizing. Examples of caramelization are the darkening of maple sugar on heating, the color on toasting bread, and the brown color of caramel candy.

Third, dark colors result from chemical interactions between sugars and proteins, referred to as the browning reaction or the Maillard reaction. In this case, an amino group from a protein combines with an aldehyde or ketone group of a reducing sugar to produce a brown color—an example is the darkening of dried milk on long storage.

Complex color changes also occur when many organic chemicals present in foods come in contact with air. Examples are the darkening of a cut surface of an apple and the brown color of tea from tea tannins. These oxidations generally are intensified by the presence of metal ions.

In many foods and in cooking, final color is the result of a combination of several of the above, which adds to the complexity of the field of food color.

Not to be overlooked is the intentional coloring of food by the addition of natural or synthetic colors as in the coloring of gelatin desserts or the addition of vegetable dyes to Cheddar cheese to make it orange.

Flavors

If food color is complex, then the occurrence and changes that take place in food flavors are certainly no less complex.

In coffee alone over 800 constituents have been identified, many of which may contribute to the flavor and aroma, although the contribution of many of them may be quite small. These organic chemicals are highly sensitive to air, heat, and interaction with one another. The flavor and aroma of coffee, milk, cooked meats, and most foods is in a continuing state of change—generally becoming less desirable as the food is handled, processed, and stored. There are exceptions, of course, as in the improvement of flavor when cheese is ripened, wine is aged, or meat is aged.

It is important to recognize that flavor often has a regional and cultural basis for acceptance. Not only do many Orientals prize the flavors of "100-year eggs" and sauces made from aged fish, but in the United States different blends of coffee are favored in the South and in the North, and sour cream is not as popular in the Midwest as in the East.

A detailed discussion of the chemistry of flavor is beyond the scope of this book. However, flavors are one of the major areas of interest among food chemists. Much progress has been made in this area from use of analytical methods such as gas chromatography and mass spectrometry. In gas chromatography, aroma compounds are separated from one another on the basis of relative volatility by a special column through which gas is passed. Each compound gives a specific peak on a recording chart.

The peaks corresponding to aroma compounds obtained from two kinds of apples are shown in Fig. 3.5. Although such methods are highly sensitive, for many flavor and aroma compounds they are not as sensitive as the human nose. Furthermore, the instrumental approach does not tell whether a flavor is liked or disliked. Therefore, subjective methods of study also are used. These employ various kinds of taste panels. Because the results are subjective, conclusions are generally based on the judgments of several people making up the panel.

Vitamins and Minerals

Vitamins and minerals are essential parts of food because they are required for normal health. There are a wide range of minerals and vitamins that are important in the diet, and the effects of food formulation and processing on vitamins and minerals must be understood. This will be discussed in detail in Chapter 4 on the nutrients of food.

Natural Toxicants

Over the centuries, plants have evolved the ability to form many compounds which play no direct biochemical role in the plant but may serve to protect the plant or help ensure reproduction. These secondary metabolites may attract pollinating insects or reject predators which attack the plant. It is not surprising that some of these compounds can be toxic. For example, some species of mushrooms have poisonous properties due to specific nitrogen-containing bases or alkaloids that, depending on concentration, can produce marked physiological effects. Many other natural foods also contain substances that can be harmful if consumed in sufficient quantities, but are not a threat at the low concentrations present in our usual diets.

Similarly, soil and water normally contain the potentially harmful metals lead, mercury, cadmium, arsenic, zinc, and selenium, and so traces of these metals occur in foods and always have. At their low levels of occurrence, however, not only are these natural components of foods harmless but zinc, selenium, and possibly others are essential nutrients.

Many harmful substances are not normal components of foods but can become part of food; these include industrial contaminants, toxins produced in food by microorganisms, and additives whose safe-use levels are exceeded. These kinds of materials are dealt with in subsequent chapters.

In addition to heavy metals, some of the better known toxicants occurring naturally in foods include low levels of the alkaloid solanine in potatoes, cyanide-generating compounds in lima beans, safrole in spices, prussic acid in almonds, oxalic acid in spinach and rhubarb, enzyme inhibitors and hemagglutinins in soybean and other legumes, gossypol in cottonseed oil, goitrogens in cabbage that interfere with iodine binding by the thyroid gland, tyramine in cheese, avidin in egg white which is antagonistic to the growth factor biotin, thiaminase in fish which destroys vitamin B_1, and several additional toxins associated with specific fish and shellfish. Vitamins A and D and essential amino acids such as methionine also exhibit toxic effects in excessive concentration.

Several of these materials and certain other natural toxicants are largely removed or inactivated when foods are processed. Thus, the heat of cooking destroys enzyme

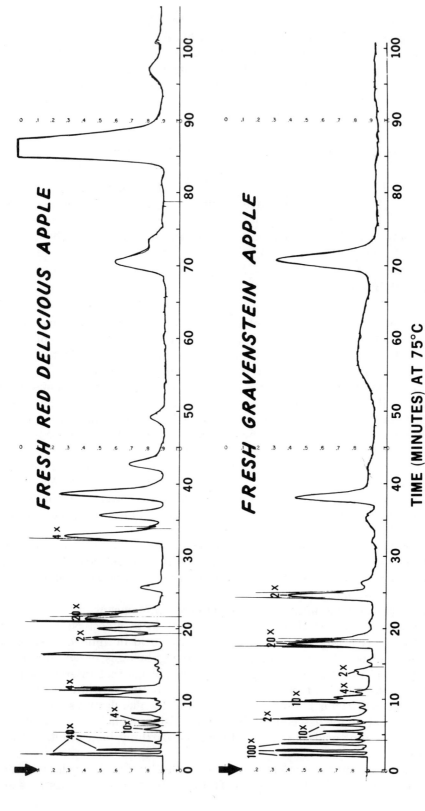

Figure 3.5. Gas chromatographic analysis of aroma volatiles from apples. *Courtesy of U.S. Dept. of Agric.*

inhibitors and hemagglutinins of beans, avidin of egg white, and thiaminase of fish. Water soaking and fermentation also remove some cyanogenic compounds. Removal of gonads, skin, and parts of certain fish eliminates toxins concentrated in these tissues. Breeding and selection also have lowered concentrations of toxicants in certain plant foods. Further, in the course of evolution, man has developed physiological mechanisms to detoxify low levels of many potentially dangerous substances and has learned to exclude clearly toxic species as food sources.

Although much more remains to be learned, a varied diet of the conventional foods of a region or culture pose small risk from natural toxicants to normally healthy individuals. Departures from conventional food sources and time-honored processes without adequate testing, microbial toxins, and harmful levels of industrial chemicals generally present greater dangers. With respect to all substances that may be normal constituents of food or become part of a food, it is important to recognize that such substances are not harmless or harmful per se but only so in terms of their concentrations.

Water

Water is present in most natural foods to the extent of 70% of their weight or greater. Fruits and vegetables may contain 90% or even 95% water. Cooked meat from which some of the water has been driven off still contains about 60% water. Water greatly affects the texture of foods—a raisin is a dehydrated grape, and a prune, a dried plum. The form in which water occurs in foods to a large extent dictates the physical properties of the food. For example, fluid milk and apples contain approximately the same amount of water but have different physical structures.

Water greatly affects the keeping qualities of food, which is one reason for removing it from foods, either partially as in evaporation and concentration, or nearly completely as in true food dehydration. When foods are frozen, water as such also is removed, since water is most active in foods in its liquid form. As a liquid in foods, it is the solvent for numerous food chemicals and thus promotes chemical reactions between the dissolved constituents. It also is necessary for microbial growth.

The other reason for removing water from food (in addition to preservation) is to reduce the weight and bulk of the food and thus save on packaging and shipping costs.

A great deal of food science and food technology can be described in terms of the manipulation of the water content of foods: its removal, its freezing, its emulsification, and its addition in the case of dissolving or reconstituting dehydrated foods.

Water exists in foods in various ways—as free water in the case of tomato juice, as droplets of emulsified water in the case of butter, as water tied up in colloidal gels in gelatin desserts, as a thin layer of adsorbed water on the surface of solids often contributing to caking as in dried milk, and as chemically bound water of hydration as in some sugar crystals.

Some of these bound water forms are extremely difficult to remove from foods even by drying, and many dehydrated foods with as little as 2–3% residual water have their storage stability markedly shortened.

Close control of final water content is essential in the production of numerous foods: as little as 1–2% of excess water can result in such common defects as molding of wheat, bread crusts becoming tough and rubbery, soggy potato chips, and caking of salt and sugar. Many skills in food processing involve the removal of these slight excesses of water without simultaneously damaging the other food constituents. On the other hand, even where a dehydrated product is involved, it is possible to remove too much water. In some cases the storage stability of a dehydrated item is enhanced by leaving a trace of moisture, equivalent to a monomolecular layer of water, to coat all internal and external surfaces. This monomolecular layer of water then may serve as a barrier between atmospheric oxygen and sensitive constituents in the food which otherwise would be more easily oxidized.

It is obvious that the purity of water used in foods or associated with the manufacture of foods is of utmost importance. It is less obvious, however, that suitable drinking water from a municipal water supply may not be of adequate purity for certain food uses. This is particularly important in the manufacture of carbonated beverages, as will be discussed in Chapter 19.

References

Aurand, L.W., Woods, A.E. and Wells, M.R. 1987. Food Composition and Analysis. Chapman & Hall, London, New York.

Binkley, R.W. 1988. Modern Carbohydrate Chemistry. Marcel Dekker, New York.

Birch, G.G. and Parker, K.J. 1982. Nutritive Sweeteners. Applied Science, London.

Birch, G.G. and Lindley, M.G. 1986. Interactions of Food Components. Elsevier Applied Science Publishers, New York.

Cherry, J.P. (Editor) 1982. Food Protein Deterioration: Mechanisms and Functionality. American Chemical Society. Washington, DC.

Fennema, O.R., Chang, W.H. and Lii, C.Y. 1986. Role of Chemistry in the Quality of Processed Foods. Food & Nutrition Press, Westport, CT.

Fox, P.F. and Condon, J.J. 1982. Food Proteins. Chapman & Hall, London, New York.

Galliard, T. 1987. Starch: Properties and potential. *In* Critical Reports on Applied Chemistry. Vol. 13. Society of Chemical Industry. Wiley, Chichester.

Gould, G.W. and Christina, J.H.B. 1988. Characterization of the state of water in foods: Biological aspects. *In* Food Preservation by Moisture Control, C.C. Seow (Editor). Elsevier Applied Science Publishers, London.

Heath, H.B. and Reineccius, G.A. 1986. Flavor Chemistry and Technology. Chapman & Hall, London.

Hudson, B.J.F. 1992. Biochemistry of Food Proteins. Elsevier Applied Science, London.

Kennedy, J.F. 1988. Carbohydrate Chemistry. Oxford University Press, New York.

Lawson, H.W. 1995. Food Oils and Fats: Technology, Utilization and Nutrition. Chapman & Hall, London, New York.

Luallen, T.E. 1985. Starch as a functional ingredient. Food Technol. *39*(1), 59–63.

Mateos, N.A. 1986. Water in foods. Alimentaria *178*, 53–60.

Meuser, F. and Suckow, P. 1986. Non-starch polysaccharides. *In* Chemistry and Physics of Baking: Materials, Processes, and Products: The Proceedings of an International Symposium Held at the School of Agriculture, April 1985. J.M.V. Blanshard, P.J. Frazier and T. Galliard (Editors). Royal Society of Chemistry, London.

Okos, M.R. 1986. Physical and Chemical Properties of Food. American Society of Agricultural Engineers, St. Joseph, MI

Olson, A., Gray, G.M. and Chiu, M. 1987. Chemistry and analysis of soluble dietary fiber. Food Technol. *41*(2), 71–80.

Phillips, R.D. and Finley, J.W. 1988. Protein Quality and the Effects of Processing. Marcel Dekker, New York.

Pomeranz, Y. 1985. Functional Properties of Food Components. Academic Press, Orlando, FL.

Regenstein, J.M., Regenstein, C.E., and Kochen, B. 1984. Food Protein Chemistry: An Introduction for Food Scientists. Academic Press, Orlando, FL.

Simatos, D. and Multon, J.L. 1985. Properties of water in foods. *In* Relation to Quality and Stability. Dordrecht, Boston.

Simatos, D. and Karel, M. 1988. Characterization of the condition of water in foods: Physicochemical aspects. *In* Food Preservation by Moisture Control, C.C. Seow (Editor). Elsevier Applied Science Publishers, London.

Soucie, W.G. and Kinsella, J.E. 1989. Food Proteins. American Oil Chemists' Society, Champaign, IL.

Tucker, G.A. and Woods, L.F.J. 1991. Enzymes in Food Processing. Chapman & Hall, London, New York.

Weiss, T.J. 1983. Food Oils and Their Uses. 2nd ed. AVI Publishing Co., Westport, CT.

Whistler, R.L., BeMiller, J.N. and Paschall, E.F. 1984. Starch: Chemistry and Technology. 2nd. ed. Academic Press, Orlando, FL.

Whistler, R.L. and BeMiller, J.N. 1992. Industrial Gums: Polysaccharides and Their Derivatives. 3rd ed. Academic Press, San Diego, CA.

4

Nutritive Aspects of Food Constituents

Food supplies both the energy for all of the body's functions and the building blocks for growth and maintenance. Even in fully grown adults there is a requirement for energy and to build and maintain body components that are being replaced. For example, the human stomach is constantly being lost and replaced. Also, there is increasing evidence that diet plays a major role in our defense against disease, including chronic diseases such as cancer and heart disease. Mental processes and behavioral attitudes appear to be influenced by nutritional status and specific nutrients.

The food scientist must consider the nutritive aspects of food from two points of view: first, what nutrients do foods contain and what is a human's requirement for these; and second, what are the relative stabilities of these nutrients and how are they affected by food processing, storage, and preparation. The science of nutrition, concerned with these broad areas, also deals with the physiological and biochemical phenomena of food utilization as related to health.

The nutrients in food, required in balanced amounts to produce and maintain optimum health, belong to the broad groups of carbohydrates, proteins, fats, vitamins, and minerals. Water, not generally classified as a nutrient, must not be overlooked, because a lack of water even for a short period is life threatening.

FOOD AND ENERGY

Food is the "fuel" which supplies chemical energy to the body to support daily activity and synthesis of necessary chemicals within the body. The major sources of energy for humans and other animals are carbohydrates, fats, and proteins. In addition to supplying energy, these nutrients have other specific functions, but their conversions to energy are of fundamental importance. The energy value of foods is measured in heat units called calories.

Calories

A calorie is the amount of heat required to raise the temperature of one gram of water one degree Celsius (from 14.5 to 15.5°C). The kilocalorie (1000 calories) is the

46

unit commonly used in expressing energy values of foods. In an effort to standardize nonmetric and metric measurement under the International System of Units (SI), the kilojoule is sometimes used in place of the kilocalorie (kilocalories × 4.2 = kilojoules). Calories remain, however, the more common unit of nutritionists, and so this term will be used in subsequent discussions.

The total potential energy of foods and food components is determined by burning the food in a steel bomb calorimeter under elevated oxygen pressure. The bomb, and water that it is immersed in, rise in temperature to an extent that is directly related to the gross energy content of the food. This is termed "calorimetry."

The total potential energy of a food as determined by calorimetry may not be equal to the energy that can be derived from it by an animal or human. If a food or food constituent is not totally digestible, or if the food is not completely oxidized within the body, then its caloric value in metabolism will be less than its theoretical total energy content.

Not all carbohydrates (apart from their relative utilization in the body) yield equal amounts of energy when burned in a calorimeter. This is true of fats and proteins also. A fat that is more highly oxidized chemically than another will yield less energy on further combustion in a calorimeter than a corresponding unoxidized fat. Nevertheless, common averages from calorimeter studies in kcal/g generally are given as 4.1 for carbohydrates, 9.5 for fats, and 5.7 for proteins.

Carbohydrates such as sugars and starches, which generally are about 98% digested and fully oxidized by humans, provide about 4 kcal/g. Most fats are generally digested to the extent of 95%, yielding 9 kcal/g. Proteins, due to incomplete digestion and oxidation, generally also yield an energy equivalent of 4 kcal/g. Thus, on an equal-weight basis, fat generally yields 2.25 times as many calories as protein or carbohydrate. These simple relationships permit approximate calculations of the caloric values of foods when their compositions are known. Calories are needed to satisfy the body's energy requirements for production of body heat, synthesis of body tissue, and performance of work. The greater part of the food we consume goes to satisfy these energy requirements. When the body performs little work, a greater proportion of this energy is conserved and stored in the form of fat. Likewise, when energy demands exceed the intake of calories, fat and other tissues are oxidized to provide this energy and the body loses weight. An excess intake of about 9 g (⅓ oz) of butter or margarine daily can result in the deposition of about 3.2 kg (7 lb) of fat in a year. This can be counteracted by walking approximately an additional 2.4 km (1.5 miles) daily.

Comprehensive calorie charts for common foods are readily available and are not included here, although the caloric content of a few representative foods are given in Table 4.1, which also includes the contents of other nutrients in these foods. While the caloric contents of foods are relatively fixed, a human's caloric requirements vary widely, depending upon such factors as physical activity, climatic conditions, weight, age, sex, and individual metabolic differences. Table 4.2 gives daily dietary allowances for each of the major nutrients, but not calories, recommended by the Food and Nutrition Board, National Academy of Sciences–National Research Council. These recommended daily dietary allowances are intended to cover most normal persons as they live in the United States under usual work and environmental stresses. Caloric needs are related to energy expenditure and are dealt with in a separate section of the recommended allowances. Especially depending on physical activity, an adult male's daily requirement can range from about 2500 to 5000 kcal. If he is a laborer and needs

Table 4.1. Approximate Calories and Amounts of Major Nutrients in Common Foods

Food	Serving[1]	Calories	Protein (g)	Calcium (mg)	Iron (mg)	Vitamins						Water (%)
						A (IU)	C (mg)	D (IU)	Thiamin (µg)	Riboflavin (µg)	Niacin (mg)	
Apple, raw	1 large	117	0.6	12	0.6	180	9	0	80	60	0.4	85
Banana, raw	1 large	176	2.4	16	1.2	860	20	0	80	100	1.4	76
Beans, green, cooked	1 cup	27	1.8	45	0.9	830	18	0	90	120	0.6	92
Beef, round, cooked	3.2 ounces	214	24.7	10	3.1	0	0	0	74	202	5.1	55
Bread, white, enriched	1 slice	63	2.0	18	0.4	0	0	0	60	40	0.5	36
Broccoli, cooked	⅔ cup	29	3.3	130	1.3	3,400	74	0	70	150	0.8	90
Butter	1 tablespoon	100	0.1	33	0.0	460	0	5	tr.	tr.	tr.	16
Cabbage, cooked	½ cup	20	1.2	39	0.4	75	27	0	40	40	0.3	92
Carrots, raw	1 cup shredded	42	1.2	39	0.8	12,000	6	0	60	60	0.5	88
Cheese, Cheedar American	1 ounce	113	7.1	206	0.3	400	0	0	10	120	tr.	36
Chicken, fried	½ breast	232	26.8	19	1.3	460	0	0	67	101	10.2	53
Egg, boiled	1 medium	77	6.1	26	1.3	550	0	27	40	130	tr.	74
Liver, beef, fried	1 slice	86	8.8	4	2.9	18,700	10	19	90	1,300	5.1	57
Margarine, fortified	1 tablespoon	101	0.1	3	0.0	460	0	0	0	0	0.0	16
Milk, whole, cow's	6 ounces	124	6.4	216	0.2	293	2	4	73	311	0.2	87
Oatmeal, cooked	1 cup	148	5.4	21	1.7	0	0	0	220	50	0.4	85
Orange, whole	1 medium	68	1.4	50	0.6	285	74	0	120	45	0.3	86
Pork, shoulder, roasted	2 slices	320	19.2	9	2.0	0	0	0	592	144	3.2	43
Tomatoes, raw	1 large	40	2.0	22	1.2	2,200	46	0	120	80	1.0	94
Potatoes, white, baked	1 medium	98	2.4	13	0.8	20	17	0	110	50	1.4	75
Rice, white, cooked	1 cup	201	4.2	13	0.5	0	0	0	20	10	0.7	71
Sugar, white, granulated	1 tablespoon	48	0.0	0	0.0	0	0	0	0	0	0.0	tr.

SOURCE: The World Book Encyclopedia.
[1] 1 cup = 237 ml; 1 ounce = 28 g; 1 tablespoon = 15 ml.

5000 kcal per day, he must eat some fat since a human stomach is not large enough to hold sufficient carbohydrates and proteins, consumed at usual mealtimes, containing this many calories.

Whereas fats are the most concentrated source of food calories, carbohydrates are the cheapest source—and proteins the most expensive. It generally is agreed that quite apart from the other nutritional demands of the body, and except for very young children and the aged, a daily intake of less than about 2000 kcal represents dietary insufficiency. It is one of the sad contrasts of our time that while so many of the world's people go hungry, in the United States and certain other countries obesity from excess caloric intake is a major nutritional disease.

ADDITIONAL ROLES OF CARBOHYDRATES, PROTEINS, AND FATS IN NUTRITION

Carbohydrates, proteins, and fats in many ways are interrelated and interconvertible in animal metabolism. Although dietary carbohydrate is an economical source of calories and provides rapidly available energy for a variety of physiologic functions, the body can fulfill its energy and carbon requirements from proteins and fats. It also can synthesize blood glucose, liver glycogen, the ribose sugar components of nucleic acids, and other important biological carbohydrates from proteins and fats.

On the other hand, carbohydrates from the foods consumed help the body use fat efficiently. They do this by supplying an organic acid formed as an intermediate in the oxidation of carbohydrates. This organic acid is required for the complete oxidation of fat to CO_2 and water. When fat is not efficiently oxidized, ketone bodies can accumulate in the blood and produce the disease condition known as ketosis.

Carbohydrates also exert a protein-sparing effect. When carbohydrates are depleted in the animal body and the animal needs additional energy, it gets this energy by oxidizing fats and proteins. In the case of proteins, this energy requirement is thus satisfied at the expense of the body's requirement for proteins and amino acids as components of body tissues, enzymes, antibodies, and other essential nitrogen-containing substances. However, if carbohydrates are supplied, the body oxidizes them for energy in preference to protein and, thus, the protein is spared. In similar fashion, fats can exert a protein-sparing effect.

The role of carbohydrates such as cellulose and hemicellulose in providing fiber and bulk is essential to a healthy condition of the intestine. In addition, the microflora of the intestine are much influenced by the nature of carbohydrates in the diet. When these carbohydrates are comparatively slow to dissolve, as in the case of starch and lactose, they remain in the intestinal tract for longer periods than the more highly soluble sugars. In this case, they serve as readily available nutrients for growth of microorganisms that synthesize several vitamins of the B complex. On the other hand, the slow rate of absorption of lactose from the intestine can cause diarrhea in some adults consuming excessive amounts of this sugar. Lactose also appears to increase calcium retention in children.

The role of protein in supplying chemical building materials for the synthesis of body tissues and other constituents of life and in providing those essential amino acids that the body cannot itself synthesize have been mentioned. The nutritional value of

Table 4.2. Food and Nutrition Board, National Academy of Sciences—National Research Council *nutrition of practically all healthy people in the United States.*

Category	Age (years) or Condition	Weight[b] (kg)	Weight[b] (lb)	Height[b] (cm)	Height[b] (in)	Protein (g)	Fat-Soluble Vitamins Vita-min A (µg RE)[c]	Vita-min D (µg)[d]	Vita-min E (mg α-TE)[e]	Vita-min K (µg)	Water-Soluble Vitamins Vita-min C (mg)	Thia-min (mg)
Infants	0.0–0.5	6	13	60	24	13	375	7.5	3	5	30	0.3
	0.5–1.0	9	20	71	28	14	375	10	4	10	35	0.4
Children	1–3	13	29	90	35	16	400	10	6	15	40	0.7
	4–6	20	44	112	44	24	500	10	7	20	45	0.9
	7–10	28	62	132	52	28	700	10	7	30	45	0.9
Males	11–14	45	99	157	62	45	1000	10	10	45	50	1.3
	15–18	66	145	176	69	59	1000	10	10	65	60	1.5
	19–24	72	160	177	70	58	1000	10	10	70	60	1.5
	25–50	79	174	176	70	63	1000	5	10	80	60	1.5
	51+	77	170	173	68	63	1000	5	10	80	60	1.2
Females	11–14	46	101	157	62	46	800	10	8	45	50	1.1
	15–18	55	120	163	64	44	800	10	8	55	60	1.1
	19–24	58	128	164	65	46	800	10	8	60	60	1.1
	25–50	63	138	163	64	50	800	5	8	65	60	1.1
	51+	65	143	160	63	50	800	5	8	65	60	1.0
Pregnant						60	800	10	10	65	70	1.5
Lactating	1st 6 months					65	1300	10	12	65	95	1.6
	2nd 6 months					62	1200	10	11	65	90	1.6

[a]The allowances, expressed as average daily intakes over time, are intended to provide for individual variations among most normal persons as they live in the United States under usual environmental stresses. Diets should be based on a variety of common foods in order to provide other nutrients for which human requirements have been less well defined. See text for detailed discussion of allowances and of nutrients not tabulated.

[b]Weights and heights of reference adults are actual medians for the U.S. population of the designated age, as reported by National Health and Nutrition Examination Survey II. The use of these figures does not imply that the height-to-weight ratios are ideal.

SUMMARY TABLE. Estimated Safe and Adequate Daily Dietary Intakes of Selected Vitamins and Minerals[a]

Category	Age (years)	Vitamins Biotin (µg)	Pantothenic Acid (mg)	Trace Elements[b] Copper (mg)	Manganese (mg)	Fluoride (mg)	Chromium (µg)	Molybdenum (µg)
Infants	0–0.5	10	2	0.4–0.6	0.3–0.6	0.1–0.5	10–40	15–30
	0.5–1	15	3	0.6–0.7	0.6–1.0	0.2–1.0	20–60	20–40
Children and	1–3	20	3	0.7–1.0	1.0–1.5	0.5–1.5	20–80	25–50
adolescents	4–6	25	3–4	1.0–1.5	1.5–2.0	1.0–2.5	30–120	30–75
	7–10	30	4–5	1.0–2.0	2.0–3.0	1.5–2.5	50–200	50–150
	11+	30–100	4–7	1.5–2.5	2.0–5.0	1.5–2.5	50–200	75–250
Adults		30–100	4–7	1.5–3.0	2.0–5.0	1.5–4.0	50–200	75–250

[a]Because there is less information on which to base allowances, these figures are not given in the main table of RDA and are provided here in the form of ranges of recommended intakes.

[b]Since the toxic levels for many trace elements may be only several times usual intakes, the upper levels for the trace elements given in this table should not be habitually exceeded.

Recommended Dietary Allowances,[a] Revised 1989. *Designed for the maintenance of good*

Water-Soluble Vitamins					Minerals						
Ribo-flavin (mg)	Niacin (mg NE)[f]	Vita-min B$_6$ (mg)	Folate (μg)	Vita-min B$_{12}$ (μg)	Cal-cium (mg)	Phos-phorus (mg)	Magne-sium (mg)	Iron (mg)	Zinc (mg)	Iodine (μg)	Sele-nium (μg)
0.4	5	0.3	25	0.3	400	300	40	6	5	40	10
0.5	6	0.6	35	0.5	600	300	60	10	5	40	5
0.8	9	1.0	50	0.7	800	800	80	10	10	70	20
1.1	12	1.1	75	1.0	800	800	120	10	10	90	20
1.2	13	1.4	100	1.4	800	800	170	10	10	120	30
1.5	17	1.7	150	2.0	1200	1200	270	12	15	150	40
1.8	20	2.0	200	2.0	1200	1200	400	12	15	150	50
1.7	19	2.0	200	2.0	1200	1200	350	10	15	150	70
1.7	19	2.0	200	2.0	800	800	350	10	15	150	70
1.4	15	2.0	200	2.0	800	800	350	10	15	150	70
1.3	15	1.4	150	2.0	1200	1200	280	15	12	150	45
1.3	15	1.5	180	2.0	1200	1200	300	15	12	150	50
1.3	15	1.6	180	2.0	1200	1200	280	15	12	150	55
1.3	15	1.6	180	2.0	800	800	280	15	12	150	55
1.2	13	1.6	180	2.0	800	800	280	10	12	150	55
1.6	17	2.2	400	2.2	1200	1200	320	30	15	175	65
1.8	20	2.1	280	2.6	1200	1200	355	15	19	200	75
1.7	20	2.1	260	2.6	1200	1200	340	15	16	200	75

[c]Retinol equivalents. 1 retinol equivalent = 1 μg retinol or 6 μg β-carotene.
[d]As cholecalciferol. 10 μg cholecalciferol = 400 IU of vitamin D.
[e]α-Tocopherol equivalents. 1 mg dα- tocopherol = 1 α-TE.
[f]1 NE (niacin equivalent) is equal to 1 mg of niacin or 60 mg of dietary tryptophan.

different proteins depends on their different amino acid compositions. A complete protein is one that contains all of the essential amino acids in amounts and proportions to maintain life and support growth when used as the sole source of protein. Such a protein is said to have high biological value. Many animal proteins such as those found in meat, poultry, fish, milk, and eggs generally are of high biological value. An exception is gelatin, which contains limited amounts of isoleucine, threonine, and methionine, and no tryptophan. Plant proteins generally are not as high in biological value as animal proteins because of amino acid limitations. Thus, for example, most varieties of wheat, rice, and corn lack lysine; corn also lacks tryptophan; legumes are of somewhat higher protein quality but have limited amounts of methionine.

Incomplete proteins can be supplemented with the missing essential amino acids either in the form of synthetic compounds or as protein concentrates from natural sources. Blends of plant and animal products also can overcome essential amino acid limitations and produce nutritional adequacy, but complementary components should best be given at the same feeding since the body has very limited protein storage capacity and all amino acids are needed for daily protein synthesis. Much protein supplementation is now being practiced to improve world food resources.

The amount of protein required daily, which beyond early childhood may range from about 40 to 60 g (Table 4.2), depends on the body demand—the demand being greatest during growth, pregnancy, and lactation.

One of the severest needs for protein on a world population basis is in infants after weaning and in young children. Protein shortage or protein malnutrition can be dramatically reversed by proper diet. However, in instances where adequate protein and proper diet are withheld too long, recovery may not be complete due to irreversible damage and possible mental retardation.

In addition to supplying calories for energy, fats supply polyunsaturated fatty acids, at least one of which, linoleic acid, is an essential fatty acid. As in the case of the essential amino acids, linoleic acid is called an essential fatty acid because animals cannot adequately synthesize it and so it must be supplied by the diet as such. In rats and in human infants, absence of linoleic acid interferes with normal growth rates and results in skin disorder. Two other polyunsaturated fatty acids, linolenic acid and arachidonic acid, formerly were listed also as essential fatty acids. However, since the body can convert linoleic acid to arachidonic acid and since linolenic acid can only partially replace linoleic acid, we now regard only linoleic acid as an essential fatty acid. Good sources of linoleic acid include grain and seed oils, fats from nuts, and fats from poultry. Linoleic and other unsaturated fatty acids when present in high proportion of dietary fats can lower blood cholesterol levels under certain dietary conditions; more will be said about this in the last section of this chapter.

Vitamins A, D, E, and K are fat soluble and so are to be found associated with the fat fractions of natural foods. Additionally, phospholipids, which are organic esters of fatty acids and also contain phosphoric acid and usually a nitrogenous base, are partially soluble in fats. The emulsifying properties of lecithin were discussed in Chapter 3. Lecithin, cephalin, and other phospholipids are found in brain, nerve, liver, kidney, heart, blood, and other tissues in addition to their presence in egg yolk. Because of their strong affinity for water, they facilitate the passage of fats in and out of the cells and play a role in fat absorption from the intestine and the transport of fats from the liver. Fat also physically insulates the body from rapid changes in temperature and helps cushion organs from sudden injury. Excess dietary fat is stored in the body's adipose (fatty) tissue, as are fats formed from the metabolism of excess carbohydrates and proteins. These stored fats can be drawn upon as a reserve source of energy. In excessive amounts they contribute to obesity.

PROTEIN QUALITY

It previously was stated that the comparative value of different proteins depends on their different amino acid compositions, especially their contents of the essential amino acids, leucine, isoleucine, lysine, methionine, phenylalanine, threonine, tryptophan, and valine, plus histidine to meet the demands of growth during childhood. Although this statement is essentially true, it requires some further consideration.

Protein quality, or the nutritional value of a protein, is meaningful only in terms of the usefulness of a protein for specific vital purposes such as growth, replacement of metabolic losses and damaged tissue, reproduction, lactation, and general well-being. The usefulness of a protein may differ for several of these functions. Further, a measure of usefulness of a protein based on chemical analysis of its amino acid makeup is complicated by a number of factors. These include accuracy of the analytical method under conditions that can preclude the detection of one or another amino acid or cause its destruction, availability and digestibility of the protein from foods that may not be readily broken down by digestive enzymes or absorbed through the intestine, and factors contributing to unpalatability of the protein-containing food. Yet another factor has to do with amino acid imbalance. It is possible to have an excess of one or more amino acids relative to others in a protein. This can have a negative effect on growth rate.

These objections to chemical analysis of amino acid content to determine protein quality are not encountered when a biological method, such as an animal feeding study, is used. However, in this case a different set of obstacles is encountered. One of the most obvious is how closely results obtained with laboratory animals apply to humans. Even where questions of digestibility and vital response are ruled out, the palatability differences between species must be considered in arriving at a valid measure of the nutritional usefulness of a protein source. Notwithstanding these difficulties, research has shown that results obtained with young rats generally are applicable to humans, and feeding tests under controlled conditions are far more easily carried out on rats than with humans. Such tests may be run under a variety of experimental conditions which then influence the interpretations that may be given the nutritional findings.

Several test methods using the rat have been developed and result in specific terms that the food scientist encounters in the area of protein evaluation. One of the commonest methods involves measurement of weight gain of rats per gram of protein eaten. This is known as Protein Efficiency Ratio (PER). Typical PER values representing protein quality from various foods are given in Fig. 4.1. One of the principal limitations of PER values is that results depend on the amount of food eaten, which can give an erroneous picture if the food is unpalatable to the test animal. The modification known as Net Protein Retention (NPR) improves upon this. If two groups of animals

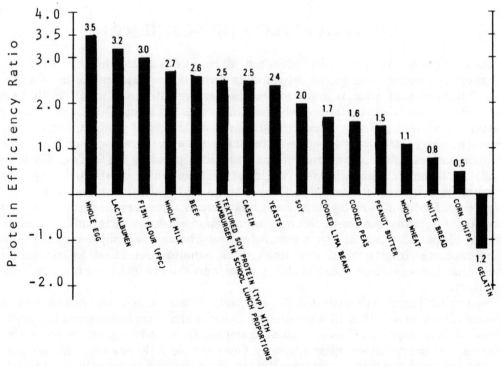

Figure 4.1. Protein quality (PER values) from various types of foods. Source: *Borgstrom and Proctor, Encyclopedia of Food Technology,* A. H. Johnson and M. S. Peterson (Editors). AUI Publishing Co., Westport, CT, *1974.*

are used and one is placed on the test protein diet and the other is placed on a protein-free diet, then the weight loss of the protein-free group can be compared with the weight gain of the protein-receiving group. Properly controlled, the test becomes independent of food intake. Another measure is the proportion of absorbed nitrogen that is retained in the body for maintenance and/or growth. This is known as Biological Value (BV). It requires measurement of protein consumed and the fraction that is excreted in the urine and feces. Since Biological Value measures percentage of the absorbed nitrogen that is retained, it does not account for the digestibility of the protein. Digestibility (D) also can be measured and is the proportion of consumed food nitrogen that is absorbed. When Biological Value (BV) is corrected for the digestibility factor, we get the proportion of nitrogen intake that is retained and this is termed Net Protein Utilization (NPU), which is BV $\times$ D. Since the nutritive value of a protein food involves both the quality and quantity of protein contained, still another measure employs NPU multiplied by the amount of protein in the food—this is known as Net Protein Value. These are but a few of the approaches for measuring protein usefulness.

The complexity of the assessment of protein quality and usefulness has had particular relevance to efforts to develop high-protein foods and supplements to correct nutritional deficiencies in underdeveloped regions of the world. Here the importance of field studies with human subjects under real life conditions has been repeatedly observed. Many new foods of excellent protein content measured by sophisticated laboratory procedures have failed in their purpose because of poor palatability, having been manufactured in a nonfamiliar physical form or having been presented in a manner in conflict with accepted custom or social status.

BIOAVAILABILITY OF NUTRIENTS

As with protein, the contents of other nutrients in foods determined by chemical or physical analysis may be quite misleading in terms of the nutrient status of a food. Apart from amount, what is important is whether the nutrient is in a form that can be utilized in metabolism; that is, whether the nutrient is bioavailable. For example, adding small iron pellets to cereals would increase their iron content, but the iron would not be very available to people eating the cereal and, therefore, be of little value.

Many factors influence a nutrient's bioavailability, including the food's digestibility and the nutrient's absorbability from the intestinal tract, which are affected by nutrient binding to indigestible constituents and nutrient–nutrient interactions in food raw materials. Processing and cooking procedures also can influence nutrient bioavailability. Apart from the food itself, different animal species exhibit variations in bioavailability of specific nutrients from a particular food. The age, sex, physiological health, consumption of drugs, general nutritional status, combinations of foods eaten together, and other factors all influence the ability of an individual to make use of a particular nutrient.

Bioavailability of carbohydrates, proteins, fats, vitamins, and minerals may be increased or decreased since all nutrients are reactive and generally present in varying amounts in food systems. There are many examples of how food composition, processing, and storage affect nutrient bioavailability. One example is the essential mineral iron. Under practical conditions its bioavailability from foods may be only 1–10% of its total level determined by chemical analysis. The recommended dietary allowances for nutrients in the United States and other countries attempt to take bioavailability into account. However, the many factors influencing nutrient bioavailability and the difficulties inherent in meaningful evaluation procedures leave much research in this area still to be done.

VITAMINS

Vitamins are organic chemicals, other than essential amino acids and fatty acids, that must be supplied to an animal in small amounts to maintain health. An exception to this is vitamin D, the only major vitamin the human body is known to be capable of manufacturing. Under certain circumstances, however, vitamin D may not be synthesized in adequate amounts and then it too must be supplied by diet or as a dietary supplement if life and health are to be sustained. Vitamins function in enzyme systems which facilitate the metabolism of proteins, carbohydrates, and fats, but there is growing evidence that their roles in maintaining health extend yet further.

The vitamins are conveniently divided into two major groups, those that are fat soluble and those that are water soluble. Fat-soluble vitamins are A, D, E, and K. Their absorption by the body depends on the normal absorption of fat from the diet. Water-soluble vitamins include vitamin C and the several members of the vitamin B complex.

Vitamin A (Retinol)

Vitamin A as such naturally occurs only in animal materials—meat, milk, eggs, and the like. Plants contain no vitamin A but contain its precursor, β-carotene. Humans and other animals need either vitamin A or β-carotene, which they easily convert to vitamin A. β-Carotene is found in orange and yellow vegetables, as well as green leafy vegetables.

A deficiency of vitamin A leads to blindness, failure of normal bone and tooth development in the young, and diseases of epithelial cells and membranes of the nose, throat, and eyes, which can decrease the body's resistance to infection. These diseases are rarely seen in the developed world but are sadly too common in some parts of the world.

Food sources rich in vitamin A are liver, fish oils, dairy products containing butterfat, and eggs. Sources of its main precursor, β-carotene, are carrots, squash, sweet potatoes, spinach, and kale. Vitamin A and β-carotene also are made synthetically, as are other vitamins.

Until recently, vitamin A activity in foods was expressed in terms of International Units (IU). The IU is a measure of a vitamin's biological activity. Because the biological activity of preformed vitamin A (retinol), β-carotene, and other carotenoids differs, confusion can be avoided by expressing total vitamin A activity in terms of the equivalent weight of pure retinol. Thus, several countries have replaced IU with "retinol equivalents." A retinol equivalent is equal to 1 μg of retinol or 6 μg of the β-carotene. It is also equal to 3.33 IU of vitamin activity from retinol and 10 IU of vitamin A activity from β-carotene. In the United States, the recommended allowance for vitamin A activity is 1000 retinol equivalents (RE) for the healthy adult male (Table 4.2). Because of the smaller size of women, their allowance is 80% of this, but it is increased during lactation. Like many other nutrients, excessive doses of preformed vitamin A can be toxic. Large intakes of carotenes are not similarly harmful since the body will limit the conversion to vitamin A; however, yellow coloration of the skin may result. To combat vitamin A deficiency, several countries in South America have passed laws that all sugar for home consumption be fortified with this vitamin.

Vitamin D

Vitamin D is formed in the skin of humans and animals by activation of sterols by ultraviolet light from the sun or by ultraviolet activation of sterols artificially. Such sterols as cholesterol and ergosterol are involved. Cholesterol is found in and under the skin of animals. Irradiated ergosterol from yeast has served as a vitamin D source for addition to milk and other foods. Vitamin D increases absorption of calcium and phosphorus from the intestinal tract and is necessary for their efficient utilization. Shortage of vitamin D results in bone defects, the principal one being rickets. This shortage may occur when exposure to the sun is limited. Most foods are low in vitamin D, although good sources are liver, fish oils, dairy products, and eggs. In children, 400 IU of vitamin D per day is considered optimum, and this is the basis of fortifying milk with added vitamin D at the level of 400 IU per 0.946 liter (1 qt). In the case of vitamin D, 400 IU is equivalent to 10 μg of the naturally occurring form of the vitamin in animal tissues. Excessive intake of vitamin D provides no benefits and is potentially harmful.

Vitamin E

Also known as α-tocopherol, vitamin E is an antisterility factor in rats and is essential for normal muscle tone in dogs and other animals, but its significance for humans is still uncertain. Vitamin E is a strong antioxidant and probably functions as such in human metabolism. Diets excessive in polyunsaturated fats can lead to the formation of peroxidized fatty acids that may reach harmful levels. There is evidence that vitamin E can prevent this. Further, vitamin E favors the absorption of iron and may play a role in maintaining stability of biological membranes. Because of its antioxidant properties, vitamin E also is able to spare carotene and vitamin A from oxidative destruction.

Vegetable oils are good sources of vitamin E, but vitamin E deficiency under practical conditions of human nutrition is rare. Vitamin E in large doses has been promoted as a remedy for numerous diseases and as an agent to prolong youth and increase sexual potency. There is little scientific evidence for such claims.

Vitamin K

Vitamin K is essential for normal blood clotting. Its deficiency generally parallels liver disease where fat absorption is abnormal. It also can be deficient in infants. This is prevented by giving infants vitamin K with their formulas. Good sources of vitamin K are green vegetables such as spinach and cabbage. Vitamin K also is synthesized by bacteria in the human intestinal tract. Thus, antibiotic therapy that destroys intestinal organisms can produce deficiencies of vitamin K and certain other vitamins synthesized by bacteria.

Vitamin C (Ascorbic Acid)

Vitamin C is the antiscurvy vitamin. Its deficiency causes fragile capillary walls, easy bleeding of gums, loosening of teeth, and bone joint diseases. It is necessary for the normal formation of the protein collagen, which is an important constituent of skin and connective tissue. Like vitamin E, vitamin C favors the absorption of iron.

Vitamin C, also known as ascorbic acid, is easily destroyed by oxidation, especially at high temperatures, and is the vitamin most easily lost during food processing, storage, and cooking. Vitamin C-containing foods must be protected against exposure to oxygen to prevent losses.

The recommended daily allowance for vitamin C in the United States for the male and female adults is 60 mg. In the United Kingdom and Canada, the recommended daily allowance has been 30 mg. This is true with other vitamin and nutrient recommendations—there is not complete international agreement.

Excellent sources of vitamin C are citrus fruit, tomatoes, cabbage, and green peppers. Potatoes also are a fair source (although the content of vitamin C is relatively low) because we consume large quantities of potatoes. Milk, cereals, and meats are poor sources.

Two of the more recent claims for vitamin C are that it removes high levels of cholesterol from the blood of rats and prevents colds in humans. The significance of the rat studies in relation to humans has not yet been established. A very high level of one or more grams of vitamin C taken daily in the form of tablets has been advocated by some as a way to prevent colds. However, the effectiveness of this treatment has not been supported by the medical profession or the FDA.

Vitamins of the B Complex Group

All members of the vitamin B complex generally are found in the same principal food sources, such as liver, yeast, and the bran of cereal grains. All are required for essential metabolic activities and several function as parts of active enzymes. Absence of a particular B vitamin results in a specific deficiency disease.

Thiamin (Vitamin B₁).

Thiamin was the first of the B vitamins to be recognized. The disease beriberi, caused by a deficiency of thiamin, is common where polished rice is a major dietary item. Fortification of rice or white bread with thiamin corrects this disease. A most important role of thiamin is in the utilization of carbohydrate to supply energy, where it functions as the coenzyme thiamin pyrophosphate, or cocarboxylase, in the oxidation of glucose.

Important to the food technologist is the sensitivity of thiamin to sulfur dioxide (SO_2), a common food preservative chemical, and to sulfite salts. Sulfur dioxide destroys the vitamin activity and should not be used to preserve foods that are major sources of thiamin—a practice that is prohibited by the FDA and the meat inspection laws. The recommended adult daily allowance for thiamin is about 1.0–1.5 mg, depending on age and sex. Best sources are wheat germ, whole cereals containing bran, liver, pork, yeast, and egg yolk. Thiamin is stable to heat in acid foods but less so in neutral and alkaline foods, and this is taken into account in food processing.

Riboflavin (Vitamin B₂)

Riboflavin is the yellow-green pigment of skim milk and whey. It functions in the oxidative processes of living cells and is essential for cellular growth and tissue maintenance.

Deficiency in humans generally results in skin conditions, such as cracking at the corners of the mouth. Recommended daily allowance for adults is 1.2–1.7 mg, depending on sex and age. Liver, milk, and eggs are good sources. Meats and green leafy vegetables are moderate sources of riboflavin.

Riboflavin is quite resistant to heat but very sensitive to light and this is why brown milk bottles have seen limited use in the past. Paper cartons, which protect milk from light, are more practical.

Niacin (Nicotinic Acid)

Niacin, also referred to as nicotinamide in the United Kingdom, is not to be confused with nicotine from tobacco. A deficiency of niacin adversely affects tissue respiration and oxidation of glucose and results in the disease known as pellagra in humans. This is characterized by skin and mucous membrane disorders as well as depression and confusion. Pellagra can be cured by feeding niacin or by feeding the essential amino acid tryptophan from which niacin can be made in the body. The adult recommended daily allowance is 13–20 mg niacin, depending on sex and age. Good sources of this vitamin are yeast, meat, fish, poultry, peanuts, legumes, and whole grain cereals. Niacin is very stable to heat, light, and oxidation, but like other water-soluble nutrients it can be leached from foods during processing and cooking.

Vitamin B₆

Vitamin B₆ is the name given to the closely related substances pyridoxine, pyridoxal, and pyridoxamine. Although essential in the human diet for specific enzyme systems and normal metabolism, a deficiency of this vitamin does not cause a well-recognized disease. Vitamin B₆ is widely distributed in foodstuffs—good sources being muscle meat, liver, green vegetables, and grain cereals with bran. The recommended daily allowance for adults is approximately 2 mg, and 2.2 mg during pregnancy and lactation. Women taking steroid contraceptive pills may require higher levels.

Pantothenic Acid

Because pantothenic acid is widespread in foods, obvious symptoms of its deficiency are rare in humans. But a deficiency may appear in experimental animals on limited diets or in severly malnourished individuals. In this case there is a general lowering in the state of well-being of the individual with signs of depression, less resistance to infection, and possibly less tolerance to stress. The human requirement for this vitamin is not well established but is believed to be about 5 mg per day, including pregnant and lactating women. This is easily supplied in a normal diet.

Vitamin B₁₂

Called the anti-pernicious anemia factor, vitamin B_{12} also is important in nucleic acid formation and in fat and carbohydrate metabolism. Vitamin B_{12}, also called cyano-cobalamin, is the largest vitamin molecule and contains cobalt in its structure, giving rise to an essential requirement for the mineral cobalt in nutrition.

Vitamin B_{12} is synthesized by bacteria and molds and is a commercial by-product of antibiotic production. Good natural sources of this vitamin are liver, meats, and seafoods. Strict vegetarians may not get sufficient vitamin B_{12} from their diets since it is virtually absent from plant tissues. The recommended daily allowance for adults is 2.0 μg. Vitamin B_{12} activity is not restricted to a single substance but is exhibited by several structurally related compounds.

Folacin

Folacin and folate are the names given to related compounds exhibiting the vitamin activity of folic acid. Like Vitamin B_{12} folacin prevents certain kinds of anemias, is involved in the synthesis of nucleic acids, and is synthesized by microorganisms. Folacin is present in animal and plant foods, especially liver, leafy vegetables, legumes, and cereal grains and nuts. The recommended daily allowance for folacin is about 200 μg for adult males, 180 μg for females, and 400 μg during pregnancy. This allowance recognizes a limited biological availability of the vitamin from certain foods of a mixed diet.

Biotin and Choline

Two additional substances that are water soluble and generally listed with the vitamins of the B complex are biotin and choline. Biotin is active in the metabolism of fatty acids and amino acids. Choline is a component of cell membranes and brain tissue, and it functions in the transmission of nerve impulses. Biotin and choline are seldom in short supply when the diet is adequate in the other B vitamins. Further, these and other growth factors, such as inositol and para-aminobenzoic acid, are produced by the normal microflora of the intestine.

Daily Allowances and Insufficiency

Recommended daily allowances for vitamins not only differ for children and adults, for different physiological states, and for different levels of physical activity, but a distinction also must be made between recommended levels and minimum acceptable levels. The recommended levels given in Table 4.2 provide a substantial margin of safety and may be as much as five times the minimum levels required to sustain life.

Although diet may provide liberal quantities of the various vitamins, several practices and life situations that have become common can result in vitamin inadequacies. As mentioned, women taking steroid contraceptive pills may require higher intakes of vitamin B_6. Oral contraceptives also lower body levels of vitamins C, B_1, B_2, B_{12}, and folacin. Heavy consumption of alcohol may result in vitamin B_1, B_6, and folacin insufficiency. Smoking reduces blood levels of vitamin C. Emotional stress can decrease absorption and increase excretion of vitamins and other nutrients. Prolonged use of certain drugs also can increase vitamin and other nutrient requirements.

MINERALS

Calcium and Phosphorus

Calcium and phosphorus are the minerals that humans require in the greatest amounts. Deficiencies result chiefly in bone and teeth diseases. Calcium also is necessary for clotting of the blood, for the function of certain enzymes, and for control of fluids through cell membranes. Phosphorus is an essential part of every living cell. It is involved in the enzyme-controlled energy-yielding reactions of metabolism. Phosphorus also helps control the acid–alkaline reaction of the blood. Highest requirements for calcium and phosphorus are for the young, and for pregnant and nursing mothers.

Not only is the dietary intake of these minerals important but also the percentage that is absorbed into the bloodstream. Because calcium and phosphorus can combine and precipitate one another, they actually interfere with the effective absorption of one another. Oxalates in foods like rhubarb also can precipitate calcium and make it unavailable for nutritional purposes. Milk and dairy products are excellent sources of calcium and phosphorus, and in normal diets there is seldom any deficiency of these minerals. In recent years the role of calcium in preventing the loss of calcium from bones (called osteoporosis) has been studied. This diesease is especially prevalent in older women. There is some evidence that increasing calcium intake, especially when young, can help reduce osteoporosis later in life. Vitamin D is essential for absorption of calcium from the intestinal tract, and lactose also is effective in promoting this absorption. This makes milk, especially milk fortified with vitamin D, a particularly valuable source of calcium.

Magnesium

Magnesium is essential to the function of several enzyme systems, is important in maintaining electrical potential in nerves and membranes, is involved with liberation of energy for muscle contraction, and is required for normal metabolism of calcium and phosphorus. Deficiency symptoms are more common in farm and experimental animals, which may have a restricted diet, than in humans whose diets generally are adequate in magnesium.

Iron and Copper

Iron is required as a component of blood hemoglobin, which carries oxygen, and muscle myoglobin, which stores oxygen. Of all required nutrients, shortages of iron may be the most common inadequacy in the diets of the industrialized world. Copper aids in the utilization of iron and in hemoglobin synthesis. The need for iron and copper is related to the rate of growth and to blood loss. Much of the iron in plant foods is bound in poorly soluble iron phytate and iron phosphates and is not bioavailable. Iron from animal sources generally is more readily absorbed in digestion, as is iron from soluble salts used in food enrichment and fortification.

Cobalt

As mentioned, cobalt is a part of vitamin B_{12}. However, cobalt will not replace the need for vitamin B_{12} in humans.

Zinc

Zinc is an essential constituent of enzymes involved in carbohydrate and protein metabolism and nucleic-acid synthesis. Its deficiency results in impaired growth and development, skin lesions, and loss of appetite.

Sodium and Chloride

Sodium and chloride are the chief extracellular ions of the body. They are involved primarily with maintaining osmotic equilibrium and body-fluid volume. Chloride ion also is necessary for the production of hydrochloric acid of gastric juice. Great losses occur in sodium and chloride during loss of body fluids, such as perspiration during exercise, and these must be replaced to prevent weakness, nausea, and muscle cramps. A human's daily intake from food of about 10 g of salt more than meets their needs, and indeed may be excessive since high sodium can contribute to elevating blood pressure. Vegetables are relatively low in salt and so vegetarians and grass-eating animals generally need salt supplementation to their diets.

Potassium

Potassium is the principal intracellular cation and with sodium helps regulate osmotic pressure and pH equilibria. It also is involved with cellular enzyme function. Potassium is essential for life but rarely is limiting even in the most meager diets.

Iodine

Iodine is part of the thyroid hormone and is essential for the prevention of goiter in humans. There is never a shortage of iodine where saltwater fish are eaten. The central United States and parts of South America, away from the ocean, are short of indigenous iodine. Today, the common use of iodized salt prevents deficiency, and in the United States there is concern that iodine levels not become excessive.

Fluorine

The fluoride ion is required for the development of sound teeth with resistance to tooth decay. Diets of growing children appear to be low in fluorine since supplementation of water with about 1 ppm reduces incidence of tooth decay. No other dietary requirement for fluorine is well documented.

Other Elements

Several other trace minerals are required by humans in at least trace amounts, but normal diets generally provide these. Thus, manganese is needed for normal bone structure, reproduction, and functioning of the central nervous system. Chromium is required for normal glucose metabolism. Molybdenum is involved in protein metabolism and oxidation reactions. Requirements also have been demonstrated in experimental animals for selenium, nickel, tin, vanadium, arsenic, and silicon, but their roles in human nutrition remain to be determined.

FIBER

The role of indigestible components of plant materials in providing roughage and bulk and in contributing to a healthy condition of the intestine has long been recognized. Celluloses, hemicelluloses, pectins, lignins, and other plant substances that are not readily digested perform this role and are collectively referred to as fiber or dietary fiber. All of these substances hold water, tend to soften stools, and decrease stool transit time through the large intestine.

In addition to these benefits of a diet adequate in fiber, research over the past decade has revealed further physiological actions of fiber under specific conditions. These include the lowering of plasma cholesterol levels, decreasing the incidence of colon cancer, lowering insulin requirements of diabetics, and others. This has led to many exaggerated health claims for fiber beyond experimental findings and the promotion of many new high-fiber food products and supplements. Although the term fiber is used inclusively, it is clear that the fiber from different food sources contains varying proportions of the different indigestible components and these components are not equal in terms of physiological effect. Further, grinding and other processing can affect the physical properties (e.g., particle size) and, in turn, the water-holding capacity of a fiber from a particular source. Fiber also may bind minerals, making them unavailable for absorption; if excessive fiber is ingested, this binding could produce essential mineral imbalance and deficiency.

Diets that contain moderate quantities of cereal grains, fruits, and vegetables are not likely to be low in fiber nor excessive in mineral binding. Persons in good health consuming such diets would not be expected to benefit from high-fiber supplements.

WATER

About 60%, by weight, of a person's body is water. A normal person experiences symptoms of dehydration when 5–10% of the body weight is lost as water and not soon replaced. Long before this occurs, thirst, weakness, and mental confusion are experienced. If the state of dehydration progresses further, the skin and lips lose elasticity, the cheeks become pale and the eyeballs sunken, the volume of urine decreases, and ultimately respiration ceases. Under certain conditions, a person may survive without food for about 5 weeks but can seldom live without water for more than a few days.

The need for water exists at the molecular level, the cellular level, and at the metabolic and functional levels. Water is the major solvent for the organic and inorganic chemicals involved in the biochemical reactions that are essential to life. Water is the principal medium that transports nutrients via body fluids to cell walls and through membranes. Water is the medium that carries nitrogenous waste products from the cells for ultimate elimination. The evaporation of water from the skin is one important mechanism for controlling and maintaining normal body temperature, essential for the controlled rate of metabolic reactions and physical comfort of the individual.

The quantitative requirement for water is directly related to the sum of water losses from the body. These include losses from excretion and elimination of body wastes, perspiration, and respiration. Any factors that increase the rates of these processes, such as exercise, excitation, elevated temperature, or low relative humidity, also increase the need for water replenishment.

An adult may consume 400 liters of water a year. About an equal amount is obtained from food. Given sufficient water, or water in excess, the body closely regulates its water content. Except in unusual cases of deprivation or illness, the body seldom suffers from a deficiency of water in the sense that there may be a deficiency of other essential nutrients. This is because, unlike many of the other nutrients, a decrease in body water causes almost immediate discomfort, driving the individual to correct the shortage.

STABILITY OF NUTRIENTS

One of the principal responsibilities of the food scientist is to preserve nutrients through all phases of food acquisition, processing, storage, and preparation. The key to doing this is a knowledge of the stability of nutrients under different conditions. As shown in Table 4.3, vitamin A is highly sensitive (i.e., is unstable) to acid, air, light, and heat; on the other hand, vitamin C is stable in acid but is sensitive to alkalinity, air, light, and heat. Because of the instability of nutrients under various conditions and their water solubility, cooking losses of some essential nutrients may be greater than 75% (Table 4.3). In modern food processing operations, however, losses seldom exceed 25%.

Where nutrient losses are unavoidably high, the law permits restoration or enrichment by the addition of vital nutrients. A common example is the enrichment of flour and white bread (Table 4.4). The standards for enrichment of food products are in a state of periodic revision as knowledge of nutrition increases.

The final nutritive value of a food reflects losses incurred throughout its history—from farmer to consumer. Nutrient value begins with the genetics of the plant and animal. The farmland fertilization program affects tissue composition of plants, and animals consuming these plants. The weather and degree of maturity at harvest also affect tissue composition. Storage conditions before processing affect vitamins and other nutrients. Washing, trimming, and heat treatments affect nutrient content. Canning, evaporating, drying, and freezing alter nutritional values, and the choices of times and temperatures in these operations must be balanced between good bacterial destruction and minimum nutrient destruction. Packaging and subsequent storage affect nutrients. One of the most important factors is the final preparation of the food in the home and the restaurant—the steam table can destroy much of what has been preserved through all prior manipulations.

DIET AND CHRONIC DISEASE

With a few specific subpopulation exceptions such as the very poor or people with other medical problems, diseases resulting from insufficient nutrients in most of the

Table 4.3. Stability of Nutrients

Nutrient	Neutral pH 7	Acid <pH 7	Alkaline >pH 7	Air or Oxygen	Light	Heat	Cooking Losses (%)
Vitamins							
Vitamin A	S	U	S	U	U	U	0–40
Ascorbic acid (C)	U	S	U	U	U	U	0–100
Biotin	S	S	S	S	S	U	0–60
Carotenes (pro-A)	S	U	S	U	U	U	0–30
Choline	S	S	S	U	S	S	0–5
Cobalamin (B_{12})	S	S	S	U	U	S	0–10
Vitamin D	S		U	U	U	U	0–40
Essential fatty acids	S	S	U	U	U	S	0–10
Folic acid	U	U	S	U	U	U	0–100
Inositol	S	S	S	S	S	U	0–95
Vitamin K	S	U	U	S	U	S	0–5
Niacin (PP)	S	S	S	S	S	S	0–75
Panthothenic acid	S	U	U	S	S	U	0–50
p-Amino benzoic acid	S	S	S	U	S	S	0–5
Vitamin B_6	S	S	S	S	U	U	0–40
Riboflavin (B_2)	S	S	U	S	U	U	0–75
Thiamin (B_1)	U	S	U	U	S	U	0–80
Tocopherols (E)	S	S	S	U	U	U	0–55
Essential amino acids							
Isoleucine	S	S	S	S	S	S	0–10
Leucine	S	S	S	S	S	S	0–10
Lysine	S	S	S	S	S	U	0–40
Methionine	S	S	S	S	S	S	0–10
Phenylalanine	S	S	S	S	S	S	0–5
Threonine	S	U	U	S	S	U	0–20
Tryptophan	S	U	S	S	U	S	0–15
Valine	S	S	S	S	S	S	0–10
Mineral Salts	S	S	S	S	S	S	0–3

SOURCE: Harris, R.S. and Kamas, E. 1975. Nutritional Evaluation of Food Processing, 2nd ed. AVI Publishing Westport, CT.

Table 4.4. Federal Standards for Flour and Bread Enrichment

	White Flour		White Bread	
	mg/100 g	mg/lb	mg/100 g	mg/lb
Thiamin	0.64	2.9	0.40	1.8
Riboflavin	0.40	1.8	0.24	1.1
Niacin	5.3	24.0	3.3	15.0
Iron	2.9–3.6	20	1.8–2.8	12.5
Calcium[a]	212	960	132	600

SOURCE: Code of Federal Regulations. 1993. 21:137–115 and 21:137–165.
[a]Enrichment with calcium optional.

developed world including the United States have nearly disappeared. This has led to a major shift in nutrition emphasis in the affluent countries. There is now much concern with the effects of consuming the wrong amounts or mix of nutrients and how this may affect one's risk from chronic diseases such as heart disease and cancer. The problems of obesity from overconsumption and its relationship to major degenerative diseases is also of concern. Increasingly, research, nutrition education, and food product development are focused on the relationship between diet and chronic disease and the problems associated with overconsumption.

Recent research indicates that diet is a significant factor in several diseases. A comprehensive report titled "Diet and Health" by the National Research Council found strong evidence that dietary patterns can influence several common diseases including atherosclerotic cardiovascular (heart) diseases and hypertension, and highly suggestive evidence that diet affects cancer. Certain dietary patterns seem to predispose for diabetes mellitus and dental caries.

The evidence for an affect on osteoporosis and chronic renal disease is insufficient. The U.S. Public Health Service has laid out a large number of health objectives for the U.S. population in the year 2000 ("Healthy People 2000"). Prominent were a number of recommendations for changing American's eating habits.

Atherosclerosis and Cardiovascular Disease

Atherosclerosis is used to describe several pathological processes occurring in a number of arteries and is responsible for coronary heart disease, stroke, and diseases of the peripheral circulatory system. Presently in the United States atherosclerosis and related cardiovascular diseases cause over one-half of all deaths, and the incidence of heart disease is greater than in any other country of the world. Atherosclerosis is a disease characterized by deposition of a fatty material on the walls of the arteries. This material consists essentially of cholesterol, triglyceride fats, fibrous tissue, and red blood cells. As the deposit continues to build, it restricts blood flow through the artery. When the coronary artery is involved, heart attack and death may follow. Coronary thrombosis refers to the presence of a blood clot in the coronary artery that blocks the normal flow of blood to the heart. Thus, atherosclerosis can contribute to a coronary thrombosis by narrowing the lumen of the coronary artery so that a clot is more likely to exert blockage.

Both animal studies and human studies have indicated a link between atherosclerosis and diet exists. The intake of saturated fats and cholesterol increases the likelihood of having elevated serum cholesterol which is associated with atherosclerosis. Other factors in addition to diet are associated with the occurrence of atherosclerosis. Among them are obesity, hypertension, diabetes, sedentary living, cigarette smoking, and high blood cholesterol levels. The latter may be caused by diet or be of hereditary origin. Although diet does appear to be involved, it must be emphasized that its relative importance in contributing to atherosclerosis is not entirely clear. Since cholesterol, a sterol found in all animal tissues, eggs, milk, and other foods of animal origin, is a component of the atherosclerotic deposit, it has been reasonable to hypothesize that foods high in cholesterol can contribute to atherosclerosis. Such foods may increase the level of cholesterol in the blood. But other components of diet—especially large quantities of saturated fats and sugars—also can result in high levels of blood cholesterol. Further, some investigators find that high levels of blood triglycerides correlate even more closely with coronary disease than do high levels of blood cholesterol. High levels of blood triglycerides also result from the consumption of large quantities of saturated fats and sugars. Whereas consumption of large quantities of saturated fats can increase levels of both cholesterol and triglycerides in the blood, liberal quantities of polyunsaturated vegetable oils tend to decrease blood cholesterol.

Such considerations have influenced the thinking of many doctors and nutritionists. Until more is learned, many are advocating a diet less rich in fats and sugars, and the substitution of polyunsaturated vegetable oils for at least part of the saturated animal fats. A reduction in quantity of foods high in cholesterol also has been generally recommended.

All of this has had an influence on manufacturers of foods. Research has made it possible to lower the cholesterol content of egg and dairy products. Margarines and other fatty foods are available that have been manufactured with a high concentration of vegetable oils rich in polyunsaturated fatty acids. Such products are said to have a high polyunsaturated to saturated fat ratio or high P/S ratio. Because the complex interrelationships between heart disease and diet have not yet been fully explained, the Food and Drug Administration has been most cautious in regulating promotional claims for such products. In this regard, recognition also must be given to possible adverse effects from excessive levels of polyunsaturated fatty acids; these have been demonstrated under experimental conditions in test animals but remain uncertain in humans. However, as is discussed in Chapter 24, certain health claims regarding heart disease can now be made on food labels.

Hypertension

Hypertension, or high blood pressure, is a major contributor to death from cardiovascular diseases as well as diseases of other organs such as the kidney. It has a genetic component but is further augmented by obesity, lack of physical activity, emotional stress, cigarette smoking, and diet. Of dietary components, sodium has been most studied. Blood pressure is positively correlated with sodium intake in populations which routinely consume larger amounts of sodium. In 1989, the Diet and Health committee of the National Academy of Sciences recommended that daily salt intake (as sodium chloride) be limited to 6 g or less. They also recommended that the use of salt in cooking be limited and that consumption of high-salt foods be limited. This has resulted in recommendations that foods be labeled with respect to their sodium content, and food processors modify products to contain less sodium. A number of low-sodium-content foods have been introduced by the food industry in response to this concern.

Cancer

Among the many causes and contributors to various kinds of cancers, diet-related factors continue to be suggested and to promote controversy. A 1982 report titled "Diet, Nutrition, and Cancer," issued by the National Academy of Sciences–National Research Council, summarized evidence leading to their conclusion that diet affects the risk of getting cancer, especially specific kinds of cancer. The 1989 report of the Committee on Diet and Health of the National Academy of Sciences found sufficient epidemiological evidence that up to one-third of all cancers were related in some way to diet. This led to recommendations that diet be modified to decrease the risk of certain cancers.

The strongest causal relationships cited were between diets high in fat and low in fresh fruits and vegetables and the incidence of gastrointestinal-tract cancers (stomach and colon). This has lead to the recommendations that fat intake should contribute no more than 30% of dietary calories. Fresh fruit and vegetable intake should be increased to at least five servings per day. Excessive consumption of cured and smoked foods, which are associated with increased incidence of cancers of the stomach and esophagus, and excessive consumption of alcohol should be avoided. Less definite were positions relative to consumption of protein, sugar, and fiber. Beneficial effects from increased consumption, but not overconsumption, of vitamins A and C and certain other nutrients were noted also. However, it is not yet possible to say how effective diet alteration might be in reducing the incidence of cancers.

Dietary Guidelines and Recommendations

In addition to the Diet and Health report cited above, several other groups have issued dietary goals in order to arrive at a national nutrition policy which promotes better health. Dietary goals and guidelines have been issued and discussed by a number of health authorities including the Surgeon General's Office, the U.S. Department of Agriculture and the Department of Health and Human Services, the National Academy of Sciences Board on Food and Nutrition, The National Cancer Institute, and others. Although there have been areas of disagreement, most reports recommend the following: avoidance of overweight; consumption of a variety of foods; reduction of total fat to less than 30% of calories and reduction in the amount of saturated fat and cholesterol consumed; moderation in the consumption of salt and alcohol; and increased consumption of fresh fruits and vegetable and other fiber-containing foods.

The general population in the United States has begun to alter their diet in response to some of these recommendations. The food industry has developed many new products in an attempt to respond to these needs. For example, a large number of reduced-fat products have been introduced in recent years.

References

Anon. 1992. Food Guide Pyramid replaces the Basic 4 circle. Food Technol. *46*(7), 64–67.

Committee on Designing Foods. 1988. Designing Foods: Animal Product Options in the Marketplace. National Academy Press, Washington, DC.

Committee on Diet, Nutrition, and Cancer. 1982. Diet, Nutrition, and Cancer. National Academy Press, Washington, DC.

Committee on Dietary Guidelines Implementation of the Food & Nutrition Board. 1991. Improving America's Diet and Health: From Recommendations to Action. National Academy Press, Washngton, DC.

Davidson, L.S.P., Passmore, R., and Eastwood, M.A. 1986. Davidson and Passmore Human Nutrition and Dietetics. 8th ed. Churchill Livingstone, Edinburgh.

FDA. 1974. Improvement of nutrient levels of enriched flour, enriched self-rising flour, and enriched breads, rolls or buns. Fed. Register *39*, 5188–5189, 20891–20892.

Friedman, M.I., Tordoff, M.G., and Kare, M.R. (Editors). Chemical Senses, Vol. 4: Appetite and Nutrition. Marcel Dekker, New York.

Gaby, S.K., Bendich, A., Singh, V.N., and Machlin, L.J. 1991. Vitamin Intake and Health: A Scientific Review. Marcel Dekker, New York.

Goldberg, I. (Editor). 1994. Functional Foods: Designer Foods, Pharmafoods, Nutraceuticals. Chapman & Hall, London.

Guthrie, H.A. 1986. Introductory Nutrition. 6th ed. Times Mirror/Mosby College, St. Louis, MO.

Harris, R.S. and Karmas, E. 1975. Nutritional Evaluation of Food Processing. 2nd ed. AVI Publishing Co., Westport, CT.

Holman, S.R. 1987. Essentials of Nutrition for the Health Professions. Lippincott, Philadelphia.

Machlin, L.J. (Editor). 1991. Handbook of Vitamins. 2nd ed. Marcel Dekker, New York.

National Research Council Committee on Diet and Health. 1989. Diet and Health: Implications for Reducing Chronic Disease Risk. National Academy Press, Washington, DC.

National Research Council Subcommittee on the Tenth Edition of the RDAs. 1989. Recommended Dietary Allowances. National Academy Press, Washington, DC.

Ory, R.L. 1991. Grandma Called it Roughage: Fiber Facts and Fallacies. American Chemical Society, Washington, DC.

Russell, P. and Williams A. 1995. The Nutrition and Health Dictionary. Chapman & Hall, London, New York.

Scherz, H., Kloos, G., and Senser, F. 1986. Food Composition and Nutrition Tables 1986/87. 3rd revised and completed ed. Wissenschaftliche Verlagsgesellschaft, Stuttgart.

United States Public Health Service. 1991. Healthy People 2000: National Health Promotion and Disease Prevention Objectives. Superintendent of Documents, U.S. GPO, Washington, DC.

Whitney, E.N., Hamilton, E.M.N., and Boyle, M.A. 1987. Understanding Nutrition. West Publishing Co., St. Paul, MN.

Woteki, C.E. and Thomas, P.R. (Editors). 1992. Eat for Life: The Food and Nutrition Board's Guide to Reducing Your Risk of Chronic Disease. National Academy Press, Washington, DC.

5

Unit Operations in Food Processing

The number of different food products and the operations and steps involved in their production are indeed very great. Further, each manufacturer introduces departures in methods and equipment from the traditional technology for that product, and processes are in a continual state of evolution. The food scientist would soon experience great frustration if there were not unifying principles and a systematic approach to the study of these operations.

The processes used by the food industry can be divided into common operations, called unit operations. Examples of unit operations common to many food products include cleaning, coating, concentrating, controlling, disintegrating, drying, evaporating, fermentation, forming, heating/cooling (heat exchange), materials handling, mixing, packaging, pumping, separating, and others. These operations are listed alphabetically, not in the order of their natural sequence or importance.

Most unit operations are utilized in the making of a variety of food products. Heat exchanging, or heating, for example, is used in the manufacture of liquid and dry food products, in such diverse operations as pasteurizing milk, sterilizing foods in cans, roasting peanuts, and baking bread.

Unit operations may include numerous different activities. The unit operation of mixing, for example, includes agitating, beating, blending, diffusing, dispersing, emulsifying, homogenizing, kneading, stirring, whipping, and working. We may want to mix to beat in air, as in making an egg white foam, or to blend dry ingredients, as in preparing a ton of dry cake mix; or we may wish to mix to emulsify, as in the case of mayonnaise, or to homogenize to prevent fat separation in milk. We may wish to mix and develop a bread dough, which requires stretching and folding, referred to as kneading.

One of the key elements to food processing is the proper selection and combination of unit operations into more complex integrated processing systems. These operations and processes consume great quantities of energy.

COMMON UNIT OPERATIONS

Materials Handling

Materials handling includes such varied operations as hand and mechanical harvesting on the farm, refrigerated trucking of perishable produce, box car transportation of live cattle, and pneumatic conveying of flour from rail car to bakery storage bins. Throughout such operations emphasis must be given to maintaining sanitary conditions, minimizing product losses (including weight loss of livestock), maintaining raw material quality (e.g., vitamin content and physical appearance), minimizing bacterial growth, and timing all transfers and deliveries so as to minimize holdup time, which can be costly as well as detrimental to product quality.

The movement of produce from farm to processing plant and of raw materials through the plant may take many forms. Oranges, for example, are moved by truck trailers to juice plants, where they are graded and washed. There is a limit to the size the trucks may be and the length of time the fruit may be held since fruits and vegetables are alive, respire, and can cause the temperature of a batch to rise to the point where complete spoilage may occur.

Bulk dry sugar delivered to confectionery and other types of food plants is conveyed from the truck to storage bins by a pneumatic lift system. Storage must not be for a period of time nor at a temperature and humidity that will allow the sugar to cake. Transfer of sugar in the plant must avoid dusting and the buildup of static electricity to prevent possible explosion of the highly combustible sugar particles. This is also true in handling finely divided flour. The pneumatic conveying of spices is shown in Fig. 5.1. This method of materials handling has the additional advantages of preventing loss of desirable volatiles from the spices, irritation to personnel, and flavor exchange between different spices.

The use of a wide variety of screw conveyors, bucket conveyors, belt conveyors, and vibratory conveyors in food plants needs no elaboration here beyond the obvious recognition that conveying and handling equipment for eggs in the shell must be different than for less fragile products.

Cleaning

Foods by the nature of the way they are grown or produced on farms in open environments often require cleaning before use. Cleaning ranges from simple removal of dirt

Figure 5.1. Air conveying systems for handling ground pepper. *Courtesy of R. T. French Co.*

from egg shells with an abrasive brush to the complex removal of bacteria from a liquid food by passing it through a microporous membrane. Grains must be cleaned of stones before use. Cleaning can be accomplished with brushes, high-velocity air, steam, water, vacuum, magnetic attraction of metal contaminants, mechanical separation, and so on, depending on the product and the nature of the dirt.

The cleanliness of water used in the soft drink bottling industry must exceed many of the standards found adequate for drinking water. If a high degree of carbonation is to be achieved, then the water used in making the drink must be remarkably free of dust particles, colloidal particles, and certain inorganic salts, since these minimize carbon dioxide solubility and promote excessive escape of gas bubbles. To adequately clean this water may require that city water receive such additional treatments as controlled chemical flocculation of suspended matter, sand filtration, carbon purification, microfiltration, and deaeration. This is no longer the unit operation of cleaning but a total cleaning process.

Some cleaning methods are dictated by surface characteristics of the product. Because pineapples have an irregular surface, the scrubbing action of high-pressure water jets is used.

Just as different food materials require special cleaning, the surfaces of food processing equipment need thorough and frequent attention. The cleaning of equipment, as well as a facility's walls and floors, must take into consideration the chemical and physical properties of both the surface to be cleaned and the type of soil. Many types of soil can be removed with mildly alkaline detergents, but strong alkali may be required for more tenacious deposits and heavy deposits of fats and oils or built-up protein deposits. Alkaline films and hard-water scales may require mildly acid detergents. Strong acids are highly corrosive to several metals, fabrics, wood, rubber, and concrete floors. Strong alkalis also are corrosive to various metals and to glass. For these reasons, moderately alkaline and neutral detergents find wide application in the food industry. Several are listed in Fig. 5.2, along with the properties that affect their cleaning efficiency. Food plant operators generally call on detergent manufacturers for expertise in establishing highly effective cleaning procedures since these further depend on detergent concentrations, temperatures of application, order of application where more than one cleaning aid is used, and other variables.

Separating

The unit operation of separating can involve separating a solid from a solid, as in the peeling of potatoes or the shelling of nuts; separating a solid from a liquid, as in the many types of filtration; or a liquid from a solid, as in pressing juice from a fruit. It might involve the separation of a liquid from a liquid, as in centrifuging oil from water, or removing a gas from a solid or a liquid, as in vacuum removal of air from canned food in vacuum canning.

One of the commonest forms of separating in the food industry is the hand sorting and grading of individual units as in the case of vegetables and fruit. However, because of the high cost of labor, mechanical and electronic sorting devices have been developed. Difference in color can be detected with a photocell and off-color products rejected. This can be done at enormous speeds with automatic rejection of discolored or moldy nuts or kernels of grain that flow past the photocell. In the case of peanuts to be made into peanut butter, each peanut individually passes through a light beam that activates a jet of air to blow the discolored peanuts from the main stream when an off-color changes the amount of reflected light. Light shining through eggs can detect blood

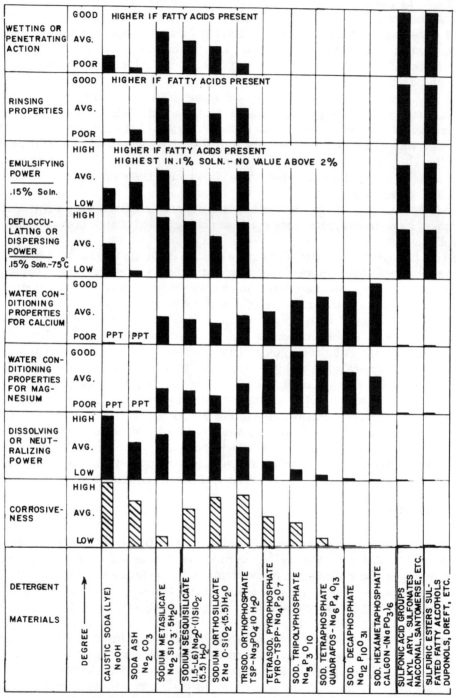

Figure 5.2. Properties of various detergent cleaning materials. Source: *Quality Control in the Food Industry*, 3rd ed., Vol. 2. AVI Publishing Co., Westport, CT, 1973.

spots and automatically reject such eggs. Automatic separation according to size is easily accomplished by passing fruits or vegetables over different size screens, holes, or slits.

The skins of fruits and vegetables may be separated using a lye peeler (Fig. 5.3). Peaches, apricots, and the like are passed through a heated lye solution. The lye or caustic softens the skin to where it can be easily slipped from the fruit by gentle action of mechanical fingers or by jets of water. Differences in the density of the fruit and skin can then be used to float away the removed skin.

To separate corn oil from corn kernels, the germ portion of the corn first is separated from the rest of the kernel by milling; then the oil is separated from the germ by applying high pressure to the germ in an oil press. Similarly, pressure is used to squeeze oil out of peanuts, soybeans, and cottonseeds. The last traces of oil can be removed from the pressed cake by the use of fat solvents. There then remains the separation of the oil from the solvent.

Crystallization is used to separate salt from sea water, or sugar from sugar cane juice. Here, evaporation of some of the water causes supersaturation, and crystals form. Since crystals are quite pure, this is also considered a purification process. The crystals are then separated from the suspending liquid by centrifugation.

Newer methods of separation include several techniques involving manufactured membranes with porosities or permeabilities capable of separations and fractionations at the colloidal and macromolecular size level. Ultrafiltration uses membranes of such porosity that water and low-molecular-weight salts, acids, and bases pass through the membrane but larger protein and sugar molecules are retained. This selective separation process, carried out at ambient temperatures, avoids the heat

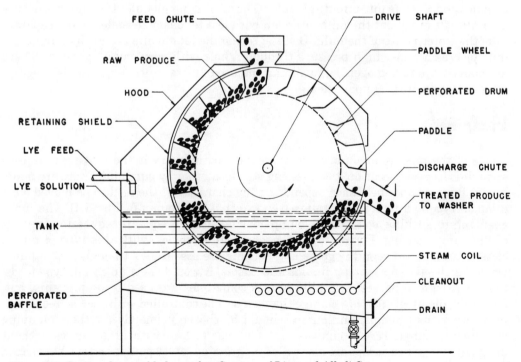

Figure 5.3. Fruit and vegetable lye peeler. *Courtesy of Diamond Alkali Co.*

damage to sensitive food constituents that is often associated with water evaporation at high temperatures. Further, removal of acids and salts with the water prevents their concentration, which would otherwise be detrimental to sensitive retained solids.

Disintegrating

Operations which subdivide large pieces of food into smaller units or particles are classified as disintegrating. It may involve cutting, grinding, pulping, homogenizing, and so on. Although the dicing of vegetables is done on automatic machines, the cutting of meat still largely represents a time-consuming, hand-labor operation. This is because skill is required to separate specific cuts of meat, and the value of the cuts can be sharply reduced by a sloppy job. However, automatic knives with a "brain" are being researched and developed. In another application, cutting of bakery products can be done cleanly and precisely with fine jets of high-pressure, high-velocity water. Laser beams also can replace knives in some cutting applications.

Disintegrating by grinding, as in the preparation of hamburger or hash, always is associated with heating of the product due to friction created in the grinding process. This can be damaging to the food product. It can partially denature proteins or it can give burned flavors to ground coffee. Some kind of cooling is therefore required. In the case of meat, this sometimes is done by grinding the meat in frozen form. Dry ice can be added to the meat or other food to chill it. Dry ice is used rather than regular ice since regular ice would melt and water the food, but dry ice goes off as carbon dioxide and so does not change the composition of the food.

Homogenizing produces disintegration of fat globules in milk or cream from large globules and clusters into minute globules. The smaller fat globules then remain evenly distributed throughout the milk or cream with less tendency to coalesce and separate from the water phase of the milk. Disintegrating the fat globule is done by forcing the milk or cream under high pressure through a hole with very small openings. There are many ways to homogenize, including the use of ultrasonic energy to disintegrate fat globules or break up particles.

Pumping

One of the most common operations in the food industry is the moving of liquids and solids from one location or processing step to another by pumping. There are many kinds of pumps and the choice depends on the character of the food to be moved. One common type is a rotary gear pump (see external gear pump in Fig. 5.4). The inner gears rotate, sucking food into the pump housing and subsequently squeezing food out of the pump housing. For reasons of mechanical efficiency, with this type of pump, close clearances between the gears and housing are essential. Although such a pump would be effective for moving liquids and pastes, it would chew up chunk-type foods, reducing them to purees. Actually, pumps sometimes are used to do just this, but, generally, disintegration is a change that can best be controlled with specialized equipment other than pumps, and pumps should be chosen primarily for their pumping efficiency. A single screw pump is best for moving food with large pieces without disintegration. Such pumps are also called progressing cavity pumps and can be selected for large clearances of the cavities between the turning center rotor and the housing. The food is gently propelled from large clearance to large clearance by the screwlike

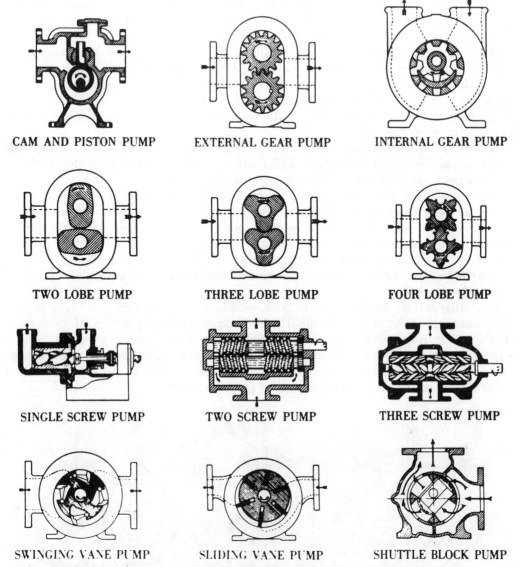

Figure 5.4. Various types of rotary positive displacement pumps. *Courtesy of Hydraulic Institute.*

action of the turning rotor. Food pieces such as corn kernels, grapes, and even small shrimp can be pumped without physical damage. In the gear-type pump these would be ground up.

An essential feature for all food pumps is ease of disassembly for thorough cleaning. Today's sanitary stainless steel pumps in many cases can be disassembled in minutes with a single tool.

Mixing

Like pumps, there are scores of kinds of mixers, depending on the materials to be mixed. One may wish to mix solids with solids, liquids with liquids, liquids with solids, gases with liquids, and so on.

For simple mixing of dry ingredients such as the components that make up a baking power, a conical blender is suitable. The bowl has a tumbling action and this may be continued for 10–20 min until the mixture is made homogenous.

If we are preparing a dry cake mix, we must cut the shortening into the flour, sugar, and other dry ingredients in order to produce a fluffy homogeneous dry mix. We may use a ribbon blender, which is a horizontal trough with one of several types of mixing elements rotating within it. The efficiency of mixing depends on the choice of the mixing element. Three types of ribbonlike elements suitable for cutting in shortening are illustrated in Fig. 5.5.

For mixing solids into liquids to dissolve them, a propeller-type agitator mounted within a stainless steel vat is best. There are a great number of propeller, turbine, and paddle types available for this kind of mixing.

All types of mixers do some work on the material being mixed and produce some increase in temperature. It is often desirable to minimize this temperature rise. However, mixers are also chosen to do special kinds of work on heavy viscous materials while they are being mixed. These mixers may have arms that knead dough, or paddles and arms that work butter. These working mixers are designed with precise geometries to maximize efficiency and minimize energy requirements to achieve the mixing–working operation.

Still other mixers are designed to beat air into a product while it is being mixed. The mixer–beater found in ice cream freezers is an example (Fig. 5.6). As the ice cream mix is being frozen within the freezer bowl, the beating element, or dasher, turns within the bowl. It not only keeps the freezing mass moving to speed freezing and make freezing more uniform, but it also beats air into the product to give the desired volume increase, or overrun, necessary for proper texture.

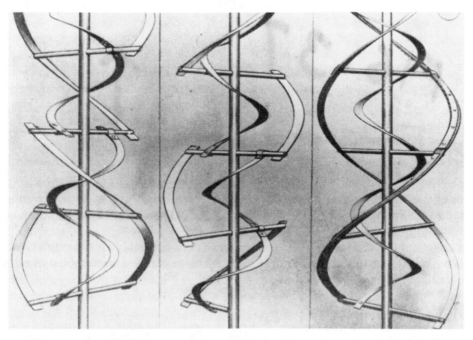

Figure 5.5. The shape of precisely designed mixing elements determine the mixing efficiency of ribbon blender. *Courtesy of J. H. Day Co.*

Figure 5.6. Mixer–beater element of continuous ice cream freezer. *Courtesy of D. K. Bandler.*

Heat Exchanging

Heating

We heat foods for many different reasons. Many foods are heated to destroy microorganisms and preserve the food, for example, during pasteurization of milk and canning of vegetables. Others are heated to drive off moisture and develop flavors, as during the roasting of coffee and toasting of cereals. Still others are heated during normal cooking to make them more tender and more palatable. Some food ingredients, such as soybean meal, are heated to inactivate natural toxic substances. Foods are heated by conduction, convection, radiation, or a combination of these.

Most foods are sensitive to heat, and prolonged heating causes burned flavors, dark colors, and loss of nutritional value. Microorganisms are more sensitive to rapid heating than are chemical reactions. Rapid heating can, therfore, destroy microorganisms faster than it causes undesirable chemical reactions. Hence, it is desirable to heat and cool foods rapidly to maintain optimal quality. Rapid heating is facilitated if the food is given maximum contact with the heating source. This may be accomplished by dividing the food into thin layers in contact with heated plates as in the plate-type heat exchanger used to pasteurize milk (Fig. 5.7). The milk flows across one side of the plates while hot water or steam heats the other side. The same equipment can be used for quick cooling with cold water or brine instead of hot water. This type of heater can be used only with liquid foods.

A jacketed tank or kettle with steam circulated in the jacket is another means of heating liquid foods. It will also heat foods with suspended solids like vegetable soup.

Figure 5.7. Plate-type heat exchanger used to heat and cool liquid foods. *Courtesy of De Laval Separator Co.*

The soup is kept in motion with a mixer propeller for uniform heating and to minimize burning onto the kettle wall.

For sterilizing foods in cans and other containers, entirely different heaters are used. The containers must be heated to temperatures higher than the boiling point of water to achieve sterility in nonacid foods, and so large pressure cookers or retorts are used (Fig. 5.8). Steam under pressure is used to obtain the high temperature needed, and the retort is of heavy construction to withstand this pressure. Another type of retort employs mechanical agitation for better convection of heat within individual cans. The outside of the cans are heated by conduction from the steam.

For roasting coffee beans or nuts, many kinds of heaters have been used. In one type, the beans or nuts move from overhead hoppers into cylindrical vessels that turn and keep the beans in constant motion for even heating. The vessels may be heated within by circulating heated air, or with radiant heat from the vessel walls, the exterior of which can be heated by contact with hot air, gas flame, or steam. In some instances

Figure 5.8. A large steam retort for heat processing packaged foods at temperatures above the boiling point of water. *Courtesy of FMC Corporation.*

this type of roaster is replaced with tunnel ovens in which the coffee beans or nuts pass on moving belts or are vibrated beneath radiating infrared rods or bulbs. Whatever the method, precise control of temperature is essential for proper roasting.

Foods may be heated or cooked using toasters, direct injection of steam, direct contact with flame, electronic energy as in microwave cookers, and so on; all of these methods are currently used in the food industry. Such processes as baking, frying, most food concentration, food dehydration, and various kinds of package closure all employ the unit operation of heating.

Cooling

While heating is the addition of heat energy to foods, cooling is the removal of heat energy. This may be done to the degree where food is chilled to refrigerator temperature, or beyond this range to where the food is frozen. Primarily, we refrigerate and freeze foods to prolong their keeping quality. But there are some foods that owe their entire character to the frozen state. A prime example is ice cream.

A great deal of milk and cream are cooled by passing them in thin layers through heat exchangers of the type shown in Fig. 5.7, or by allowing the liquids to run down over the surface of a hinged leaf cooler. Within the leaves are pipes through which cold water or refrigerant is pumped.

Liquid egg, apple slices, and other fruits in 13.6-kg cans are commonly frozen solid in an air-blast freezer or sharp freezer room maintained at about −26°C. The cans are spaced to allow the cold air, which is circulated by fans and blowers, to get between them and speed the freezing operation. Products in this form go mainly to bakeries for use after thawing.

There are many kinds of commercial air-blast freezers designed to freeze peas, beans, and other vegetables as individual pieces. In one type the peas are loaded on trays that are automatically moved upward through a cold air blast. After freezing, the peas are dislodged from the trays and conveyed under cold air to packaging equipment. The trays return to a position under the pea hopper to receive additional product and the cycle is repeated. Freezing of canned or packaged foods may be done by direct immersion in a refrigerant. Here the cans may be agitated as they pass through the refrigerant within a cylindrical shell, or tube. Agitation increases the efficiency of heat transfer.

The value of quick freezing to food quality, which is discussed more fully in Chapter 9, has led to the use of liquid nitrogen with its extremely low temperature of −196°C. Many food plants have installed large liquid nitrogen tanks and pump liquid-nitrogen to freezers where it is sprayed directly onto foods to be frozen. Delicate products such as mushrooms are frozen this way.

Evaporation

Evaporation in the food industry is used principally to concentrate foods by the removal of water. It is also used to recover desirable food volatiles and to remove undesirable volatiles.

The simplest kind of evaporation occurs when the sun evaporates water from sea water and leaves behind salt; this process is used commercially. Grapes and other fruits can be dried using the energy from the sun. Another simple form of evaporation occurs when a heated kettle is used to boil water from a sugar syrup, as is common in some kinds of candy-making. However, this requires considerable energy in the form of heat for a long period of time, which would cause heat damage to such products as milk or orange juice if an attempt were made to concentrate them this way.

All liquids boil at lower temperatures under reduced pressure and this is the key to modern evaporation. If a heated kettle is enclosed and connected to a vacuum pump, one has a simple vacuum evaporator. Such evaporators are used to remove water from sugar cane press juice in the early stages of crystalline sugar production.

Evaporators differ widely in their design and can be connected in series as in the triple-stage system diagrammed in Fig. 5.9. In this device, progressively higher degrees of vacuum are maintained in the subsequent stages through which the liquid food passes. Regardless of design, however, a principal objective of vacuum evaporators is to remove water at temperatures low enough to avoid heat damage to the food. Multiple-stage evaporators can easily remove water at 50°C and some are designed to boil off water at temperatures as low as 21°C.

Drying

In drying, the object also is to remove water with minimum damage to the food. Whereas evaporators will concentrate foods twofold or threefold, driers will take foods very close to total dryness—in many cases less than 2% or 3% water. Driers are used

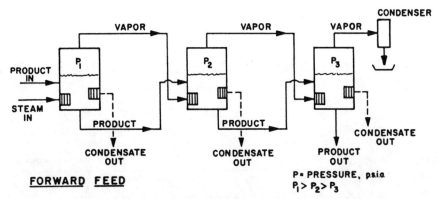

Figure 5.9. Schematic for forward-feed triple effect evaporator. *Courtesy of Hall and Hedrick.*

to prepare such well-known products as dried milk powder and instant coffee. Although food traditionally has been dried to preserve it from spoilage and to reduce its weight and bulk, some foods are dried as convenience items and for their novelty appeal; an example has been freeze-dried fruits for cereals. Drying, as well as several of the other unit operations, is treated in detail in later chapters, and so only a few brief comments are needed here.

Liquid foods such as milk and foods in chunk form like shrimp or steak may be dried. It is generally much easier to dry liquid foods because these are easier to subdivide, either as a spray or a film, and in a subdivided form, the moisture can be removed more quickly.

Subdivision of a liquid is the principle behind the widely used spray drier (Fig. 5.10). Liquid food such as milk, coffee, or eggs is pumped into the top of the large tower, at which point the liquid is atomized by a spray nozzle or equivalent device. At the same time, heated air is introduced to the tower. The heated air in contact with the fine droplets of food dries the droplets, and dehydrated particles fall to the bottom of the tower and are drawn off into collectors. The moisture removed during drying is exhausted separately. Most commercially dried liquid foods are made this way.

Drying by subdividing food as a thin film is commonly done on a drum or roller drier of the kind illustrated in Fig. 5.11. The drum is heated by steam from within, and the applied layer of food flashes off its moisture on contact with the heated drum. The dried food is then mechanically scraped from the drum with long knives. Mashed potatoes, tomato puree, and several milk products are frequently dried this way.

Small food pieces such as peas and diced onions can be dried by moving them through a long tunnel-oven, and many types are in use. However, overheating and shrinkage in the course of water removal may give poor quality products in the case of particularly sensitive foods. For chunk foods a milder method is vacuum freeze-drying. There are many kinds of vacuum freeze-driers. In all types the food pieces first are frozen and then dehydrated under vacuum from the frozen state. The ice does not melt but under the conditions of high vacuum goes off directly as gaseous water vapor, a process known as sublimation. This very gentle kind of drying protects all food quality attributes such as texture, color, flavor, and nutrients. Freeze-drying is by no means restricted in its use to solid and particulate foods. Brewed coffee and quality juice products are dehydrated by freeze-drying.

Figure 5.10. Spray drying tower and associated equipment. *Courtesy of De Laval Separator Co.*

Figure 5.11. Sheets of dehydrated potato coming off commercial drum drier. *Courtesy of Burr and Reeve.*

Forming

Foods must often be formed into specific shapes. For example, hamburger patties are formed by gently compacting ground beef into a disk shape in various types of patty-making machines, which apply controlled pressure to the beef within an appropriate form. Excessive pressure is avoided or the hamburger will be overly tough after cooking. Uniform pressure is essential or patties will vary in weight. Pressure extrusion through dies of various shapes forms doughs into spaghetti and other pasta shapes for subsequent oven drying.

In the confectionery industry, besides pressure extrusion, candies are formed by depositing fondants, chocolate, and jellies into appropriate molds where they cool and harden. Here edible release agents may be sprayed on the mold to help separate the hardened confection. Other confections and food tablets may be formed from powdered ingredients by the application of intense pressure in specially designed tableting machines. Sometimes, as in the case of malted milk tablets, an edible binding agent is required to hold the malted milk powder together. When the powders are high in certain sugars or other thermoplastic food constituents, an additional binding agent is not required. Here high pressure during tablet forming produces heat, melting some of the sugar or other thermoplastic material, which on cooling helps fuse the powdered mass together. This is one way to form fruit juice tablets from dehydrated fruit juice crystals. Some tablet-forming machines also may employ additional heat beyond that generated by pressure.

Forming is an important unit operation in the breakfast cereal and snack food industries. The characteristic shapes of several popular breakfast cereals are the result of pressure extrusion through meticulously designed dies, together with adherence to precisely controlled operating conditions of temperature, pressure, dough consistency, cutoff, and other variables. One special kind of forming is known as extrusion cooking. In this case a formulated dough or mash is extruded under high pressure with or without supplemental heat. The heat, largely from pressure, causes gelatinization of starch and other cooking effects while the material is forced through the extruder. In some cases pressure and temperature are so regulated that the food rises in temperature above that required to boil water. Then when the shaped food emerges from the extrusion cooker, the heated water rapidly boils as pressure is relieved at the exit nozzle. This causes puffing of the formed piece. Such formed and puffed items may then receive additional oven drying. A versatile cooker extruder is diagrammed in Fig. 5.12.

Further examples of forming include the shaping of butter and margarine bars, the pressing of cheese curd into various shapes, the many manipulations given to bread dough to produce variety breads, and the shaping of sausage products in natural and artificial flexible casings.

Packaging

Food is packaged for several purposes, including containment for shipping, dispensing, and unitizing into appropriate sizes, and improving the usefulness of the product. A primary reason is to protect it from microbial contamination, physical dirt, insect invasion, light, moisture pickup, flavor pickup, moisture loss, flavor loss, and physical abuse.

Foods are packaged in metal cans, glass and plastic bottles, paper and paperboard, a wide variety of plastic and metallic films, and combinations of these. Packaging is done by continuous automatic machines sometimes at speeds of more than 1000 units per minute. Many items formerly filled into rigid containers of metal and glass are being increasingly packaged in flexible and formable materials, and filling and capping machines are being joined by more sophisticated systems. Much of the consumer milk

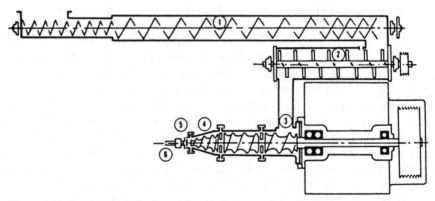

Figure 5.12. Continuous cooker–extruder: (1) Steam conditioning section; (2) mixing zone; (3) kneeding zone; (4) heating zone; (5) die for shaping; (6) cutoff knife. *Courtesy of Wenger Mixer Manufacturing Co.*

supply is packaged in paper cartons. Containers are automatically formed from stacked paper flats, volumetrically filled, and sealed by passing the upper flaps through heated jaws, which melt the plastic coating and thus provide adhesion. In recent years, paperboard cartons which have been coated with special plastics which inhibit oxygen from entering the carton have become widely used for orange juice and similar products.

Other machines form pouches from rolls of plastic film, fill them, and seal them. This is the way many popular snack food items are packaged. Still more complete systems (Fig. 5.13) form the container from roll stock film, fill the container to exact weight, draw a vacuum on the package to remove oxygen, back flush the package with inert nitrogen gas, seal the package, and finally stack the packages into cardboard cartons. This is the way some dessert powders and dehydrated soups are commonly packaged.

The container-forming step is not limited to the use of paper flats or films of various materials. Some food-packaging machines start with plastic resins in granular form, melt these, and blow-mold or otherwise form rigid or semirigid containers for immediate filling and sealing. Two advantages of such a system are the savings of space in food plants that otherwise would have to store great numbers of empty containers, and the in-line production of virtually sterile containers, since the heat to melt the plastic resins also kills microorganisms.

Figure 5.13. Corn chips being fed into a 12-bucket filling system. Bags are formed below the buckets from continuous rolls of material and filled after the correct amount is weighed in each bucket. Source: Anon *Packaging World.* 1(11): 46. 1994.

Controlling

With all of these and several additional unit operations combined into complex processing operations there have to be ways of measuring and controlling them to obtain the desired food product quality. Controlling may be considered a unit operation in itself. Its tools are valves, thermometers, scales, thermostats, and a wide variety of other components and instruments to measure and adjust such essential factors as temperature, pressure, fluid flow, acidity, specific gravity, weight, viscosity, humidity, time, liquid level, and so on.

Figure 5.14 shows a retort commonly used in the canning industry for processing food in metal cans. It is equipped to heat cans of food to the proper sterilization temperature, hold them for the required time, and then cool them. It has controls for steam flow, steam pressure, air pressure, water temperature, water level, and holding time. These controls can be manually operated or can be designed for automatic operation. In modern food plants most instrumentation and controls are automatic, and the plant operator or supervisor directs the process from a remote panelboard or integrates several processes under microprocessor-based computer control from a programmable console.

Overlapping Unit Operations

The division or grouping of food processing steps into unit operations is not perfect and there can be overlapping. For example, filtering bacteria out of beer might logically

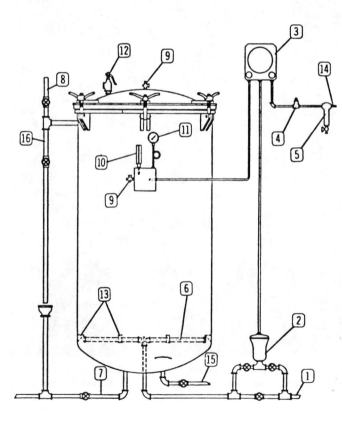

Figure 5.14. Vertical retort with several elements for control of processing. (1) Steam; (2) regulating valve; (3) controller; (4) reducing valve, air; (5) air filter; (6) steam distributor; (7) drain; (8) vent; (9) bleeders; (10) indicating thermometer; (11) pressure gauge; (12) pop safety valve; (13) basket supports; (14) air for controller; (15) water; (16) overflow; ⊗ manual valves. *Courtesy of Continental Can Co.*

be considered cleaning or it might be considered separating. Moving milk to a cheese vat might be viewed as pumping or it might be considered materials handling. Milling grain to yield flour might be considered disintegrating or separating, although actually it is both of these unit operations—disintegrating followed by separating.

Overlapping does not detract from the value of the unit operations concept. This concept permits one to think in an orderly fashion. What is more, some food texts and most food equipment catalogs are divided by unit operations. One may have the problem of blending fragile stuffed olives into sausage meat emulsion with minimum breakage as in the manufacture of certain table-ready meats, or of incorporating a whip improver into commercial liquid egg white without foaming the white. Applicable available equipment generally will not be found in reference sources under food commodity headings but will be grouped in the mixing sections of equipment and engineering references.

If one considers any total food process, such as the manufacture of bread or frozen orange juice concentrate, it is seen that the process is always a series of unit operations performed in a logical sequence. In modern food processing these operations are so connected as to commonly permit smooth, continuous, automatically controlled production.

Energy Conservation

Many food processing unit operations require considerable amounts of energy. Thus, the cost of energy is a significant part of the cost of producing foods. This has focused attention on unit operations, equipment design, and overall processes from the standpoint of optimizing energy use. There is now much interest in the analysis of heating and cooling processes and the recovery and reuse of heat units that formerly were sewered or vented to the atmosphere. Dehydration, concentration, freezing, sterilization, and other operations are being reevaluated in terms of times and temperatures, which affect energy expenditures but also product properties and safety. The energy requirements for materials handling and cleaning often are influenced by varietal types of fruits and vegetables and by agricultural practices before commodities reach processing plants. This also is true for the energy required in peeling, cutting, and disintegrating. The energy requirements for producing the papers, tinplate, aluminum, glass, and plastics of modern packaging, as well as the various package forms, differ, but so do their protective properties, contributions to litter and waste, and recycle values. Preservation by methods that remove water affect product weight and volume and therefore the energy requirements of subsequent transportation and climate-controlled storage. Often the energy required to produce processed foods industrially is far less than that needed to prepare similar foods at home, but careful analyses in this area have been few and the variables that can influence results are many.

There are countless simple measures to conserve energy throughout the food production chain that often are overlooked. These include common everyday practices such as increasing boiler and steam efficiency, optimizing refrigeration and space conditioning through improved temperature control and insulation, closing unnecessary building openings and reducing excessive ventilation rates, reducing lighting excessive to the task performed, using optimum-sized equipment, scheduling regular maintenance including periodic checks on sensors and controlling devices, and so on. Today it is common practice to employ energy conservation specialists and involve plant engineers in general energy management.

New Processes

New processing technologies are constantly being developed which increase the range of options within each unit operation. Major goals of food scientists and processing engineers are to develop new methods which improve quality or increase efficiency. Not surprisingly, there has been considerable research in this area. Newer processing methods that hold promise for improved products over the coming years include super-critical fluid extraction, ohmic heating, and high hydrostatic pressure.

Supercritical fluid extraction uses gases such as carbon dioxide at high pressures to extract or separate food components. At high pressures carbon dioxide behaves as a liquid and can selectively dissolve portions of a food product. For example, supercritical fluid extraction is already commercially used to extract caffeine from coffee to produce a decaffeinated product. The advantage is that the extraction can be made highly selective and occurs under mild conditions which result in a high quality product. This process also has the advantage that the solvent (i.e., carbon dioxide) is not toxic and can be disposed of easily. Supercritical fluid processing can also be used to extract and concentrate delicate flavor compounds from biological materials such as spices and herbs. Research has further shown that supercritical extraction can be used to fraction-ate the fats from dairy products, lower their cholesterol content, and produce lipids with desirable functional characteristics. Large continuous systems are being devel-oped which may lower the cost of such applications and increase efficiency.

Heat to destroy microorganisms can cause undesirable effects in food texture, flavor, and color. The more quickly heat can be applied and removed, the less detrimental are the changes that occur, and so it is not surprising that considerable effort has been directed at developing high-temperature–short time thermal processes.

When foods containing particulates such as beef in a stew are heated by conventional heat exchange systems, the liquid portion becomes overprocessed by the time the inner portion of the beef is sufficiently heated. One proposed solution to this problem is termed ohmic heating. In ohmic heating, the temperature of particulates in a conduct-ing medium such as a salt brine is raised quickly. The food is pumped between two electrodes which are charged with an alternating current similar to common household electricity. The temperature of both particles and liquid increases rapidly. Although this process was first applied to fruit juices, it is particularly useful for foods which consist of particles suspended in liquids like soups or stews.

A third new technology utilizes high hydrostatic pressure. Liquid foods such as fruit juices and beverages or particulate foods suspended in liquid are subjected to pressures as high as several thousand atmospheres. These high pressures can inactivate microor-ganisms and in some cases enzymic activity. In a suggested application, foods are packaged in flexible pouches which are loaded into vessels capable of withstanding the high pressures. Water is then pumped into the vessel and all air removed. The water transmits the pressure to the package. A system for increasing the pressure inside the vessel is activated after the vessel is closed. The major advantage of such a system is that foods may be preserved without the input of large amounts of heat, thus improving their quality. The commercial feasibility of such a system, however, remains to be determined.

References

Batty, J.C. and Folkman, S.L. 1983. Food Engineering Fundamentals. Wiley, New York.
Earle, R.L. 1983. Unit Operations in Food Processing. Pergamon Press, Oxford

Fellows, P. 1988. Food Processing Technology: Principles and Practice. E. Horwood, Chichester.

Fellows, P.J. 1990. Food Processing Technology: Principles and Practice. Prentice-Hall, Englewood Cliffs, NJ.

Giese, J. 1993. On-line sensors for food processing. Food Technol. *47*(5), 88, 90–95.

Hall, C.W., Farrall, A.W., and Rippen, A.L. 1986. Encyclopedia of Food Engineering. 2nd ed. AVI Publishing Co., Westport, CT.

Hayes, G.D. 1987. Food Engineering Data Handbook. Wiley, New York.

Heldman, D.R. and Lund, D.B. 1992. Handbook of Food Engineering. Marcel Dekker, New York.

Karmas, E. and Harris, R.S. 1988. Nutritional Evaluation of Food Processing. Chapman & Hall, London, New York.

Knorr, D. 1993. Effects of high-hydrostatic-pressure processes on food safety and quality. Food Technol. *47*(6), 156, 158–161.

LeMaguer, M. and Jelen, P. 1986. Food Engineering and Process Applications. Elsevier Applied Science Publishers, London.

McLellan, M.R. 1985. Introduction to computer-based process control in a food engineering course. Food Technol. *39*(4), 96–97.

Mertens, B. and Deplace, G. 1993. Engineering aspects of high-pressure technology in the food industry. Food Technol. *47*(6), 164–169.

Paine, F.A. 1987. Modern Processing, Packaging and Distribution Systems for Food. Chapman & Hall, London, New York.

Parrott, D.L. 1992. Use of ohmic heating for aseptic processing of food particulates. Food Technol. *46*(12):68–72.

Rizvi, S.S.H. 1986. Engineering Properties of Foods. Marcel Dekker, New York.

Rizvi, S.S.H. and Mittal, G.S. 1992. Experimental Methods in Food Engineering. Chapman & Hall, London, New York.

Singh, R.P. and Wirakartakusumah, M.A. 1992. Advances in Food Engineering. CRC Press, Boca Raton, FL.

Toledo, R. (Editor). ed. 1991. Fundamentals of Food Process Engineering. 2nd ed. Chapman & Hall, London, New York.

United National Economic Commission for Europe, Geneva. 1991. Food-Processing Machinery, United Nations, New York.

Watson, E.L., Harper, J.C., and Harper, J.C. 1988. Elements of Food Engineering. Chapman & Hall, London, New York.

6

Quality Factors in Foods

In countries where food is abundant, people choose foods based on a number of factors which can in sum be thought of as "quality." Quality has been defined as degree of excellence and includes such things as taste, appearance, and nutritional content. We might also say that quality is the composite of characteristics that have significance and make for acceptability. Acceptability, however, can be highly subjective. Quality and price need not go together, but food manufacturers know that they generally can get a higher price for or can sell a larger quantity of products with superior quality. Often "value" is thought of as a composite of cost and quality. More expensive foods can be a good value if their quality is very high. The nutrient value of the different grades of canned fruits and vegetables is similar for all practical purposes, yet the price can vary as much as threefold depending on other attributes of quality. This is why processors will go to extremes to control quality.

When we select foods and when we eat, we use all of our physical senses, including sight, touch, smell, taste, and even hearing. The snap of a potato chip, the crackle of a breakfast cereal, and crunch of celery are textural characteristics, but we also hear them. Food quality detectable by our senses can be divided into three main categories: appearance factors, textural factors, and flavor factors.

Appearance factors include such things as size, shape, wholeness, different forms of damage, gloss, transparency, color, and consistency. For example, apple juice is sold both as cloudy and clear juice. Each has a different appearance and is often thought of as a somewhat different product.

Textural factors include handfeel and mouthfeel of firmness, softness, juiciness, chewiness, grittiness. The texture of a food is often a major determinant of how little or well we like a food. For example, many people do not like cooked liver because of its texture. Texture of foods can be measured with sophisticated mechanical testing machines such as the one shown in Fig. 6.1.

Flavor factors include both sensations perceived by the tongue which include sweet, salty, sour, and bitter, and aromas perceived by the nose. The former are often referred to as "flavors" and the latter "aromas," although these terms are often used interchangeably. Flavor and aroma are often subjective, difficult to measure accurately, and difficult to get a group of people to agree. A part of food science called sensory science is dedicated to finding ways to use humans to accurately describe the flavors and other sensory properties of foods. There are hundreds of descriptive terms that have been invented to describe flavor, depending on the type of food. Expert tea tasters have a language all of their own, which has been passed down to members of their guild from generation to generation. This is true of wine tasters as well.

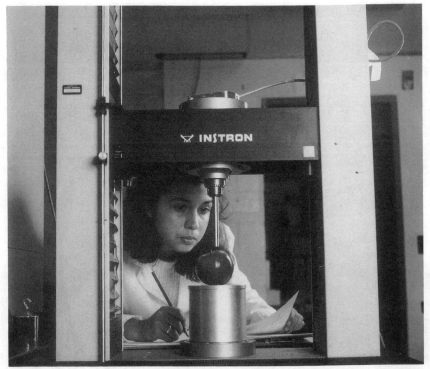

Figure 6.1. Instron Universal Testing Machine being used to test the mechanical/physical properties of an apple. *Courtesy of Cornell University Photo Services.*

Since we generally experience the properties of food in the order of (1) appearance, (2) texture, and (3) flavor, it is logical to discuss quality factors in this order now.

APPEARANCE FACTORS

In addition to size, shape, and wholeness, pattern (e.g., the way olives are laid out in a jar or sardines in a can) can be an important appearance factor. Wholeness refers to degree of whole and broken pieces; the price of canned pineapple goes down from the whole rings, to chunks, to bits. Appearance also encompasses the positive and negative aspects of properly molded blue-veined cheeses, and the defect of moldy bread, as well as the quality attribute of ground vanilla bean specks in vanilla ice cream, and the defect of specks and sediment from extraneous matter. Although some ice cream manufacturers have added ground vanilla bean as a mark of highest quality, others have concluded that as often as not a less-sophisticated consumer misinterprets these specks and rejects the product.

Size and Shape

Size and shape are easily measured and are important factors in federal and state grade standards. Fruits and vegetables can be graded for size by the openings they will pass through. The simple devices shown in Fig. 6.2 were the forerunners of current

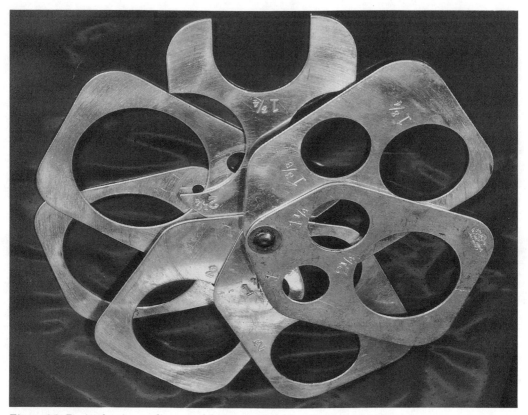

Figure 6.2. Device for size grading round fruits or vegetables. *Courtesy of A. Kramer.*

high-speed automatic separating and grading machines, although they are still used to some extent in field grading and in laboratory work. Size also can be approximated by weight after rough grading, for example, determining the weight of a dozen eggs.

Shape may have more than visual importance, and the grades of certain types of pickles include the degree of curvature (Fig. 6.3). Such curiosities can become quite important, especially in the design of machines to replace hand operations. When an engineer attempts to design a machine for automatically filling pickles into jars at high speeds, it must be recognized that all pickles are not shaped the same, and a machine that will dispense round objects like olives or cherries can be totally inadequate. Mechanized kitchen, restaurant, and vending systems for rapid mass feeding

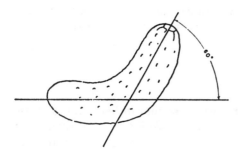

Figure 6.3. Measurement of curvature in pickles as an indicator of quality. *Courtesy of U.S. Department of Agriculture.*

have become commonplace. Some of the most difficult engineering problems encountered in such facilities were in designing equipment that would dispense odd-shaped food pieces into moving dishes.

Color and Gloss

Food color not only helps to determine quality, it can tell us many things. Color is commonly an index of ripeness or spoilage. Potatoes darken in color as they are fried—and we judge the endpoint of frying by color. The bleaching of dried tomato powder on storage can be indicative of too high an oxygen level in the headspace of the package, whereas the darkening of dried tomato can reflect too high a final moisture level in the powder. The color of a food foam or batter varies with its density and can indicate a change in mixing efficiency. The surface color of chocolate is a clue to its storage history. These and many other types of color changes can be accurately measured in the laboratory and in the plant—all influence or reflect food quality.

If the food is a transparent liquid such as wine, beer, or grape juice, or if a colored extract can be obtained from the food, then various types of colorimeters or spectrophotometers can be used for color measurement. With these instruments, a tube of the liquid is placed in a slot and light of selected wavelength is passed through the tube. This light will be differentially absorbed depending on the color of the liquid and the intensity of this color. Two liquids of exactly the same color and intensity will transmit equal fractions of the light directed through them. If one of the liquids is a juice and the other is the same juice somewhat diluted with water, the latter sample will transmit a greater fraction of the incoming light and this will cause a proportionately greater response on the instrument. Such an instrument can also measure the clarity or cloudiness of a liquid depending on the amount of light the liquid lets pass. There are several other methods for measuring the color of liquids.

If the food is liquid or a solid, we can measure its color by comparing the reflected color to defined colored tiles or chips. The quality control inspector changes tiles until the closest color match is made and then defines the color of the food as being identical to the matching tile or falling between the two nearest tiles. Working with tomato products, one would need to have only a few green and red disks to cover the usual range of tomato color. The grade standards for tomatoes have been based on such a method.

Color measurement can be further quantified. Light reflected from a colored object can be divided into three components, which have been termed value, hue, and chroma. Value refers to the lightness or darkness of the color or the amount of white versus black; hue to the predominant wavelength reflected, which determines what the perceived color is (red, green, yellow, blue, etc); and chroma refers to the intensity strength of the color. The color of an object can be precisely defined in terms of numerical values of these three components. Another three-dimensional coordinate scale for describing color utilizes the attributes of lightness–darkness, yellowness–blueness, and redness–greenness. These dimensions of color, used in tri-stimulus colorimetry, can be quantified by instruments such as the Hunterlab Color and Color Difference Meter (Fig. 6.4). Food samples having the same three numbers have the same color. These numbers, as well as numbers representing value, hue, and chroma, vary with color in a systematic fashion that can be graphed to produce a chromaticity diagram (Fig. 6.5). The color chemist and quality controller can relate these numbers to color, and through changes in the numbers can follow gross or minute changes in products that may occur during ripening, processing, or storage. In similar fashion a quality controller can define the

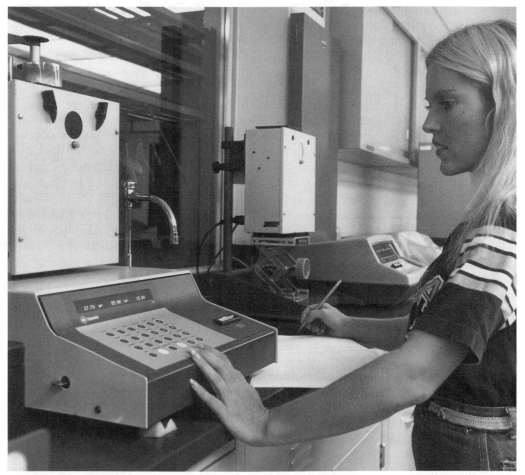

Figure 6.4. Hunterlab color and color-difference meter used to objectively measure the color of food products. *Courtesy of Hunterlab, Inc.*

color of a product and relate this information to distant plants to be matched at any future date. This is particularly useful where the food color is so unstable as to make the forwarding of a standard sample unfeasible.

As with color, there are light-measuring instruments that quantitatively define the shine, or gloss, of a food surface. Gloss is important to the attractiveness of gelatin desserts, buttered vegetables, and the like.

Consistency

Although consistency may be considered a textural quality attribute, in many instances we can see consistency and so it also is another factor in food appearance. A chocolate syrup may be thin-bodied or thick and viscous; a tomato sauce can be thick or thin. Consistency of such foods is measured by their viscosity, higher viscosity products being of higher consistency and lower viscosity being lower consistency.

The simplest method to determine consistency is to measure the time it takes for

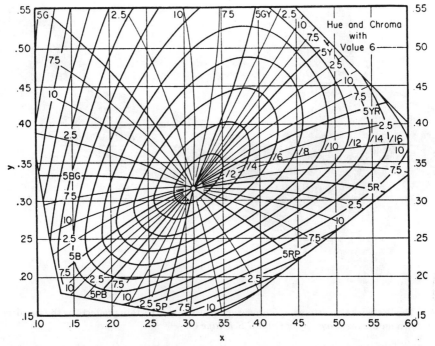

Figure 6.5. One type of chromaticity diagram for objectively describing exact colors. Source: Kramer and Twiss *Quality Control in the Food Industry, 3rd ed., Vol. 2. AVI Publishing Co., Westport, CT,* 1973.

the food to run through a small hole of a known diameter; or one can measure the time it takes for more viscous foods to flow down an inclined plane using the Bostwick Consistometer (Fig. 6.6). This device might be used for ketchup, honey, or sugar syrup. These devices are called viscometers. There are several other types of viscometers using such principles as the resistance of the food to a falling weight such as a ball, and the time it takes the ball to travel a defined distance; and resistance to the rotation of a spindle, which can be measured by the power requirements of the motor or the amount of twist on a wire suspending the spindle. Viscometers range from the quite simple as shown above to highly sophisticated electronic instruments.

TEXTURAL FACTORS

Texture refers to those qualities of food that we can feel either with the fingers, the tongue, the palate, or the teeth. The range of textures in foods is very great, and a departure from an expected texture is a quality defect.

We expect chewing gum to be chewy, crackers and potato chips to be crisp, and steak to be compressible and shearable between the teeth. The consumer squeezes melons and bread as a measure of texture which indicates the degree of ripeness and freshness. In the laboratory, more precise methods are available. However, the squeezing device in Fig. 6.7 gives only an approximation of freshness, since the reading also depends on the stiffness of the wrapping and the looseness with which the bread slices are packed.

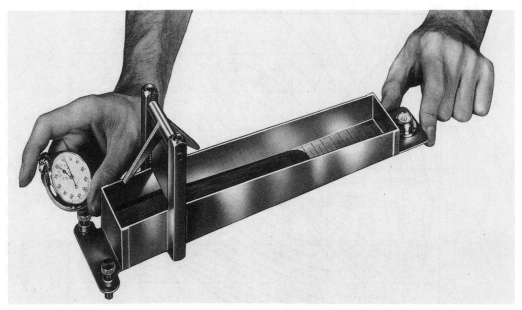

Figure 6.6. The Bostwick Consistometer for measuring viscosity of certain products by the speed with which they flow down an incline. *Courtesy of Central Scientific Co.*

Measuring Texture

Food texture can be reduced to measurements of resistance to force. If food is squeezed so that it remains as one piece, this is compression—as with the squeezing of bread. If a force is applied so that one part of the food slides past another, it is shearing—as in the chewing of gum. A force that goes through the food so as to divide it causes cutting—as in cutting an apple. A force applied away from the material results in tearing or pulling apart, which is a measure of the food's tensile strength—as in pulling apart a muffin. When we chew a steak, what we call toughness or tenderness is really the yielding of the meat to a composite of all of these different kinds of forces. There are instruments to measure each kind of force, many with appropriate descriptive names but none exactly duplicate what occurs in the mouth.

Many specialized test instruments have been devised to measure some attribute of texture. For example, a succulometer (Fig. 6.8) uses compression to squeeze juice out of food as a measure of succulence. A tenderometer applies compression and shear to measure the tenderness of peas. A universal testing machine fitted with the appropriate devices can measure firmness and crispness and other textural parameters (Fig. 6.9). This and similar instruments frequently are connected to a moving recording chart. The time–force curve traced on the chart gives a graphic representation of the rheological properties of the food item. When an apple half is tested, the tracing would show an initial high degree of force required to break the skin, and then a change in force as the compressing–shearing element enters and passes through the apple pulp.

Various forms of penetrometers are in use. These generally measure the force required to move a plunger a fixed distance through a food material. A particular penetrometer used to measure gel strength is the Bloom Gelometer. In this device, lead shot is automatically dropped into a cup attached to the plunger. The plunger positioned above the gel surface moves a fixed distance through the gel until it makes contact

Figure 6.7. Dalby–Hill squeeze-tester for measuring bread stiffness. *Courtesy of G. Dalby.*

with a switch that cuts off the flow of lead shot. The weight of shot in grams, which is proportional to the firmness of the gel, is reported as degrees Bloom. This is one way of measuring the "strength" of gelatin and the consistency of gelatin desserts. Another kind of penetrometer, also referred to as a tenderometer, utilizes a multiple-needle probe that is pressed into the rib eye muscle of raw beef (Fig. 6.10). The force needed is sensed by a transducer and displayed on a meter. The carefully engineered needle probe was designed to give readings that correlate with the tenderness of the meat after cooking, while at the same time not altering the raw meat for further use.

Several of the above methods for measuring texture alter the food sample being tested, so that it cannot be returned to a production batch. Since there are correlations between color and texture in some instances, there are applications where color may be used as an indication of acceptable texture. Under controlled conditions automatic color measurement may then be used as a nondestructive measure of texture; this is done in the evaluation of the ripeness of certain fruits and vegetables moving along conveyor belts. Another nondestructive indication of texture is obtained by the experi-

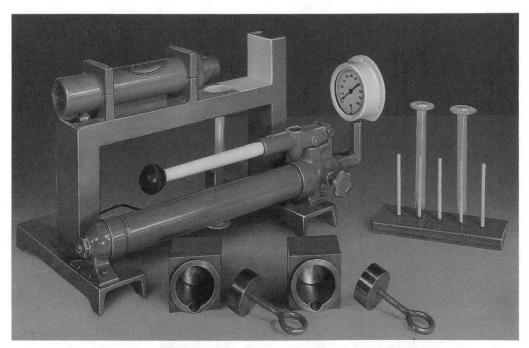

Figure 6.8. The succulometer places a fruit under a constant pressure and allows the juice, which is measured as an indicator of succulence, to run out. *Courtesy of the United Co.*

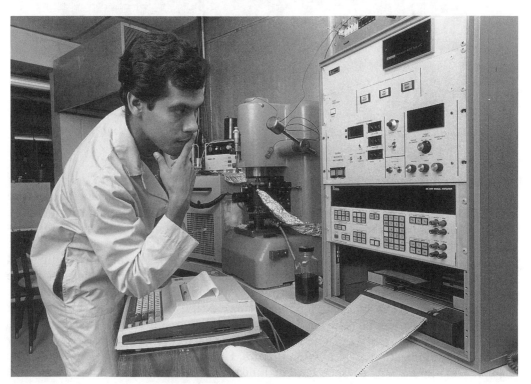

Figure 6.9. Testing the rheological properties of foods using a sophisticated mechanical spectrometer. *Courtesy of Cornell University Photo Services.*

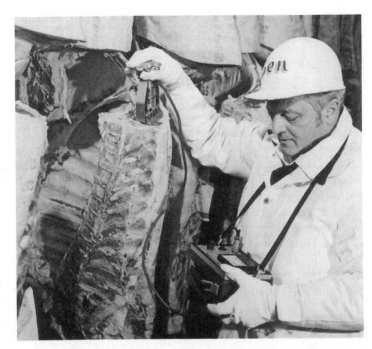

Figure 6.10. Evaluating the tenderness of beef with the Armour meat tenderometer. Source: Anon. *Food Technology* 27:1. 1973.

enced cheesemaker who thumps the outside of a cheese and listens to the sound. This gives a rough indication of the degree of eye formation during ripening of Swiss cheese. One of the newer methods of nondestructive texture measurement makes use of sonic energy, which is absorbed to different extents depending on the firmness of an object.

Texture Changes

The texture of foods, like shape and color, does not remain constant. Water changes play a major role. Foods also can change texture on ageing. Texture of fresh fruit and vegetables becomes soggy as the cell walls break down and the cells lose water. This is referred to as loss of turgor. As more water is lost from the fruit, it becomes dried out, tough, and chewy. This is desirable in the case of dried apricots, prunes, and raisins. Bread and cake in the course of becoming stale lose some water and this is a quality defect. Steaming the bread refreshes it somewhat by softening the texture. Crackers, cookies, and pretzels must be protected against moisture pickup that would soften texture.

Quite apart from changes in the texture of unprocessed foods, there are the textural aspects of processed foods. For example, lipids are softeners and lubricants that the baker blends into a cake formula to tenderize cake. Starch and numerous gums are thickeners; they increase viscosity. Protein in solution can be a thickener, but if the solution is heated and the protein coagulates, it can form a rigid structure as in the case of cooked egg white or coagulated gluten in baked bread. Sugar affects texture differently depending on its concentration. In dilute solution it adds body and mouthfeel to soft drinks. In concentrated solution it adds thickening and chewiness. In still higher concentrations it crystallizes and adds brittleness as in hard candies.

The food manufacturer not only can blend food constituents into an endless number of mixtures but may use countless approved ingredients and chemicals to help modify texture.

FLAVOR FACTORS

As noted already, flavor is a combination of both taste and smell and is largely subjective and therefore hard to measure. This frequently leads to differences of opinion between judges of quality. This difference of opinion is to be expected since people differ in their sensitivity to detect different tastes and odors, and even where they can detect them, people differ in their preference. In some cultures, strong smelling fish is desirable, whereas in others such fish would be unacceptable.

The flavor of a given food is determined by both the mixture of salt, sour, bitter, and sweet tastes and by the endless number of compounds which give foods characteristic aromas. Thus, the flavor of a food is quite complex and has not been completely described for most foods. Adding to this complexity is the fact that the same food is often perceived differently by different individuals. This difference is due to cultural and biological differences between people.

Influence of Color and Texture on Flavor

Judgments about flavor often are influenced by color and texture. For example, we associate such flavors as cherry, raspberry, and strawberry with the color red. Actually, the natural flavor essences and the chemicals they contain are colorless. But in nature they occur in foods of typical color and so we associate orange flavor with the orange color, cherry with red, lime with green, chicken flavor with yellow, and beef flavor with brown.

If gelatin-type desserts are prepared without color, inexperienced tasters will find it hard to distinguish lime from cherry. If we color the lime-flavored item red and the cherry-flavored item green, then the challenge becomes still greater. Butter and margarine may be colored by the addition of a dye. Many consumers will agree that of two samples, the yellow one has the stronger butter flavor, but this may not actually be the case. This is the reason "blind" testing is often employed in flavor evaluation, colored lighting being the means of masking out an influencing color.

Texture can be equally misleading. When one of two identical samples of gravy is thickened with a tasteless starch or gum, many will judge the thicker sample to have the richer flavor. This can be entirely psychological. However, the line between psychological and physiological reactions is not always easy to draw. Our taste buds respond in a complex fashion not yet fully understood. Many chemicals can affect taste response to other compounds. It is entirely possible for texturizing substances to influence taste and flavor in a fashion that is not imaginary. If a thickener affects the solubility or volatility of a flavor compound, its indirect influence on the nose or tongue could be very real.

Taste Panels

We can measure flavor in various ways depending on our purpose. The use of gas chromatography to measure specific volatile compounds has already been mentioned. Some flavor-contributing substances can be measured chemically or physically with other instruments. Examples are salt, sugar, and acid. Salt concentration can be measured electrically by its effect on the conductivity of a food solution. Sugar in solution can be measured by its effect on refractive index. Acid can be measured by titration with alkali, or by potentiometric determination of hydrogen ion concentration as in determining pH (negative logarithm of the hydrogen ion concentration). All of these are largely research or quality control tools. When it comes to consumer quality acceptance, there is still no substitute for measurements made by having people taste products.

We may use individuals, but groups are better because differences of opinion tend to average out. We may use trained individuals as is common in federal and state grading of agricultural products such as butter and cheese. We may employ consumer preference groups—panels that are not specifically trained but can provide a good insight into what customers generally will prefer. We can use panels of highly trained people who are selected on the basis of their flavor sensitivity and trained to recognize attributes and defects of a particular product such as coffee or wine.

A typical taste panel room is provided with separate booths to isolate tasters so that they do not influence one another with conversation or by facial expressions. The booths may be equipped with colored lights when appropriate. The food sample is given to the taster through a closed window so the taster will not see how it was prepared and thus, be influenced. The samples are coded with letters or numbers to avoid terms or brand names that might be influential.

The tasters are given an evaluation form, of which there are many kinds. One kind has columns for samples with descriptive terms such as like definitely, like mildly, neither like nor dislike, dislike mildly, and dislike definitely. The taster checks an opinion for each sample and may make additional comments. The terms are given number rankings by the taste panel leader, such as 5 for like definitely down to 1 for dislike definitely. When all evaluation forms are complete, the taste panel leader tabulates and averages the results. A number ranking scale for flavor or for other quality factors is known as a hedonic scale.

Often taste panelists are asked to choose between two samples in a preference test. Given just two samples, tasters may choose one even though they are really unable to distinguish between them. Given the same two samples again, they might choose in reverse order quite by chance. To avoid this and gain more meaningful data on samples that are quite close in the attribute being studied, preference tests often involve three samples. In this case, tasters are given two samples that are identical and one that is different, all at the same time and appropriately coded. Tasters are asked which two samples are similar, which sample is different, and which is preferred. If a taster cannot correctly pick the odd sample, then his or her preference loses significance. This is known as a triangle test. There are various ways of interpreting triangle tests and different kinds of preference tests; statistical analysis of results is commonly employed.

The number of samples tasters can reliably judge at one sitting without their taste perception becoming dulled is quite limited and depends on the product; generally, no more than about four or five samples can be reliably tested at one time. Taste panel booths often are provided with facilities for rinsing the mouth between samples, or unsalted crackers may be eaten to accomplish a similar effect.

Taste panels used in research, product development, and for purposes of evaluating new and competitive products are not restricted to evaluating flavor. Texture, color, and many other quality factors can be meaningfully measured with this technique.

Further, evaluation forms can be devised to measure the reactions of the very young (Fig. 6.11), or any other special group.

ADDITIONAL QUALITY FACTORS

Three very important quality factors that may not always be apparent by sensory observation are nutritional quality, sanitary quality, and keeping quality.

Nutritional quality frequently can be assessed by chemical or instrumental analyses for specific nutrients. In many cases this is not entirely adequate and animal feeding tests or equivalent biological tests must be used. Animal feeding tests are particularly common in evaluating the quality of protein sources. In this case, the interacting variables of protein level, amino acid composition, digestibility, and absorption of the amino acids all contribute to determine biological value. Whereas the commercial feeding of livestock is done very largely on a nutritional quality basis, unfortunately people do not choose their food on this basis.

Sanitary quality usually is measured by counts of bacteria, yeast, mold, and insect fragments, as well as by sediment levels. X-rays can be used to reveal inclusions like glass chips, stones, and metal fragments in raw materials and finished products moving at high speeds through a plant.

Keeping quality or storage stability is measured under storage and handling conditions that are set up to simulate or somewhat exceed the conditions the product is expected to encounter in normal distribution and use. As normal storage tests may require a year or longer to be meaningful, it is common to design accelerated storage tests. These usually involve extremes of temperature, humidity, or other variables to show up developing quality defects in a shorter time. Accelerated storage tests must be chosen with considerable care because an extreme temperature or other variable frequently will alter the pattern of quality deterioration.

The major quality factors of appearance, texture, and flavor are referred to as organoleptic, or sensory, properties since they are perceived by the senses. There are hundreds

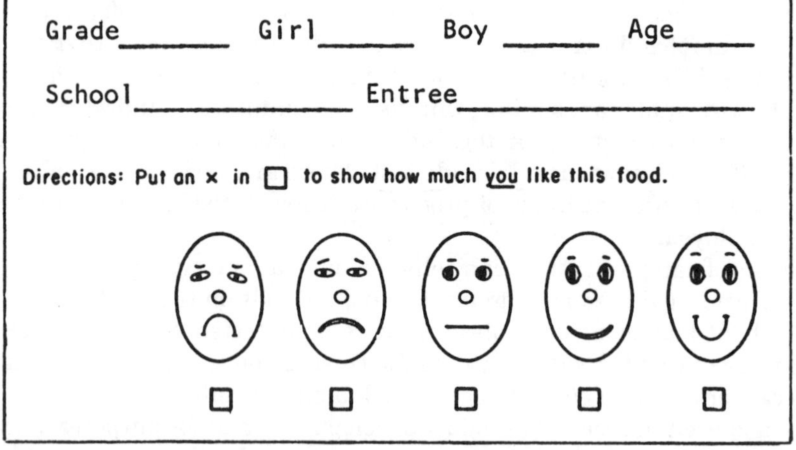

Figure 6.11. Facial hedonic scale used to determine likes and dislikes of children. *Courtesy of E. F. Eckstein.*

of specific quality attributes unique to particular foods and sometimes they do not seem to make much sense, unless we accept the fact that they are traditional and people have become used to them and expect them.

The foam or head on a glass of beer is a quality factor, and its size, bubble structure, and stability are all important quality attributes. But the slightest head foam at the top of a glass of wine or a cup of tea is a quality defect. A slight cloudiness or turbidity is desirable in orange juice and is a quality attribute, but cloudiness in apple juice may be considered a defect by some but desirable by others. The quality of Swiss cheese is judged by the size, shape, gloss, and distribution of its eyes or holes, but eyes in Cheddar cheese is a defect.

These and many other quality attributes make our foods different and interesting, and in many cases the quality attribute that seems arbitrary really is associated with a more fundamental quality factor. The eyes in Swiss cheese, for example, are an indication of flavor through proper bacterial fermentation and of a proper cheese texture capable of holding the carbon dioxide formed during fermentation.

QUALITY STANDARDS

To help ensure food quality, many types of quality standards have come into existence. These include research standards, trade standards, and various kinds of government standards.

Research standards are internal standards set up by a company to help ensure the excellence of its products in a highly competitive market. Trade standards generally are set up by members of an industry on a voluntary basis to assure at least minimum acceptable quality and to prevent the lowering of standards of quality for the products of that industry. Governments set many types of standards. Some are mandatory; these are the standards developed to protect health and prevent deception of the consumer. More will be said about them in later chapters. Other government standards known as Federal Grade Standards are largely optional and have been set up mainly to help producers, dealers, wholesalers, retailers, and consumers in marketing food products. The Federal Grade Standards provide a common language among producers, dealers, and consumers for trading purposes.

Federal Grade Standards

The Federal Grade Standards are standards of quality administered by the USDA Agricultural Marketing Service and the Food Safety and Inspection Service. To give meaning and uniformity to the standards, the USDA established an official system of food inspection and grading. The difference between inspection and grading must be understood. Inspection is often mandatory and assures the wholesomeness of products, particularly meats. Grading is voluntary and determines the quality of products such as meats. Inspectors and graders are trained in the accepted quality factors, and there are inspectors and graders for each major food category. Uniform grades of quality have been established for over 100 foods, including meat, dairy, poultry, fruit, vegetable, and seafood products.

Taking meat products as an example, a federal meat grader evaluates the overall quality of beef, taking into account such factors as shape of the animal carcass, quality and distribution of the fat, age of the animal, firmness and texture of the flesh, and color of the lean meat. The grader stamps the meat in such a repetitive way that the grade stamp will be present on all cuts even after the carcass is butchered for retail sale.

The federal grade marks for beef (Fig. 6.12), in order of decreasing quality, are Prime, Choice, Select, Standard, Commercial, Utility, Cutter, and Canner. The first three are generally found on meat cuts in retail stores or used in restaurants; the other grades are more commonly used in processed meat products. These grade standards are quality standards that do not reflect differences in wholesomeness, cleanliness, or freedom from disease. All meat must pass such inspection regardless of Federal Grade Standards. The quality level indicated by a standard sometimes is changed for reasons of limited availability or economics. The Federal Grade Standards for beef were first instituted in the 1920s and the most recent changes in the beef grading system occurred in 1980.

In similar fashion, other foods for which grade standards have been established are graded. Detailed brochures are published on each food, and quality control tests for the different quality factors are precisely defined. Liberal use of pictures is made where the quality factor is difficult to describe in words.

In the grading of eggs, for example, the freshness quality is fairly accurately measured by the visual condition of the egg white and the egg yolk (Fig. 6.13). A fresh egg has a high percentage of thick white next to the yolk and a small amount of thin white beyond the thick white. As the egg ages, the thick white breaks down and the proportion of the thick white to thin white decreases. Ultimately all of the white is

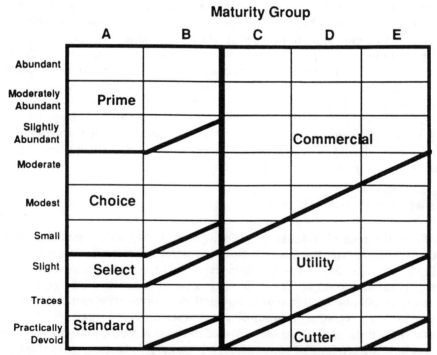

Figure 6.12. The relationship between fat marbling, animal maturity (i.e., age) and the USDA federal grade standard for beef. "A" maturity animals are 9–30 months of age, whereas "E" are more than 96 months. Source: *Muscle Foods: Meat, Poultry and Seafood Technology.* D. M. Kinsman, A. N. Kotula, and B. C. Breidenstein (editors), 1994. *Chapman & Hall*, New York.

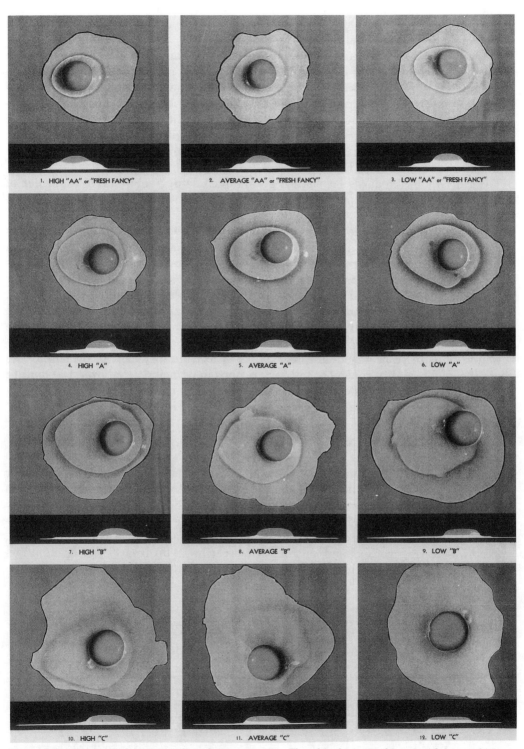

Figure 6.13. Changes in interior quality of eggs on ageing. The white becomes thinner with increasing age. *Courtesy of U.S. Department of Agriculture.*

thin watery white and no thick white will be seen. There is also flattening of the yolk with ageing and loss of freshness.

Quality defects of fruits and vegetables are of many kinds. The grade standards for asparagus recognize slight differences in development in the head and bracts. The standards for sweet potatoes consider the shape that yields "usable pieces" for processing. A usable piece means a segment of such size and shape that it will be recognizable as a sweet potato after canning or freezing. The quality attributes for fresh whole tomatoes (ripeness, color, freedom from cracks and blemishes, size, etc.) also include limited "puffiness." This is the amount of air void or open space between the tomato wall and the central pulp (Fig. 6.14).

Federal Grade Standards for shelled pecans include degree of shriveling, color of nut kernels, degree of chipping of nut halves, moldiness, decay, rancidity, broken shells, and so on.

Typically, the final grade of a product is given after weighing each of the quality factors and giving each a numerical value. The values are then added up to give a total score. The Federal Grade Standard for canned concentrated orange juice allows 40 points for color, 40 points for flavor, and 20 points for absence of defects; the latter includes freedom from seeds, excessive orange oil, proper reconstitution with water, absence of settling out, and so on. The relative importance given in Federal Grade Standards to different quality factors for a wide range of processed fruit and vegetable products is listed in Table 6.1.

Many kinds of score sheets are devised for special quality control purposes. The score sheets for quality control of military rations may include well over 100 quality factors, ranging from specific requirements for packaging materials to storage stability and performance of the packaged food under unique environmental conditions.

Planned Quality Control

Regardless of whether quality is to be maintained on agricultural raw materials or on manufactured food products, a systematic quality control program is essential. This program begins with customer specifications and market demand. What level of quality is demanded and can be produced for the price the customer can afford to pay? Additionally, what legal requirements must be met? With these specifications agreed upon, appropriate testing methods and control stations can be set up. Nearly all food manufacturing facilities have a formal quality control or quality assurance department.

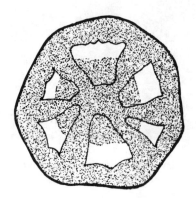 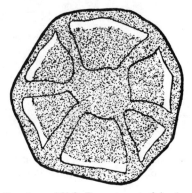

Figure 6.14. The defect of puffiness in tomatoes. *Courtesy of U.S. Department of Agriculture.*

Table 6.1. Relative Importance of Factors Involved in USDA Standards for Processed Fruit and Vegetable Products

Product	Absence of Defects	Color	Flavor	Character	Consistency	Uniformity	Texture	Tenderness and Maturity	Clearness of Liquor
Apples	20	20	—	40	—	20 siz.	—	—	—
Apple butter	20	20	20	20 fin.	20	—	—	—	—
Apple juice	20	20	60	—	—	—	—	—	—
Apple sauce	20	20	20	20 fin.	20	—	—	—	—
Apricots	30	20	—	30	—	20 siz.	—	—	—
Asparagus	30	20	—	—	—	—	—	40	10
Green & wax beans	35	15	—	—	20	—	—	40	10
Dried beans	40	—	—	40	20	—	—	—	—
Lima beans	25	35	—	30	—	—	—	—	10
Beets	30	25	—	—	—	15	30	—	—
Berries	30	25	—	30	—	20	—	—	—
Blueberries	40	20	—	40	—	—	—	—	—
Carrots	30	25	—	—	shape	15 siz.	30	—	—
Cherries, sweet	30	30	20	20	—	20 siz.	—	—	—
Cherries, sour	30	20	20	30	—	20 pits	—	—	—
Corn, cream	20	10	20	—	20	—	—	30	—
Corn, whole	20	10	20	—	40	—	—	40	10
Cranberry sauce	20	20	20	—	40	—	—	—	—
Figs, kadota	30	20	—	35	—	15 siz.	—	—	—
Frozen apples	20	20	—	40	—	20	—	—	—
Fruit cocktail	20	20	—	20	—	20	—	—	20
Fruit jelly	—	20	40	—	40	—	—	—	—
Fruit preserv. (Jam)	20	20	40	30	20	20 siz.	—	—	—
Fruit salad	30	20	—	20	—	20 siz.	(Wholeness 20)	(Drained Wt. 20)	—
Grapefruit	20	20	40	—	—	—	—	—	—
Grapefruit juice	40	20	40	—	—	—	—	—	—
Grape juice	20	40	40	—	—	—	—	—	—
Lemon juice	35	35	30	—	—	—	—	—	—
Mushroom	30	30	—	20	—	20 siz.	—	—	—
Olives, green	30	30	—	20	—	20 siz.	—	—	—

Continued

Table 6.1. (Continued)

Product	Absence of Defects	Color	Flavor	Character	Consistency	Uniformity	Texture	Tenderness and Maturity	Clearness of Liquor
Olives, ripe	10	15	30	25	—	20	—	—	—
Orange juice	20	40	40	—	—	—	—	—	—
Orange juice con.	20	40	40	—	—	—	—	—	—
Orange marm.	20	20	40	—	20	—	—	—	—
Okra	20	15	15	—	—	10	—	35	5
Peaches	30	20	—	30	—	20 siz.	—	—	—
Peanut butter	30	20	30	—	20	—	—	—	—
Pears	30	20	—	30	—	20 siz.	—	—	—
Peas	30	10	—	—	—	—	—	50	10
Peas, field	40	20	—	40	—	—	—	—	—
Cucumber pick.	30	20	—	—	—	20	30	—	—
Pimientos	40	30	—	10	—	20 siz.	—	—	—
Pineapples	30	20	—	30	—	20	—	—	—
Pineapple juice	40	20	40	—	—	—	—	—	—
Plums	30	20	—	30	—	20 siz.	—	—	—
Potatoes, peeled	40	20	—	—	—	20	20	—	—
Prunes, dr.	30	20	—	35	—	15	—	—	—
Pumpkins & squash	30	20	—	20 fin.	30	—	—	—	—
Raspberries	20	25	—	35	—	20 siz.	—	—	15
Sauerkraut	10	15	45	15 crisp	—	—	—	—	15
Sauerkraut, bulk	10	15	45	15 crisp	—	—	—	—	15
Spinach	40	30	—	30	—	—	—	—	—
Sw. potatoes	40	20	—	20	—	20 siz.	—	—	—
Tomatoes	30	30	—	—	—	(Wholeness 20)	(Drained Wt. 20)	—	—
Tomato juice	15	30	40	—	15	—	—	—	—
Tomato paste	40	60	—	—	—	—	—	—	—
Tomato pulp—pure	50	50	—	—	—	—	—	—	—
Tom. sauce-catsup	25	25	25	—	25	—	—	—	—
Chili sauce	20	20	20	20	20	—	—	—	—

SOURCE: Kramer, A. and Twigg, B.A. 1970. Quality Control for the Food Industry 3rd ed. Vol. 2. AVI Publishing, Westport, CT.

The functions of a quality control department are diverse and far ranging, as can be seen in Fig. 6.15 for a department organized to control quality of tomato products. Such a department not only is charged with quality control, which implies detection and correction of defects, but also with the broader concept of quality assurance, which encompasses anticipation and prevention of potential defects. In a food processing or manufacturing plant, quality control testing must start with the raw materials. Sampling and testing of the raw materials will provide a basis for accepting or rejecting these raw materials and will give useful information on how to handle the material in order to obtain a finished product of the desired quality and shelf life. Quality control tests on the processed products through manufacturing, packaging, and warehousing operations are then essential to ensure that customer demands and legal requirements are satisfied.

The diversity of testing indicated in Fig. 6.15 is further complicated by the variations that may be expected between units or batches of raw materials, as well as the fluctuations that occur when any processing condition is repeatedly measured over time. Are variations of a magnitude that is within an acceptable range and within the intrinsic capability of the manufacturing process or is the variation indicative of a process truly out of control? Such questions are answered by applying statistics to repeated measurements of a given quality attribute or processing condition. Repeated measurement will provide data for determination of the mean, range, and normal frequency distribution of the variable under consideration. Such data then can be used to develop quality control charts, one type of which is shown in Fig. 6.16. This chart graphs the variation of a measured attribute from a mean value ($\bar{x}$) with time. The attribute might be weight of filled cans coming from the filling machine prior to can seaming. The chart also provides a range bounded by an upper control limit (UCL) and a lower control limit (LCL) based on the measured attribute's normal frequency distribution. Filled cans with weights beyond this range would be unacceptable and could be removed for adjustment prior to seaming. This type of chart also reveals trends during processing that may call for filler machine adjustment depending on the direction of the drift. Statistical quality control charts are of many kinds and can be devised to monitor specific measurements of size, color, texture, flavor, ingredient composition, nutrients, microbial counts, and numerous types of processing variables. These and other measurements are increasingly being performed continuously by automated instrumentation systems, which can detect deviations from specification and, through computer controls, initiate process adjustment. Thus, on-line measurement of carbon dioxide levels in soft drinks, and of alcohol and carbohydrates as an index of calories in "light" beer, is currently being performed by infrared analyzers as part of the quality control programs for these products.

The influence of market demand on quality specifications cannot be overemphasized. Whereas such factors as nutritional quality and sanitary quality ought not be permitted to vary from well-established standards, organoleptic quality factors are by no means rigid. White eggs are preferred in New York, but brown eggs have been very popular in Boston. Quality bread means different formulas, textures, and shapes in many markets of the United States and indeed all over the world.

Two newer concepts in product quality and safety have been developed and widely adopted in recent years. Total Quality Management (TQM) and Hazard Analysis and Critical Control Point (HACCP) are quality and management "systems" designed to assure the highest quality and safest foods possible. TQM is a management system which strives to continuously improve the quality of products by making small but

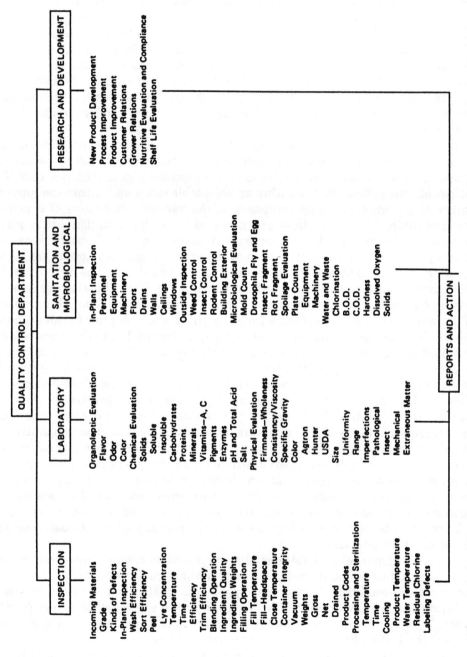

Figure 6.15. Some functions of a quality control department. Source: *Tomato Production, Processing and Quality Evaluation, 2nd ed. AVI Publishing Co., Westport, CT, 1983.*

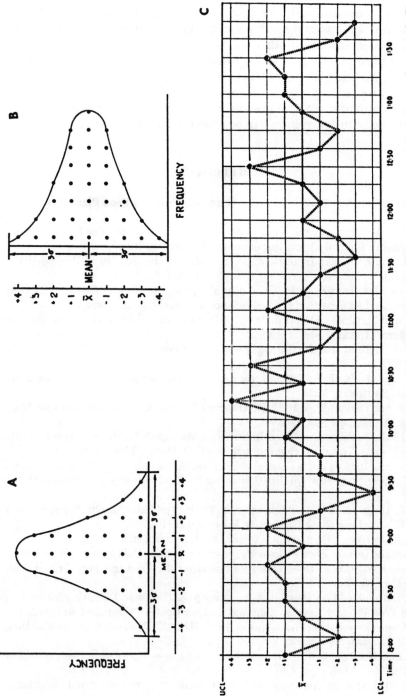

Figure 6.16. Relation of the $\bar{x}$ control chart to a normal frequency distribution: (A) frequency distribution with frequency scale vertical; (B) frequency distribution with frequency scale horizontal; (C) control chart. Frequency distribution extended into a time series. *Source: Kramer and Twigg, Quality Control in the Food Industry, 3rd ed. AVI Publishing Co., Westport, CT, 1973.*

incremental changes in a product's ingredients, manufacture, handling, or storage which result in overall improvement. All workers in plants which employ TQM techniques have joint responsibility for product quality and routinely meet in "quality circles" or similar groups to discuss potential improvements. This concept has been widely applied with success in other industries such as automobile manufacturing.

HACCP is a preventative food safety system in which a process for manufacturing, storing, and distributing a food product is carefully analyzed step-by-step. Points at which tight control of the process will result in elimination of a potential hazard are identified and appropriate control measures taken before a problem occurs. The principles and practices of such a system will be presented in greater detail in Chapter 23.

References

Bianco, L.J., & Associates. 1992. GMP–GSP Guideline Rules for Food Plant Employees. Northbrook, IL.

Bianco, L.J., & Associates. 1992. GMP–GSP Guideline Rules for Food Plant Management (HACCP–Quality Control–Audit Inspections). Northbrook, IL.

Chang, W.-H. and Lii, C.-Y. 1986. Role of Chemistry in the Quality of Processed Food. O.R. Fennema (Editor). Food & Nutrition Press, Westport, CT.

Food and Drink Manufacture. 1989 .Good Manufacturing practice: A Guide to its Responsible Management. 2nd ed. Institute of Food Science and Technology, London.

Fung, D.Y.C. and Matthews, R.F. (Editors). 1991. Instrumental Methods for Quality Assurance in Foods. Marcel Dekker, New York.

Gould, W.A. 1992. Total Quality Management for the Food Industries. CTI Publications, Inc., Baltimore, MD.

Gould, W.A. and Gould, R.W. 1988. Total Quality Assurance for the Food Industries. CTI Publications, Baltimore, MD.

Harrigan, W.F. and Park, R.W.A. 1991. Making Safe Food: A Management Guide for Microbiological Quality, Academic Press, London.

Hayes, G.D. 1991. Quality in the Food Industry: A Manager's Guide to Current Developments and Future Trends. Technical Communications/G.D. Hayes, Manchester.

Herschdoerfer, Ed., 1984. Quality Control in the Food Industry. 2nd ed. Academic Press, London.

Hubbard, M.R. 1990. Statistical Quality Control for the Food Industry. Chapman & Hall, London, New York.

Hubbert, W.T., Hagstad, H.V., and Spangler, E. 1991. Food Safety & Quality Assurance: Foods of Animal Origin. Iowa State University Press, Ames, IA.

Ji, R.I., Dav, I., Jan Vel, I.S., and Jan Pokorn, Y. (Editors). 1990.

Chemical Changes During Food Processing. Elsevier Science Publications, New York.

Kramer, A. and Twigg, B.A. 1970. Quality Control for the Food Industry. 3rd ed. Vol. 1. AVI Publishing Co., Westport, CT.

Lyon, D.H., Francombe, M.A., Hasdell, T.A., Lawson, K. (Editors). 1992. Guidelines for Sensory Analysis in Food Product Development and Quality Control. Chapman and Hall, New York.

Pierson, M.D. and Corlett, Jr., D.A. (Editors). 1992. HACCP: Principles and Applications. Chapman & Hall, London, New York.

Pinder, A.C. and Godfrey, G. (Editors). 1993. Food Process Monitoring Systems. Chapman & Hall, London, New York.

Stauffer, J.E. 1988. Quality Assurance of Food: Ingredients, Processing and Distribution. Food & Nutrition Press, Westport, CT.

Zeuthen, P., et al. (Editor). 1990. Processing and Quality of Foods. Chapman & Hall, London, New York.

7

Food Deterioration and its Control

All foods undergo varying degrees of deterioration during storage. Deterioration may include losses in organoleptic desirability, nutritional value, safety, and aesthetic appeal. Foods may change in color, texture, flavor, or another quality attribute as discussed in Chapter 6.

Food is subject to physical, chemical, and biological deterioration. The highly sensitive organic and inorganic compounds which make up food and the balance between these compounds, and the uniquely organized structures and dispersions that contribute to texture and consistency of unprocessed and manufactured products are affected by nearly every variable in the environment. Heat, cold, light and other radiation, oxygen, moisture, dryness, natural food enzymes, microorganisms and macroorganisms, industrial contaminants, some foods in the presence of others, and time—all can adversely affect foods. This range of potentially destructive factors and the great diversity of natural and processed foods is why so many variations of several basic food preservation methods find application in modern food technology.

The rapidity with which foods spoil if proper measures are not taken is indicated in Table 7.1, which lists the useful storage life of typical plant and animal tissues at 21°C. Meat, fish, and poultry can become inedible in less than a day at room temperature. This is also true for several fruits and leafy vegetables, raw milk, and many other products. Room temperature or field temperature can be much higher than 21°C during much of the year in many parts of the world. Typically, slower rates of deterioration occur with foods that are low in moisture, high in sugar, salt, or acid, or modified in other ways. Nevertheless, even in our modern and efficient warehouses and supermarkets, shelf-stable, refrigerated, and frozen foods undergo continuous change, necessitating stock rotation and product removal at definite intervals, which may be days, weeks, or months for such products as baked goods, soft cheeses, and frozen specialties, respectively.

Rapid spoilage has significance in less developed areas as well as in the most highly advanced and organized societies. In less developed areas starvation because of spoilage has been know to occur in villages only 20–30 km from locations of a lush harvest. In highly advanced societies food production generally is centralized in areas where food can be most efficiently grown or processed. These areas in the United States can be half a continent or more distant from a population center where the food will be consumed. Unless the deteriorative factors are controlled, there would be no food for these population centers and, indeed, there could be no highly advanced society.

History has been made and wars won or lost over food deterioration and its control. Wars, and the need to provide food for armies in regions remote from areas of food production, have always focused attention on the problems of food deterioration, and

Table 7.1. Useful Storage Life of Plant and Animal Tissues

Food Product	Generalized Storage Life 21°C (days)
Meat	1–2
Fish	1–2
Poultry	1–2
Dried, salted, smoked meat and fish	360 and more
Fruits	1–7
Dried fruits	360 and more
Leafy vegetables	1–2
Root crops	7–20
Dried seeds	360 and more

SOURCE: Desrosier and Desrosier (1977).

this is still very true today. It is interesting to note that some of the most important advances in preventing food deterioration have been made in time of war.

At the close of the eighteenth century France was at war and Napoleon's armies were doing poorly on inadequate rations that frequently included spoiled meat and other unwholesome or unpalatable items. Similar problems, including elimination of scurvy, were facing the navy and merchant shipping. Prizes were offered as incentive to encourage development of useful methods of preserving food. From this came the discovery by Nicolas Appert that if food was sufficiently heated in a sealed container and the container not opened, the food would be preserved. Appert was awarded 12,000 francs and honored in 1809, and the world gained the art of food canning. It was not until the work of Pasteur some 50 years later, that growth of microorganisms was shown to be a major cause of food spoilage; this provided an explanation for Appert's method of preservation.

One of the most important aspects of food science is an understanding of food deteriorative factors and their control. Commonly, various forms of preservation were developed long before an understanding of the principles involved were known; and many of the foods we prize today developed out of attempts to prevent deterioration and prolong storage life. One might not ordinarily think of butter as a means of preserving food, but long ago it was discovered that while milk deteriorated in a day or two, clumps of butter fat that formed when milk was agitated could be removed from the milk and would store for weeks or months. Similarly, cheese, smoked fish, dried fruits, and many fermented foods had their beginnings in attempts to slow down deteriorative processes.

SHELF LIFE AND DATING OF FOODS

The shelf life of a food is sometimes defined as the time it takes a product to decline to an unacceptable level. Of course, what is "acceptable" varies from one person to another. To some, the slightest odor in fish is unacceptable, whereas others may prefer fish which has developed considerable odor. In many cases, shelf life is taken as the time a product remains salable. In the final analysis, shelf life is a judgment that must be made by the food manufacturer or retailer. In many cases, the manufacturer must define a minimum acceptable quality (MAQ) for the product. The MAQ will depend on what degree of degradation will be allowed before the manufacturer no longer wants to sell the product.

The actual length of the shelf life of any given product will depend on a number of factors such as processing method, packaging, and storage conditions. For example, one cannot accurately say what the shelf life of fresh milk is without indicating the conditions. Milk held at room temperature will have a different shelf life than milk held under good refrigeration. Of course, canned and processed milk stored at room temperature will outlast pasteurized milk held at refrigeration temperature.

It has become a widespread practice to add some form of dating system to retail packages of foods so that consumers may have some indication of the shelf life or freshness of the products they buy. Several types of code dates have emerged including the date of manufacture ("pack date"), the date the product was displayed ("display date"), the date by which the product should be sold ("sell by date"), the last date of maximal quality ("best used by date"), and the date beyond which the product is no longer acceptable ("use by date" or "expiration date").

Considerable effort has gone into predicting and monitoring shelf life in recent years. Methods of predicting are particularly useful for new products which do not have a history of distribution. Predictions are based on the mechanism of deterioration and the rate at which these deteriorations occur. One recent system for monitoring shelf life uses labels or tags on foods which respond to a combination of time and temperature to which the product has been exposed. These indicators are, hence, called "time–temperature" indicators and are based on the principle that both time and temperature are important in the spoilage of foods.

MAJOR CAUSES OF FOOD DETERIORATION

The major factors affecting food deterioration include the following: (1) growth and activities of microorganisms, principally bacteria, yeasts, and molds; (2) activities of food enzymes and other chemical reactions within food itself; (3) infestation by insects, parasites, and rodents; (4) inappropriate temperatures for a given food; (5) either the gain or loss of moisture; (6) reaction with oxygen; (7) light; (8) physical stress or abuse; and (9) time. These factors can be divided into biological, chemical, and physical factors.

Often these factors do not operate in isolation. Bacteria, insects, and light, for example, can all be operating simultaneously to spoil food in the field or in a warehouse. Similarly, heat, moisture, and air simultaneously affect the multiplication and activities of bacteria, as well as the chemical activities of food enzymes. At any one time, many forms of deterioration may take place, depending on the food and environmental conditions. Effective preservation must eliminate or minimize all of these factors in a given food. For example, in the case of canned meats the meat is sealed in a metal can, which protects it from insects and rodents as well as from light, which could affect its color and possibly its nutritive value. The can also protects the meat from drying out. Vacuum is applied or the can is flushed with nitrogen to remove oxygen before sealing. The sealed can is then heated to kill microorganisms and to destroy meat enzymes. The processed cans are stored in a cool room and the length of time the cans are held in supermarkets and in our homes is limited. In this case the preservation method takes into account all of the major factors in food deterioration. It is well to consider these factors individually.

Bacteria, Yeasts, and Molds

There are thousands of genera and species of microorganisms. Several hundred are associated in one way or another with food products. Not all cause disease or food spoilage, and the growth of several types is actually desirable because they are used to make and preserve foods. The lactic acid-producing organisms used to make cheese, sauerkraut, and certain types of sausage are examples. Others are used for alcohol production in making wine or beer, or for flavor production in other foods. However, except where these microorganisms are especially cultivated by selective inoculation or by controlled conditions to favor their growth over the growth of less desirable types, microorganism multiplication on or in foods frequently is the major cause of food deterioration.

Microorganisms capable of spoiling food are found everywhere—in the soil, water, and air; on the skins of cattle and the feathers of poultry; and in the intestines and other cavities of the animal body. They are found on the skins and peels of fruits and vegetables, and on the hulls of grain and the shells of nuts. They are found on food processing equipment that has not been sanitized, as well as on the hands, skin, and clothing of food-handling personnel.

A most important point, however, is that microorganisms generally are not found within healthy living tissue—such as within the flesh of animals, or the flesh or juice of plants. But they are always present to invade the flesh of plants or animals through a break in the skin, or if the skin is weakened by disease or death. In this case they may digest the skin and penetrate through it to the tissue below. In nearly all cases, the presence of spoilage organisms in foods is a result of contamination. Therefore, one of the major strategies in reducing food spoilage due to microorganisms is to reduce contamination by ensuring good sanitation practices.

Milk from a healthy cow is sterile as secreted, but becomes contaminated as it passes through the teat canals, which are body cavities. Milk becomes further contaminated from dirt on the cow's hide, from the air, from dirty utensils and containers, and so on. Beef becomes contaminated when the animal is slaughtered and the protective skin is broken, especially during cutting. Fruits, vegetables, grains, and nuts become contaminated when the skins or shells are broken or weakened. This is also true of healthy eggs. The inside of a healthy egg is sterile, but the shell of the egg can be highly contaminated from passage through the chicken's body cavity at time of laying.

Bacteria are single-celled organisms, many of which can be classified into one of three types based on the shape of individual cells (Fig. 7.1) These are the spherical shapes represented by several forms of cocci, the rod shape of the bacilli, and spiral forms possessed by the spirilla and vibrios. Many bacteria can move by means of whiplike flagella. Some bacteria, yeast, and all molds produce spores, which are seedlike packets and which under proper conditions can germinate into full-sized cells called "vegetative cells." Spores are remarkably resistant to heat, chemicals, and other adverse conditions (Fig. 7.2). Bacterial spores are far more resistant than yeast or mold spores, and more resistant to most processing conditions than vegetative cells. Sterilization processes are designed specifically to inactivate these highly resistant bacterial spores.

All bacteria associated with foods are small. Most are of the order of one to a few micrometers (μm) in length and somewhat smaller than this in diameter. All bacteria can penetrate the smallest openings; many can pass through the natural pores of an egg shell once the natural bloom of the shell is worn or washed away. Yeasts are

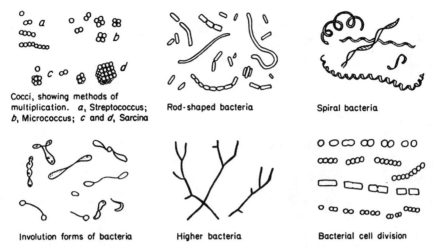

Figure 7.1. Morphology of some common bacteria when viewed under a microscope. Source: *Weiser et al., Practical Food Microbiology and Technology, 2nd ed., AVI Publishing Co., Westport, CT, 1971.*

somewhat larger, of the order of 20 μm or so in individual cell length and about a third this size in diameter. Most yeasts are spherical or ellipsoidal.

Molds are still larger and more complex in structure (Fig. 7.3). They grow by a network of hairlike fibers called mycelia and send up fruiting bodies that produce mold spores referred to as conidia. The blackness of bread mold and the blue-colored veins of blue cheese are due to conidia; beneath the fruiting heads, the hairlike mycelia anchor the mold to the food. Mycelia are 1 μm or so in thickness and, like bacteria, can penetrate the smallest opening; in the case of a weakened fruit skin or egg shell, they can digest the skin and make their own route of penetration.

Bacteria, yeasts, and molds can attack virtually all food constituents. Some ferment sugars and hydrolyze starches and cellulose; others hydrolyze fats and produce rancidity; some digest proteins and produce putrid and ammonia like odors. Other types form acid and make food sour, produce gas and make food foamy, and form pigments and discolor foods. A few produce toxins and give rise to food poisoning. When food is contaminated under natural conditions, several types of organisms will be present

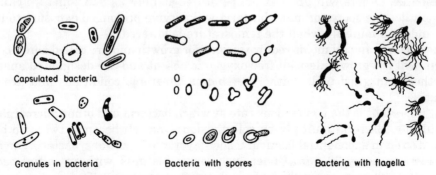

Figure 7.2. Common bacterial structures. Source: *Weiser et al., Practical Food Microbiology and Technology, 2nd ed. AVI Publishing Co, Westport, CT, 1971.*

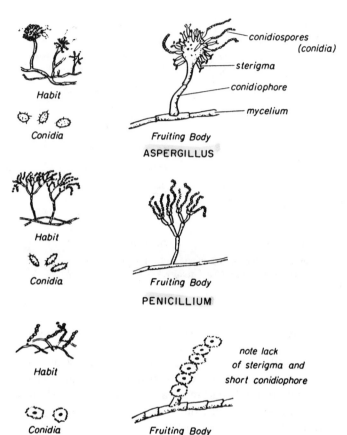

Figure 7.3. Structures of some common molds. Source: *Weiser et al., Practical Food Microbiology and Technology, 2nd ed. AVI Publishing Co., Westport, CT,* 1971.

together and contribute to a complex of simultaneous or sequential changes, which may include production of acid and gas, putrefaction, and discoloration.

Bacteria, yeasts, and molds like warm, moist conditions. Most bacteria multiply best at temperatures between 16°C and 38°C; these are termed mesophilic. Some will grow at temperatures down to the freezing point of water and are called psychrotrophic or psychrophilic. Others will grow at temperatures as high as 82°C, and we call these thermophilic. The spores of many bacteria will survive prolonged exposure to boiling water and then multiply when the temperature is lowered.

Some bacteria and all molds require oxygen for growth and are called aerobic. Other bacteria will not grow unless all free oxygen is absent and are designated anaerobic. Still others can grow under either aerobic or anaerobic conditions and are called facultative.

Most important is the tremendous rate at which bacteria and other microorganisms can multiply. Bacteria multiply by cell division. One cell becomes two, two become four, and so on in exponential fashion. Under favorable conditions, bacteria can double their numbers every 30 min. Under such conditions milk with an initial bacterial count of 100,000 or so per milliliter, which is not uncommon before pasteurization, if left standing at room temperature can reach a bacterial population of about 25 million in 24 h, and over 5 billion per milliliter in 96 h.

Food-Borne Disease

A special kind of food deterioration that may or may not alter a food's organoleptic properties has to do with food-borne disease. Food-borne diseases are commonly classified as food infections or food intoxications. Whereas the distinction is sometimes imperfect, food infections involve microorganisms present in the food at time of consumption which then grow in the host and cause illness and disease. Food intoxications involve toxic substances produced in foods as by-products of microorganisms prior to consumption and cause disease upon ingestion. Where the toxin producer is a microorganism, it need not grow in the host to produce disease or even be present in the food.

Staphylococcus aureus and *Clostridium botulinum* produce bacterial food poisoning by intoxication through the production of specific bacterial toxins. In fact, the toxin produced by *C. botulinum* is one of the most toxic substances known. Certain molds produce mycotoxins, the best known being the aflatoxins of *Aspergillus flavus*. Unlike the toxins of *S. aureus* and *C. botulinum*, which are highly toxic to man, aflatoxins may be more toxic to domestic animals than to man. However, their carcinogenic properties are cause for much concern since aflatoxins can be produced in a wide range of cereals, legumes, nuts, and other products allowed to become moldy. When such products occur in feeds, aflatoxins may subsequently be detected in the milk of animals consuming the feed and in cheese made from such milk.

Many bacteria can transmit food-borne infections capable of causing human disease. These include *Clostridium perfringens*, numerous members of the genus *Salmonella*, *Shigella dysenteriae*, *Vibrio parahaemolyticus*, *Streptococcus pyogenes*, *Bacillus cereus*, *Campylobacter jejuni*, and others. A number of viral infections also may be contracted by man through contaminated food that has not been adequately processed or handled, including infectious hepatitis, poliomyelitis, and various respiratory and intestinal disorders. Microorganisms which cause disease in humans are known as pathogenic or pathogens.

Scientists are still learning about food-borne diseases. Over the last decade or so, several bacteria that had not been thought to be transmitted by food and cause human disease have been found to do just that. Chief among these "newer" pathogens are *Aeromonas hydrophila*, *Yersinia enterocolitica*, *Listeria monocytogenes*, *Vibrio parahaemolyticus* and a particular type of *Escherichia coli* called 0157:H7. Of particular importance is the recent discovery that some food-borne pathogenic bacteria can multiply at temperatures as low as 3.3°C (38°F). This means that temperatures which have been considered good for refrigerated storage may not always keep food from becoming a hazard. Considerable research is ongoing into ways to further protect foods from these psychrotrophic pathogens. Some of the causes of food intoxications and infections are listed in Table 7.2, along with the types of foods usually involved and general comments on corrective practices.

Insects, Parasites, and Rodents

Insects are particularly destructive to cereal grains and to fruits and vegetables. Both in the field and in storage it has been estimated that insects destroy 5–10% of the U.S. grain crop annually. In some parts of the world the figure may be in excess of 50%. The insect problem is not just one of how much an insect can eat, but when insects eat, they damage the food and open it to bacterial, yeast, and mold infection,

Table 7.2. Major Food-Borne Pathogenic Microorganisms Causing food-borne Illness

Microorganism	Source in Nature	Characteristics of Illness	Associated Foods
Clostridium botulinum	Soil, sediment intestinal tracts of fish, mammals, gills, viscera of fish crabs, seafood	Neurotoxicity; shortness of breath, blurred vision; loss of motor capabilities; death; onset ranges between 12 and 36 h	Low-acid foods, especially home canned; meats, fish, smoked/fermented fish, vegetables, other marine products
Clostridium perfringens	Soil and sediment (widespread), water, intestinal tracts of humans and animals	Nausea, occasional vomiting, diarrhea and intense abdominal pain; onset ranges from 8 to 22 h; short duration (24 h)	Improperly prepared roast beef, turkey, pork, chicken, cooked ground meat and other meat dishes, gravies, soups and sauces
Salmonella spp.	Water, soil, mammals, birds, insects, intestinal tracts of animals, especially poultry and swine	Nausea, vomiting, abdominal cramps, diarrhea, fever, and headache; normal incubation period 6–48 h	Beef, turkey, pork, chicken eggs and products, meat salads, crabs, shellfish, chocolate, animal feeds, dried coconut, baked goods and dressings
Listeria monocytogenes	Soil, silage, water, and other environmental sources, birds, mammals, and possibly fish and shellfish	Healthy individuals generally have mild flulike symptoms; severe forms of listeriosis include septicemia, meningitis, encephalitis, and abortion in pregnant women	Raw milk, soft cheese, cole slaw, ice cream, raw vegetables, raw meat sausages, raw and cooked poultry, raw and smoked fish
Campylobacter jejuni	Soil, sewage, sludge, untreated waters, intestinal tracts of chickens, turkeys, cattle, swine, rodents, and some wild birds	Fever, headache, nausea, muscle pain, and diarrhea (sometimes watery, sticky, or bloody); onset time 2–5 days and duration 7–10 days; relapses common	Raw milk, chicken, other meats and meat products

Organism	Source	Symptoms	Foods
Staphylococcus aureus	Hands, throats, and nasal passages of humans; common on animal hides	Nausea, vomiting, diarrhea, abdominal cramps, and prostration. Symptoms may be severe; normal onset from 30 min to 8 h; duration usually 24–48 h	Ham, turkey, chicken, pork, roast beef, eggs, salads (e.g., egg, chicken, potato, macaroni), bakery products, cream-filled pastries, luncheon meats, milk and dairy products
Shigella spp.	Polluted water and intestinal tracts of humans and other primates	Diarrhea with bloody stools, abdominal cramps, and fever; Severe cases caused by *S. dysenteriae* may result in septicemia, pneumonia, or peritonitis; onset averages 0.5–2 days but may be as high as 7 days; recovery is slow	Milk and dairy products, raw vegetables, poultry, and salads (e.g., potato, tuna, shrimp, macaroni, and chicken)
Vibrio parahaemolyticus	Estuarine and marine waters	Abdominal cramps, nausea, vomiting, headache, and diarrhea (with occasional blood and mucus in stools) and fever; onset from 4–96 h; symptoms last an average of 2.5 days	Raw, improperly cooked, or cooked, recontaminated fish, shellfish, or crustacea
Vibrio cholerae O1	Untreated water, intestinal tracts of humans	Copious watery stools, vomiting, prostration, dehydration, muscular cramps, and occasionally death; incubation from 1 to 5 days	Shell fish, raw fish, and crustacea
Bacillus cereus	Soils, sediments, dust, water, vegetation, and a variety of foods, notably cereals, dried foods, spices, milk and dairy products, meat products, and vegetables	TYPE 1: Diarrheal type food poisoning—watery diarrhea, abdominal cramps, nausea, usually no fever or vomiting; onset time 6–15 h; short duration (24 h)	Meats, vegetable dishes, milk, cream pastries, soups, and puddings

Table 7.2. (Continued)

Microorganism	Source in Nature	Characteristics of Illness	Associated Foods
Bacillus cereus (con't.)		TYPE II: Emetic type food poisoning—nausea and vomiting within 0.5–6 h; abdominal cramps and diarrhea occasionally occur; short duration (less than 24 h)	Fried, boiled; or cooked rice, and other starchy foods (e.g., potatoes and pasta)
Yersinia enterocolitica	Soil, natural waters, intestinal tracts of various animals; (pigs, birds, beavers, dogs and cats)	Diarrhea, and/or vomiting, fever and abdominal pain are hallmark symptoms; pseudo-appendicitis; onset 24–48 h; recovery in 1–2 days	Fresh meat and meat products (particularly swine), fresh vegetables, milk, and milk products
Escherichia coli (Enterovirulent types)	Intestinal tracts of humans and animals	Mild to severe bloody diarrhea, vomiting, cramping, dehydration, and shock; can result in more serious symptoms; some illnesses may last up to 8 days	Raw or rare meats and poultry, raw milk and milk products, unprocessed cheese, salads

SOURCE: Pierson and Corlett (1992).

causing further destruction. A small insect hole in a melon, not so bad in itself, can result in the total decay of the melon from bacterial invasion. Insects have been controlled in stored grain, fruit, and spices by the use of pesticides, inert atmospheres, or cold temperatures. The use of chemical pesticides on foods continues to raise questions of possible toxic effects and maximum safe levels, and there are currently active programs to increase plant resistance and other biological-based methods of insect control. Genetic engineering, for example, offers the possibility of producing food plants which have the genes to make chemicals which are toxic to insects. When the pest eats the plant, the toxicant kills the insect. Many of these toxicants are only effective on insects and of little concern for humans or the environment.

Insect eggs may persist, or be laid, in food after processing, as for example in flour. An interesting method of destroying insect eggs is to throw the flour with high impact against a hard surface as in a centrifuge-type machine known as an Entoleter (Fig. 7.4). The impact destroys the eggs. They remain in the flour, but no further insect multiplication results.

Inspection of foods for insect contamination, which the Federal Food, Drug, and Cosmetic Act defines as a form of adulteration, requires some knowledge of insect life cycles and insect morphology since fragments of insects from their various stages of development also constitute adulteration. The life cycle of the common Drosophila fruit fly progresses through the egg, larval, pupal, and adult fly stages. It is virtually impossible to produce and transport grains and other food commodities completely devoid of insects and insect parts and so the Food and Drug Administration recognizes certain low levels of insect contamination as tolerable and takes action when these levels are exceeded. For example, the acceptable levels for Drosophila eggs and larvae in tomato products are indicated in Table 7.3.

Commodities containing highly destructive insects are prohibited from import, ex-

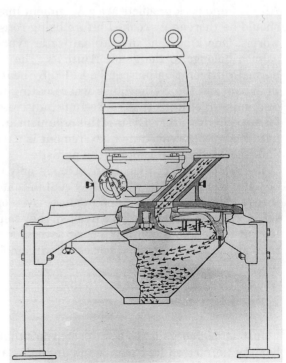

Figure 7.4. Entoleter machine capable of destroying insect eggs by impact. *Courtesy of Entoleter, Inc.*

Table 7.3. FDA Defect Action Level for *Drosophila* Eggs and Larvae

Product	Sample Size (g)	No. of Eggs	No. of Larvae
Tomatoes	500	10 or	
		5 and	1 or
			2
Tomato juice	100	10 or	
		5 and	1 or
			2
Tomato purée	100	20 or	
		10 and	1 or
			2
Tomato paste, pizza, and other sauces	100	30 or	
		15 and	1 or
			2

SOURCE: Gould (1983).

port, and sometimes transport across state lines. In the United States, the Animal and Plant Health Inspection Service of the USDA is responsible for such regulations. Some states and many countries prohibit fresh fruits and vegetables from being imported in order to try and prevent spread of the insects.

An important food-borne parasite is the trichinosis nematode, *Trichinella spiralis*, which can enter hogs eating uncooked food wastes. The nematode penetrates the hog's intestines and finds its way into the pork. If the meat is not thoroughly cooked, the live worm can infect man. It also is possible to destroy the nematode by frozen storage. All pork and pork products are government inspected, but as a further safeguard they should be thoroughly cooked before being consumed. Fish may also harbor parasitic worms. One kind that invades saltwater types such as herring, cod, mackerel, and salmon belongs to the genus *Anisakis*. The worm, which can infect man, survives normal refrigeration but can be killed by heating or freezing. It has been a problem in Japan and the Netherlands where eating raw fish is common. Another parasitic contaminant of foods that causes much distress is *Entamoeba histolytica*, responsible for amoebic dysentery. Cysts of this organism are transmitted in feces and may contaminate foods where raw human excrement is used as fertilizer for crops. Infected water and poor hygiene also spread the parasite.

The problems with rodents involve not only the quantity of food they may consume but also the filth with which they contaminate food. Rats live 2–3 years, may have three to five litters a year, and produce seven or eight young per litter. In parts of the world where they are poorly controlled, they contribute substantially to food shortages. Rodent urine and droppings harbor several kinds of disease-producing bacteria, and rats spread such human diseases as salmonellosis, leptospirosis, typhus fever, and plague.

Food Enzymes

Just as microorganisms possess enzymes that ferment, rancidify, and putrefy foods, healthy uninfected food plants and animals have their own enzyme complement, the activity of which largely survives harvest and slaughter. Cereal grains and seeds recovered after 60 years of storage still possessed the properties of respiration, germination, and growth—all enzyme-controlled functions. Not only can enzyme activity persist throughout the entire useful life of many natural and manufactured foods, but this activity often is intensified after harvest and slaughter. This is because enzymatic reactions are delicately balanced in the normally functional living plant and animal; but the balance is upset when the animal is killed or the plant removed from the field. Thus, although pepsin helps digest protein in the animal intestine, it does not digest the intestine itself in the healthy living animal. However, when the body defenses cease on slaughter, pepsin does contribute to proteolysis of the organs containing it. A great many similar "runaway" enzymatic reactions can be found in plants.

Unless these enzymes are inactivated by heat, chemicals, radiation, or some other means, they continue to catalyze chemical reactions within foods after slaughter or harvest. Some of these reactions are highly desirable if not allowed to go too far—like continued ripening of tomatoes after they are picked, and natural tenderizing of beef on ageing. But ripening and tenderizing beyond an optimum point becomes deterioration; the weakened tissues fall subject to microbial infections and the deterioration reaches the point of rotting. This can happen in the field, supermarket, and home refrigerator, given sufficient time.

Heat and Cold

Quite beyond their effects on microorganisms, heat and cold can cause deterioration of food if not controlled. Within the moderate temperature range over which most food is handled, such as 10–38°C, the rate of chemical reaction is approximately doubled for every 10°C rise in temperature. This includes the rates of many enzymatic as well as nonenzymatic reactions. Excessive heat, of course, denatures proteins, breaks emulsions, dries out foods by removing moisture, and destroys vitamins.

Uncontrolled cold also will damage foods. If fruits and vegetables are allowed to freeze, they suffer discoloration, changes in texture, or cracked skins, leaving the food susceptible to attack by microorganisms. Freezing may also cause deterioration of liquid foods. If a container of milk is allowed to freeze, the emulsion will be broken and fat will separate. Freezing also will denature milk protein and cause it to curdle. Carefully controlled freezing on the other hand need not cause these defects.

Cold damage to foods does not necessarily require the extreme of freezing. Fruits and vegetables after harvest, like other living systems, have optimum temperature requirements. When held at refrigeration temperatures of about 4°C, some are weakened or killed and deteriorative processes follow. This is termed "chill injury." Table 7.4 gives a partial listing of chill injury of some fruits and vegetables held cold but above the freezing point. The deterioration includes off-color development, surface pitting, and various forms of decay. Bananas, lemons, squash, and tomatoes are examples of products that should be held at temperatures no lower than about 10°C for maximum quality retention.

Moisture and Dryness

As already pointed out, excessive moisture pickup or loss causes substantial deteriorative changes in foods. Moisture is required for chemical reactions and for microorgan-

Table 7.4. Damage to Several Fruits and Vegetables by Cold Temperatures Above Freezing Zone

Commodity	Approximate Lowest Safe Temperature, °C	Character of Injury When Stored Between 0°C and Safe Temperature
Apples, certain varieties	1–2	Internal browning, soggy breakdown
Avocados	7	Internal browning
Bananas		
green or ripe	13	Dull color when ripened
Beans (snap)	7–10	Pitting increasing on removal, russeting on removal
Cranberries	1	Low-temperature breakdown
Cucumbers	7	Pitting, water-soaked spots, decay
Eggplants	7	Pitting or bronzing, increasing on removal
Grapefruit	7	Scald, pitting, watery breakdown, internal browning
Lemons	13–14	Internal discoloration, pitting
Limes	7	Pitting
Mangoes	10	Internal discoloration
Melons		
cantaloupes	7	Pitting, surface decay
Honey Dew	5–10	Pitting, surface decay
Casaba	5–10	Pitting, surface decay
Crenshaw and Persian	5–10	Pitting, surface decay
watermelons	2	Pitting, objectionable flavor
Okra	5	Discoloration, water-soaked areas, pitting, decay
Olives, fresh	7	Internal browning
Oranges, California	1.5–2.5	Rind disorders
Papayas	7	Breakdown
Peppers, sweet	7	Pitting, discoloration near calyx
Pineapples		
mature-green	7	Dull green when ripened
Potatoes		
Chippewa and Sebago	5	Mahogany browning
Squash, winter	10–13	Decay
Sweet potatoes	13	Decay, pitting, internal discoloration
Tomatoes		
mature-green	13	Poor color when ripe; tendency to decay rapidly
ripe	10	Breakdown

SOURCE: USDA (1954).

ism growth; excessive moisture can accelerate these types of deterioration. Excessive loss of moisture can also have detrimental affects particularly on appearance and texture. Moisture need not be present throughout the food to exert major effects.

Surface moisture resulting from slight changes in relative humidity can cause lumping and caking, as well as such surface defects as mottling, crystallization, and stickiness. The slightest amount of condensation on the surface of food can become a virtual pool for the multiplication of bacteria or the growth of mold.

This condensation need not come from the outside. In a moisture-barrier package, fruits or vegetables can give off moisture from respiration and transpiration. This moisture is then trapped within the package and can support the growth of destructive microorganisms. Nonrespiring foods in a moisture-barrier package also can give up moisture and change the relative humidity of the package headspace. This moisture can then condense on the surface of the food, particularly when storage temperature is permitted to decrease.

Oxygen

Whereas nitrogen, which makes up 79% of air, is inert from the perspective of foods, the 20% oxygen in the air is quite reactive and causes substantial deteriorative effects in many foods. Besides the destructive effects due to chemical oxidation of nutrients (especially vitamins A and C), food colors, favors, and other food constituents, oxygen is also essential for mold growth. All molds are aerobic and this is why they are found growing on the surface of foods and other substances or within cracks in these materials.

Atmospheric oxygen is excluded from foods by vacuum deaeration or inert gas purging in the course of processing, by vacuum-packaging or by flushing containers with nitrogen or carbon dioxide, and in some instances by adding to foods and containers oxygen scavengers, which promote removal of residual trace oxygen through chemical reaction.

In recent years, positive use has been made of the fact that certain gases influence the rate of deterioration of foods. Several products which are often stored under refrigeration have been packaged in containers in which the air has been removed and replaced with some other gas or gas mixture. Often this mixture is made up of all nitrogen or a mixture of nitrogen and carbon dioxide. These mixtures can reduce the rate of deterioration and substantially increase shelf life. This is known as modified atmosphere packaging and will be discussed in more detail in the chapter on packaging. The high-moisture-content pastas that are now sold are packaged in modified atmospheres that are devoid of oxygen which inhibits the growth of mold.

Light

As mentioned in Chapter 4, light destroys some vitamins, notably riboflavin, vitamin A, and vitamin C, and causes deterioration of many food colors. Milk in bottles exposed to the sun develops "sunlight" flavor due to light-induced fat oxidation and changes in the protein. Not all wavelengths making up natural or artificial light are equally absorbed by food constituents or are equally destructive. Surface discolorations of sausages and meat pigments are different under natural light and under fluorescent light that may be encountered in display cases. Sensitive foods often can be protected from light by opaque packaging or by incorporating compounds into glass and transparent films that screen out light of specific wavelengths.

Time

After slaughter, harvest, or food manufacture there is a time when the quality of food is the highest. In many products this quality peak can be passed in the field in a day or two, or after harvest in a matter of hours. Fresh corn and peas are notable examples. The growth of microorganisms, destruction by insects, action of food enzymes, nonenzymatic interaction of food constituents, loss of flavor volatiles, and the effects of heat, cold, moisture, oxygen, and light all progress with time. This is not to say that certain cheeses, sausages, wines, and other fermented foods are not improved with ageing up to a point. But for the vast majority of foods, quality decreases with time and major goals of food-handling and preservation practices are to capture and maintain freshness. Adequate processing, packaging, and storage may prolong the shelf life of foods considerably but cannot extend it indefinitely. Eventually, the quality of any food product decreases. This is the reason for considerable interest in shelf life dating of processed foods, a subject that will be considered further in Chapter 24.

SOME PRINCIPLES OF FOOD PRESERVATION

In subsequent chapters several of the more important food preservation methods are treated individually. For now we will enlarge somewhat on some of the general principles underlying control of food deterioration. These are the principles on which food preservation methods are largely based.

If foods are to be kept only for short periods of time, then there are two very simple rules:

1. Keep food alive as long as possible; kill the animal or plant just before it is to be used. A good example of this is keeping lobsters alive in a tank in a supermarket or restaurant—while alive and healthy, they do not seriously deteriorate. This is also practiced with fish, poultry, fruits, and vegetables where possible. Unfortunately, the possibilities are limited.
2. If the food must be killed, clean it, cover it, and cool it as quickly as possible. However, cleaning, covering, and cooling will only delay deterioration for a short time, for hours or perhaps at most for a few days. Microorganisms and natural food enzymes will not be destroyed or totally inactivated and so will take over very quickly.

For longer-term and practical preservation, as required for most of our food supply, further precautions are necessary. These are largely directed at inactivating or controlling microorganisms, enzymes, and reducing or eliminating chemical reactions which cause food spoilage.

CONTROL OF MICROORGANISMS

The most important means of controlling bacteria, yeasts, and molds are heat, cold, drying, acid, sugar, salt, smoke, air, chemicals, and radiation. Any one of these can also cause deterioration of foods and so it is a matter of balance: an amount of heat that will kill microorganisms but still leave the food acceptable; a level of a preservative chemical additive that will inhibit microbial growth but have minimum adverse effects on food components or human health. Indeed, the entire science of food preservation is one of compromise with respect to dosage or treatment.

Heat

Most bacteria, yeasts, and molds grow best in the temperature range of about 16–38°C. Thermophiles will grow in the range 66–82°C. Most bacteria are killed in the range 82–93°C, but many bacterial spores are not destroyed even by boiling water at 100°C for 30 min. To ensure sterility, that is, total destruction of microorganisms including spores, a temperature of 121°C (wet heat) must be maintained for 15 min or longer. This is generally done with steam under pressure, as in a laboratory autoclave or commercial retort. These and other temperature effects on microorganisms are listed in Table 7.5. Commercial pressure retorts of the kind used in the canning industry operate at temperatures and for time intervals adequate to destroy large numbers of highly resistant bacterial spores within the canned food. Sterility or "commercial sterility," to be defined later, is essential because the food may be stored in the can for a year or longer.

Not all foods require the same amount of heat for sterilization. When foods are high in acid, such as tomatoes or orange juice, the killing power of heat is increased. A temperature of 93°C for 15 min may be enough to gain sterility if sufficient acid is present. Safe temperatures and times for different foods have been well established and are published in handbooks used by the canning industry.

Another fundamental point on the use of heat and other means of preservation: it is not always necessary to kill all microorganisms and produce a sterile product. It may be necessary to employ only sufficient heat to destroy disease-producing organisms in the food. This is done in the case of pasteurized milk. Most of the bacteria and all of the disease-producing organisms that might be present in milk are destroyed by pasteurization at 63°C for 30 min, but the milk is not sterile. Nor need it be, since it will be held in a refrigerator and consumed generally within a few days. However,

Table 7.5. Effects of Temperature on Microorganisms

°C	°F	Temperature Effects
121	250	Steam temperature at 15 lb[a] pressure kills all forms including spores in 15–20 min
116	240	Steam temperature at 10 lb pressure kills all forms including spores in 30–40 min
110	230	Steam temperature at 6 lb pressure kills all forms including spores in 60–80 min
104	220	Steam temperature at 2 lb pressure
100	212	Boiling temperature of pure water at sea level; kills in vegetative stage quickly but not spores after long exposure
82–93	179–200	Growing cells of bacteria, yeasts, and molds usually killed
66–82	151–180	Thermophilic organisms grow
60–77	140–171	Pasturization of milk in 30 min kills all important bacteria pathogenic to humans except spore-forming pathogens
16–38	61–100	Active growing range for most bacteria, yeasts, and molds
10–16	50–61	Growth retarded for most organisms
4–10	39–50	Optimum growth of psycrophilic organisms (10–4°C). Some food-borne pathogens still grow
0	32	Freezing; usually the growth of all organisms stopped
−18	0	Bacteria preserved in latent state
−251	−420	Many species of bacteria not killed by the temperature of liquid hydrogen

Source: Adapted from Weiser (1971). Updated 1994.
[a] A pressure of 1 lb/in.^{2} = 0.07 kg/cm^2 = 6895 pascals.

evaporated milk, which is intended to remain in a can at room temperature for months or even years must receive a greater heat treatment to ensure sterility or commercial sterility.

Cold

As stated earlier, most bacteria, yeasts, and molds grow best in the temperature range 16–38°C. Psychrotrophs will grow down to 0°C, the freezing point of water, and below. At temperatures below 10°C, however, growth is slow and becomes slower the colder it gets (see Table 7.5). When the water in food is completely frozen, there is no multiplication of microorganisms. But in some foods all of the water is not frozen until a temperature of −10°C or lower is reached. This is because of dissolved sugars, salts, and other constituents, which depress the freezing point.

The slowing of microbial activity with decreased temperatures is the principal behind refrigeration and freezing preservation. However, it is important to note that although cold temperatures will slow microbial growth and activities and may kill a certain fraction of the bacterial population, cold, including severe freezing, cannot be depended on to kill all bacteria. This is illustrated by data in Table 7.6. An ice cream mix inoculated with typhoid bacteria still retained over 600,000 live bacteria per milliliter after 1 year in frozen storage. Not only do cold storage and freezing fail to sterilize foods but when the food is taken from cold storage and thawed, the surviving organisms often resume rapid growth since the food may have been damaged from the cold or frozen storage. As indicated above, recent studies have shown that some disease-causing bacteria can grow at refrigeration temperatures of 3.3°C.

Drying

Microorganisms in a healthy growing state may contain in excess of 80% water. They get this water from the food in which they grow. If the water is removed from the food, water will also be removed from the bacterial cells and multiplication will stop. Partial drying will be less effective than total drying, although for some microorganisms, partial drying as in concentration may be quite sufficient to arrest bacterial

Table 7.6. Effect of Freezing Ice Cream Mix on Typhoid Bacilli

Samples Taken	Typhoid Bacilli per ml of Ice Cream Mix
5 days old	51,000,000
20 days old	10,000,000
70 days old	2,200,000
342 days old	660,000
430 days old	51,000
648 days old	30,000
2 years old	6,300
2 years, 4 months old	Viable typhoid bacilli

SOURCE: Weiser (1971). Mountney, Gould

growth and multiplication. Bacteria and yeasts generally require more moisture than molds, and so molds often will be found growing on semidry foods where bacteria and yeasts find conditions unfavorable. Examples are mold growing on stale bread and partially dried fruits.

Slight differences in relative humidity within the room in which food is held, or within a food package, can greatly affect the rate at which microorganisms multiply. Figure 7.5 depicts the number of organisms present on meat stored cold for 20 days at 75% and 95% relative humidity (RH). At each temperature, the higher humidity produced the higher population of microorganisms.

Since microorganisms can live in one part of a food that may differ in moisture and other physical and chemical conditions from the food just millimeters away, the "microenvironment" of the microorganisms must be considered. Thus, it is common to refer to water conditions in terms of specific activity. The term "water activity" is related to relative humidity. Relative humidity is defined as the ratio of the partial pressure of water vapor in the air to the vapor pressure of pure water at the same temperature. Relative humidity refers to the atmosphere surrounding a material or solution. Water activity is a property of solutions and is the ratio of vapor pressure of the solution to the vapor pressure of pure water at the same temperature. Under equilibrium conditions, water activity equals RH/100. Moisture requirements of microorganisms really means water activity in the immediate environment, whether it be in solution, in a particle of food, or at a surface in contact with the atmosphere. At the usual temperatures permitting microbial growth, most bacteria require a water activity in the range 0.90–1.00. Some yeasts and molds grow slowly at a water activity as low as 0.65. More will be said about water activity in Chapter 10.

As stated previously, food is dried either partially or completely for several reasons. One of the most important is to preserve it against microbial spoilage. However, partial or complete drying of food does not kill all microorganisms. It may actually preserve

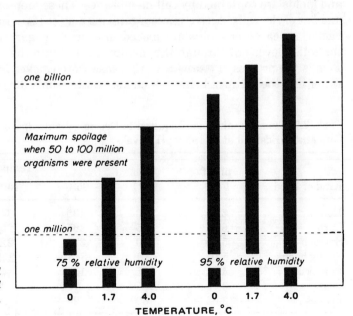

Figure 7.5. Effect of relative humidity on growth of microorganisms in meat stored 20 days. Source: *Weiser et al., Practical Food Microbiology and Technology,* 2nd ed., *AVI Publishing Co., Westport, CT,* 1971.

microorganisms as it preserves the food, and dried food usually is far from sterile. Although bacteria may not grow in the dried food, when the food is remoistened or reconstituted, bacterial growth may resume.

Acid

In sufficient strength acid modifies bacterial proteins as it denatures food proteins and so microorganisms are sensitive to acid. Some are much more sensitive than others, and the acid produced by one organism during fermentation often will inhibit another type of organism. This is one of the principles of controlled fermentation as a means of preserving foods against growth of proteolytic and other types of spoilage organisms.

Acid may be produced in foods by adding acid-producing bacterial cultures, or acids may be added directly to foods; examples are citric acid and phosphoric acid added to soft drinks. Foods such as tomatoes, citrus juices, and apples contain natural acidity with varying degrees of preservative power. Much of this is due directly to the hydrogen ion concentration (expressed in terms of pH), but two acids producing the same pH may not be equally preservative since the anions of certain acids also exert an effect. The degree of acidity tolerable in foods from the standpoint of palatability is never sufficient to ensure food sterility.

As mentioned earlier, acid combined with heat makes the heat more destructive to microorganisms. This is seen in Table 7.7 with respect to requirements for destruction of the spores of the anaerobic toxin-producing bacterium, *C. botulinum*.

Sugar and Salt

Fruits are preserved by placing them in a sugar syrup, and certain meat products are preserved by placing them in a salt brine. How does this work? Bacteria, yeasts, and molds are contained by cell membranes. These membranes allow water to pass in and out of the cells. Active microorganisms may contain in excess of 80% water. When bacteria, yeasts, or molds are placed in a heavy sugar syrup or salt brine, water in the cells moves out through the membrane and into the concentrated syrup or brine. This is the process of osmosis; in this case water moves from the cell containing 80% water into the syrup or brine, which may contain only 30% or 40% water. The tendency

Table 7.7. Time (in minutes) Required to Destroy Viable Spores of *Clostridium botulinum* in Different Foods and at Various pH Levels

Kind of Food	pH of Food	Temperature				
		90°C	95°C	100°C	105°C	108°C
Hominy	6.95	600	495	345	34	10
Corn	6.45	555	465	255	30	15
Spinach	5.10	510	345	225	20	10
String beans	5.10	510	345	225	20	10
Pumpkin	4.21	195	120	45	15	10
Pears	3.75	135	75	30	10	5
Prunes	3.60	60	20	—	—	—

SOURCE: Weiser (1971). Moutney, Gould

to equalize water concentration inside and outside the cell in this case causes a partial dehydration of the cell, referred to as plasmolysis, which interferes with microorganism multiplication. Quite the opposite can be accomplished by placing microorganisms in distilled water. In this case water enters the cells and causes them to burst. This process, known as plasmoptysis, rarely occurs in food products. All of this is closely related to the water activity of solutions and foods. Solutions high in solute concentration have a high osmotic pressure and a low water activity. Dilute solutions are low in osmotic pressure and have a high water activity. The quantitative contribution of a specific solute to osmotic pressure and water activity depends on the solute's molecular weight and the number of ions it produces in solution. An equal weight of a low-molecular-weight solute has a greater effect on increasing a solution's osmotic pressure and decreasing its water activity than the same weight of a high-molecular-weight solute.

Different organisms have various degrees of tolerance to osmosis and to sugar and salt. Yeasts and molds are more tolerant than most bacteria. This is why yeasts and molds often are found growing on high sugar or salt products; for example, fruit jam or bacon, where bacteria are inhibited.

Smoke

As with most preservative methods, smoke was used long before the reasons for its effectiveness were understood. In preserving foods such as meats and fish with smoke, the preservative action generally comes from a combination of factors. Smoke contains preservative chemicals such as small amounts of formaldehyde and other materials from the burning of wood. In addition, smoke generally is associated with heat which helps kill microorganisms. This heat also tends to dry out the food, which further contributes to preservation. Smoking over a fire may be quite effective in preserving certain foods; on the other hand, today smoke may be added merely to flavor food, that is, without heat from burning. In this case the smoke may be a very poor preservative. In meat products such as those illustrated in Fig. 7.6, smoke combined with other preservatives is used more for its flavor than for its preservative action.

Atmosphere Composition

The different requirements of microorganisms for oxygen and some of the means for removing air and oxygen already have been cited. To control organisms that require it, air is removed; for organisms that cannot tolerate it, air is provided. It is often fairly easy to exclude air from aerobes such as molds. Wax coating of cheeses or oxygen-impermeable skin-tight plastic films can be quite effective. Controlling strict anaerobes by providing air can be more difficult and dangerous, especially with large food pieces. This is because the center of the food piece may remain anaerobic even if air is left in the headspace of a package. Further, some organisms consume oxygen and thus convert a formerly aerobic microenvironment into an anaerobic one favorable to other organisms. Thus, in preserving food against *C. botulinum* which is a strict anaerobe, measures in addition to inclusion of air are essential.

Chemicals

Many chemicals will kill or inhibit the growth of microorganisms, but most of these are not permitted in foods. A few that are permitted, in prescribed low levels in certain

Figure 7.6. Smoking combined with other preservatives in preparation of various sausage products. *Courtesy of Vilter Corp.*

foods, include sodium benzoate, sorbic acid, sodium and calcium propionate, ethyl formate, and sulfur dioxide. The use of chemicals to control microorganisms, and for any other purpose in foods, is part of the broader subject of food additives, which will be dealt with in a subsequent chapter. At the present time the Food, Drug, and Cosmetic Act, administered by the Food and Drug Administration, regulates what chemicals may be used in foods, and the conditions of their use. New chemicals to be approved must undergo rigorous toxicological testing, and the burden of proof of safety resides with the producer or user of the chemical. The FDA also retains the right to reverse former decisions and prohibit use of approved chemicals if new knowledge relative to safety warrants such action.

These chemicals can inhibit the growth of specific microorganisms. For example, sorbic acid is effective at inhibiting the growth of many molds. Thus, the surface of cheese may be treated with sorbic acid or cottage cheese may have sorbic acid mixed directly in the product.

Radiation

Microorganisms are inactivated to various degrees by different kinds of radiation. X-Rays, microwaves, ultraviolet light, and ionizing radiations are each a type of electromagnetic radiation, differing in wavelength and energy, that have been used to preserve food. The effectiveness of each type differs and each imparts different changes in foods. For all types of radiation, the doses required to effectively sterilize most foods, and inactivate their natural enzymes, are generally excessive or borderline from the standpoint of food quality; and all may cause flavor, color, texture, or nutritional defects. Doses less than sterilizing appear more generally useful to extend storage life. These substerilization doses can inactivate enzymes responsible for initiating vegetable sprouting or can kill human disease-causing bacteria and insects.

Today, foods usually are irradiated with ionizing radiation, obtained from radioactive isotopes or electron accelerators. There is no significant temperature rise from this form of irradiation and so the term "cold sterilization" has been used. Several foods such as spices, vegetables and fruits, pork, and poultry have been approved for irradiation pasteurization at specific doses in the United States and other countries. Safety of irradiated foods in the United States comes under the jurisdiction of the FDA. Food irradiation, which is receiving renewed interest in the United States, will be considered in more detail in Chapter 11.

CONTROL OF ENZYMES
AND OTHER FACTORS

Most of what has been said in this chapter regarding control of deteriorative factors has dealt specifically with microorganisms. This emphasis is justified because by far the greatest losses to the food supply are due to microorganisms. Further, preservation of foods against deterioration from inherent food enzymes, probably the second greatest cause of spoilage, follows many of the same principles and methods that apply to preservation against microbial deterioration.

Just as microorganisms are controlled with heat, cold, drying, certain chemicals, and radiation, these are the principal means used to control and inactivate damaging inherent food enzymes. Indeed, when foods are sterilized or pasteurized to inactivate microorganisms, natural food enzymes are simultaneously partially or completely destroyed. Likewise, when cold is employed to slow down microbial activity, it can also retard the activity of natural food enzymes.

It is important to recognize that some natural food enzymes may be more resistant to the effects of heat, cold, drying, radiation, and other means of preservation than microorganisms. Thus, a heat or a radiation treatment may effectively destroy bacteria but leave damaging food enzymes intact to carry on food deterioration. Although a few broad generalizations can be made, specific conditions for preservation must be selected in accordance with the unique spoilage patterns of individual foods. Neverthe-

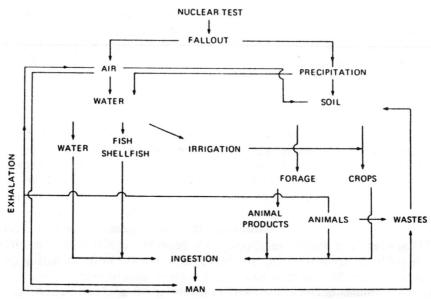

Figure 7.7. Pathways by which atmospheric nuclear materials enter the food chain. *Courtesy of J. M. DeMan.*

less, in two areas of preservation especially, irradiation and freezing, conditions that destroy or retard microorganism activity frequently leave enzymes active.

As for preservation against such other factors as moisture, dryness, air, and light, protective packaging is the major means employed to retard losses. Insects and rodents are also largely controlled by protective packaging combined with a high degree of sanitation.

Many harmful substances can enter the food supply for which the control measures discussed here are of little value. Among these are industrial pollutants and pesticides. For example, radioactive nucleotides were found in foods in Europe as a result of the releases at the Russian nuclear powerplant at Chernoble. Many of these substances are not readily degraded, and if permitted to contaminate the environment, they find their way into foods by numerous routes (Fig. 7.7). These pose especially difficult problems for the Environmental Protection Agency and the FDA. These agencies have developed elaborate quarantine and other plans in the event of disasters, but greater protection resides in strict preventative regulation and surveillance.

References

Anon. 1979. Basic Food Plant Sanitation Manual. 3rd ed. American Institute of Baking, Manhattan, KS.

Anon. 1988. Current Concepts in Food Protection. U.S. Dept. of Health and Human Services, Public Health Service, Food and Drug Administration, State Training Branch, Rockville, MD.

Aurand, L.W., Woods, A.E., and Wells, M.R. 1987. Food deterioration, preservation, and contamination. *In* Food Composition and Analysis. L.W. Aurand, A.E. Woods, and M.R. Wells, (Editors). Chapman & Hall, London, New York. pp. 621–663.

Board, R.G. 1983. A Modern Introduction to Food Microbiology. Blackwell Mosby Books, St. Louis, MO.

Bryan, F.L. 1992. Hazard Analysis Critical Control Point Evaluations: A Guide to Identifying Hazards and Assessing Risks Associated with Food Preparation and Storage. WHO, Geneva.

Charalambous, G. 1986. Handbook of Food and Beverage Stability. Chemical, Biological, Microbiological, and Nutritional Aspects. Academic Press, Orlando, FL.

Doyle, M.P. 1992. A New Generation of Foodborne Pathogens. Dairy Food Environ. Sanitat. *12*(8), 490, 492–493.

Desrosier, N.W. and Desrosier, J.N. 1977. Technology of Food Preservation. 4th ed. AVI Publishing Co., Westport, CT.

Gould, W.A. 1983. Tomato Production, Processing and Quality Evaluation. 2nd ed. AVI Publishing Co., Westport, Ct.

Guthrie, R.K. 1988. Food Sanitation. 3rd ed. Chapman & Hall, London, New York.

Jarvis, B. 1989. Progress in Industrial Microbiology. Vol. 21. Statistical Aspects of the Microbiological Analysis of Foods. Elsevier Science Publishers, New York.

Jay, J.M. 1992. Microbiological food safety. Crit. Rev. Food Sci. Nutr. *31*(3) 177–190.

Jay, J.M. 1992. Modern Food Microbiology. 4th ed. Chapman & Hall, London, New York.

John, G. 1987. Objective Methods in Food Quality Assessment. CRC Press, Boca Raton, FL.

King, P. 1992. Implementing a HACCP program. Food Manag. *27*(12), 54, 56, 58.

Labuza, T.P. 1982. Shelf-life Dating of Foods. Food & Nutrition Press, Inc., Westport, CT.

Marriott, N. 1994. Principles of Food Sanitation. 3rd ed. Chapman & Hall, London, New York.

Mossel, D.A.A. 1988. Impact of foodborne pathogens on today's world, and prospects for management. Anim. Hum. Hlth. *1*(1), 13–23.

Mountney, G.J., Gould, W.A., and Weiser, H.H. 1988. Practical Food Microbiology and Technology. Chapman & Hall, London, New York.

National Research Council. 1985. An Evaluation of the Role of Microbiological Criteria for Foods and Food Ingredients. National Academy Press, Washington, DC.

National Research Council. 1984. Insect Management for Food Storage and Processing. American Association of Cereal Chemists, St. Paul, MN.

Paine, F.A. 1987. Modern Processing, Packaging and Distribution Systems for Food. Chapman & Hall, London, New York.

Palumbo, S.A. 1986. Is refrigeration enough to restrain foodborne pathogens? J. Food Protect. *49*(12), 1003–1009.

Pierson, M.D. and Corlett, D.A. Jr. 1992. HACCP: Principles and Applications. Van Nostrand Reinhold, N.Y. 212 pp.

Troller, J.A. 1993. Sanitation in Food Processing. 2nd ed. Academic Press, New York.

USDA. 1954. The Commercial Storage of Fruits, Vegetables, and Florist and Nursery Stock. Their Epidemiologic Characteristics. AVI Publishing Co., Westport, CT.

Weiser, H.H., Mountney, G.J., and Gould, W.A. 1971. Practical Food Microbiology and Technology. 2nd ed. AVI Publishing Co., Westport, Ct.

Wolf, I.D. and Lechowich, R.V. 1989. Current issues in microbiological food safety. Pediatrics *84*(3), 468–472.

8

Heat Preservation
and Processing

Of the various means of preserving foods, the use of heat finds very wide application. The simple acts of cooking, frying, broiling, or otherwise heating foods prior to consumption are forms of food preservation. In addition to making foods more tender and palatable, cooking destroys a large proportion of the microorganisms and natural enzymes in foods; thus, cooked foods generally can be held longer than uncooked foods. However, cooking generally does not sterilize a product, so even if it is protected from recontamination, food will spoil in a comparatively short period of time. This time is prolonged if the cooked foods are refrigerated. These are common household practices.

Another feature of cooking is that it is usually the last treatment food receives prior to being consumed. The toxin that can be formed by *Clostridium botulinum* is destroyed by a 10-min exposure to moist heat at 100°C. Properly processed commercial foods will be free of this toxin. Cooking provides a final measure of protection in those unfortunate cases where a processing error does occur, or a faulty food container becomes contaminated. However, heat preservation of food generally refers to controlled processes that are performed commercially, such as blanching, pasteurizing, and canning.

DEGREES OF PRESERVATION

It is important to recognize that there are various degrees of preservation by heating, and that commercial heat-preserved foods are not truly sterile. A few terms must be defined and understood.

Sterilization

Sterilization refers to the complete destruction of microorganisms. Because of the resistance of certain bacterial spores to heat, this frequently requires a treatment of at least 121°C of wet heat for 15 min or its equivalent. It also means that every particle of the food must receive this heat treatment. If a can of food is to be sterilized, then immersing it into a 121°C pressure cooker or retort for 15 min will not be sufficient because of the relatively slow rate of heat transfer through the food into the can. Depending on the size of the can, the effective time to achieve true sterility may be several hours. During this time there can be many changes in the food which reduce its quality. Fortunately, many foods need not be completely sterile to be safe and have keeping quality.

Commercially Sterile

The term commercially sterile or the word "sterile" (in quotes), sometimes seen in the literature, means that degree of sterilization at which all pathogenic and toxin-forming organisms have been destroyed, as well as all other types of organisms which if present could grow in the product and produce spoilage under normal handling and storage conditions. Commercially sterilized foods may contain a small number of heat-resistant bacterial spores, but these will not normally multiply in the food supply. However, if they were isolated from the food and given special environmental conditions, they could be shown to be alive.

Most canned and bottled food products are commercially sterile and have a shelf life of 2 years or more. Even after longer periods, so-called deterioration is generally due to texture or flavor changes rather than to microorganism growth.

Pasteurization

Pasteurization involves a comparatively low order of heat treatment, generally at temperature below the boiling point of water. Pasteurization treatments, depending on the food, have two different primary objectives. In the case of some products, notably milk and liquid eggs, pasteurization processes are specifically designed to destroy pathogenic organisms that may be associated with the food and could have public health significance. The second, more general, objective of pasteurization is to extend product shelf life from a microbial and enzymatic point of view. This is the objective when beer, wine, fruit juices, and certain other foods are pasteurized. In the latter case, these foods would not be expected to be a source of pathogens, or would be protected by some other means of control. Pasteurized products will still contain many living organisms capable of growth—of the order of thousands per milliliter or per gram—limiting the storage life compared to commercially sterile products. Pasteurization frequently is combined with another means of preservation, and many pasteurized foods must be stored under refrigeration. Pasteurized milk may be kept stored in a home refrigerator for a week or longer without developing significant off-flavors. Stored at room temperature, however, pasteurized milk may spoil in a day or two. Pasteurization is not limited to liquid foods. A newer application is the steaming of oysters in the shell to reduce bacterial counts.

Blanching

Blanching is a kind of pasteurization generally applied to fruits and vegetables primarily to inactivate natural food enzymes. This is common practice when such products are to be frozen, since frozen storage in itself would not completely arrest enzyme activity. Blanching, depending on its severity, also will destroy some microorganisms, as pasteurization will inactivate some enzymes.

SELECTING HEAT TREATMENTS

Heat sufficient to destroy microorganisms and food enzymes also generally affects other properties of foods adversely. The mildest heat treatments that guarantee freedom from pathogens and toxins and produce the desired storage life will be the heat treatments of choice. How then do processors choose the optimal heat treatment for a particular food? To select a safe heat-preservation treatment, the following must be known:

1. Time-temperature combination required to inactivate the most heat-resistant pathogens and spoilage organisms in a particular food
2. heat-penetration characteristics in a particular food, including the can or container of choice if it is packaged

Processors must provide the heat treatment which will ensure that the remotest particle of food in a batch or within a container will receive sufficient heat, for a sufficient time, to inactivate both the most resistant pathogen and the most resistant spoilage organisms if they are to achieve sterility or commercial sterility, and to inactivate the most heat-resistant pathogen if pasteurization for public health purposes is the goal. Different foods will support growth of different pathogens and different spoilage organisms, and so the targets will vary depending on the food to be heated.

HEAT RESISTANCE OF MICROORGANISMS

The most heat-resistant pathogen found in foods, especially those that are canned and held under anaerobic conditions, is *Clostridium botulinum*. However, there are nonpathogenic spore-forming spoilage bacteria, such as Putrefactive Anaerobe 3679 (PA 3679) and *Bacillus stearothermophilus* (FS 1518), which are even more heat resistant than *C. botulinum*. If a heat treatment inactivates these spoilage organisms, *C. botulinum* and all other pathogens in the food also will be destroyed.

Thermal Death Curves

Bacteria are killed by heat at a rate that is very nearly proportional to the number present in the system being heated. This is referred to as a logarithmic order of death, which means that under constant thermal conditions the same percentage of the bacterial population will be destroyed in a given time interval, regardless of the size of the surviving population. In other words, if a given temperature kills 90% of the population in the first minute of heating, 90% of the remaining population will be killed in the second minute, 90% of what is left will be killed in the third minute, and so on. This principle is illustrated in Fig. 8.1. The logarithmic order of death also applies to bacterial spores, but the slope of the death curve will differ from that of vegetative cells, reflecting the greater heat resistance of spores.

Figure 8.1 also illustrates the concept of the "D value," which is defined as the time in minutes at a specified temperature required to destroy 90% of the organisms in a population. Thus, the D value, or decimal reduction time, decreases the surviving population by one log cycle. If a quantity of food in a can contained one million organisms and it received heat for a time equal to four D values, then it would still contain 100 surviving organisms. If there were 100 such cans in a retort initially and the retort provided heat for a period equivalent to 7 D values, then it would be expected that the 100 cans with a total initial bacterial population of 100 million organisms would still contain 10 surviving organisms. Statistically, these 10 organisms should

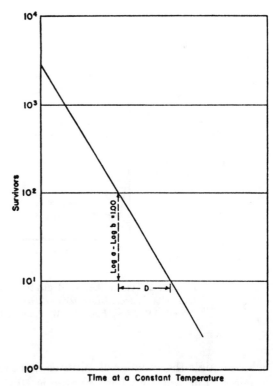

Figure 8.1. Bacterial destruction rate curve showing logarithmic order of death. Source: Stumbo, *Thermobacteriology in Food Processing, 2nd ed.,* Academic Press, New York, 1973.

be distributed among the cans. Obviously, no container can have a fraction of an organism although the 100 cans will average 0.1 organism per can. In this case, 10 of the cans probably will have one organism each and could possibly ultimately spoil, while 90 of the cans would be sterile.

Figure 8.1 is one kind of thermal death *rate* curve. It provides data on the rate of destruction of a specific organism in a specific medium or food at a *specific* temperature. From thermal death rate curves determined at different temperatures, a thermal death *time* curve (Fig. 8.2) can be constructed. A thermal death time curve for a specific organism in a specific medium or food provides data on the destruction times for a defined population of that organism at *different* temperatures.

Figure 8.2 illustrates two terms to characterize thermal death time curves. These are the "z value" and the "F value." The z value is the number of degrees required for a specific thermal death time curve to pass through one log cycle (change by a factor of 10). It is also the negative slope index of the thermal death time curve. Different organisms in a given food will have different z values, which characterize resistance of the populations to changing temperature. Similarly, a given organism will have different z values in different foods. The F value is defined as the number of minutes at a specific temperature required to destroy a specified number of organisms having a specific z value. Thus, the F value is a measure of the capacity of a heat treatment to sterilize.

Since F values represent the number of minutes to diminish a population with a specific z value at a specific temperature, and z values as well as temperatures vary, it is convenient to designate a reference F value. Such a reference is the F_0 value,

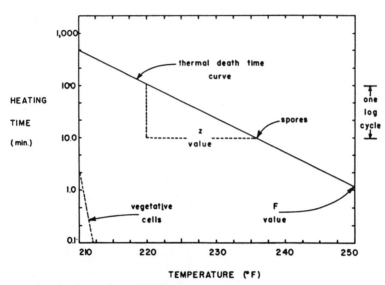

Figure 8.2. Typical thermal death time curves for bacterial spores and vegetative cells. Source: *Desrosier and Desrosier, Technology of Food Preservation,
4th ed. AVI Publishing Co., Westport, CT, 1977.*

which equals the number of minutes at 121°C (250°F) required to destroy a specified
number of organisms whose z value is 10°C (18°F). If such a population is destroyed
in 6 min at 121°C, then the heat treatment was equal to an F_0 of 6. Other temperatures
for different times can have the same lethality as this heat treatment. If they do, then
they also can be described as having an F_0 value of 6. If they have less lethality, they
have an F_0 value of less than 6 and vice versa. The F_0 value of a heat treatment is
thus a measure of its lethality, and the F_0 value is also known as the "sterilization
value" of the heat treatment. F_0 is a common term in the canning industry and other
areas utilizing heat processes. Not only do different amounts of heating provide different F_0 values but the F_0 requirements of various foods differ and are a measure of the
ease or difficulty with which these foods can be heat-sterilized.

Thermal death time curves have been carefully determined for many important
pathogens and food spoilage organisms. Two such curves for Putrefactive Anaerobe
3679 and *Bacillus stearothermophilus* (FS 1518) are shown in Fig. 8.3. They tell us how
long it takes to kill these organisms (under defined conditions) at a chosen temperature.
Thus, for example, it would take about 60 min at 104°C (220°F) to kill a specified
number of spores of PA 3679. On the other hand, at a temperature of 121°C (250°F),
these spores are killed in little over 1 min.

The conditions that must be defined to make a thermal death curve meaningful and
applicable to food processing are many. The requirement for a greater heat treatment
the larger the initial microbial population is inherent in the logarithmic order by which
bacteria die. In addition, the sensitivities of microorganisms to heat (and therefore the
characteristics of the thermal death curve) are markedly affected by the composition of
the food in which the heating is done. It already has been pointed out that acid increases
the killing power of heat. As will be enlarged upon shortly, many food constituents have
an opposite effect on heat sensitivity of microorganisms and protect them against heat.
Thus, a thermal death curve established in a synthetic medium or in a given food gener-

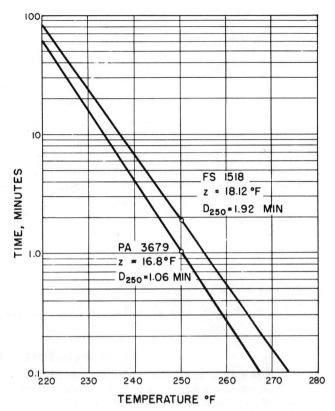

Figure 8.3. Thermal death time curves for test microorganisms PA 3679 and FS1518. Source: *Pflug and Esselen, Fundamentals of Food Canning Technology, Jim Jackson and B. M. Shinn (Editors). AVI Publishing Co., Westport, CT, 1979.*

ally is not applicable to a different food, and thermal death curves, to be valid, should be established in the specific food for which a heat process is being designed.

Margin of Safety

Data from thermal death curves can be plotted in various ways. In Fig. 8.4, data are plotted to show the heat resistance of bacterial spore suspensions as a function of initial spore concentration. Regardless of the temperature chosen, the greater the number of microorganisms or spores, the greater the heat treatment that will be required to destroy them.

We generally do not know how many organisms are present in food to be commercially sterilized, or indeed which specific types of organism are present. To provide a substantial margin of safety in low-acid foods, we may assume that a highly heat-resistant spore-former such as *C. botulinum* is present and that its population is large. From its thermal death rate curve established in the same food (or established in a medium giving no greater protection against heat destruction), we may take its D value at the temperature we choose to employ and heat for a time such that every particle of food in the can is exposed to this temperature for a period equal to 12 D values. This is sufficient to decrease any population of *C. botulinum* through 12 log cycles. Since even highly spoiled food rarely supports a bacterial population greater than a billion organisms per can, 12 D values will bring the microbial population of the can to a condition of sterility. Had a great number of such cans originally contained 1 billion

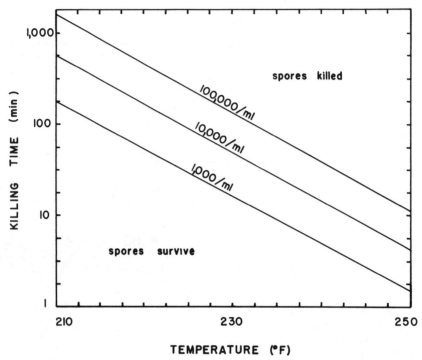

Figure 8.4. Thermal death curves for bacterial spore suspensions of different initial concentrations. Source: *Desrosier and Desrosier, Technology of Food Preservation, 4th ed., AVI Publishing Co., Westport, CT, 1977.*

C. botulinum organisms, then statistically, after a 12 *D* heat treatment, only 1 can in 1000 would be expected to still harbor 1 living organism or spore; the other 999 cans would be sterile. Had the food contained 1 million organisms per can before heating (which is still unusually high), then the same 12 *D* heat treatment would be expected to render 999,999 cans out of 1 million sterile. Since the 12 *D* heat treatment was based also on destroying *C. botulinum*, it would be still more effective against less heat-resistant spore-formers and other far less heat-resistant non-spore-forming pathogens or spoilage organisms that might be present. When organisms still more heat-resistant than *C. botulinum* are chosen as targets for destruction, then a heat treatment of less than 12 *D* values against these may be adequate. Thus, against Putrefactive Anaerobe 3679 (PA 3679) or *Bacillus stearothermophilus* (FS 1518) in low-acid foods, a 5 *D* heat treatment is considered essentially equal to a 12 *D* value against *C. botulinum* and quite sufficient to eliminate microbial spoilage and render the product pathogen-free. However, at present there is no general agreement on what theoretical number of survivors of different microorganisms is best for process calculations.

These heat treatments, which are commonly employed in the canning industry for low-acid foods, would be excessive and unnecessary for acid foods. Acid foods are currently defined as foods having a pH of 4.6 or less. Low-acid foods are those foods with a pH greater than 4.6. Table 8.1 gives the pH values for a number of foods, common spoilage agents associated with these foods, and an indication of the degree of heat processing required for their treatment. With many acid foods, temperatures at or below 100°C for a few minutes constitute adequate heat treatment.

Table 8.1. Classification of Canned Foods on Basis of Processing Requirements

Acidity Classification	pH Value	Food Item	Food Groups	Spoilage Agents	Heat and Processing Requirements
Low acid	7.0	Lye hominy	Meat	Mesophilic spore-forming anaerobic bacteria	High temperature processing 116°–121°C (240°–250°F)
		Ripe olives, crabmeat, eggs, oysters, milk, corn, duck, chicken, codfish, beef, sardines	Fish, Milk, Poultry		
	6.0	Corned beef, lima beans, peas, carrots, beets, asparagus, potatoes	Vegetables	Thermophiles, Naturally occurring enzymes in certain processes	
	5.0	Figs, tomato soup	Soup		
Medium acid	4.5	Ravioli, pimientos	Manufactured foods	Lower limit for growth of *C. botulinum*	
Acid		Potato salad	Fruits	Nonspore-forming aciduric bacteria	Boiling water processing 100°C (212°F)
		Tomatoes, pears, apricots, peaches, oranges			
	3.7	Sauerkraut, pineapple, apple, strawberry, grapefruit	Berries	Aciduric spore-forming bacteria, Naturally occurring enzymes	
High acid	3.0	Pickles	High acid foods (Pickles)	Yeasts Molds	
		Relish	High acid-high solids foods (jam-jelly)		
		Cranberry juice			
		Lemon juice			
	2.0	Lime juice	Very acid foods		

SOURCE: Desrosier and Desrosier (1977).

Certain spices and food chemicals also combine with heat in killing microorganisms and so reduce the heat treatment that must be used. Still another factor in permitting lesser heat treatments to be used with acid foods is the sensitivity of *C. botulinum* to acid. *Clostridium botulinum* will not grow in foods at pH 4.6 or below. Therefore, such foods even if unheated would not constitute a health hazard from the standpoint of this heat-resistant organism.

HEAT TRANSFER

Even after the time and temperature required to destroy target organisms are known from thermal death curves and a sufficient margin of safety has been calculated, a problem remains: how to ensure that every particle of food (within the container if the food is canned) receives the required heat treatment. This becomes a problem of heat transfer, that is, heat penetration into and throughout the can or mass of food.

If cans are heated from the outside, as would be the case if they are submerged within a retort, the larger the can, the longer it will take to heat the center portion of the can to any desired temperature. However, there are several other factors besides the size and shape of a can that affect heat penetration into the food within it. Principal among these factors is the nature and consistency of the food itself. This will determine, for example, whether heat will reach the center by straight conduction or will be speeded by some convection within a can.

Conduction and Convection Heating

Heat energy is transferred by conduction, convection, and radiation. In retorts used in canning, conduction and convection are important. Conduction is the method of heating in which the heat moves from one particle to another by contact, in more or less straight lines. In the case of conduction, the food does not move in the can and there is no circulation to stir hot food with cold food. Convection, on the other hand, involves movement in the mass being heated. In natural convection the heated portion of the food becomes lighter in density and rises; this sets up circulation within the can. This circulation speeds temperature rise of the entire contents of the can. Forced convection occurs when circulation is promoted mechanically.

A liquid food such as canned tomato juice can be readily set into convection heating motion in addition to the heating by conduction it receives through the can wall. On the other hand, a solid food such as corned beef hash is too viscous to circulate, and so it will be virtually completely heated by conduction through the can wall and through itself. A product containing free liquid and solid, such as a can of pears within a sugar syrup, will be intermediate and will rise in temperature from a combination of conduction and convection; conduction through the fruit and convection from the moving syrup. Convection heating is far more rapid than conduction heating and so, other things being equal, if cans of these three products were placed in the same retort, uniform complete heating would be expected to be reached first in the tomato juice, second in the canned pears, and last in the corned beef hash.

Cold Point in Food Masses

When heat is applied from the outside, as in retorting, the food nearest the can surfaces will reach sterilization temperature sooner than the food nearer the center of the can. The point in a can or mass of food which is last to reach the final heating temperature is designated the "cold point" within the can or mass.

In a can of solid food heated by conduction the cold point is located in the very center of the can. However, in foods that undergo convection heating, unless the cans are agitated, the cold point is somewhat below the dead center of the can. To ensure that commercial sterilization is achieved, sufficient time must be allowed for the cold point of cans to reach the sterilization temperature and remain there for the required time interval to destroy the most resistant bacterial spores. If 12 D values are indicated and this corresponds, for example, to 121°C for 2.5 min in a particular system, and if we ensure that the cold point of cans receives 121°C for 2.5 min, or an equivalent heat treatment, we are assured that every other region within the can has been adequately heated.

Determining Process Time and Process Lethality

The time required in a retort to produce lethal temperatures at the cold point can be determined with a heat-sensing thermocouple. Figure 8.5 indicates proper placement of the thermocouple to measure temperature at the cold points in canned foods that heat by conduction and natural convection. The cans with thermocouples are filled with the particular food under study, sealed, and placed in the retort. After steam is admitted to the retort, the temperature rise is recorded with time. In a particular retort filled with a given number of cans of specified size and contents, it may require 30–40 min for the cold point of cans to reach a lethal temperature close to 121°C. This is due to the "come-up" time required for the retort to reach processing temperature plus the time for heat penetration into the cans. The addition of the needed holding time completes the sterilization requirement.

Although a lethal effect at the cold point equivalent, for example, to 121°C for 2.5 min, may be called for, this degree of lethality can be achieved by various equivalent time–temperature exposures. Further, since the temperature rise during heat penetration of the can also accomplishes a degree of microbial destruction, this is commonly accounted for by decreasing the required holding time accordingly. Once treated suffi-

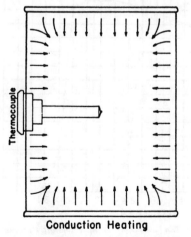 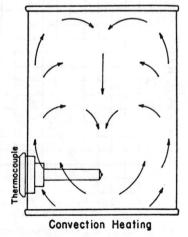

Conduction Heating   **Convection Heating**

Figure 8.5. Proper thermocouple placement in a can when heating is primarily by conduction or convection. *Courtesy of American Can Co.*

ciently, cans are quickly cooled to prevent additional heat damage to the food. Because cooling is not instantaneous, some additional microbial destruction also occurs during the cooling period. Thus, to calculate effective retort process treatments, accurate heat penetration and cooling curves must be established. Total lethality of the process then represents a summation of the lethal effects of changing temperatures with time during the entire retort operation.

To perceive how total lethality of a process is calculated, one must first understand what is meant by the term "unit of lethality." For heat process calculations, a unit of lethality has been defined as the heat kill equivalent to 1 min at 121°C against an organism of a given z value. All equally destructive heat treatments provide a unit of lethality. Further, fractions of a minute at 121°C, or their equivalents, represent corresponding fractions of a unit of lethality. These fractions are referred to as "lethal rates." We can calculate the lethal rate of any temperature reached at the cold point of a can being retorted, for any target organism, from the following relationship: lethal rate = antilog $[(T-250)/z]$, where T is the temperature of the cold point in the container in degrees Fahrenheit and z for the target organism also is in degrees Fahrenheit. Similarly, lethal rate = antilog $[(T-121)/z]$, where T and z are in degrees Celsius. These lethal rates, corresponding to successive temperatures taken from the heat penetration and cooling curves of a retort process are integrated to determine the total lethality of the process, which is its sterilization value or F_0 value. This may be done by plotting lethal rates against corresponding time from the heat penetration and cooling curves, as in Fig. 8.6. The resulting total area under this lethal rate curve divided by the area corresponding to one unit of lethality gives total lethality or F_0. In Fig. 8.6, F_0 equals 9.74; thus, the retort process was equivalent to a heat treatment of 121°C for 9.74 min against an organism with a z value of 10°C. In Fig. 8.6, lethal rates increase and then begin to decrease after about 30 min, which is the time when retort steam was turned off and cooling water turned on. The dotted line traced parallel to the descending line encloses an area corresponding to the retort process had the steam been turned off after 25.5 min (vertical dotted line). In this case, F_0, the total lethality or sterilization value of the process, would equal 6.3.

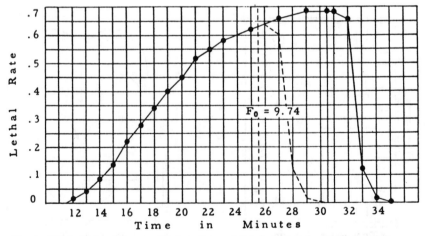

Figure 8.6. Lethal rate curve. Source: *Canned Food—Principles of Thermal Process Control and Container Closure Evaluation.* Source: *National Canners Association, Berkeley, CA, 1973.*

Since come-up time and penetration time will vary between different retorts, different size and shape of cans or bottles, and different compositions of foods, it is obvious that the required heat treatment will be different for each specific case. More advanced mathematical methods than those discussed here have been devised for the calculation of safe but not excessive process times and the effects of changes in processing on lethality. These calculations may be performed by computers which further control retort processes in highly instrumented canning plants. In all cases, however, bacterial death curve data, heat penetration properties of the food, and certain characteristics of the retort must be known if an optimum process is to be calculated. Of course, a great deal of experience has been gained by the canning industry over the years and simple tables of heat treatments for well-known foods in common can sizes can be found in appropriate canning references (Table 8.2). However, when a new product is developed or when new packaging shapes or materials are employed, then specific determinations of effective heat treatments must be made.

PROTECTIVE EFFECTS
OF FOOD CONSTITUENTS

Several constituents of foods protect microorganisms to various degrees against heat. For example, sugar in high concentration protects bacterial spores, and canned fruit in a sugar syrup generally requires a higher temperature or longer time for sterilization than the same fruit without sugar. Starch and protein in foods generally act somewhat like sugar. Fats and oils have a great protective effect on microorganisms and their spores by interfering with the penetration of wet heat. As has been noted, wet heat at a given temperature is more lethal than dry heat, because moisture is an effective conductor of heat and penetrates into microbial cells and spores. If microorganisms are trapped within fat globules, then moisture can less readily penetrate into the cells and heating becomes more like dry heat. In the same can or food mass, organisms in the liquid phase may be quickly killed while more heating time is required for inactiva-

Table 8.2. Proceess Time for Vegetables in 307 × 409 Cans and No. 303 Glass Jars

Product	Initial Temp		307 × 409 Cans		No. 303 Jars	
	(°C)	(°F)	Min at 116°C 240°F	Min at 121°C 250°F	Min at 116°C 240°F	Min at 121°C 250°F
Green beans, whole or cut	21	70	21	12	25	—
Lima beans, succulent	21	70	40	20	45	—
Beets, whole, cut, diced	21	70	35	23	35	—
Carrots, whole, cut	21	70	35	23	30	—
Corn, cream style	71	160	100	80	105	80
Corn, whole kernel in brine	38	100	55	30	50	30
Peas in brine	21	70	36	16	45	25
Peas and carrots	21	70	45	20	45	—
Potatoes, white, small whole	21	70	35	23	35	25
Pumpkin or squash	71	160	80	65	80	65

SOURCE: National Canners Assoc. (1966, 1971).

tion of the oil-phase flora. This makes sterilization of meat products and fish packed in oil very difficult; the severe heat treatments required often adversely affect other food constituents. Likewise, because there is more fat and more sugar in ice cream mix than there is in milk, ice cream mix must be pasteurized at a higher temperature or for a longer time than milk to accomplish equivalent bacterial destruction.

In addition to any direct protective effects food constituents may have on microorganisms, there are indirect effects related to differences in heat conductivity rates through different food materials. Fat, for example, is a poorer conductor of heat than is water. Further, and often more important, are the effects related to food consistency and its influence on whether conduction or convection heating will take place. If sufficient starch or other thickener is added to a food composition to convert it from a convection heating system to a conduction heating system, then in addition to any direct microbial protection, there will be a slowing down of the heat penetration rate to the cold point within the container or food mass, and this will protect microorganisms. Because common starches in solution thicken upon heating, foods supplemented with starch have a reduced rate of convection within cans during retorting and require longer retort times. Special starches have been developed which do not thicken on early heating but instead thicken on later heating or on cooling. Foods supplemented with these starches retain maximum convection rates in the retort, permitting shortened retort times and less heat damage. Then upon cooling, the starch imparts the desired thickening. In a typical application, a product like chow mein can be heated with less softening of the vegetables from excessive heating yet possess the desired viscosity in the liquid phase.

The size and type of container that holds the food during thermal processing can also affect the sterilization process. Thin flexible pouches allow faster heat penetration into the cold center when compared to a cylindrical shape of a can, for example. This means that less heat is required for an equivalent lethality in pouches. Often higher quality product can be achieved from pouches than cans but at a higher container cost. The heat transfer properties of the container can also affect processing time. Thus, metal cans transfer heat more readily than plastic cans, resulting in shorter processing times.

INOCULATED PACK STUDIES

The many variables discussed so far make the determination of safe heat treatments by calculation alone difficult and sometimes risky, especially when applied to new products. In practice, therefore, formulas based on thermal death curves, heat penetration rates, and properties of specific retorts are used to gain an approximation of the safe heat treatment, but results are checked by what are termed inoculated pack studies.

In inoculated pack studies, a substantial population of a heat-resistant food spoilage organism such as PA 3679 is inoculated into cans of food, which are then processed in a retort. If formulas call for 60 min of heat, representative cans may be heated for 50, 55, 60, 65, and 70 min. The cans are then stored at a temperature that would be favorable for growth of any surviving spores. The cans are periodically examined for evidence of growth and spoilage, such as bulging from gas production (Fig. 8.7). Samples of nonbulging cans also are examined bacteriologically. The shortest heat treatment

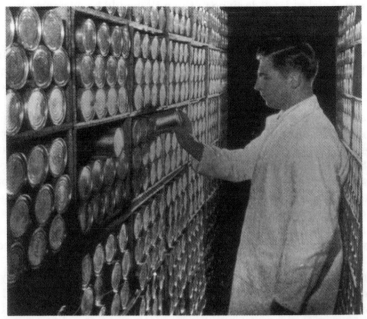

Figure 8.7. Cans being examined for bulging as evidence of spoilage. *Courtesy of American Can Co.*

that consistently produces commercial sterility is then taken as the effective heat treatment to be used for subsequent commercial packs.

DIFFERENT TEMPERATURE–TIME COMBINATIONS

Different temperature–time combinations that are equally effective in microbial destruction can differ greatly in their damaging effect on foods. This is of the greatest practical importance in modern heat processing and is the basis for several of the more advanced heat preservation methods. If the time–temperature combinations required for destruction of *C. botulinum* in low-acid media are taken from thermal death curves, the following will be found to be equally effective:

0.78 min at 127°C	10 min at 116°C
1.45 min at 124°C	36 min at 110°C
2.78 min at 121°C	150 min at 104°C
5.27 min at 118°C	330 min at 100°C

This illustrates the simple relationship that the higher the temperature the less time is required for microbial destruction. This principle holds true for all types of microorganisms and spores. On the other hand, foods are not equally resistant to these combinations, and the more important factor in damaging the color, flavor, texture, and nutritional value of foods is long time rather than high temperature. If we were to inoculate milk with *C. botulinum* and then heat samples for 330 min at 100°C, 10 min at 116°C, and less than 1 min at 127°C, equal microbial destruction would occur in all three samples, but heat damage to the milk would be enormously different. The sample heated for 330 min would be thoroughly cooked in flavor and brown in color. The 10-min sample would be almost as bad. The 1-min sample, although still somewhat overheated, would not be far different from unheated milk. This difference in sensitivity to time and temperature between microorganisms and various foods is a general phenomenon. It applies to milk, meat, juices, and generally all other heat-sensitive food materials.

The greater relative sensitivity of microorganisms than food constituents to high temperatures can be quantitatively defined in terms of different temperature coefficients for their destruction. Thus, whereas each increase of 10°C in temperature approximately doubles the rate of chemical reactions contributing to food deterioration, each 10°C increase, above the maximum temperature for growth, produces approximately a tenfold increase in the rate of microbial destruction. Since higher temperatures permit use of shorter times for microbial destruction, and shorter times favor food quality retention, high-temperature–short-time heating treatments rather than low-temperature–long-time heating treatments are used for heat-sensitive foods whenever possible.

In pasteurizing certain acid juices, for example, the industry formerly used treatments of about 63°C for 30 min. Today, flash pasteurization at 88°C for 1 min, 100°C for 12 sec, or 121°C for 2 sec is the common practice. Although bacterial destruction is very nearly equivalent, the 121°C 2-sec treatment gives the best quality juice with respect to flavor and vitamin retention. Such short holding times, however, require special equipment which is more difficult to design and generally is more expensive than that needed for processing at 63°C.

HEATING BEFORE OR AFTER PACKAGING

The foregoing principles very largely determine the design parameters for heat preservation equipment and commercial practices. The food processor will employ no less than that heat treatment which gives the necessary degree of microorganism destruction. This is further ensured by periodic inspections from the FDA or equivalent local authorities. However, the food processor also will want to use the mildest effective heat treatment to ensure highest food quality, as well as to conserve energy.

It is convenient to separate heat preservation practices into two broad categories: one involves heating of foods in their final containers, the other employs heat prior to packaging. The latter category includes methods that are inherently less damaging to food quality, especially when the food can be readily subdivided (such as liquids) for rapid heat exchange. However, these methods then require packaging under aseptic or nearly aseptic conditions to prevent or at least minimize recontamination. On the other hand, heating within the package requires less technical sophistication and produces quite acceptable quality with the majority of foods; most canned foods are heated within the package.

Heating Food in Containers

Still Retort

One of the simplest applications of heating food in containers is sterilization of cans in a still retort, that is, the cans remain still while they are being heated. In this type of retort, temperatures above 121°C generally may not be used or foods cook against the can walls. This is especially true of solid foods that do not circulate within the cans by convection, but it also can be a problem with liquid foods. Because 121°C is the upper temperature, and there is relatively little movement in the cans, the heating time to bring the cold point to sterilizing temperature is relatively long; for a small can of peas it may be 40 min.

Agitating Retorts

Processing time can be markedly reduced by shaking the cans during heating, especially with liquid or semiliquid foods. Not only is processing time shortened, but food quality is improved. This is accomplished with various kinds of agitating retorts, one type of which is shown in Fig. 8.8. Part of the wall has been cut away to show the cans resting in reels which rotate and thereby shake the contents. Forced convection within cans also depends on the degree of can filling, since some free headspace within cans is necessary for optimum food turnover within the cans. In addition to faster heating, there is less chance for food to cook onto the can walls since the can contents are in motion. Different types of agitation are possible; for example, cans may be made to turn end over end or to spin on their long axis. Depending on the physical properties of the food, one method may be more effective.

The reduction in processing times possible with agitating retorts compared with still retorts is seen in Table 8.3. These substantial reductions in time with associated quality advantages would not be realized in foods that heat primarily by conduction; for such foods the simpler and generally less costly still retorts may be quite satisfactory.

Pressure Considerations

Whether still or agitating retorts are used, the high temperatures required for commercial sterilization commonly are obtained from steam under pressure. Steam

Figure 8.8. Cutaway view of a continuous agitating retort. *Courtesy of FMC Corp.*

Table 8.3. Comparison of Process Times in Agitating and Conventional Retorts

Product	Can Size	Process-agitated			Conventional		
		Time (min)	Temp (°C)	Temp (°F)	Time (min)	Temp (°C)	Temp (°F)
Peas	307×409	4.90	127	260	35	116	240
Carrots	307×409	3.40	127	260	30	116	240
Beets, sliced	307×409	4.10	127	260	30	116	240
Asparagus spears	307×409	4.50	132	270	16	120	248
Asparagus, cuts and tips	307×409	4.00	132	270	15	120	248
Cabbage	307×409	2.75	132	270	40	116	240
Asparagus, spears							
brine packed	307×409	5.20	127	260	50	116	240
brine packed	603×700	10.00	127	260	80	116	240
vacuum packed	307×306	5.00	127	260	35	121	250
Mushroom soup	603×700	19.00	127	260	—	—	—
Evaporated milk	300×314	2.25	93	200	18	116	240

Courtesy of L. E. Clifcorm.

pressures of approximately 10, 15, and 20 psi (above atmospheric pressure) are required for heating at 116°C, 121°C, and 127°C (1 psi = 0.07 kg/cm^2 = 6895 pascals).

Moist foods in cans have part of their moisture converted to steam at these temperatures and produce equivalent pressures within the cans although there is little pressure difference between the inside and outside of the can when temperatures are equal. A special case is foods canned under vacuum. Then the initial pressure within a can will be less than the pressure in the retort to an extent determined by the degree of vacuum used at time of can closure. Control of pressure differences inside and outside of cans and other containers during and following heat treatment are of obvious importance to prevent mechanical damage to containers. Several techniques are employed to prevent such damage.

If the vacuum within cans is such that retort pressures cause can collapse, a heavier gauge of steel may be required. More commonly, pressure problems are due to greater pressures within the container than on its outside. This occurs when steam pressure is too rapidly released in closing down a batch-type retort or when heated containers are too suddenly conveyed from a continuous pressure retort to atmospheric pressure. The problem is greater in the case of glass jars than with cans; excessive internal pressure can easily blow the lids from glass jars since these generally have a weaker seal than the lids of cans. During retorting of glass jars, provisions are made for air pressure over a layer of water to balance internal and external pressures. Partial cooling of containers before releasing them from retorts is the common way to decrease internal container pressure. Many continuous retorts, such as the agitating type in Fig. 8.8, provide semipressurized cooling zones following the heating zone just prior to release of cans to atmospheric pressure.

With the increasing use of flexible packaging materials has come the sterilization of foods in flexible plastic pouches. Here, pressure problems can be still greater than with glass jars. Overriding air pressure must be applied when the pouches are cooled after retorting so that the steam pressure inside the pouch does not cause the pouch to burst. In addition, a uniform heat treatment requires that the pouches be evenly exposed to the heating medium rather than be allowed to contact each other and pile up. One means of better controlling pouches during retorting is to sandwich the pouches between rigid supports (Fig. 8.9). Plastic pouches require shorter retort times since heat penetration through the thin pouches is quite rapid. This, in turn, can produce high quality products and save on energy costs.

All plastic rigid "cans" are also being used to retort foods. These cans have the advantage that they can be reheated in a microwave oven. They require many of the same types of handling requirements as flexible pouches.

Hydrostatic Cooker and Cooler

Continuous retorts (usually of the agitating type) are pressure-tight and built with special valves and locks for admitting and removing cans from the sterilizing chamber. Without these, pressure conditions would not be held constant and sterilizing temperatures could not be closely controlled. Another type of continuous pressure retort, which is open to the atmosphere at the inlet and outlet ends, is the hydrostatic pressure cooker and cooler.

This type of heating equipment consists essentially of a "U" tube with an enlarged lower section. Steam is admitted to the enlarged section and hot water fills one of the legs of the "U" while cool water fills the other leg (Fig. 8.10). Cans are carried by a chain conveyor down the hot water leg, through the steam zone, which may involve an undulating path to increase residence time, and up the cool water leg. These legs are sufficiently high to produce a hydrostatic head pressure to balance the steam pressure in the sterilizing zone. If a temperature of 127°C is used in the sterilizing zone, then this would be equal to a pressure of about 140,000 pascals (20 psi) above

Figure 8.9. Racks of flexible retortable pouches in retort ready for processing. *Courtesy of Magic Pantry Foods, Inc.*

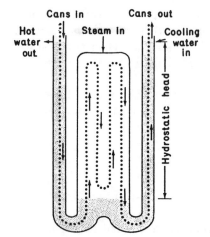

Figure 8.10. Hydrostatic cooker and cooler illustrating how steam pressure is balanced by water heads. *Courtesy of Food Processing.*

atmospheric pressure which would be balanced by water heights of about 14 m (46 ft) in the hot and cold legs.

As cans descend the hot water leg and enter the steam zone, their internal pressure increases as food moisture begins to boil. But this is balanced by the increasing external hydrostatic pressure. Similarly, as high-pressure cans pass through the water seal and ascend the cool water leg, their gradually reduced internal pressure is balanced by the decreasing hydrostatic head in this cool leg. In this way cans are not subjected to sudden changes in pressure. For this reason the system also is well suited to the retorting of foods and beverages in jars and bottles.

Direct Flame Sterilization

Where sterilizing temperatures above 100°C are needed, steam under pressure generally is the heat exchange medium, and vessels capable of withstanding pressure add to the cost of equipment. Another method, introduced from France, employs direct flame to contact cans as the cans are rotated in the course of being conveyed past gas jets. Excellent rates of heating are achieved with high product quality and reduced costs, but commercial experience with this type of system is still somewhat limited.

In-package Pasteurization

In-package heating need not be to the point of sterilization or commercial sterilization. Tunnels of various designs are used to pasteurize food and beverages in cans, bottles, and jars. Hot water sprays or steam jets are directed at the containers and varying temperature zones progressing to cooling temperatures are commonly employed. Temperature changes must be gradual to prevent thermal shock to glass. Such systems are operated at atmospheric pressure. This is one of the methods of pasteurizing beer containers.

Heating Food Prior to Packaging

As already stated, there are advantages to heating heat-sensitive foods prior to packaging. These are related to the ability to heat rapidly by exposing food in a subdivided state to a heat exchange surface or medium, rather than having to allow appreciable time for heat penetration into a relatively large volume of food in a container.

Batch Pasteurization

One of the earliest and simplest methods of effectively pasteurizing liquid foods, such as milk, is to heat the food in a vat with mild agitation. Raw milk commonly is pumped into a steam-heated jacketed vat, brought to temperature, held for the prescribed time, and then pumped over a plate-type cooler prior to bottling or cartoning. Milk must be quickly brought to 62.8°C (145°F), held at this temperature for 30 min, and rapidly cooled. In addition to destroying common pathogens, this heat treatment also inactivates the enzyme lipase, which otherwise would quickly cause the milk to become rancid. Batch pasteurization, also known as the holding method of pasteurization, is still widely practiced in some parts of the world, but it has largely given way to high-temperature–short-time continuous pasteurization.

High-Temperature–Short-Time Pasteurization

High-temperature–short-time (HTST) pasteurization of raw milk employs a temperature of at least 71.7°C (161°F) for at least 15 sec. This is equivalent in bacterial destruction to the batch method. In HTST pasteurization (Fig. 8.11), raw milk held in a cool storage tank is pumped through a plate-type heat exchanger and brought to temperature. The key to the process rests in ensuring that every particle of the milk

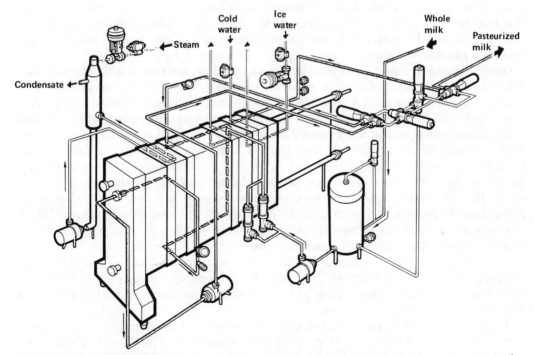

Figure 8.11 Flow diagram of a high-temperature—short-time plate-type pasteurization system commonly used for milk. Source: Anon *Dairy Handbook, Alfa-Laval, Inc. Lund, Sweden.*

remains at not lower than 71.7°C for no less than 15 sec. This is accomplished by pumping the heated milk through a holding tube of such length and diameter that it takes every milk particle at least 15 sec to pass through the tube. At the end of the tube is an accurate temperature-sensing device and valve. Should any milk reach the end of the holding tube and be down in temperature even one degree, a flow diversion valve checks this flow of milk and sends it back through the heat exchanger once again to be reheated. In this way no milk escapes the required heat treatment. Frequent checks of the equipment are made by authorized milk inspectors to help ensure its proper operation. After emerging from the holding tube, the milk is cooled and may be cartoned or bottled. Cooling not only prevents further heat damage to the milk but also retards subsequent bacterial multiplication since the milk is not sterile.

Pasteurization by the HTST method is not limited to milk and is widely used in the food industry. However, times and temperatures vary in accordance with the effects of different foods on microorganism survival and the heat sensitivities of these foods.

Aseptic Packaging

Aseptic packaging is a method in which food is sterilized or commercially sterilized outside of the can, usually in a continuous process, and then aseptically placed in previously sterilized containers which are subsequently sealed in an aseptic environment.

The most commercially successful form of aseptic packaging utilizes paper and plastic materials which are sterilized, formed, filled, and sealed in continuous operation. The package may be sterilized with heat or a combination of heat and chemicals. In some cases, the disinfectant property of hydrogen peroxide (H_2O_2) is combined with heated air or with ultraviolet light to make lower temperatures effective in sterilizing these less heat-resistant packaging materials. Coffee cream and coffee whiteners, for example, are packaged in small single-service paper packets this way, as are larger volume-size milk and juice products.

Quick heating of liquid foods may be done in a plate-type heat exchanger (see Fig. 5.7) or in a tubular scraped-surface heat exchanger (Fig. 8.12). This latter type consists essentially of a tube within a tube. Steam flows through the space between the tubes while food flows through the inner tube. The inner tube also is provided with a rotating shaft or mutator equipped with scraper blades to prevent food from burning onto the heat exchange surface. In contact with the hot surface, the thin layer of food may be brought to sterilization temperature in 1 sec or less. Food temperatures employed may be as high as 150°C and sterilization takes place in 1 or 2 sec, yielding food products of the highest quality, and often with significant energy savings. If it is desired to prolong residence time beyond this, then a holding tube is added as in the case of HTST pasteurization. Such rapid sterilization at extremely high temperatures is referred to as ultrahigh-temperature (UHT) sterilization.

The sterile food must be quickly cooled to room temperature, because at these high temperatures product quality can be impaired in seconds. Quick cooling can be accomplished with the same types of plate or tubular scraped-surface heat exchangers, used with refrigerants instead of steam.

Aseptic packaging is also used with metal cans as well as large plastic and metal drums or large flexible pouches (Fig. 8.13). Great quantities of food materials are used as intermediates in the production of further processed foods. This frequently requires packaging of such items as tomato paste or apricot puree in large containers such as

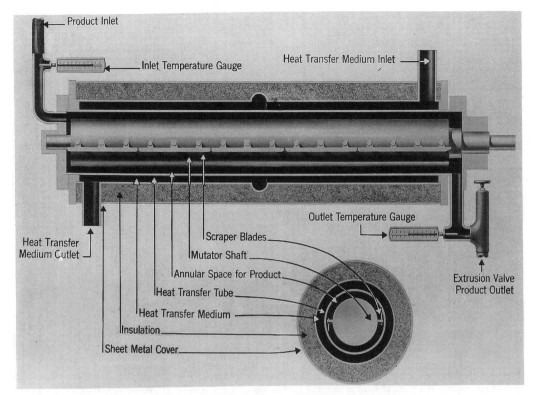

Figure 8.12. Tubular-scraped surface-type heat exchanger. *Courtesy of Chemetron Corp.*

55-gal drums, as smaller units involve greater expense. The food manufacturer then may use the tomato paste in the production of ketchup or the apricot puree in bakery products. If such large volumes were to be sterilized in drums, by the time the cold point reached sterilization temperature the product nearer the drum walls would be excessively burned. Such items can be quickly sterilized in efficient heat exchangers and aseptically packaged. In this case, large chambers have been developed in which the drums and lids are sterilized under superheated steam and then filled with the product and sealed aseptically within the chamber. This technology has advanced to the point where sterile food can be aseptically filled into previously sterilized silo tanks and tank cars. More will be said about aseptic packaging in Chapter 21.

Hot Pack or Hot Fill

The terms hot pack or hot fill refer to the packing of previously pasteurized or sterilized foods, while still hot, into clean but not necessarily sterile containers, under clean but not necessarily aseptic conditions. The heat of the food and some holding period before cooling the closed container is utilized to render the container commercially sterile.

Hot pack, as distinguished from aseptic packaging, is most effective with acid foods since lower temperatures in the presence of acid are lethal; further, at a pH of 4.6 *Clostridium botulinum* will not grow or produce toxin, so this health hazard is not present. Hot pack with low-acid foods (above pH 4.6) is not feasible unless the product

Figure 8.13. Aseptically packaged liquid eggs can be packaged in 220- or 330-gal bag-in-box containers for use in large-scale bakeries or food processing plants. Source: J. Giese *Food Technology 48*(9)95. 1994.

is recognized as being only pasteurized and will be stored under refrigeration or unless the hot pack treatment is combined with some additional means of preservation such as a very high sugar content. This is because the residual heat of the food in the absence of appreciable acid is not sufficient to guarantee destruction of spores that may be present on container surfaces or that may enter containers during filling and sealing. Even with acid foods, very definite food temperatures and holding times in the sealed containers before they are cooled for warehouse storage must be adhered to for hot pack processing to be effective. These temperatures and times depend on the specific product's pH and other food characteristics.

In home canning, when fruit and sugar are boiled together to make jam and the hot jam is poured into jars that have been previously boiled, the principle of hot pack is being employed. Home canning instructions further call for inverting the filled jars after a short time. This is to ensure that the hot acid product contacts all surfaces of the jar lid for sterilization. However, for home canning of meats and other low-acid foods, directions always call for pressure cooking of closed containers as is done in conventional commercial retorting.

In commercial practice, acid juices such as orange, grapefruit, grape, tomato, and various acid fruits and vegetables, such as sauerkraut, commonly are hot packed following prior pasteurization or sterilization. Typically, acid fruits and juices are first heated in the range of about 77–100°C for about 30–60 sec, hot filled at no lower than 77°C and often closer to 93°C, and held at this temperature for 1–3 min, including an inversion before cooling. In the case of tomato juice, a common practice is HTST heating of juice at 121°C for 0.7 min, cooling below the boiling point but not below 91°C for hot fill can sterilization, and can holding for 3 min, including an inversion before final cooling. Precise times and temperatures depend on the pH of the particular tomato juice batch and can be confirmed by inoculated pack studies.

Microwave Heating

Microwave energy produces heat in materials that absorb it. Microwave energy and energies of closely related frequencies are finding ever-increasing applications in the food industry. These include heat preservation. Microwave energy heats foods in a unique fashion that largely eliminates temperature gradients between the surface and center of food masses. Foods do not heat from the outside to the inside as with conventional heating since microwave penetration can generate heat throughout the food mass simultaneously. In this case the concept of cold point and the limitations of conventional heat penetration rates are not directly applicable. The use of microwaves can result in very rapid heating but requires special equipment, and often specific packaging materials, since microwaves will not pass through metal cans or metal foils. Microwave heating also can produce major differences in food appearance and other properties compared with the more conventional methods of heating. More will be said about microwave heating in Chapter 11.

GOVERNMENT REGULATIONS

In the United States, as elsewhere, the Food and Drug Administration requires thermal processing to be carried out under what is known as Good Manufacturing Practices (GMPs) to help assure food safety and wholesomeness. Among these GMPs are specific regulations pertaining to low-acid canned foods (foods that are thermally processed, have pH values greater than 4.6 and water activity greater than 0.85, are packaged in hermetically sealed containers, and are not stored under refrigeration). FDA has issued a number of regulations for low-acid foods. Additional regulations for acidified foods have also been released. The primary purpose of these regulations is to describe safe procedures for manufacturing, processing, and packing of foods that could otherwise support the growth of and toxin production by *Clostridium botulinum*. The safety of low-acid and acidified foods is further ensured through the Emergency Permit Control regulations, which require manufacturers to register their processing plants and file their processes with the FDA. These regulations also require firms to adhere to their filed and approved processes, to maintain detailed records, and to make these records available to authorized FDA personnel. Since differences in processing equipment, operating conditions, container type or size, kind of food, and food form constitute different processes, presently over 100,000 processes have been filed with the FDA under these regulations.

References

Anon. 1982. Canned Foods: Principles of Thermal Process Control, Acidification and Container Closure Evaluation. 4th ed. The Food Processors Institute, Washington, DC.

Desrosier, N. W. and Desrosier, J. N. 1977. Technology of Food Preservation. 4th ed. AVI Publishing Co., Westport, Conn.

Dietz, J.M. and Erdman, J.W. 1989. Effects of thermal processing upon vitamins and proteins in foods. Nutr. Today *24*(4), 6–15.

Heldman, D. R. and Singh, P. 1981. Food Process Engineering. 2nd ed. Chapman & Hall, London.

Le Maguer, M. and Jelen, P. 1986. Food Engineering and Process Applications. Elsevier Applied Science Publishers, London.

Lopez, A. 1987. A Complete Course in Canning. Books I, II, III. The Canning Trade Inc., Baltimore, MD.

Mohsenin, N. N. 1980. Thermal Properties of Foods and Agricultural Materials. Gordon and Breach, New York.

National Canners Assoc. 1966. Processes for Low-Acid Canned Foods in Metal Containers. Natl. Canners Assoc. Bull. 26-L.

National Canners Assoc. 1971. Processes for Low-Acid Canned Foods in Glass Containers. Natl. Canners Assoc. Bull. 30-L.

Pflug, 1. J. and Esselen, W. B. 1979. Heat sterilization of canned food. *In* Fundamentals of Food Canning Technology. J. M. Jackson and B. M. Shinn (Editors). AVI Publishing Co., Wesport, CT.

Polvino, D. 1992. Thermal processing and the role it plays in quality of food. Act Rep R&D Assoc. *44*(1), 116–129.

Rees, J.A.G. and Bettison, J. 1991. Processing and Packaging of Heat Preserved Foods. Chapman & Hall, London, New York.

Stumbo, C. R., Purohit, K. S., Ramakrishnan, T. V., Evans, D. A., and Francis, F. J. 1983. Handbook of Lethality Guides for Low-Acid Canned Foods. Vols. I and 2. CRC Press, Boca Raton, FL.

USDA. 1984. Guidelines tor aseptic processing and packaging systems in meat and poultry plants. U.S. Department of Agriculture, Washington, D.C.

Woodruff, J.G. and Luh, B.S. 1986. Commercial Fruit Processing. 2nd ed. AVI Publishing, Wesport, CT.

Zeuthen, P. 1984. Thermal Processing and Quality of Foods. Elsevier Applied Science, London.

9

Cold Preservation and Processing

Freezing and refrigeration (i.e., cold storage) are among the oldest methods of food preservation, but it was not until 1875 that a mechanical ammonia refrigeration system capable of supporting commercial refrigerated warehousing and freezing was invented. This major advance was hampered by the lack of proper facilities, a prime requirement for any refrigerated or frozen food industry. Thus, as late as the 1920s, food delivered to a market in a frozen state commonly thawed before it could be brought home or else thawed in household ice boxes and generally was of marginal to poor quality. Starting in the 1920s, Clarence Birdseye pioneered research on quick-freezing processes, equipment, frozen products, and frozen food packaging. As household refrigerators and freezers became more common, the modern frozen food industry grew rapidly.

Refrigeration today markedly influences the practices of agriculture and marketing and sets the economic climate of the food industry. Without mechanical refrigeration in transit, much of world trade in perishable food commodities would be impossible. Large cities that are distant from growing areas would cease to enjoy abundant fruits and vegetables. Refrigeration and cold storage equalize food prices throughout the year and make products available year round. Without them, prices would be very low at time of harvest and extremely high later on, if indeed the foods were available at all.

DISTINCTION BETWEEN REFRIGERATION AND FREEZING

The difference between refrigeration and cool storage on the one hand and freezing and frozen storage on the other should be noted. Cool storage generally refers to storage at temperatures above freezing, from about 16°C down to −2°C. Commercial and household refrigerators are usually operated at 4.5–7°C. Commercial refrigerators sometimes are operated at a slightly lower temperature when a particular food is being favored. Whereas pure water will freeze at 0°C, most foods will not begin to freeze until about −2°C or lower. Frozen storage refers to storage at temperatures that maintain food in frozen condition. Good frozen storage generally requires temperatures of −18°C or below. Refrigerated or cool storage generally will preserve perishable foods for days or weeks, depending on the food. Frozen storage will preserve foods for months or even years if properly packaged.

Further distinctions between refrigeration and freezing temperatures are related to microorganism activity (Fig. 9.1). Most food spoilage microorganisms grow rapidly at temperatures above 10°C but some grow at temperatures below 0°C as long as there

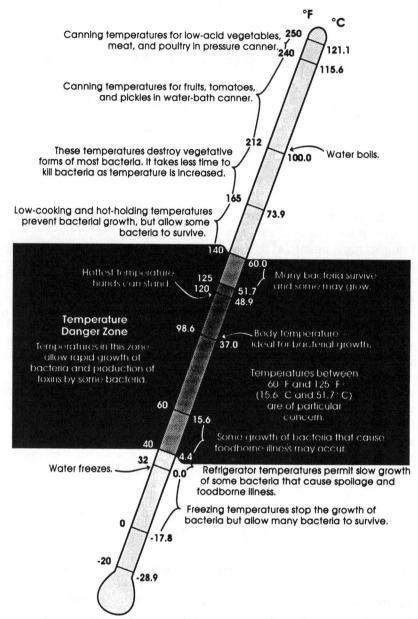

Figure 9.1. Some relationships between temperature and microbial growth in foods. Source: *Preventing Foodborne Illness: A Guide to Safe Food Handling. USDA Food Safety and Inspection Service, Washington, DC, 1990.*

is unfrozen water available. Food held under good refrigeration temperatures can still spoil due to microorganism growth. However, until recently, it was thought that although properly refrigerated food could spoil due to undesirable changes in odor, flavor, and appearance, this did not cause a safety problem because disease-causing organisms did not grow appreciably at these low temperatures. Indeed, this is true for many such microorganisms, but in recent years, food scientists have found that some food-poisoning microorganisms grow, albeit slowly, at temperatures as low as 3.3°C. These microorganisms are know as psychrotrophic pathogens (i.e., cold tolerant disease causing). This is a serious concern because it means that even good refrigeration cannot always be assumed to protect foods. Below −9.5°C there is no significant growth of spoilage or pathogenic microorganisms in food; instead there is a gradual decrease in the numbers of living organisms. But, as pointed out previously, destruction of microorganisms by freezing is not complete; when the food is thawed, there can be rapid multiplication and spoilage.

REFRIGERATION AND COOL STORAGE

Refrigeration and cool storage in general is the gentlest method of food preservation. It has relatively few adverse effects on the taste, texture, nutritive value, and other attributes of foods, provided simple rules are followed and storage periods are not excessive. One cannot say this of heat, dehydration, irradiation, and other methods of preservation, which often immediately result in changes in food, however small.

Although refrigeration and cool storage reduces the rate of food deterioration, with most foods it will not prevent deterioration to anywhere the same degree as does heat, dehydration, irradiation, fermentation, or true freezing. Table 9.1 indicates the generally useful storage life of plant and animal tissues at various temperatures. At 0°C, which is lower than most commercial or household refrigerators, the life of such perishables as animal flesh, fish, poultry, and many fruits and vegetables is generally less than 2 weeks. At the more common refrigerator temperature of 5.5°C, storage life is often less than 1 week. On the other hand, these products held at 22°C or above may spoil in 1 day or less.

Ideally, refrigeration of perishables starts at time of harvest or slaughter and is maintained throughout transportation, warehousing, merchandising, and storage prior to ultimate use. This is not required from the standpoint of microbial spoilage alone but also to maintain the flavor, texture, and other quality attributes of many foods.

A few hours delay between harvest or slaughter and refrigeration is sufficient to permit marked food deterioration to occur. This is particularly true with certain metabolically active fruits and vegetables. These not only will generate heat from respiration

Table 9.1. Useful Storage Life of Plant and Animal Tissues at Various Temperatures

Food	Average Useful Storage Life (days)		
	0°C (32°F)	22°C (72°F)	38°C (100°F)
Meat	6–10	1	less than 1
Fish	2–7	1	less than 1
Poultry	5–18	1	less than 1
Dry meats and fish	1000 and more	350 and more	100 and more
Fruits	2–180	1–20	1–7
Dry fruits	1000 and more	350 and more	100 and more
Leafy vegetables	3–20	1–7	1–3
Root crops	90–300	7–50	2–20
Dry seeds	1000 and more	350 and more	100 and more

Courtesy of N. W. Desrosier.

but will convert metabolites from one form to another. The loss of sweetness from sweet corn is an example of the latter. At 0°C, sweet corn can still metabolize its own sugar, but generally less than 10% is lost in 1 day and 20% in 4 days. At 20°C, however, these losses can amount to 25% in 1 day and can far exceed this on a hot summer afternoon. To minimize such losses, cooling systems are brought into the harvest field. Figure 9.2 shows one type of portable cooler. As they are picked, fruits or vegetables pass through this hydro-cooler where they are sprayed with jets of cold water. The water also may contain a germicide to inactivate surface microorganisms. The cooled produce is then loaded into refrigerated trucks or railroad cars in route to refrigerated warehouses.

Quick cooling does not simply mean immediate placement of bulk foods into a refrigerated railroad car or warehouse in all cases. Cooling is the taking of heat out of a body. If the body is large, the time to remove sufficient heat can be so long as to permit considerable food spoilage before effective preservation temperatures are reached. The hydrocooler of Fig. 9.2 accomplishes rapid cooling aided by subdivision of the produce as it is fed through the machine. Similarly, subdivision of bulk produce favors cold air circulation in refrigerated storage rooms. Leafy vegetables may be quickly cooled by spraying them with water and creating a vacuum to promote evaporative cooling. Current use of cold nitrogen gas from evaporating liquid nitrogen in refrigerated trucks, railcars, and shipholds also aids in providing intimate cold contact and quick cooling of produce. This has the further advantage of displacing air from the refrigerated area, which can be beneficial for certain products. Bulk liquids are best cooled rapidly by passing them through an efficient heat exchanger before putting them into refrigerated storage. Animal carcasses at time of slaughter are at a temperature of about 38°C, which must be lowered to about 2°C in less than 24 h if quality is to be maintained.

Requirements of Refrigerated Storage

The principal requirements for effective refrigerated storage are controlled low temperature, air circulation, humidity control, and modification of gas atmospheres.

Figure 9.2. Portable hydro-cooler used to cool produce in the field. *Courtesy of FMC Corp.*

Controlled Low Temperature

Properly designed refrigerators, refrigerated storage rooms, and warehouses will provide sufficient refrigeration capacity and insulation to maintain the room within about ± 1°C of the selected refrigeration temperature. In order to design a refrigerated space capable of maintaining this temperature, it is necessary to know, in addition to the insulation requirements, all factors that may generate heat within this space or influence ease of removal of heat from the space. These factors include the number of heat-generating electric lights and electric motors that may be operating, the number of people that may be working in the refrigerated space, how often doors to the area will be open to permit entrance of warm air, and the kinds and amounts of food products that will be stored in the refrigerated area.

This latter item is of importance for two major reasons: First, the quantity of heat that must be removed from any amount of food to lower it from one temperature to another is determined by the specific heat of the particular food; and second, during and after cooling, such foods as fruits and vegetables respire and produce their own heat at varying rates. Both the specific heats and respiration rates of all important foods are known or can be closely estimated. These values, in addition to the items mentioned above, are necessary to calculate the "refrigeration load," which is the quantity of heat that must be removed from the product and the storage area in order to go from an initial temperature to the selected final temperature and then maintain this temperature for a specified time.

The heat evolved during respiration by representative fruits and vegetables is listed in Table 9.2. The amount of heat produced varies with each product, and like all metabolic activities decreases with storage temperature. Products with particularly high respiration rates, such as snap beans, sweet corn, green peas, spinach, and strawberries, are particularly difficult to store. Such products, if closely packed in a bin, can rot in the center even when the surrounding air is cool due to the heat generated by the product. The relationships between specific heats of foods and calculation of refrigeration load will be discussed in the section on freezing and frozen storage.

Air Circulation and Humidity

Proper air circulation helps move heat away from the vicinity of food surfaces toward refrigerator cooling coils and plates. But the air that is circulated within a cold storage room must not be too moist or too dry. Air of high humidity can condense moisture on the surface of cold foods. If this is excessive, molds will grow on these surfaces at common refrigeration temperatures. If the air is too dry, it will cause drying out of foods. All foods are different with respect to supporting mold growth and tendency to dry out, and so for each, an optimum balance must be reached. The optimum relative humidity (RH) to be maintained in cool storage rooms for most foods is known. Table 9.3 summarizes the best storage temperatures and relative humidities for many food items and their approximate storage life. (This table also includes data necessary for calculation of refrigeration loads.) Most foods store best at refrigeration temperatures when the relative humidity of air is between about 80% and 95%. The optimal relative humidity for a particular food is generally related to its moisture content and the ease with which it dries out. For example, celery and several other crisp vegetables require a relative humidity of 90–95%, whereas nuts may do well at only 70%. On the other hand, dry and granular products such as powdered milk and eggs, which have extended

Table 9.2. Heat Evolved During Respiration of Fruits and Vegetables

Commodity	Btu[a] per Ton per 24 h		
	0°C (32°F)	4.4°C (40°F)	16°C (60°F)
Apples	500–900	1,100–1,600	3,000–6,800
Beans, snap	4,400	7,700	20,500
Cabbage	3,000	4,700	12,600
Carrots	—	4,300	8,700
Celery	1,600	2,400	8,200
Corn, sweet	—	17,100	35,800
Onions, green	4,200	6,200	19,600
Oranges	400–1,100	800–1,600	2,800–5,200
Peaches	900–1,400	1,400–2,000	7,300–9,300
Pears, Bartlett	700–1,500	1,100–2,200	3,300–13,200
Peas, unshelled	8,500	—	25,700[b]
Potatoes, mature	—	1,300	2,600
Spinach	—	10,100	39,300
Strawberries	2,700–3,900	3,600–7,300	15,600–20,300
Tomatoes			
mature green	—	1,540	4,500
ripening	—	3,100[b]	5,900

SOURCE: Adapted from Lutz and Hardenburg (1968), Ryall and Lipton (1979), Ryall and Pentzer (1982).

[a] Btu = 252 cal = 1055 joules.

[b] At 50°F (10°C).

storage lives at refrigeration temperatures, are favored by very dry atmospheres, and relative humidities above about 50% can cause excessive lumping and caking if the packaging is not moisture tight.

When refrigerated storage is to be for prolonged periods, various techniques are used to maintain quality. Foods that tend to lose moisture can be protected by several packaging methods. This is important since otherwise there would be a continual migration of moisture from the food to the storage atmosphere and onto refrigerated coils and plates since moisture vapor tends to condense on cold surfaces.

Large cuts of meat are often packaged in sealed plastic bags or they may be sprayed with various moisture-resistant coatings. Cheeses that are ripened for many months in cold warehouses are also packaged in plastic films. An older method is to coat the cheese with wax. This not only minimizes moisture loss but affords protection against contamination and growth of surface molds. Eggs in the shell tend to lose moisture as well as carbon dioxide. This can be retarded by coating the eggs with a thin edible oil, such as mineral oil, to seal the minute pores of the egg shell.

Beef that is tenderized by ageing in cool rooms often presents a problem. Conventionally, ageing is done at about 2°C for a period of several weeks. If the relative humidity of the storage room is much below 90%, the beef dries out; if it is above 90%, the beef will mold. Precise control of relative humidity is difficult to achieve. To retard mold growth and the development of surface bacterial slime, ultraviolet light is sometimes employed. In one type of particular accelerated ageing process beef is aged in 2 or 3 days by combining high humidity with a temperature of about 18°C. This also speeds surface microbial growth, which is kept in check by ultraviolet light. In applications

Table 9.3. Storage Requirements and Properties of Perishable Foods

Commodity	Storage Temp (°F)[a]	Relative Humidity (%)	Approximate Storage Life	Water Content (%)	Average Freezing Point (°F)[a]	Specific Heat above Freezing[b]	Specific Heat below Freezing[b]	Latent Heat of Fusion[c] (Btu/lb)	Heat of Respiration[d] (Btu/ton/24 hr)
Apples	30–32	85–90	—	84.1	28.2	0.87	0.45	121	1,500–12,380(70)
Apricots	31–32	85–90	1–2 wk	85.4	29.6	0.88	0.46	122	
Asparagus	32	90–95	3–4 wk	93.0	30.4	0.94	0.48	134	
Avocados	45–55	85–90	4 wk	65.4	30.0	0.72	0.40	94	
Bananas	—	85–95		74.8	29.6	0.80	0.42	108	6,160–52,950(70)
Beans (green or snap)	45	85–90	8–10 days	88.9	30.2	0.91	0.47	128	
Blackberries	31–32	85–90	7 days	84.8	29.4	0.88	0.46	122	
Bread	0	—	Several wk	32–37	—	0.70	0.34	46–53	
Broccoli, sprouting	32	90–95	7–10 days	89.9	30.3	0.92	0.47	130	7,450–100,000
Cabbage, late	32	90–95	3–4 mo	92.4	30.5	0.94	0.47	132	1,200–6,120(70)
Carrots (topped)	32	90–95	4–5 mo	88.2	28.8	0.90	0.46	126	2,130–8,080
Cauliflower	32	85–90	2–3 wk	91.7	30.2	0.93	0.47	132	
Celery	31–32	90–95	2–4 mo	93.7	30.9	0.95	0.48	135	1,620–14,150(70)
Cherries	31–32	85–90	10–14 days	83.0	27.7	0.87	0.45	120	1,249–13,200
Corn, sweet	31–32	85–90	4–8 days	73.9	30.8	0.79	0.42	106	6,560–61,950(80)
Cranberries	36–40	85–90	1–3 mo	87.4	30.0	0.90	0.46	124	720–1,800(50)
Cucumbers	45–50	90–95	10–14 days	96.1	30.5	0.97	0.49	137	1,690–10,460
Dairy products									
butter	32–36	80–85	2 mo	15.5–16.5	—	0.33	—	23	
butter	−10–−20	80–85	1 yr	15.5–16.5	—	—	0.25	23	
cheese	35	65–70	—	37–38	28.0	0.50	0.31	54	
cream (sweetened)	−15	—	Several mo	—	—	—	—	—	
ice cream	−15	—	Several mo	—	22–29	0.80	0.45	96	
skim milk (dried)	40	—	Several mo	3.5	—	0.23	—	5	
Dried fruits	32	50–60	9–12 mo		—	0.30–0.32	—	17–21	
Eggplant	45–50	85–90	10 days	92.7	30.4	0.94	0.48	132	

(Continued)

169

Table 9.3. (Continued)

Commodity	Storage Temp (°F)[a]	Relative Humidity (%)	Approximate Storage Life	Water Content (%)	Average Freezing Point (°F)[a]	Specific Heat above Freezing[b]	Specific Heat below Freezing[b]	Latent Heat of Fusion[c] (Btu/lb)	Heat of Respiration[d] (Btu/ton/24 hr)
Eggs									
dried spray albumen	35	Low as possible	6 mo	Up to 6.0	—	0.25	—	9	
dried, whole	35	Low as possible	6 mo–1 yr	5.0	—	0.25	0.21	9	
dried, yolk	35	Low as possible	6 mo–1 yr	3.0	—	0.22	0.21	4	
frozen	–10–0	Low as possible	1 yr, plus	73	28.0	0.74	0.42	104	
shell	29–31	85–90	8–9 mo	67.0	28.0	0.74	0.40	96	
Fish									
fresh	33–40	90–95	5–20 days	62–85	28.0	0.80	0.40	89–122	
frozen	–10–0	90–95	8–10 mo	62–85	—	0.80	0.40	115	
Grapefruit	32–50	85–90	4–8 wk	88.8	28.6	0.91	0.46	126	950–6.840(90)
Grapes									
American type	31–32	85–90	3–8 wk	81.9	29.4	0.86	0.44	116	
European type	30–31	85–90	3–6 mo	81.6	27.1	0.86	0.44	116	
Honey	—	—	1 yr	18.0	—	0.35	0.26	26	
Kale	32	90–95	3–4 wk	86.6	30.7	0.89	0.46	124	
Lemons	32,55–58	85–90	1–4 mo	89.3	29.0	0.92	0.46	127	900–5,490(80)
Lettuce	32	90–95	3–4 wk	94.8	31.2	0.96	0.48	136	11,320–45,980
Mangoes	50	85–90	2–3 wk	81.4	29.4	0.85	0.44	117	
Meat									
beef, fresh	32–34	88–92	1–6 wk	62–77	28–29	0.70–0.84	0.38–0.43	89–110	
frozen	–10–0	90–95	9–12 mo	—	—	—	—	—	
hams and shoulders, fresh	32–34	85–90	7–12 days	47–54	28–29	0.58–0.63	0.34–0.36	67–77	
cured	60–65	50–60	0–3 yr	40–45	—	0.52–0.56	0.32–0.33	57–64	
frozen	–10–0	90–95	6–8 mo	—	—	—	—	—	

Commodity									
lamb, fresh	32–34	85–90	5–12 days	60–70	28–29	0.68–0.76	0.38–0.51	86–100	1,230–8,500
frozen	−10–0	90–95	8–10 mo	—	—	—	—	—	
pork, fresh	32–34	85–90	3–7 days	35–42	28–29	0.48–0.54	0.30–0.32	50–60	
Melons									
cantaloupe and Persian	45–50	85–90	1–2 wk	92.7	29.9	0.94	0.48	132	
watermelons	36–40	85–90	2–3 wk	92.1	30.6	0.97	0.48	132	
Mushrooms	32–35	85–90	3–5 days	91.1	30.0	0.93	0.47	130	6,160–58,000(70)
Nuts	32–50	65–75	8–12 mo	3–6	—	0.22–0.25	0.21–0.22	4–8	
Oil (vegetable salad)	35	—	1 yr	0	—	—	—	—	
Oleomargarine	35	60–70	1 yr	15.5	—	0.32	0.25	22	
Olives, fresh	45–50	85–90	4–6 wk	75.2	28.5	0.80	0.42	108	
Onions and onion sets	32	70–75	6–8 mo	87.5	30.1	0.90	0.46	124	1,100–4,180(70)
Oranges	32–34	85–90	8–12 wk	87.2	30.6	0.90	0.46	124	1,030–9,420(90)
Papayas	45	85–90	2–3 wk	90.8	30.1	0.82	0.47	130	
Peaches	31–32	85–90	2–4 wk	86.9	29.6	0.90	0.46	124	1,370–22,460(80)
Pears	29–31	85–90	—	82.7	27.7	0.86	0.45	118	880–13,200
Peas, green	32	85–90	1–2 wk	74.3	30.1	0.79	0.42	106	8,360–82,920(80)
Peppers, chili (dry)	32–40	65–75	6–9 mo	12.0	30.9	0.30	0.24	17	
Peppers, sweet	45–50	85–90	8–10 days	92.4	30.5	0.94	0.47	132	2,720–8,470
Pineapples									
mature green	50–60	85–90	3–4 wk	85.3	29.1	—	—	—	
ripe	40–45	85–90	2–4 wk	—	29.7	0.88	0.45	122	
Popcorn, unpopped	32–40	85	—	13.5	—	0.31	0.24	19	
Potatoes									880–3,530(70) (Irish Potatoes)
early crop	50–55	85–90	—	—	30.0	—	—	—	
late crop	38–50	85–90	—	77.8	29.8	0.82	0.43	111	
Poultry									
fresh	32	—	1 wk	74.0	27.0	0.79	—	106	
frozen, eviscerated	−10–0	—	9–10 mo	—	—	—	—	—	
Pumpkins	50–55	70–75	2–6 mo	90.5	29.9	0.92	0.47	130	
Rhubarb	32	90–95	2–3 wk	94.9	29.9	0.96	0.48	134	
Spinach	32	90–95	10–14 days	92.7	31.3	0.94	0.48	132	4,860–38,000
Squash									
acorn	45–50	75–85	4–5 wk	—	30.0	—	—	—	
summer	32–40	85–95	10–14 days	95.0	30.4	0.96	—	135	

(Continued)

171

Table 9.3. (Continued)

Commodity	Storage Temp (°F)[a]	Relative Humidity (%)	Approximate Storage Life	Water Content (%)	Average Freezing Point (°F)[a]	Specific Heat above Freezing[b]	Specific Heat below Freezing[b]	Latent Heat of Fusion[c] (Btu/lb)	Heat of Respiration[d] (Btu/ton/24 hr)
Strawberries									
fresh	31–32	85–90	7–10 days	89.9	30.2	0.92	—	129	3,800–46,400(80)
frozen	−10–0	—	1 yr	72.0	—	—	0.42	103	
Sugar, granulated	50–100	Below 60	1–3 yr	0.5	—	0.20	0.20	72	
Sweet potatoes	55–60	90–95	4–6 mo	68.5	29.2	0.75	0.40	97	2,440–6,300
Tomatoes									
mature green	55–70	85–90}[a]	2–5 wk	94.7	30.4	0.95	0.48	134	580–6,230
ripe	32}[a]	85–90}	7 days	94.1	30.4	0.95	0.48	134	1,020–5,640
Yeast, compressed baker's	31–32}	—	—	70.9	—	0.77	0.41	102	

SOURCE: McCoy (1963). Additional data in Lutz and Hardenburg (1968) and ASHRAE (1978, 1981).

[a] °C = 5/9 (°F—32).

[b] Calculated by Siebel's formula. For values above freezing point, Specific Heat = 0.008 (% water) + 0.20; for values below freezing point, Specific Heat = 0.003 (% water) + 0.20.

[c] Calculated by multiplying the % of water content by the latent heat of fusion of water, 143.4 Btu/lb.

[d] 1 Btu = 252 cal = 1055 joules.

172

such as this, the dosage of ultraviolet irradiation must be controlled, as excessive exposure to ultraviolet light can cause surface fat to become rancid.

Modification of Gas Atmospheres

Controlled-atmosphere (CA) storage is used for apples and other fruits in order to inhibit overripening in cold storage. Stored fruits and vegetables consume oxygen and give off carbon dioxide during storage. Three ways to slow down this respiration and the physiological changes that accompany it are reducing temperature, reducing but not eliminating oxygen, and increasing the carbon dioxide. The optimum temperatures, relative humidities, and gas compositions of the atmosphere differ for different fruits and even for varieties of the same fruit. In the case of McIntosh apples, optimal conditions include a temperature of 3°C, 87% relative humidity (RH), and an atmosphere containing 3% oxygen (normally, 21%), 3% carbon dioxide (normally 0.03%) for about 1 month then 5% carbon dioxide, and nitrogen to make up the balance. In practice, a cold storage warehouse is made gas-tight, brought to temperature, filled with fruit, and sealed. Commercial gas generators then replace air with the chosen gas atmosphere and may also introduce water vapor to maintain the desired relative humidity. The warehouse is usually sealed for months until it is to be emptied. If someone must enter the warehouse to make repairs, an oxygen mask is required. Under these conditions, apples retain quality in storage for better than 6 months.

Controlled-atmosphere storage is not limited to storage in a warehouse. In a sense, CA storage is practiced whenever food is sealed in a package under vacuum, nitrogen, carbon dioxide, or any other departure from the composition of air. In recent years, the concept of controlled atmosphere packaging has become widely used for individually packaged foods. Perishable foods such as meats, high-moisture pasta, fish, and fresh fruits and vegetables can be sealed in packages in which the air has been replaced with some mixture which extends shelf life. High-moisture-content pasta products are common examples of such technologies. These technologies require the use of strict control of quality and storage temperatures in order to ensure the safety and quality of such products.

Other examples of controlled- or modified-atmosphere storage include use of antimicrobial vapors or fumigants to control molds and use of ethylene gas to speed ripening and color development of citrus fruits and bananas. The fact that liquid-nitrogen cooling displaces air with nitrogen gas has already been mentioned, and considerable research has been done to determine the full potential of this kind of modified-atmosphere storage.

Since animal as well as plant tissues consume and give off gases, it would be expected that gas equilibria would affect many food properties. This certainly is so with respect to the pigment changes of red meat, the growth and metabolic patterns of surface ripening as well as spoilage microorganisms, and the staling rate of cold storage eggs. In the latter case, besides oil coating to minimize water and carbon dioxide losses, eggs have been stored in warehouses enriched with carbon dioxide to minimize loss of this gas which is associated with egg pH and freshness.

The term *hypobaric storage* has been used to describe another type of CA storage. In this case the refrigerated storage area is maintained under reduced pressure and high humidity. This decreases the amount of air, and with it the amount of oxygen in the area, while the high humidity prevents product dehydration. Hypobaric storage has been used in warehouses as well as enclosed truck bodies. Since altered gas equilibria can change the rates of microbial and enzymatic spoilage, the possibilities for energy savings (less intense refrigeration) through use of modified atmospheres is of increasing interest.

Changes in Food During Refrigerated Storage

The deterioration of foods during cool storage are influenced by the growing conditions and varieties of plants, feeding practices of animals, conditions of harvest and slaughter, sanitation and damage to tissues, temperature of cool storage, mixture of foods in storage, and other variables.

For example, Florida grapefruits store well at 0°C, whereas Texas Marsh grapefruits store better at 11°C. McIntosh apples store well at 2–5°C, but Delicious apples do better at 0°C. Pigs fed on substances high in unsaturated fats such as peanuts and soybeans produce softer pork and lard than the same animals fed on cereal grains; meat from the latter keeps better in cold storage. Animals permitted to rest before slaughter build up glycogen (animal starch) reserves in their muscles. Following slaughter, this is converted to lactic acid, which is a mild preservative and enhances the keeping quality of meat in cold storage. Animals that are exercised or excited before slaughter use up their glycogen reserves, less is available for conversion to lactic acid, and keeping quality is impaired.

Too low refrigeration temperature can cause damage called "chill injury" to fruits and vegetables even when these are not physically damaged by freezing (see Table 7.4). This is not surprising since living plants would be expected to have optimum temperature requirements just as animals do. Many of the defects listed in Table 7.4 are of microbial origin, reflecting a weakened physiological state and a decrease in resistance to this kind of deterioration. In the case of bananas and tomatoes on the other hand, storage temperatures below about 13°C slow down the activities of natural ripening enzymes and result in poor colors. Nevertheless, for the majority of perishable foods, no cooling at all generally would be far worse than refrigeration temperatures that are somewhat too low.

Refrigerated storage permits exchange of flavors between many foods. Butter and milk will absorb odors from fish and fruit, and eggs will absorb odors from onions. It is best to store different foods, especially odorous ones, separately, but this is not always economically feasible. In many instances, odor exchange can be prevented by effective packaging.

The previously mentioned losses of sugar in sweet corn stored at refrigeration temperatures are due to synthesis of starch from the sugar and, thus, do not represent actual nutrient losses. However, some changes that occur in foods during refrigerated storage represent true nutrient losses. An important example is loss of vitamin C and other vitamins which is common in many foods held for relatively short periods under refrigeration (Table 9.4).

Still other common changes during refrigerated storage involve loss of firmness and crispness in fruits and vegetables, changes in the colors of red meats, oxidation of fats, softening of the tissues and drippage from fish, staling of bread and cake, lumping and caking of granular foods, losses of flavor, and a host of microbial deteriorations often unique to a specific food and caused by the dominance of a particular spoilage organism. Some foods should not be refrigerated. Bread is an example. The rate of staling of bread is greater at refrigeration temperatures than it is at room temperature.

Table 9.4. Losses of Vitamin C in Selected Vegetables During Cold Storage

Produce	Storage Conditions			Losses, %
	Days	°C	°F	
Asparagus	1	1.7	35	5
	7	0	32	50
Broccoli	1	7.8	46	20
	4	7.8	46	35
Green beans	1	7.8	46	10
	4	7.8	46	20
Spinach	2	0	32	5
	3	1.1	34	5

Courtesy of N. W. Desrosier.

Staling can be arrested by freezing. These and other differences between foods at refrigerated temperatures cause the storage requirements indicated in Table 9.3.

Benefits Other Than Preservation

In the food industry, cooling generally is used for its preservation value. There are many situations, however, where cooling provides other advantages and improves the processing properties of foods. Cooling is employed to control the rates of certain chemical and enzymatic reactions as well as the rates of growth and metabolism of desirable food microorganisms. This is the case in the cool ripening of cheeses, cool ageing of beef, and cool ageing of wines. Cooling also improves the ease of peeling and pitting peaches for canning. Cooling citrus fruits reduces changes in flavor during extraction and straining of juice. Cooling improves the ease and efficiency of meat cutting and bread slicing. Cooling precipitates waxes from edible oils. Water for soft drinks is cooled before carbonating to increase the solubility of carbon dioxide. More will be said about these applications of cooling in later chapters.

Economic Considerations

Where cooling is used for preservation in a multiproduct warehouse, supermarket, or household refrigerator, it is not always economical or practical to separate foods and provide each with the optimum temperature and humidity. As a compromise, the refrigerated area often is held somewhere within the range 2–7°C with no special provisions made to control humidity. Even under these conditions, refrigeration significantly improves the safety, appearance, flavor, and nutritional value of our food supply. It further reduces losses from insects, parasites, and rodents.

FREEZING AND FROZEN STORAGE

As a preservation method, freezing takes over where refrigeration and cool storage leave off. Freezing has been a major factor in bringing convenience foods to the home, restaurant, and institutional feeding establishments. Because freezing, properly done, preserves foods without causing major changes in their size, shape, texture, color, and flavor, freezing permits much of the work in preparing a food item or an entire meal to be done prior to the freezing step. This transfers operations that formerly had to be done in the home or restaurant to the food processor. Such diverse items as chicken pot pie, breaded fish sticks, whole entrees, whipped topping, chiffon and fruit pies, and complete dinners are, today, commonly frozen. The wide array of available frozen food products, many sold in their final serving dishes, represents a major revolution in the food industry and reflects gross changes in eating habits. There are more meals being eaten outside the home than ever before. This includes meals in restaurants, colleges, school lunch programs, hotels, airplanes, hospitals, and so on. Labor costs are steadily rising, forcing maximum use of convenience foods in food-handling establishments.

At present, no form of food preservation is as well suited to provide maximum convenience as freezing. Although dehydrated foods offer convenience, they require reconstitution on an individual component basis to satisfy varying water needs, and then also require heating. Not so with frozen foods. Many items can be completely prepared and assembled together for a single thawing–heating operation. Quality, of course, rests on well-developed scientific principles.

Initial Freezing Point

It is a basic property of aqueous solutions that increasing their concentrations of dissolved solids will lower their freezing points. Thus, the more salt, sugar, minerals, or proteins in a solution, the lower its freezing point and the longer it will take to freeze when put into a freezing chamber. If water and fruit juice, for example, are placed in a freezer, the water will freeze first. Further, unless the temperature is considerably below the freezing point of pure water, the juice will never freeze completely but rather will become icy and slushy. What is happening is that the water component of the juice freezes first and leaves the dissolved solids in a more concentrated solution which requires a still lower temperature to freeze it.

Since different foods have quite different water contents and kinds and amounts of solids dissolved in the water, they have different initial freezing points (Table 9.3) and, under a given freezing condition, require different times to reach a solidly frozen state. This alone provides much of the explanation of why cultivars of the same fruit or vegetable, which have somewhat different compositions, behave differently on freezing. Even the same cultivar grown under different irrigation and fertilization practices will exhibit variations in composition, including differences in mineral content absorbed from the fertilizers. For this reason, frozen food producers who want to have strict control over the freezing processes specify the cultivar to be grown and may even supply seed and fertilizer to help guarantee controlled composition and other properties of raw materials.

Freezing Curve

A given unit of food will not freeze uniformly; that is, it will not suddenly change from liquid to solid. In the case of milk placed in a freezer, the liquid nearest the container wall will freeze first, and the first ice crystals will be pure water. As water continues to be frozen out, the milk will become more concentrated in minerals, proteins, lactose, and fat. This concentrate, which gradually freezes, also becomes more concentrated as freezing proceeds. Finally, a central core of highly concentrated unfrozen liquid remains; if the temperature is sufficiently low, this central core also will ultimately freeze.

The freezing point of pure water is 0°C, but, actually, water does not begin to freeze at 0°C. Instead, it generally becomes supercooled to a temperature several degrees below 0°C before some stimulus such as crystal nucleation or agitation initiates the freezing process. When this occurs, there is an abrupt rise from the supercooled temperature to 0°C due to the evolution of the latent heat of crystallization. Even if the water is in an environment far below 0°C, as long as free water is freezing and giving up latent heat of crystallization or fusion the temperature of a pure water–ice mixture will not drop below 0°C. Only after all of the water has frozen will the system drop below the equilibrium temperature of 0°C, and then rapidly approach the temperature of the freezing environment.

Much of this is true also for food systems containing water, but since foods contain dissolved solids, progressive freezing is somewhat more complex. The freezing curve for a thin section of beef shown in Fig. 9.3 was obtained by placing the beef in a freezing chamber which was below −18°C and recording the temperature of the beef with time as it froze. At the same time, the percentage of water that was converted to ice was determined as a function of temperature or time. As the beef is chilled, it first drops from its initial temperature to a supercooled temperature below its freezing point. Nucleation or agitation initiates formation of the first ice crystal, and latent heat of fusion causes the temperature to rise to the freezing point, which is just below 0°C because of the dissolved solids in the water phase.

If this were the freezing curve of pure water, it would not drop below the freezing point as long as liquid water remained. In the case of beef and other foods, however, the temperature continues to drop as more and more water is frozen. This is largely because as more water is frozen out, the concentration of solutes in the remaining water progressively increases and exerts a greater and greater freezing point depression on the remaining solution.

It also should be noted from Fig. 9.3 that at about −4°C some 70% of the water is frozen, and the beef would appear solidly frozen; however, at −9.4°C, about 3% of the

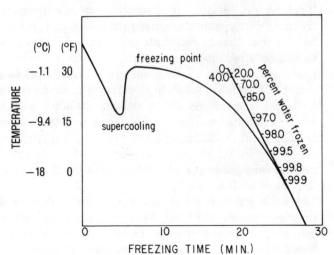

Figure 9.3. Freezing curve for thin sections of beef. Source: *Desrosier and Desrosier, Technology of Food Preservation, 4th ed., AVI Publishing Co., Westport, CT, 1977.*

water still remains unfrozen, and even at $-18°C$, not all of the water is completely frozen. These small quantities of unfrozen water are highly significant, particularly since within them are dissolved food solutes which are concentrated and are therefore more prone to reaction with one another and with other food constituents.

Since the compositions of foods differ, they have characteristic freezing curves differing somewhat in shape. Generally, one can identify the zone of supercooling, the inflection up to the freezing point, and the subsequent drop in temperature if there is a sufficient temperature differential between the freezing food and the freezer environment. This differential provides the driving force for continued heat transfer out of the food.

Changes During Freezing

Water when frozen can burst iron pipes, so it should not be surprising that unless properly controlled, freezing can disrupt food texture, break emulsions, denature proteins, and cause other changes of both physical and chemical nature. Many of these changes are related to food composition, which, in turn, is influenced by agricultural practices long before the freezing process.

Concentration Effects

For quality to be maintained in frozen storage, most foods must be solidly or very nearly solidly frozen. An unfrozen core or a partially frozen zone will deteriorate in texture, color, flavor, and other properties. In addition to the possible growth of psychrotrophic microorganisms and to the greater activity of enzymes when water remains unfrozen, a major reason for deterioration in partially frozen food is due to the high concentration of solutes in the remaining water. Thus, when milk is slowly frozen, as can occur outdoors in winter, the concentration of minerals and salts can denature proteins and break fat emulsions, to cause curdling and butter granules. Flavor changes also occur.

Damage from the concentration effect can be of various kinds:

• If solutes precipitate out of solution, as do excessive levels of lactose in freezing ice cream, a gritty, sandy texture to the food can occur.
• Solutes that do not precipitate but remain in concentrated solution can cause proteins to denature because of a "salting out" effect.
• Acidic solutes on concentration can cause the pH to drop below the isoelectric point (point of minimum solubility), causing proteins to coagulate.
• Colloidal suspensions are in delicate balance with respect to the concentration of anions and cations. Some of these ions are essential to maintain colloids, and concentration or precipitation of these ions can disturb this balance.
• Gases in solution also are concentrated when water freezes. This can cause supersaturation of the gases and ultimately force them out of solution. Frozen beer or soda pop may have such a defect.
• The concentration effect can also cause a dehydration of adjacent tissues at the microenvironmental level. Thus, when ice crystals form in extracellular liquid and solutes are concentrated in the vicinity of the ice crystals, water will diffuse from within the cells through the membranes into the region of high solute concentration

to restore osmotic equilibrium. This shift of moisture is rarely completely reversed on thawing and can result in loss of tissue turgor.

Ice Crystal Damage

Solid foods from living tissues such as meats, fish, fruits, and vegetables are of cellular structure with delicate cell walls and cell membranes. Within and between the cells is water. When water freezes rapidly, it forms minute ice crystals; when it freezes slowly, it forms large ice crystals and clusters of crystals. Large ice crystals forming within or between cells cause much more physical rupture and separation of cells than do small crystals. The changes in the texture of strawberries resulting from freezing is an example. Large ice crystal damage is detrimental not only to cellular foods, but it also can disrupt emulsions such as butter, frozen foams like ice cream, and gels such as puddings and pie fillings. In the case of butter, ice crystals that grow within individual water droplets dispersed in the continuous fat phase can penetrate through the fat and merge. When such butter is later thawed, water pockets and water drippage result. In the case of ice cream, large ice crystals can puncture frozen foam bubbles. Such a condition leads to loss of volume on storage and during partial melting. Gels behave somewhat like butter, often exhibiting syneresis or water separation.

Rate of Freezing

Whether the concentration effect or physical damage from large ice crystals is the more detrimental during freezing and frozen storage depends on the particular food system under consideration. However, in either case, fast freezing is necessary for high quality.

Fast freezing produces minute ice crystals. It also minimizes concentration effects by decreasing the time concentrated solutes are in contact with food tissues, colloids, and individual constituents during the transition from the unfrozen to the fully frozen state. For these reasons, modern methods of freezing and freezing devices are designed for very rapid freezing where high food quality can justify the cost. Generally, the faster the freezing rate, the better the product quality. However, from a practical standpoint, progressive freezing equivalent to about 1.3 cm (0.5 in.) per hour are satisfactory for most products. This would mean that a flat package of food 5 cm in thickness and frozen from both major surfaces should be frozen (to −18°C or below) at its center in about 2 h. Plate freezers easily do this, and liquid-nitrogen freezers may cut down the time to a few minutes. Unfortunately, home freezers can have one of the slowest rates of freezing.

Choice of Final Temperature

A consideration of all the factors—textural changes, enzymatic and nonenzymatic chemical reactions, microbiological changes, and costs—leads to the general conclusion that foods should be frozen to an internal temperature of −18°C (0°F) or lower and kept at −18°C or lower throughout transport and storage. Economic considerations generally preclude temperatures below about −30°C during transport and storage, although many foods commonly are frozen to temperatures below this in an effort to achieve the advantages of rapid freezing.

The choice of −18°C or below as the recommended temperature for freezing and storage is based on substantial data and represents a compromise between quality and cost. Microbiologically, −18°C storage would not be strictly required since pathogens do not grow below about 3.3°C and normal food spoilage organisms do not grow below −9.5°C. On the other hand, transportation and frozen storage facilities are expected to vary somewhat at any chosen temperature setting. Therefore, the use of −18°C provides a reasonable measure of safety against food spoilage organisms and a still greater margin of safety against pathogens; and, indeed, frozen foods have enjoyed an excellent public health record over the years.

For control of enzymatic reactions, −18°C is not exceptionally low since some enzymes retain activity even at −73°C, although reaction rates are extremely slow. Enzyme reaction rates are faster in supercooled water than in frozen water at the same temperature. In most foods there remains considerable unfrozen water at −9.5°C, and long-term storage at this temperature results in severe enzymatic deterioration of food quality, especially of an oxidative nature. Storage at −18°C sufficiently retards the activity of many food enzymes to prevent significant deterioration. In the case of fruits and vegetables, the enzymes are inactivated prior to freezing by heat blanching or chemical treatment.

Nonenzymatic chemical reactions are not entirely stopped at −18°C, but proceed very slowly. In the freezing zone, the generalization that reaction rates are approximately halved for every 10°C drop in temperature does not hold well, since many reactions proceed in solution, and in this zone, solution concentration is changing rapidly with the freezing of water. Nevertheless, the lower the temperature, the slower the reaction rates and the less unfrozen water present to serve as solvent for chemical reactants.

The overall effects of low temperature on long-term storage of various foods are indicated in Table 9.5. Many vegetables, fruits, and nonfatty meats, if properly packaged and frozen, retain good quality during storage at −18°C for 12 months or longer.

Table 9.5. Approximate Number of Months of High-Quality Storage Life

Product	Storage Temperature		
	−18°C (0°F)	−12°C (10°F)	−6.7°C (20°F)
Orange juice (heated)	27	20	4
Peaches	12	<2	6 days
Strawberries	12	2.4	10 days
Cauliflower	12	2.4	10 days
Green beans	11–12	3	1
Green peas	11–12	3	1
Spinach	6–7	<3	¾
Raw chicken (adequately packaged)	27	15½	<8
Fried chicken	<3	<30 days	<18 days
Turkey pies or dinners	>30	9½	2½
Beef (raw)	13–14	5	<2
Pork (raw)	10	<4	<1.5
Lean fish (raw)	3	<2¼	<1.5
Fat fish (raw)	2	1½	0.8

Source: U.S. Dept. of Agriculture.

Most fish are less stable. At higher temperatures of −9 to −7°C, quality may be retained for periods of only days or a few weeks, depending on product.

Quality and subsequent storage life would be still better for many foods if they were frozen and stored well below −18°C. It is relatively easy to freeze foods to −30°C and even lower by several methods, and the costs are not excessive. What is more difficult and very expensive, however, is maintaining food at or below −30°C during transport and warehouse or supermarket cabinet storage. Many refrigerated trucks on the road today are not capable of holding a temperature of even −18°C, and supermarket display cases often are above −18°C near the top, although they may be colder below.

Damage from Intermittent Thawing

The kinds of damage that can occur to foods during slow freezing also occur during slow thawing. Repeated freezing and thawing cycles are very detrimental to stored foods. Repeated thawing need not be complete for damage. Complete thawing in storage is rare and generally occurs only when there is complete breakdown of the cold storage equipment. This is easily recognized and quickly corrected. However, all commercial frozen food distribution or storage systems have a measurable temperature cycle. Such cycles are part of the temperature control systems, and it is not uncommon for a frozen storage chamber to go from its maximum to its minimum temperature and back again on roughly a 2-h cycle. This could mean 360 cycles a month and over 4000 cycles a year.

As little as a 3°C fluctuation in freezer storage temperature above and below the −18°C mark can be damaging to many foods. Above −12°C, thawing intensifies the concentration effect. Upon refreezing, water melted from small ice crystals tends to bathe unmelted crystals, causing them to grow in size. Whatever the temperature fluctuation in the storage area, because heat transfer has a finite rate there will be a lag effect in the food itself and the food generally will experience less of a temperature range than the room or cabinet. Nevertheless, room or cabinet temperature variations of greater than a few degrees from the −18°C mark over a period of weeks or months will noticeably damage the quality of most frozen foods, and such freezer storage facilities should be repaired.

Frozen foods being thawed for ultimate use also are subject to quality loss, especially if thawing is slow. Once again, concentration effects can occur. Brought to the same temperature, the most concentrated solutions that freeze last are the first to thaw. These commonly are eutectic mixtures. A eutectic mixture is a solution of such composition that it freezes (or thaws) as such rather than becoming more concentrated due to further separation of pure ice. In other words, the eutectic mixture freezes (and thaws) as a mixture, and the frozen eutectic will have a constant proportion of ice crystals intermingled with solute crystals. The temperature at which a eutectic mixture is formed is called the eutectic temperature or eutectic point. A dilute solution of NaCl in water under freezing conditions will first freeze out pure water and become more concentrated in NaCl. At −21°C, the remaining water and salt would consist of a mixture of 23% NaCl and 77% water. This would freeze into eutectic ice of the same composition. Where ice from pure water has a melting point of 0°C, eutectic ice could be removed from the system and would have a melting point of −21°C, and, therefore, such ice would be a better refrigerant than pure water ice and has been used as such commercially.

Food materials are complex mixtures and pass through several eutectic compositions in the course of becoming solidly frozen and on thawing. If thawing is slow, there is more time for food constituents to be in contact with concentrated eutectic mixtures and damaging concentration effects are intensified.

Another reason why quick final thawing is superior to slow thawing is illustrated in Table 9.6. Large volumes of frozen food, such as a 30-lb can of frozen whole egg, can take 20–60 h to thaw in air, depending on the air temperature. Cool running water and other techniques can markedly reduce this time. Since bacteria survive the freezing process, when thawing times are long and temperatures of the product rise, there is an opportunity for bacterial multiplication.

Refrigeration Requirements

A product's refrigeration load is the quantity of heat that must be removed to reduce the temperature of the product from its initial temperature to the temperature consistent with good frozen food storage. If the food is cooled from above its freezing point to a storage temperature below its freezing point, then quantitatively this refrigeration load is made up of three parts: the number of heat units that must be removed to (1) cool the food from its initial temperature to its freezing point, (2) cause a change of state at the freezing point, and (3) lower the temperature of the frozen product to the specified storage temperature. These heat units may be calculated in terms of British Thermal Units (Btu), calories (cal), or joules (J). Much of the original work in refrigeration and freezing was done using the British system of units, which has dominated the literature. This will gradually be replaced by the International System of Units (SI), which employs joules and grams as the units of heat and weight.

Definitions and Heat Constants

A Btu is the quantity of heat that will raise or lower the temperature of 1 lb of water 1°F through the range 32–212°F at normal atmospheric pressure. A calorie is the amount of heat that will raise or lower the temperature of 1 g of water 1°C (from 14.5 to 15.5°C) at normal atmospheric pressure. (1 Btu = 252 calories = 1055 joules or 1.055 kJ.)

Different substances can absorb different amounts of heat and are said to have different heat capacities. The heat capacity of water is 1 Btu/(lb °F) or 1 cal/(g °C).

Table 9.6. Effect of Thawing on the Microbiology of Frozen Whole Eggs After Removal from the Shell

Method	Hours Required	% Increase in Microbial Count During Thawing
In air at 27°C (80°F)	23	1000
In air at 21°C (70°F)	36	750
In air at 7.2°C (45°F)	63	225
In running water, 16°C (60°F)	15	250
In running water, 21°C (70°F)	12	300
Agitated water, 16°C (60°F)	9	40
Dielectric heat	0.25	Negligible

Source: Weiser (1971). Mountney, and Gould

The specific heat of any substance is the ratio of its heat capacity to that of water. Specific heat is a ratio like specific gravity and is independent of whether measurements are made in Btu or calories. In either case the specific heat of water is taken as the standard and is given a value of 1. Thus, in the British system the specific heat of a substance is the ratio of the heat required to raise or lower the temperature of a unit mass of the substance 1°F compared to the heat required to raise or lower the temperature of a unit mass of water 1°F. Since the specific heat of water is 1, it follows that the specific heat of any substance is the amount of heat in Btu required to raise or lower the temperature of 1 lb of the substance 1°F. In like fashion, the specific heat of any substance is the amount of heat in calories required to raise or lower the temperature of 1 g of the substance 1°C.

There are two types of heat: sensible heat and latent heat. Sensible heat is readily perceived by the sense of touch and produces a temperature rise or fall as heat is added or removed from a substance. Latent heat is the quantity of heat required to change the state or condition under which a substance exists, without changing its temperature. Thus, a definite quantity of heat must be removed from water at 0°C to change it to ice at 0°C; the same amount of heat must be added to ice at 0°C to change it to water at 0°C. This is known as the latent heat of fusion or crystallization. Similarly, in going from water at 100°C to steam at 100°C, the latent heat of evaporation must be added to the system. The latent heat of fusion, which must be removed during freezing, is 144 Btu/lb for water. The specific heat of a material is different in the liquid state and in the frozen state; that is, a different number of Btu/lb is required to raise or lower the temperature of a material 1°F depending on whether the material is above or below its freezing point.

The specific and latent heats of foods are used to determine refrigeration requirements for cooling, freezing, and storage. Typical values for a few foods are given in Table 9.7. Additional values can be found in several handbooks and in Table 9.3.

Calculation of Refrigeration Load

As indicated in Table 9.7, water has a specific heat before freezing of 1.00, a latent heat of fusion of 144, and a specific heat after freezing of 0.48. Therefore, if we wanted to freeze a pound of water from 60°F to 0°F, we would have to remove 1 Btu/°F from 60°F down to 32°F, plus 144 Btu to go from water to ice at 32°F, plus 0.48 Btu/°F from 32°F down to 0°F. This would be equal to $28 + 144 + 15$ or 187 Btu.

Using the same method, the following general equations can be used to calculate the heat to be removed to cool and freeze any quantity of food material from any starting temperature to any frozen storage temperature, provided the freezing point of the material is known:

$$H_1 = S_L W(T_i - T_f), \qquad H_3 = S_S W(T_f - T_s)$$
$$H_2 = H_F W, \qquad H_{fs} = H_1 + H_2 + H_3$$

Thus, H_1 is the number of Btu required to cool the food from its initial temperature to its freezing point and is equal to S_L (the specific heat of the food above its freezing point) $\times$ W (its weight in pounds) $\times$ $T_i - T_f$ (the difference between the initial temperature and the freezing point in °F).

H_2 is the number of Btu required to change the food from the "liquid" state to the frozen state at its freezing point and is equal to H_F (the latent heat of fusion of the food) $\times$ W (its weight).

Food Science

Table 9.7. Specific and Latent Heats of Selected Foods

| Food Product | Specific Heat (Btu/lb)[a] | | Latent Heat of Fusion (Btu/lb) |
	Above Freezing	Below Freezing	
Asparagus	0.94	0.48	134
Bacon	0.50	0.30	29
Beef			
lean	0.77	0.40	100
fat	0.60	0.35	79
dried	0.34	0.26	22
Cabbage	0.94	0.47	132
Eggs	0.76	0.40	100
Fish	0.76	0.41	101
Lamb	0.67	0.30	83
Milk	0.93	0.49	124
Oysters, shelled	0.90	0.46	125
Poultry, fresh	0.79	0.37	106
Pork, fresh	0.68	0.38	86
Veal	0.71	0.39	91
Water	1.00	0.48	144

[a] 1 Btu/lb = 0.556 cal/g = 2.326 joules/g.

H_3 is the number of Btu required to lower the frozen food from its freezing point to the desired storage temperature and is equal to S_S (the specific heat of the food below its freezing point) × W (the weight of the food) × T_f-T_s (the difference between the freezing point and the desired storage temperature).

H_{fs}, the total Btu requirement, is equal to $H_1+H_2+H_3$.

These same relationships will hold for calculations in SI units if temperatures are converted to degrees Celsius, weights to grams, and latent heats of fusion from British thermal units per pound to calories per gram. Then the total amount of heat to be removed, or the refrigeration load, would be in calories or joules (1 cal = 4.187 J) rather than British thermal units.

Table 9.7 also indicates some important differences between foods. Foods that are high in water, such as asparagus, cabbage, and milk, have heat constants very close to those of water, and foods low in water have lower heat constants. This says that foods low in water require removal of less heat per unit weight to cool and freeze them than do foods higher in water. Comparison of the constants for lean beef, fatty beef, and dried beef clearly shows that as the water content of food decreases, the specific heats before and after freezing, as well as latent heats of fusion, go down.

To calculate refrigeration loads using specific heat data, the initial freezing points of foods must be known. These are usually given in handbooks, along with moisture contents of foods and the specific and latent heat constants (Table 9.3). The satisfactory use of such data depends on the purpose for which calculations are being made and the degree of accuracy that is needed.

Specific and latent heat values given in handbooks strictly apply only at the moisture content of the food for which they were established. Thus, the value listed in Table

9.3 for the latent heat of fusion of sweet corn with 74% moisture is 106 Btu/lb. But sweet corn can vary by more than 10% in water content. If sweet corn has only 64% moisture, the latent heat of fusion would be found on careful calorimetric measurement to be not 106 Btu/lb but 91.7 Btu/lb.

For research purposes it sometimes is necessary to know refrigeration requirements with a high degree of accuracy; then heat constants based on calorimetric measurements at specific moisture contents are required. But, generally, in commercial practice, refrigeration requirements may be estimated quite well from average handbook values. Further, for practical purposes, where the actual freezing point of a food is not known, the value of $-2.2°C$ (28°F) is assumed to be the freezing point of most foods.

Where handbook values are not available, as would be the case for a new product of a proprietary nature, or a natural product of altered moisture content, three simple formulas are commonly used to estimate heat constants.

Specific heat of a food before freezing = 0.008(% water) + 0.20

Specific heat after freezing = 0.003(% water) + 0.20

Latent heat of fusion in Btu/lb = 144(% water in food)/100

When a product refrigeration load is calculated in Btu by the methods described here, it often is expressed in terms of the standard refrigeration unit, tons of refrigeration. A ton of refrigeration is the number of Btu required to convert 1 ton of water at 32°F to 1 ton of ice at 32°F in 24 h. Since the latent heat of fusion for water is 144 Btu/lb, a ton of refrigeration is 144 times 2000 lb, or 288,000 Btu. Therefore, the product refrigeration load in Btu divided by 288,000 equals the tons of refrigeration required to bring the food batch to its storage temperature. The refrigeration engineer then adds to this the additional refrigeration required to cool down the cold storage room or cabinet and to maintain it at proper storage temperature against heat pickup through poor insulation, food respiration, opening of doors, electric motors, and so on.

Factors Determining Freezing Rate

Apart from the absolute refrigeration requirements to freeze foods, there are other factors that affect freezing rates and thereby help determine quality.

The rate of cooling and freezing food may be expressed generally as a function of two variables; the driving force divided by the sum of resistances to heat transfer. The driving force is the temperature difference between the product and the cooling medium. The resistances depend on such factors as air velocity, thickness of product, geometry of the system, composition of the product, and the resistance to heat transfer of the food package. Geometry of the system could include such factors as degree of contact of the refrigerant with the food to be cooled, extent of agitation, in a continuous cooling and freezing system whether the refrigerant is circulated in the same direction or counter current to the food, and so on. Composition of the product not only involves the chemical composition and the thermal conductivities of various constituents but also the physical arrangement of constituents—for example, the way the fat is distributed within a cut of meat and the direction of orientation of muscle fibers in relation to a refrigerating surface. Different packaging materials have different degrees of resistance to heat transfer.

Food Composition

Like metals and other materials, food constituents have different thermal conductivity properties which change with temperature. The greater the conductivity, the greater the cooling and freezing rates—all other things being equal. In the cooling and freezing temperature range, heat conductivities of water change little until the phase change from water to ice occurs. Since the thermal conductivity of ice is far greater than that of water, the thermal conductivity of a food increases rapidly as it passes from the unfrozen to the frozen state. It should also be noted that fat has a much lower thermal conductivity than water, and air has a thermal conductivity far less than that of water or fat.

Several generalizations can be made about the effect of the composition of a food product on freezing rate under controlled freezing conditions. First, high levels of fat or entrapped air tend to reduce the rate of freezing. Second, rates of cooling and freezing are not constant during these processes since thermal conductivities change as water changes to ice. In addition, the physical structure of foods influences freezing rates. For example, if two food systems both contained 50% fat and 50% moisture, but one was an oil-in-water emulsion and the other was a water-in-oil emulsion, the two foods would be expected to have different thermal conductivity properties. The oil-in-water emulsion, with water being the continuous phase, should have greater thermal conductivity values at different temperatures than the corresponding water-in-oil emulsion of the same chemical composition. Other things being equal, the oil-in-water emulsion should freeze at a faster rate than the water-in-oil emulsion. Similarly, cuts of meat should conduct heat at different rates depending on whether the meat is in contact with a refrigerated surface in a direction parallel or perpendicular to the layers of fat and to the direction of orientation of the muscle fibers. We can make educated guesses as to how heat transfer will be affected by these variations, but unfortunately there is little published work on these food systems.

Noncompositional Influences

The effects of various other factors (e.g., air velocity, product thickness, agitation, degree of contact between food and cooling medium, and packaging) on freezing rate are well known, follow the simpler rules of heat transfer, and largely determine the design of freezer systems. It is easy to cite the directional effects of such variables on freezing rate, and these will apply in virtually any system design. However, a quantitative measure of the effects of these variables generally must be established experimentally for each different food and system geometry to have validity.

The following principles apply in any system: (1) The greater the temperature difference between the food and the refrigerant, the faster the freezing rate; (2) the thinner the food piece or greater the heat transfer rate of the food package, the faster the freezing rate; (3) the greater the velocity of refrigerated air or circulating refrigerant, the faster the freezing rate; (4) the more intimate the contact between the food and the cooling medium, the faster the freezing rate; and (5) the greater the refrigerating effect or heat capacity of the refrigerant, the faster the freezing rate.

In this latter case, if the refrigerant is a liquid that expands to a gas, then the refrigerating effect is determined largely by its particular latent heat of vaporization. If the refrigerant does not undergo phase change, such as a salt brine, then its refrigerating effect is determined by its heat capacity or specific heat. More will be said about the refrigerating effect under liquid-nitrogen freezing.

The magnitude of the influence of these major variables on freezing rate can be substantial. Thus, lowering air temperature in a tunnel-type freezer from −18° to −30°C can shorten freezing time of small cakes from 40 min to about 20 min. Spraying with liquid nitrogen at −196°C would cut freezing time to under 2 min. Lowering of freezer temperatures to very low levels, however, does not give a straight-line freezing rate response; thus, the increase in freezing rate as temperatures decrease tends to drop off, especially at freezer temperatures below about −45°C.

In still air at −18°C, small items such as individual fruits or small fish fillets may freeze in about 3 h. Increasing the air velocity at this temperature to 1.25 m/sec (250 ft/min) will decrease freezing time of these items to about 1 h, whereas an air velocity of 5 m/sec will further decrease freezing time to about 40 min. Whether in cold air or in any other flowing refrigerant, increased velocity speeds freezing by carrying heat away from the surface of food and rapidly replacing warmed refrigerant with cold refrigerant to maintain the maximum temperature difference between the food and the refrigerant. However, freezing rate does not increase linearly with air velocity.

The effect of food or package thickness on freezing rate is such that when common packages are doubled from about 5 to 10 cm (2–4 in.) in thickness, freezing time increases about 2.5-fold. The slope of the thickness curve is such that as product thickness is increased further, the rate of increase in freezing time goes up faster than the rate of change in thickness. Thus, 13.6-kg (30-lb) cans of eggs or fruit may require 48–72 h to become thoroughly frozen. Commercial 0.21-m^3 (55-gal) drums of fruit juices may require more than a week to freeze unless the juice is first slush-frozen by being passed through a heat exchanger prior to being filled into the large drums.

Freezing Methods

There are three basic freezing methods in commercial use: freezing in air, freezing by indirect contact with the refrigerant, and freezing by direct immersion in a refrigerating medium. Each method can be subdivided in various ways, as indicated in Table 9.8.

Cold air may be used with various degrees of velocity, progressing from still-air "sharp" freezing to the high-velocity blast freezer tunnel. The velocity of air also may be used to subdivide and move particles of materials to be frozen, as in the case of fluidized-bed freezing.

Indirect contact freezing includes those methods in which the food or the food package is in contact with a surface that, in turn, is cooled by a refrigerant, but the food or food package does not contact the refrigerant directly. In the case of solid foods or foods in containers, this most commonly involves providing a flat surface in contact with refrigerated plates, which may contact one or two surfaces of the food or package.

Table 9.8. Commercial Freezing Methods

Air Freezing	Indirect Contact Freezing	Immersion Freezing
Still-air "sharp" freezer	Single plate	Heat exchange fluid
Blast freezer	Double plate	Compressed gas
Fluidized-bed freezer	Pressure plate	Refrigerant spray
	Slush freezer	

In the case of liquid foods and purees, the food is pumped through a cold wall heat exchanger and frozen to the slush condition.

Immersion freezing involves direct contact of the food or package with the refrigerant either by submerging the food or spraying the cold liquid onto the food or package surface.

With the exception of still-air "sharp" freezing, all of these methods are referred to as fast freezing methods and may be engineered for batch, semicontinuous, or continuous operation.

Air Freezing

The oldest and least expensive air freezing method from an equipment standpoint is still-air "sharp" freezing. Here the food is simply placed in an insulated cold room at a temperature usually maintained in the range of $-23°$ to $-30°C$. The method was introduced around 1860 and became termed "sharp" freezing, since anything below $-18°C$ was then considered a very low temperature. Although there is some air movement by natural convection, and in some cases gentle air movement is promoted by placing circulating fans in the room, the method is essentially a still-air freezing method to be clearly distinguished from air-blast freezing, which employs air velocities that may exceed hurricane speeds. Depending on size of food items or packages and degree of separation between units, freezing time can be several hours to several days. Sharp freezing remains, today, a very important freezing method. Sharp freezing conditions also are essentially the freezing conditions that exist in home freezers, except temperatures commonly are closer to $-18°C$ than to $-23°C$ or $-30°C$. Commercially, sharp freezers often double as frozen storage rooms where space permits.

Air-blast freezers typically are operated at temperatures of -30 to $-45°C$ with forced air velocities of 10–15 m/sec (2000–3000 ft/min). Under such conditions, 30-lb cans of eggs or fruit, which take 72 h to freeze thoroughly in a sharp freezer, may be frozen in 12–18 h. Blast freezers are of many designs, from rooms where food is frozen as a batch to tunnels through which carts or a belt may be moved continuously. The tunnel blast freezer shown in Fig. 9.4 combines an overhead conveyor belt with a lower level track for moving food carts. Particulate unpackaged foods, such as loose vegetables, are automatically fed onto the moving belt whose speed is adjustable according to the required freezing time. Frozen product is dropped from the belt into a collection hopper at the opposite end of the tunnel. Independently, loose foods or packaged cartons loaded on the carts can be moved through the tunnel at a selected rate.

Other designs make use of vertical movement of food on trays. Trays of particulate products such as peas or beans are automatically moved upward through a cold air blast. Freezing time of such particulates in thin layers may be on the order of 15 min. Today, along with foods frozen as a single block in a package, vegetables and other particulate items, such as shrimp, are individually quick frozen (IQF) so that they can be poured from a bag for greater convenience. Since there is some tendency for such particulates to stick together during freezing, product mechanically dislodged from trays may be passed through a breaker devise to disaggregate large clusters before being conveyed under cold air to packaging.

In air-blast systems, manufacturers have developed numerous patterns of cold air flow to pass over, under, or through the product. Frequently, the principle of countercurrent air flow is employed to bring the coldest air into contact with the already frozen product as it is about to leave the tunnel or column. In this way, freezing is progressive

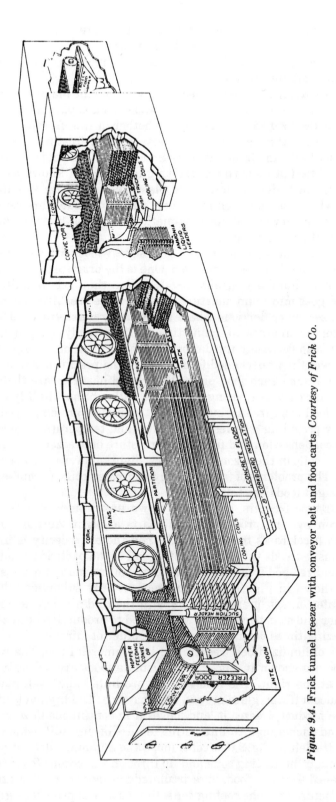

Figure 9.4. Frick tunnel freezer with conveyor belt and food carts. *Courtesy of Frick Co.*

and there is no tendency for product to rise in temperature and partially thaw through the freezing process, as would be possible in a cocurrent system, where coldest air enters with unfrozen product and tends to rise in temperature through the tunnel as product gives up heat and freezes.

Modern blast freezers also provide means to overcome the major drawbacks inherent in high-velocity air freezers. Whenever unwrapped food is placed in a cold zone, there is a tendency for the food to lose moisture, whether during freezing or after the food is frozen. In a freezer this can have two consequences: frosting over of refrigerated coils or plates, necessitating frequent defrosting to maintain heat exchanger efficiency; and drying out of food at its surface, resulting in the defect known as "freezer burn." This situation is markedly accelerated in blast freezers where the air is moving at high velocity. When food is frozen and water vapor molecules are removed from the frozen surface by the dryer cold air, sublimation, a form of freeze-drying occurs; this is essentially what freezer burn is.

To minimize freezer burn, which produces unsightly food surfaces, nutrient loss, and other defects, two techniques are employed. One is the practice of prechilling food with air of high relative humidity at about $-4°C$. Here the food partially frozen in the humid air undergoes minimum moisture loss. Then the prechilled surface-frozen food is moved into a second colder zone where it is quickly finish-frozen. The rapid finish-freezing in colder air provides minimum time for the already cold product to lose more moisture, which also decreases the defrosting requirement on freezer coils. The other technique is to wet the unpackaged food pieces in the prechilling zone so as to freeze a thin ice glaze around each food piece. The glazed particles are then moved to the colder zone for quick finish-freezing. The glaze will sublime slightly but will protect the underlying food from freezer burn. These techniques are employed in continuous blast freezers by providing a series of humidity- and temperature-controlled zones.

To minimize condensation and freezing of moisture on freezer coils, it is common to maintain freezer coils in these zones only very slightly lower in temperature than the circulating air. The problems of freezer burn and frosted coils, of course, are markedly less when packaged foods are frozen.

In various air-type freezers, cold air is blown up through a wire mesh belt that supports and conveys the product. This imparts a slight vibratory motion to food particles, which accelerates freezing rate. When the air velocity is increased to the point where it just exceeds the velocity of free-fall of the particles, fluidization occurs; this is called fluidized-bed freezing. The dancing-boiling motion of peas in a fluidized-bed freezer is shown in Fig. 9.5. This motion not only subdivides the product and provides intimate contact of each particle with the cold air, but keeps clusters from freezing together and so is particularly well suited to production of frozen items in the IQF form. Freezing times commonly are in the order of minutes.

Several types of fluidized-bed freezers are in commercial use. The workings of one continuous type are shown in Fig. 9.6. Particulate foods are fed by a shaker onto a porous trough at the right. The food may be prechilled and even moistened if it is desired to produce a frozen glaze around each particle. The high-velocity refrigerated air fluidizes the product, freezes it, and moves it in continuous flow from right to left for collection and packaging. An interesting feature of this unit, which can be seen in the diagram at the left in Fig. 9.6, is continuous and automatic defrosting. Air is blown via the fan through the cooling coils and up through the porous food trough. The cold air, now moistened from the food, is recirculated to conserve refrigeration. It tends to condense its moisture onto the cooling coils. But a spray of propylene glycol antifreeze

Figure 9.5. "Boiling" motion of peas during fluidized-bed freezing. *Courtesy of Frigoscandia AB, Sweden.*

is maintained over the cooling coils to melt ice as it would be formed. In this way the cooling coils are maintained at maximum operating efficiency. The glycol solution is bled off to an evaporator where the accumulated water is easily removed.

Indirect Contact Freezing

Although it is of course possible to place solid food directly onto the surface of a block of ice or a block of dry ice, this is rarely done commercially. Rather, the food is placed on plates, trays, belts, or other cold walls which are chilled by circulating refrigerant but separate the food from the refrigerant. Thus, the food or its package is in direct contact with the cold wall but in indirect contact with the refrigerant. This permits the use of refrigerants that might otherwise adversely affect the food or its package.

The more important indirect contact freezers are typified by the Birdseye Multiplate Freezer. This consists of a number of metal shelves or plates through which refrigerant is circulated. The food, usually as flat packages, is placed between shelves and there is provision after loading for applying pressure to squeeze the shelves into more intimate contact with the top and bottom of the packages for faster freezing. All is enclosed within an insulated cabinet. Depending on refrigerant temperature, package size, degree of contact, and the type of food, freezing time is 1–2 h for commercial packages 4–5 cm (1.5–2.0 in.) thick. This unit is a batch freezer.

Quite similar are various automatic plate freezers. These have provisions for automatically loading shelves from the packaging line. As a shelf is loaded, it is moved into pressure contact with the preceding shelf and into an insulated zone where freezing proceeds. At the rear of the freezing zone, frozen packages are discharged one shelf at a time, and the empty shelves return to the loading position. The plate freezer shown in Fig. 9.7 operates in this fashion.

In all indirect contact freezing machines, efficiency is dependent on the extent of contact between the plates and the food. For this reason, packages should be well filled or slightly overfilled to make good pressure contact with the plates; solid compact products such as meat or fish fillets freeze more rapidly than shrimp or vegetables where the individual pieces are separated from one another by small air spaces.

Indirect contact freezers for liquid foods and purees are quite different. They generally take the form of the tubular scraped-surface heat exchangers previously described (see Fig. 8.12), with refrigerant rather than steam on the side of the wall opposite the food. As is the case when this equipment is used for heating, the liquid food is pumped through the inside tube with its rotating shaft or mutator. This occupies most of the inner tube space, so the food is forced to pass through the remaining annulus as a thin layer in contact with the cold wall. Scraper blades attached to and rotating with the mutator continually scrape the cold wall as food tends to freeze onto it. This speeds the freezing rate three ways: (1) It keeps the cold wall free of a frozen insulating food layer, which would minimize the temperature difference between the unfrozen food and the effective cold wall; (2) it shaves ice crystals from the cold wall as freezing progresses and these ice crystals seed the unfrozen portion of the food, promoting further freezing within the mass; and (3) the scraper blades (and mutator) keep the mass in motion, continually bringing new portions of food into contact with the cold wall. Freezing is virtually instantaneous, occurring in a matter of seconds. In this type of unit, freezing is never carried to completion, or else the frozen product would freeze in the tube and choke off continuous flow. Instead, the product is frozen to the slush condition, packaged, and then hard-frozen in an air-blast or immersion-type freezer.

Immersion Freezing

In a strict sense, air freezing is a kind of immersion freezing. However, the term immersion freezing generally is applied to refrigerants other than cold air. The advantages of direct immersion freezing include the following:

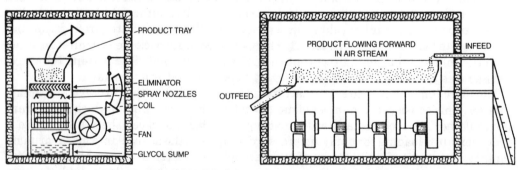

Figure 9.6. Diagram of continuous fluidized-bed freezer and method for defrosting. *Courtesy of Frigoscandia AB, Sweden.*

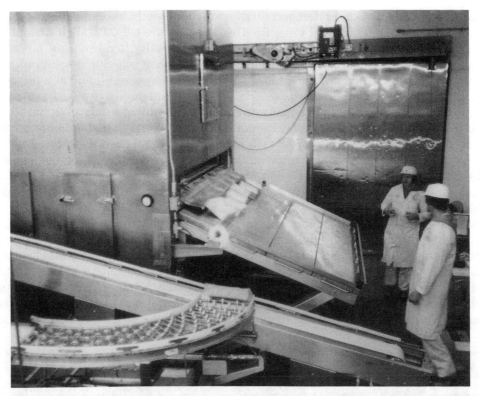

Figure 9.7. Cartons of hamburger patties emerge frozen from automatic plate freezer. *Courtesy of Crepaco, Inc.*

1. There is intimate contact between the food or package and the refrigerant; therefore, resistance to heat transfer is minimized. This is particularly important with irregularly shaped food pieces to be frozen very rapidly, such as loose shrimp, mushrooms, and the like.
2. Although loose food pieces can be frozen individually by immersion freezing and air freezing, immersion freezing minimizes their contact with air during freezing, which can be desirable for foods sensitive to oxidation.
3. For some foods, the speed of immersion freezing with cryogenic liquids produces quality unattainable by any other currently known freezing method.

Direct immersion freezing places limitations on the refrigerants that may be used, especially if they are to come into contact with unpackaged foods. These refrigerants must be nontoxic, pure, clean, free from foreign tastes, odors, colors or bleaching agents, and so on. Similarly, when the food is packaged, nontoxicity and noncorrosiveness to the packaging material are important. The refrigerants for immersion freezing are of two broad classes: low-freezing-point-liquids, which are chilled by indirect contact with another refrigerant; and cryogenic liquids, such as compressed liquified nitrogen, which owe their cooling effect to their own evaporation.

The low-freezing-point liquids that have been used for contact with nonpackaged foods include solutions of sugars, sodium chloride, and glycerol. These must be used at sufficient concentration to remain liquid at −18°C or lower to be effective. In the case of sodium chloride brine, for example, this requires a concentration of about 21%. Temperatures as low as −21°C can be achieved with a 23% brine, but this is the eutectic point, and at lower temperatures a salt and water mixture freezes out of solution, so the lowest practical brine freezing temperature is −21°C. Brine cannot be used with unpackaged foods that should not become salty; today, brine for direct immersion freezing is largely restricted to freezing of fish at sea. Sugar solutions have been used to freeze fruits, but the difficulty here is that to remain liquid at a temperature of −18°C, a solution of approximately two-thirds sucrose is required, which becomes very viscous at the low temperature. Glycerol and water mixtures have been used to freeze fruit, but, like sugar, cannot be used for foods that should not become sweet. One can get down to −47°C with a 67% glycerol solution in water. Another low-freezing-point liquid related to glycerol is propylene glycol. A 60% propylene glycol–40% water mixture freezes at −51°C. Propylene glycol is nontoxic but has an acrid taste. For this reason, its use in immersion freezing is generally limited to packaged foods.

Commercial equipment for immersion freezing with low-freezing-point liquids is typified by the continuous round shell direct immersion freezer which is well suited to the freezing of food in cans (Fig. 9.8). The shell has a large-diameter closed tubular reel that rotates within it. The cans are positioned at the periphery of this rotating reel, and refrigerant is circulated in the annulus between the outer shell and the inner reel. Since the inner reel is closed at its ends, refrigerant flows only through the space occupied by the cans, reducing the volume of refrigerant needed. In a typical operation,

Figure 9.8. Continuous round shell direct immersion freezer. *Courtesy of FMC Corp.*

about 400 (6 oz) cans are frozen per minute. Residence time or freezing time can be 30 min. Rotation of the cans imparts motion to fluid foods in the cans as they freeze, contributing to a faster, more uniform, small ice crystal type of freezing.

Immersion Freezing with Cryogenic Liquids. Cryogenic liquids are liquefied gases of extremely low boiling point, such as liquid nitrogen and liquid carbon dioxide, with boiling points of $-196°C$ and $-79°C$, respectively. Today, liquid nitrogen is the most commonly used cryogenic liquid in immersion freezing of foods.

The major advantages of liquid nitrogen freezing are the following:

1. It undergoes slow boiling at $-196°C$, which provides a great driving force for heat transfer.
2. Liquid nitrogen, like other immersion fluids, intimately contacts all portions of irregularly shaped foods, thus minimizing resistance to heat transfer.
3. Since the cold temperature results from evaporation of liquid nitrogen, there is no need for a primary refrigerant to cool this medium.
4. Liquid nitrogen is nontoxic and inert to food constituents. Moreover, by displacing air from the food it can minimize oxidative changes during freezing and through packaged storage.
5. The speed of liquid-nitrogen freezing produces frozen foods with a quality unattainable by noncryogenic freezing methods. Although many products do not require such fast freezing for good quality, some products such as mushrooms cannot be frozen by other methods without excessive tissue damage.

The major disadvantage of liquid-nitrogen freezing generally cited is its high cost.

Some additional properties of liquid nitrogen deal with its heat capacity or refrigerating effect. In being vaporized from a liquid at $-196°C$ to a gas at $-196°C$ each kilogram of liquid nitrogen absorbs 200 kJ, the latent heat of vaporization. Then each kilogram of gas at $-196°C$ absorbs another 186 kJ in rising in temperature to $-18°C$; this is the specific heat of the gaseous nitrogen times the temperature rise from $-196°$ to $-18°C$. Thus, the total heat uptake of the liquid at $-196°C$ going to $-18°C$ is 386kJ/kg.

This is very important in the design of spraying types of liquid-nitrogen freezers. To get the maximum freezing effect from spraying, the nitrogen should impinge on the food surface as liquid droplets rather than as a cold gas in order to achieve the cooling effect of the latent heat of vaporization plus the sensible heat of gas temperature rise, or 386kJ/kg of liquid. If the spray permits the liquid to vaporize before contacting the food, then a cooling effect of only 186kJ/kg of gas is achieved as the gas goes from a temperature of $-196°$ to $-18°C$.

The principles of manufacture and handling of liquid nitrogen should be understood. Liquid nitrogen is manufactured by compressing air and simultaneously removing the heat of compression. The cooled compressed air is then allowed to expand through specially designed valves. This expansion causes further chilling of the air to the point of liquefaction, producing a mixture principally of liquid nitrogen and liquid oxygen. Since liquid oxygen has a higher boiling point than liquid nitrogen, namely, $-183°C$, the oxygen and nitrogen can be separated by distillation. Liquid nitrogen with its lower boiling point of $-196°C$ comes off first as a gas, which can be recompressed into a liquid. In recent years, technological advances in the production of liquid gases have stemmed from the needs of the space program where liquid oxygen is an important fuel.

At $-196°C$ and at atmospheric pressure, liquid nitrogen boils gently. It does not produce an excessive pressure of nitrogen gas above the liquid if placed in a vessel and maintained at $-196°C$; this is the key to liquid-nitrogen storage and handling. Liquid nitrogen is transported and stored at $-196°C$ by housing it in large insulated tanks of the thermos bottle or Dewar flask type. As long as insulation maintains this low temperature, pressure developed in the tank is comparatively small and not dangerous. In contrast, compressed nitrogen gas at room temperature is under great pressure and requires storage in the familiar high-pressure steel gas cylinders.

Manufacturers of liquid nitrogen deliver it in tank truck quantities to insulated storage tanks at food plants. Filling is done simply by hose connection from the tank truck, often making use of the slight pressure of nitrogen gas above the liquid in the tank truck to force the liquid nitrogen through the hose into the storage vessel. The same procedure is used to deliver liquid nitrogen from the storage vessel to the liquid-nitrogen food freezer or to a refrigerated truck for transporting frozen food (Fig. 9.9). The food freezing equipment commonly is of tunnel construction with a continuous mesh belt (Fig. 9.10). The earlier practice of submerging the food under liquid nitrogen has largely given way to more efficient use of liquid-nitrogen sprays. Design features are aimed at depositing the spray as a liquid onto the food surface and minimizing sensible heat loss through insulation; in very large installations, provisions may be engineered to recover and recompress the vaporized nitrogen for reuse. When liquid nitrogen contacts the relatively warm food, it boils violently. In most installations it is not recompressed, but the spent nitrogen gas, which still may be in the range of about -18 to $4°C$, is vented to contact and prechill incoming food, or to cool a refrigerated storage room.

Although liquid nitrogen is capable of freezing food down to $-196°C$, this is virtually never done, because it entails unnecessary cost and could even be damaging to some foods. The food is seldom frozen to a temperature below $-45°C$, and quality results largely from the speed at which this temperature is reached. In the case of many fruit, vegetable, meat, and fish items this may require 1–3 min.

Much shrimp is liquid-nitrogen frozen in the IQF form. Typically, the shrimp enter the freezing tunnel at one end and a liquid-nitrogen spray is directed at the conveyor belt at the opposite end of the tunnel. The spray vaporizes and cold nitrogen gas is directed through the tunnel to meet the incoming shrimp in a countercurrent fashion. The incoming shrimp thus are first chilled by cold gas to about $0°C$ before they reach the spray. Then they are passed under the liquid spray and are frozen to a surface temperature of about $-185°C$. The shrimp are then moved through an equilibration zone where the cold surface and warmer core equilibrate to a uniform temperature of about $-45°C$. The shrimp are then passed through a controlled water spray, where the stored refrigeration of the shrimp freezes a thin ice glaze protective against subsequent dehydration in storage. The shrimp then emerge from the glaze spray at about $-30°C$ for packaging and storage at about $-23°C$. Generally, liquid-nitrogen freezing will produce less dehydration during freezing and less drip loss during thawing than other freezing methods. This can amount to as much as 5% of the weight of some foods.

Cryogenic freezing with carbon dioxide has taken two forms. In one, powdered dry ice, which sublimes at $-79°C$, is mechanically mixed with the food to be frozen. In the other, liquid carbon dioxide under high pressure is sprayed onto the food surface. As pressure is released in the course of spraying, the liquid carbon dioxide becomes dry ice snow at $-79°C$. There are applications where frozen food quality with this refrigerant is equal to that obtained with liquid-nitrogen freezing. In such cases, since a given weight

Figure 9.9. Liquid nitrogen is stored in large tanks prior to use for freezing foods. *Courtesy of D.C. Brown, Air Products and Chemicals, Inc.*

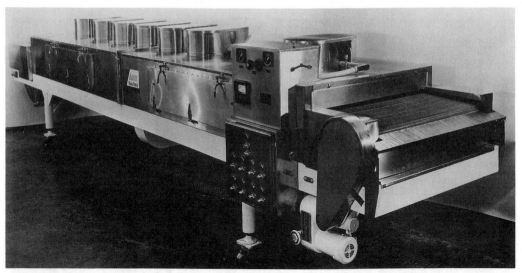

Figure 9.10. Liquid-nitrogen spray freezing unit. *Courtesy of Airco, Inc.*

of dry ice will absorb over twice as much heat in vaporizing as does liquid nitrogen, the use of carbon dioxide may have economic advantages over liquid-nitrogen freezing.

Packaging Considerations

The packaging of frozen foods imposes certain special requirements. Because of the tendency of water vapor to sublime from frozen food surfaces to colder surfaces in freezers and storage rooms, packaging materials for frozen foods should have a high degree of resistance to the permeation of water vapor. Most foods expand on freezing, some to the extent of 10% of their volume. Therefore, packages in which food is frozen should be strong and flexible. As with all foods that may be stored for months or years, packages should be protective against light and air. Because frozen foods generally will be thawed at time of use in their containers, packages should be liquid-tight to prevent leaking on thawing. Many packages and packaging materials such as cans, foils, waxed papers, plastic-coated paperboard, and plastic films are all satisfactory for frozen foods. Glass generally is not satisfactory due to breakage from expansion and from thermal shocks. Packaging will be considered in more detail in Chapter 21.

SOME ADDITIONAL DEVELOPMENTS

As in all industries, favorable economics encourage alternative methods of food processing. Freezing and frozen transportation and storage are highly energy intensive. Other methods of processing that may yield products equal or nearly equal in quality to those obtained by freezing have received increasing attention with rising energy costs. Currently, citrus juice concentrates are commonly frozen for high quality. However, juices and certain other foods may also be produced by HTST heating followed by aseptic packaging, as has been done in Europe for some years. Food irradiation, to be discussed in Chapter 11, also is capable of quality preservation of some products now routinely frozen. These and other methods can be less energy intensive and less costly than freezing and frozen storage but also have limitations. Frozen storage and display at the retail store is also expensive and retailers are looking for ways to reduce these costs.

References

Ashrae. 1983. Handbook. Equipment. American Society of Heating, Refrigerating, and Air-Conditioning Engineers, Atlanta, GA.

Ashrae. 1985. Handbook. Fundamentals Inch-pound edition. American Society of Heating, Refrigerating, and Air-Conditioning Engineers, Atlanta, GA.

Ashrae. 1984. Handbook. Systems. American Society of Heating, Refrigerating, and Air-Conditioning Engineers, Atlanta, GA.

Cano-Munoz, G. 1991. Manual on Meat Cold Store Operation and Management. Food and Agriculture Organization of the United Nations, Rome.

Cleland, A.C. 1990. Food Refrigeration Processes. Analysis, Design and Simulation. Elsevier Science Publishers Ltd., Barking, UK.

Dennis, C. and Stringer, M. (Editors). 1992. Chilled Foods: A Comprehensive Guide. Ellis Horwood Ltd., Chichester.

Fennema, O. 1982. Effect of processing on nutritive value of food: freezing. *In* Handbook of Nutritive Value of Processed Food, M. Rechcigl, Jr. (Editor). CRC Press, Boca Raton, FL, pp. 31–43.

Fennema, O. 1993. Frozen foods: Challenges for the future. Food Australia *45*(8), 374–380.

George, R.M. Freezing processes used in the food industry. Trends Food Sci. Technol. *4*(5), 134–138.

Hallowell, E.R. 1980. Cold and Freezer Storage Manual. 2nd ed. AVI Publishing Co., Westport, CT.

Heldman, D.R. 1982. Food properties during freezing. Food Technol. *36*(2), 92–96.

Heldman, D.R. 1983. Factors influencing food freezing rates. Food Technol. *37*(4), 103–109.

Heldman, D.R. and Singh, R.P. 1986. Thermal properties of frozen foods. In Physical and Chemical Properties of Food, M.R. Okos (Editor). American Society of Agricultural Engineers, St. Joseph, MO, pp. 120–137.

Kornacki, J.L. and Gabis, D.A. 1990. Microorganisms and refrigeration temperatures. Dairy, Food Environ. Sanitat. *10*(4), 192, 194–195.

Lutz, J.M. and Hardenburg, R.E. 1968. The Commercial Storage of Fruits, Vegetables, and Florist and Nursery Stocks. Agricultural Handbook 66, U.S. Department of Agriculture, Washington, DC.

Mallett, C.P. (Editor). 1993. Frozen Food Technology. Chapman & Hall, London, New York.

McCoy, D.C. 1963. Refrigeration in food processing. *In* Food Processing Operations, Vol. 1, M.A. Joslyn and J.L. Heid (Editors). AVI Publishing Co. Westport, CT.

Moberg, L. 1989. Good manufacturing practices for refrigerated foods. J. Food Protect. *52*(5), 363–367.

Renaud, T., Briery, P., Andrieu, J. and Laurent, M. 1992. Thermal properties of model foods in the frozen state. J. Food Eng. *15*(2), 83–97.

Ryall, A.L. and Lipton, W.J. 1979. Handling, Transportation and Storage of Fruits and Vegetables. 2nd ed. Vol. 1. AVI Publishing Co. Westport, CT.

Ryall, A.L. and Pentzer, W.T. 1982. Handling, Transporation and Storage of Fruits and Vegetables. 2nd ed. Vol. 2. AVI Publishing Co. Westport, CT.

Weiser, H.H., Mountney, G.J., and Gould, W.A. 1971. Practical Food Microbiology and Technology. 2nd ed., AVI Publishing Co., Westport, CT.

10

Food Dehydration and Concentration

Water is removed from foods under natural field conditions, by a variety of controlled dehydration processes and during such common operations as cooking and baking. However, in modern food processing, the terms *food dehydration* and *food concentration* have acquired rather special meanings.

Grains in the field dry on the stalk by exposure to the sun. Often a sufficient degree of dryness is achieved (approximately 14% moisture) to require no further drying for effective preservation. This also is true of many plant seeds and spices, and is approached by certain fruits, such as dates and figs, that develop high sugar contents as they dry out on the tree. Centuries ago, humans learned to copy this natural sun-drying process to dry fish and thin slices of meat by hanging them in the air and sun. Where drying of these animal products took a long time, bacterial spoilage during the slow operation occurred, so the use of smoke and salt as further preservative agents in combination with drying gradually evolved.

Sun drying is still in use in many parts of the world, but although sun drying in some parts of the world and for certain products is the most economical kind of drying, it has several obvious disadvantages. Sun drying is dependent on the elements; it is slow and not suitable for many high quality products; it generally will not lower moisture content below about 15%, which is too high for storage stability of numerous food products; it requires considerable space; and the food being exposed is subject to contamination and losses from dust, insects, rodents, and bird droppings.

Efforts at artificial drying with heated air date back to the close of the eighteenth century. The term *food dehydration* refers generally to artificial drying under controlled conditions. However, in modern food processing the term does not refer to all processes that remove water from foods. For example, water is removed when potatoes are fried, cereals toasted, and steak broiled. But these operations do much more than simply remove water and are not considered as a form of food dehydration. Likewise, concentration processes that remove only part of the water from foods (e.g., in the preparation of syrups, evaporated milk, condensed soups) do not come under the currently accepted meaning of the term food dehydration.

In a strict sense, then, food dehydration refers to the nearly complete removal of water from foods under controlled conditions that cause minimum or ideally no other changes in the food properties. Such foods, depending on the item, commonly are dried to final moisture within the range of about 1–5%. Examples are dried milk and eggs, potato flakes, instant coffee, and orange juice crystals. Such products will have storage stability at room temperature of a year or longer. A major criterion of the quality of dehydrated foods is that when reconstituted by the addition of water they be very close to, or virtually indistinguishable from, the original food material used in their preparation. In food dehydration, the technological challenge is especially difficult since very low moisture levels for maximum product stability are not easily obtained with minimum change to food materials. Further, such optimization frequently can be approached only at the expense of increased drying costs. With sensitive foods, product quality and processing costs usually are correlated also in the case of concentration processes.

FOOD DEHYDRATION

Preservation is the principal reason but not the only reason for dehydrating foods. Foods may be dehydrated to decrease weight and bulk (Fig. 10.1). Since orange juice contains approximately 12% solids, removal of all the water leaves one-eighth the weight; that is, 237 ml (8 fl oz) of orange juice yield approximately 28 g (1 oz) of solids. To reconstitute, 207 ml (7 fl oz) of water are added prior to consumption. In the case of juices, the volume of the powders is less than the original juices, although rarely are the powders decreased in volume to the same extent that they are reduced in weight. These reductions can result in lower shipping and container cost, but this is not always the case with dehydrated foods.

Figure 10.1. Weight and volume relationships in dehydrated juices. *Courtesy of U.S. Army Natick Laboratories.*

Some drying processes are chosen to retain the size and shape of the original food. Freeze-drying of large food pieces is such a process. The freeze-dried steak on the left in Fig. 10.2 has essentially the same volume as the original steak. Savings may be made in shipping costs from reduced weight, but not in size of containers in this instance. Further, sometimes shipping costs are not based on weight but are based on volume. In such a case, freeze-dried steaks would not be cheaper to package or ship than their original counterparts.

A third reason for dehydration is the production of convenience items. Good examples of this are instant coffee and instant mashed potatoes. In both cases all brewing or cooking steps are completed before the products are dried. The consumer simply adds water and stirs or mixes. Regardless of the reasons for water removal, food dehydration processes are based on sound scientific principles.

Heat and Mass Transfer

Whatever method of drying is employed, food dehydration involves getting heat into the product and getting out moisture. These two processes are not always favored by the same operating conditions. For example, pressing food between two heated plates would give close contact and improve heat transfer into the food through the top and bottom surfaces but would interfere with the escape of free moisture. It might be better

Figure 10.2. Freeze-dried steaks which have been rehydrated and cooked. *Courtesy of Food Processing.*

to use one bottom hot plate and get heat in, and a free surface on top of the food to let out moisture. In food dehydration, we generally are interested in a maximum drying rate, and so every effort is made to speed heat and mass transfer rates. The following considerations are important in this regard.

Surface Area

Generally, food to be dehydrated is subdivided into small pieces or thin layers to speed heat and mass transfer. Subdivision speeds drying for two reasons. First, a larger surface area provides more surface in contact with the heating medium and more surface from which the moisture can escape. Second, smaller particles or thinner layers reduce the distance heat must travel to the center of the food and reduce the distance through which moisture must travel to reach the surface and escape. Nearly all types of food driers ensure a large surface area of the food to be dried.

Temperature

The greater the temperature difference between the heating medium and the food, the greater will be the rate of heat transfer into the food; this provides the driving force for moisture removal. When the heating medium is air, temperature plays a second important role. As water is driven from the food as water vapor, it must be carried away or it will create a saturated atmosphere at the food's surface, which will slow down the rate of subsequent water removal. The hotter the air, the more moisture it will hold before becoming saturated. Obviously, a greater volume of air also can take up more moisture than a lesser volume.

Air Velocity

Not only will heated air take up more moisture than cool air, but air in motion, that is, high-velocity air, will sweep it away from the drying food's surface, preventing the moisture from creating a saturated atmosphere. Thus, clothes dry more rapidly on a windy day.

Humidity

When air is the drying medium, the drier the air, the more rapid is the rate of drying. Moist air is closer to saturation and so can absorb and hold less additional moisture than if it were dry.

But the dryness of the air also determines to how low a moisture content the food product can be dried. Dehydrated foods are hygroscopic. Each food has its own equilibrium relative humidity. This is the humidity at a given temperature at which the food will neither lose moisture to the atmosphere nor pick up moisture from the atmosphere. Below this atmospheric humidity level, food can be further dried; above this humidity, it cannot, rather it picks up moisture from the atmosphere. The equilibrium relative humidity at different temperatures can be determined by exposing the dried product to different humidity atmospheres in bell jars and weighing the product after several hours of exposure. The humidity at which the product neither loses nor gains moisture is the equilibrium relative humidity. Plots of such data yield water sorption isotherms

(Fig. 10.3). We can see from Fig. 10.3 that at 100°C and 40% relative humidity (RH), potato comes into equilibrium at 4% moisture; if we wish to dry it down to 2% moisture with 100°C air, then this air must be at about 15% RH. Similar water sorption isotherms have been established for a wide variety of food products and can be found in appropriate references. But for a new product, a mixture of ingredients, such as a dehydrated soup, or a new variety of fruit or vegetable, it usually is necessary to experimentally determine the isotherms for the specific product. With this information, the best temperature and humidity of the drying air can be selected. Equilibrium relative humidity data also are important when we consider storage of a dried product. If the food is packaged in a container that is not moisture-tight and is stored in an atmosphere above the dried food's equilibrium relative humidity, then the food will gradually pick up moisture and may cake or otherwise deteriorate.

Atmospheric Pressure and Vacuum

At a pressure of 1 atm (760 mm Hg) water boils at 100°C. As the pressure is lowered, the boiling temperature decreases. At constant temperature, a decrease in pressure increases the rate of boiling. Thus, food in a heated vacuum chamber will lose moisture at a lower temperature or at a faster rate than it would in a chamber at atmospheric pressure. Lower drying temperatures and shorter drying times are especially important in the case of heat-sensitive foods.

Evaporation and Temperature

As water evaporates from a surface, it cools the surface. This cooling is largely the result of absorption by the water of the latent heat of phase change from liquid to gas,

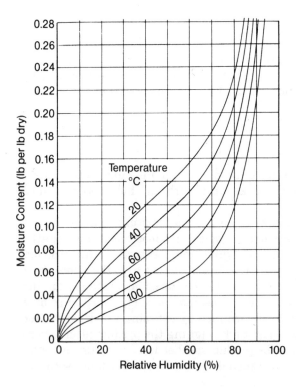

Figure 10.3. Water sorption isotherms for dried potatoes. *Courtesy of P. Görling.*

that is, the heat of vaporization going from water to water vapor. This heat is taken from the drying air or the heating surface and from the hot food, and so the food piece or droplet is cooled.

The same thing happens with a wet bulb thermometer. A sling psychrometer consists of two identical thermometers, except that the bulb of one is immersed in a wet wick. If we sling this around in the air to speed evaporation, the temperature of the wet bulb thermometer drops compared to the dry bulb thermometer if the relative humidity of the air is less than 100%. The wet bulb continues to be lower than the dry bulb as long as the wet bulb can give off moisture to the atmosphere. If the wick becomes dry, the temperature ceases to drop. If the air is at a high humidity, then the rate of evaporation and the amount of water vapor evaporated from the wet wick are less than that evaporated at lower humidity. The extent of the wet bulb temperature depression is a measure of relative humidity.

A particle or piece of solid food, or a droplet of liquid food, while it is being dehydrated acts as a wet bulb so long as it still contains free water. Regardless of the temperature of the drying air or heating surface, the temperature of the food will not be substantially higher than the temperature of a wet bulb so long as water is evaporating rapidly. Thus, in a spray drier the incoming air may be at 200°C and the exit air at perhaps 120°C, but a food particle while drying may be no higher than about 70°C. As the moisture content of the food particle decreases and evaporation slows down, the particle rises in temperature. When there is virtually no more free water, the food particle rises in temperature to that of the incoming air, and the exit air also approaches that temperature if there are no other heat losses through the drier. Because most foods are heat sensitive, they generally are removed from high-temperature driers before they reach the maximum temperatures possible or are exposed to the highest temperatures for only a very short time.

Unless heated specifically for the purpose, foods are not sterile at the end of dehydration. Although a large proportion of the microbial load is killed during most drying operations, many bacterial spores are not. This becomes still more significant if the dehydration method is designed to be gentle to protect delicate foods. In freeze-drying, for example, comparatively few microorganisms are destroyed, and indeed freeze-drying has been used for many years as a method of preserving the viability of bacterial cultures. The nonsterilizing aspects of food dehydration also apply to certain natural food enzymes that may survive drying conditions.

Time and Temperature

Since all important methods of food dehydration employ heat, and food constituents are sensitive to heat, compromises must be made between the maximum possible drying rate and maintenance of food quality. With few exceptions, drying processes that employ high temperatures for short times do less damage to food than drying processes employing lower temperatures for longer times. Thus, vegetable pieces dried in a properly designed oven in 4 h would retain greater quality than the same product sun-dried over 2 days. Several drying processes achieve dehydration in a matter of minutes or even less if the food is sufficiently subdivided.

Freeze-drying, which is discussed later in this chapter, may appear to contradict the high-temperature–short-time principle, since drying may take 8 h or more and still produce excellent quality. However, in this case the product is dried directly from the frozen state, and under such conditions there is little deterioration.

Normal Drying Curve

When foods are dried, they do not lose water at a constant rate all the way down to bone dryness. As drying progresses, the rate of water removal under any set of fixed conditions drops off. This is seen in Fig. 10.4 for carrot dice. In practice, if 90% of a product's water is removed in 4 h, it may require another 4 h to remove most of the remaining 10%. Since the removal rate becomes asymptotic, zero moisture is never reached under practical operating conditions. At the beginning of drying, and for some time thereafter, water generally continues to evaporate from a food piece at a rather constant rate, as if it were drying from a free surface. This is referred to as the constant rate period of drying; in Fig. 10.4 it extends for 4 h. This is followed by an inflection in the drying curve, which leads into the falling rate period of drying.

These changes during dehydration can be largely explained in terms of heat and mass transfer phenomena. A cube of food in the course of dehydration will lose moisture from its surfaces and gradually develop a thick dried layer, with remaining moisture largely confined to its center. From the center to the surface, a moisture gradient will be established. As a result, the outside dried layer will form an insulation barrier against rapid heat transfer into the food piece, especially since the evaporating water leaves air voids behind it. In addition to less driving force from decreased heat transfer, water remaining in the center has farther to travel to get out of the food piece than did surface moisture at the start of drying. Further, as the food dries, it approaches its normal equilibrium relative humidity. As it does, it begins to pick up molecules of water vapor from the drying atmosphere as fast as it loses them. When these rates are equal, drying ceases.

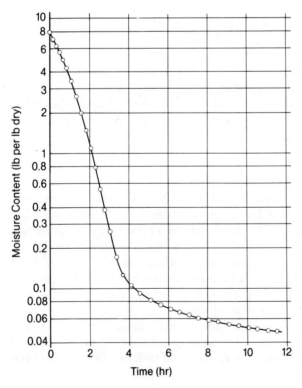

Figure 10.4. Changes in moisture content of diced carrot during dehydration. Source: *Van Arsdel et al., Food Dehydration, 2nd ed., Vol. 1, W.B. Van Arsdel, M.J. Coply, and A.I. Morgan, Jr. (Editors). AVI Publishing Co., Westport, CT, 1973.*

These are not the only food changes that contribute to the shape of the typical drying curve, although they are major factors. The precise shape of the normal drying curve varies with different food materials, for different types of driers, and in response to varying drying conditions such as temperature, humidity, air velocity, direction of the air, thickness of the food, and other factors. But the drying of most food materials generally shows periods of constant and falling rate, and the removal of water below about 2% without damage to the product is exceedingly difficult.

Effects of Food Properties on Dehydration

The physical factors affecting heat and mass transfer such as temperature, humidity, air velocity, surface area, and the like are usually relatively easy to optimize and control, and largely determine drier design. Far more subtle are the properties of food materials that may change during dehydration and affect drying rates and final product quality. In the case of food freezing, it was pointed out that various food properties affect heat transfer. The picture is more complex with regard to dehydration, since food raw material properties affect both heat and mass transfer, and both can have gross effects on characteristics of the dried products.

Constituent Orientation

Few foods approach homogeneity at the molecular level. A piece of meat, for example, will have lean and fat interlaced or marbled together. A piece of meat being dried will give up water at different rates in the regions of fat and lean, especially if the water must escape through a fat layer. This suggests that where fat occurs in layers, faster drying will occur if the meat is oriented relative to the source of heat so that moisture escapes in a line parallel to the layers of fat rather than having to pass through them. The same principle applies to layers of muscle fibers. The rate of drying will differ depending on whether orientation with respect to the heat source encourages moisture to escape parallel with or transverse to the stratification of muscle fibers. Parallel escape generally gives faster drying.

Constituent orientation also applies in food emulsions. If in a food piece or droplet, water is emulsified in oil so that the oil is the continuous phase and coats the moisture droplets, then dehydration should be slower than if the emulsion is reversed and water is the continuous phase. Sometimes this can be controlled in a manufactured food being dried, but more often we must take what nature gives us.

Solute Concentration

Solutes in solution elevate the boiling point of water systems. This occurs in food dehydration processes. Foods high in sugar or other low-molecular-weight solutes dry more slowly than foods low in these solubles. What is more, the concentration of solutes becomes greater in the remaining water as drying progresses. This is another factor that slows drying and contributes to the falling rate period in the drying of many foods.

Binding of Water

Water escapes freely from a surface when its vapor pressure is greater than the vapor pressure of the atmosphere above it. But as a product dries and its free water is progressively removed, the vapor pressure of a unit area of the product decreases. This is because there is less remaining water per unit volume and per unit area and also because some of the water is held or bound by chemical and physical forces to solid constituents of the food.

Free water is easiest to remove and evaporates first. Additional water may be loosely held by forces of adsorption to food solids. More difficult to remove is water that enters into colloidal gels such as when starch, pectin, or other gums are present. Still more difficult is the removal of chemically bound water in the form of hydrates (e.g., glucose monohydrate or hydrates of inorganic salts). These phenomena also contribute to the flattening of normal drying curves with time.

Cellular Structure

Solid foods of natural tissue have a cellular structure with moisture between and within the cells. When the tissue is alive, the cell walls and membranes hold moisture within the cells. Such cells have turgor, rather than exhibiting leakage or bleeding.

When an animal or plant is killed, its cells become more permeable to moisture. When the tissue is blanched or cooked, the cells may become still more permeable to moisture. Generally, cooked vegetables, meat, or fish will dry more easily than their fresh counterparts, provided cooking does not cause excessive toughening or shrinking.

Shrinkage, Case Hardening, Thermoplasticity

Even dead cells retain varying degrees of elasticity and will stretch or shrink under stress. If the stress is excessive, then their elastic limit is exceeded and they will not return to their original shape on removal of the stress. One of the most obvious changes during dehydration of cellular, as well as noncellular foods is shrinkage.

If moisture was removed evenly throughout the mass of a perfectly elastic material under turgor, then the material would shrink in an even linear fashion with removal of moisture. This uniform shrinkage is rarely seen in food materials being dehydrated since the food pieces generally do not have perfect elasticity and water is not removed evenly throughout the food piece as it is dried. Different food materials exhibit different shrinkage patterns in the course of dehydration. Typical changes of vegetable dice during dehydration are indicated in Fig. 10.5. The original piece before drying is represented in Fig. 10.5a. The effect of surface shrinkage is seen in Fig. 10.5b, where the edges and corners gradually pull in giving the cube a more rounded appearance in the early stages of drying. Continued dehydration gradually removes water from deeper and deeper layers and finally from the center. This causes continued shrinkage toward the center and the concave cube appearance as in Fig. 10.5c.

Often with quick high-temperature drying of food pieces, the surface becomes dry and rigid long before the center has dried out. Then when the center dries and shrinks, it pulls away from the rigid surface layers causing internal splits, voids, and honeycomb effects. Such differences in shrinkage patterns can affect the bulk density of the dried product, that is, the weight per unit volume. Products dried rapidly have a rigid, less concave surface, and more internal shrinkage and air voids. Products dried slowly are more concave and dense.

Both kinds of dried product have advantages and disadvantages. For example, a less dense product will absorb water and reconstitute quicker, is more attractive and more

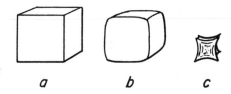

Figure 10.5. Changes in shape of diced vegetables during drying. *Courtesy of W. B. Van Arsdel.*

a   b   c

closely resembles the original material, and may be psychologically more acceptable to consumers, who frequently interpret greater volume as more substance even though weight is the same. On the other hand, a less dense product is more expensive to package, ship, and store, and because of its air voids, may be more easily oxidized or otherwise have shortened storage stability. A less dense product often is favored when it is to be sold directly to consumers, who value appearance and quick reconstitution of the product, whereas a more dense product frequently is preferred by food manufacturers who purchase dehydrated ingredients for further processing and have reconstitution kettles and mixers; processors are likely to be less concerned with reconstitution rate than with container, shipping, and storage costs.

A special condition related to shrinkage and sealing of the surface of a food piece is known as case hardening. This may occur when there is a very high surface temperature and unbalanced drying of the piece so that a dry skin forms quickly, before most of the internal moisture has had opportunity to migrate to the surface. The rather impermeable skin then traps much of the remaining water within the particle, and the drying rate drops off severely.

Case hardening is particularly common with foods that contain dissolved sugars and other solutes in high concentration. This can be explained from the various ways water may escape from a product during dehydration. Some of the water moves through cell walls and membranes of cellular foods by molecular diffusion. If the membranes are highly selective against solutes, the water will leave dissolved substances behind. Also, water may be heated to water vapor within a food piece and escape as water vapor molecules free of solute. But food pieces and food purees being dried also contain voids, cracks, and pores of various diameters down to minute capillary size. Water in foods rises in these pores and capillaries, many of which lead to the food surface. Capillary water carries sugars, salts, and other materials in solution to the surface of food pieces during dehydration. Then at the surface, the water is evaporated and the solutes are deposited. This is what causes a sticky, sugary exudate on the surface of some fruits in the early stages of drying. This can seal off the surface pores and cracks, which also are shrinking during drying. The combined effects of shrinking and pore clogging from solutes contributes to case hardening. Where case hardening is a problem, it generally can be minimized by lower surface temperatures to promote a more gradual drying throughout the food piece.

Many foods are thermoplastic, that is, they soften on heating. A cellular food, such as plant and animal tissue, has structure and some rigidity even at drying temperatures. A fruit or vegetable juice, on the other hand, lacks structure and is high in sugars and other materials that soften and melt at the drying temperature. Thus, if orange juice or a sugar syrup is dried on a pan or on a heated belt, even after all of the water has been removed the solids will be in a thermoplastic tacky condition, giving the impression that they still contain moisture. They also will stick to the pan or belt and be difficult to remove. However, on cooling, the thermoplastic solids harden into a crystalline or amorphous glass form. In this more brittle condition they generally are more easily removed from the pan or belt. Most belt type driers are equipped with a cooling zone just prior to a scraper knife to facilitate removal of this type of material from the drier.

Porosity

Many drying techniques or treatments given to food before drying are aimed at making the structure more porous so as to facilitate mass transfer and thereby speed drying rate. But in some instances, even though potential mass transfer rates are increased by puffing or otherwise opening the structure, the drying rate is not increased. Porous spongelike structures are excellent insulating bodies and will slow down the rate of heat transfer into the food. The net result depends on whether the change in porosity has a greater effect on the rate of mass transfer or heat transfer in the particular food material and drying system.

Porosity may be developed by creating steam pressure within a product during drying. The escaping steam tends to puff such a product as in the case of the potato puffs of Fig. 10.6. Porosity can be developed also by whipping or foaming a food liquid or puree prior to drying. A stable foam that resists collapse during drying is then

Figure 10.6. Potato puffs made under conditions of rapid drying and internal steam escape. *Courtesy of U.S. Department of Agriculture.*

desired. Porosity can be developed in a vacuum drier by rapid escape of water vapor into the high vacuum, and by still other means.

Quite apart from its effect on drying rate, any process that retains or creates a highly porous structure does many of the other things discussed in relation to internal voids. A porous product has the advantages of quick solubility or reconstitution and greater volume appearance, but the disadvantages of increased bulk and generally shorter storage stability because of increased surface exposure to air, light, and so on.

Chemical Changes

A great range of chemical changes can take place during food dehydration along with the physical changes already described, and these contribute to the final quality of both the dried items and their reconstituted counterparts in terms of food color, flavor, texture, viscosity, reconstitution rate, nutritional value, and storage stability. These changes frequently are product-specific, but a few major types occur in virtually all foods undergoing dehydration. The extent of these changes depends on the composition of the food and the severity of the drying method.

Browning reactions may be caused by enzymatic oxidations of polyphenols and other susceptible compounds if the oxidizing enzymes are not inactivated. Drying temperatures, because of the water evaporation cooling effect, often are not sufficient to inactivate these enzymes during drying, so it is common to pasteurize or blanch foods with heat or chemicals prior to drying. Caramelization of sugars and scorching of other materials if heat is excessive is another common type of browning. Highly important in food dehydration are nonenzymatic or Maillard browning products from the reaction of aldehydes and amino groups of sugars and proteins. Maillard-type browning, like other chemical reactions, is favored by high temperature and by high concentrations of reactive groups in the presence of some water. In the course of dehydration, reactive groups are concentrated. Maillard browning generally proceeds most rapidly during drying when the moisture content is decreased to the range of about 20–15%. As moisture content drops further, the rate of Maillard browning slows, so that in dried products below 2% moisture further color change from this kind of browning is minimally perceptible even on long-range storage. Drying systems or heating schedules generally are designed to dehydrate rapidly through the 20–15% moisture range so as to minimize time for Maillard browning at this optimum condition.

Another common consequence of dehydration is some loss in the ease of rehydration. Some of this is caused by physical shrinkage and distortion of cells and capillaries, but much also results from chemical or physicochemical changes at the colloidal level. Heat and the salt concentration effects from water removal can partially denature proteins, which cannot then fully reabsorb and bind water. Starches and gums also may be altered and become less hydrophilic. Sugars and salts escape from damaged cells into the water used to reconstitute dehydrated foods, resulting in loss of turgor. These and other chemical changes make reabsorption of water by dried products somewhat less than equal to the original water content and contribute to altered texture.

Still another common chemical change associated with dehydration is some loss of volatile flavor constituents. This invariably occurs to at least a slight degree. Complete prevention of flavor loss has as yet proven virtually impossible, and so methods of trapping and condensing the evolved vapors from the drier and adding them back to the dried product are sometimes employed. Additional techniques involve addition to dried products of essences and flavor preparations derived from other sources, as well as methods of minimizing flavor loss by incorporating gums and other materials into certain liquid foods prior to drying. Some of these materials have flavor fixative properties; others work by coating dried particles and providing a physical barrier against loss of volatile substances.

Optimization of Variables

In the design of food dehydration equipment, efforts are made to produce maximum drying rate with minimum product damage at the most economical drying cost. This requires a balancing of the various factors discussed so far; food dehydration is truly an area where the food scientist and engineer must work together to achieve optimum results.

Mathematical relationships exist between each of the major controllable drying variables and heat and mass transfer. Because of the peculiarities of food materials, optimum drying conditions are seldom the same for two different products. Engineering calculations based on model systems can go a long way toward selecting favorable drying conditions, but seldom are sufficient in themselves to accurately predict drying behavior. This is because food materials are highly variable in initial composition, in amounts of free and bound water, in shrinkage and solute migration patterns, and, most important, in how properties change throughout the drying operation. This is especially so in the falling rate period of the drying curve, where quality and economics are most affected. For these reasons, in selecting and optimizing a drying process, experimental tests with the food to be dried must always supplement engineering calculations based on less variable model systems.

Drying Methods and Equipment

There are several basic drying methods and a far greater number of modifications of the basic methods. The method of choice depends on the type of food to be dried, the quality level that must be achieved, and the cost that can be justified. Because orange juice crystals command a much higher price than starch, a processor can afford to use a more delicate and generally more expensive drying method to dehydrate orange juice, which needs a milder drying method since it is far more sensitive than starch.

Some of the more common drying methods include drum drying, spray drying, vacuum shelf drying, vacuum belt drying, atmospheric belt drying, freeze-drying, fluidized-bed drying, rotary drying, cabinet drying, kiln drying, tunnel drying, and others. Some of these methods are particularly suited to liquid foods and cannot handle solid food pieces; others are suitable for solid foods or mixtures containing food pieces.

One useful division of drier types separates them into air convection driers, drum or roller driers, and vacuum driers. Using this breakdown, Table 10.1 indicates the suitability of more common drier types for liquid and solid foods. In air convection driers, heated air is put into intimate contact with the food material and supplies a major source of the heat for evaporation. If liquid, the food may be sprayed or poured into pans or on belts. Pieces may be supported in any number of ways. Although heated moving air is common to this group of driers, additional heat also may be supplied by heated tray or belt supports. Drum or roller driers are limited to use with purees, mashes, and liquid foods that can be applied as thin films. Vacuum driers may employ any degree of vacuum to lower the boiling point of water. Freeze-driers are special

Table 10.1. Common Drier Types Used for Liquid and Solid Foods

Drier Type	Usual Food Type
Air convection driers	
kiln	Pieces
cabinet, tray, or pan	Pieces, purees, liquids
tunnel	Pieces
continuous conveyor belt	Purees, liquids
belt trough	Pieces
air lift	Small pieces, granules
fluidized bed	Small pieces, granules
spray	Liquids, purees
Drum or roller driers	
atmospheric	Purees, liquids
vacuum	Purees, liquids
Vacuum driers	
vacuum shelf	Pieces, purees, liquids
vacuum belt	Purees, liquids
freeze-driers	Pieces, liquids

kinds of vacuum driers generally operated at extremely low internal pressures so as to sublime water vapor directly from ice without going through the liquid phase. This classification is not rigid, since many driers are combinations. Thus, we can place a drum drier in a vacuum chamber or blow high-velocity heated air over the drum to speed drying; both practices are done commercially.

Air Convection Driers

All air convection driers have some sort of insulated enclosure, a means of circulating air through the enclosure, and a means of heating this air. They also have various means of product support and special devices for collecting dried product; some have air driers to lower drying air humidity. Movement of air generally is controlled by fans, blowers, and baffles. Air volume and velocity affect drying rate, but its static pressure also is important since products being dried become very light and can be blown off trays or belts. Airflow patterns are complex when they encounter surfaces, and their velocities and pressures in contact with food are seldom comparable with measurements made on the main airstream, but such measurements usually can be correlated with drying behavior. Even if two driers have an air velocity of 5 m/sec (1000 ft/min), the surface of the food in the two driers probably encounters different velocities when the driers are of different geometries.

The air may be heated by direct or indirect methods. In direct heating the air is in direct contact with a flame or combustion gases. In indirect heating the air is in contact with a hot surface, such as pipes or fins heated by steam, flame, or electricity. The important point is that indirect heating leaves the air uncontaminated. On the other hand, in direct heating the fuel is seldom completely oxidized to carbon dioxide and water. Incomplete combustion leaves gases and traces of soot, which are picked up by the air and can be transferred to the food product. Direct heating of air also contributes small amounts of moisture to the air since moisture is a product of combustion, but this is usually insignificant except with very hygroscopic foods. These disadvantages are balanced by the generally lower cost of direct heating of air compared to indirect heating, and both methods are widely used in food dehydration.

Kiln Drier.   One of the simplest kinds of air convection drier is the kiln drier. Kiln driers of early design were two-story constructions. A furnace or burner on the lower floor generated heat, and warm air would rise through a slotted floor to the upper story. Foods such as apple slices would be spread out on the slotted floor and turned over periodically. This kind of drier will not reduce moisture to below about 10%. It is still in use for apple slices.

Cabinet, Tray, and Pan Driers.   A step more advanced is the cabinet drier in which food may be loaded on trays or pans in comparatively thin layers up to a few centimeters. A typical construction for this type of drier is shown in Fig. 10.7. Fresh air enters the cabinet (B), is drawn by the fan through the heater coils (C), and is then blown across the food trays to exhaust (H). In this case, the air is heated by the indirect method. Screens filter out any dust that may be in the air. The air passes across and between the trays in this design. Other designs have perforated trays and the air may be directed up through these. In Fig. 10.7, the air is exhausted to the atmosphere after one pass rather than being recirculated within the system. Recirculation is used to conserve heat energy by reusing part of the warm air. In recirculating designs, moist air, after evaporating water from the food, may have to be dried before being recirculated to prevent saturation and slowing down of subsequent drying. In such a case, this air could be dried by passing through a desiccant such as a bed of silica gel, or the moisture could be condensed out by passing the moist air over cold plates or coils. But when the exhaust air is not dried for recirculation, then the exhaust vent should not be close to the fresh air intake area, otherwise the moist exhaust air will be drawn back through the drier and drying efficiency will be lost.

Cabinet, tray, and pan driers are usually for small-scale operations. They are comparatively inexpensive and easy to set in terms of drying conditions. They may run up to 25 trays high and operate with air temperatures of about 95°C dry bulb and with air velocities of about 2.5–5 m/sec across the trays. They commonly are used to dry

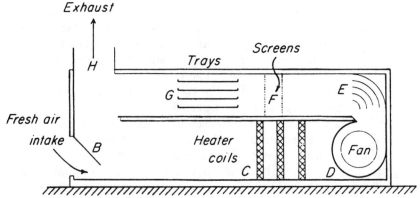

Figure 10.7. One type of cabinet or tray drier. Source: *Van Arsdel et al., Food Dehydration, 2nd ed., Vol. 1, W.B. Van Arsdel, M.J. Copley, and A.I. Morgan, Jr. (Editors). AVI Publishing Co., Westport, CT, 1973.*

fruit and vegetable pieces, and depending on the food and the desired final moisture, drying time may be of the order of 10 or even 20 h.

Tunnel and Continuous Belt Driers. For larger operations, tunnel driers with elongated cabinets, through which trays on carts pass, are used (Fig. 10.8). If drying time to the desired moisture is 10 h, each wheeled cart of trays will take 10 h to pass through the tunnel. When a dry cart emerges, it makes room to load another wet cart into the opposite end of the tunnel. Such an operation then becomes semicontinuous.

A main construction feature by which tunnel driers differ has to do with the direction of airflow relative to tray movement. In the drier shown in Fig. 10.8, wet food carts move from left to right. The drying air moves across the trays from right to left. This is a counterflow, or countercurrent, pattern in which the hottest and driest air contacts the nearly dry product, whereas the initial drying of entering carts gets cooler, moister air that has cooled and picked up moisture going through the tunnel. This means that initial product temperature and moisture gradients will not be as great, and the product is less likely to undergo case hardening or other surface shrinkage, leaving wet centers. Further, lower final moisture can be reached because the driest product encounters the driest air. In contrast, cocurrent flow tunnels have the incoming trays and incoming hottest driest air traveling in the same direction. In this case, rapid initial drying and slow final drying can cause case hardening and internal splits and porosity as centers finally dry, which sometimes is desirable in special products.

Just as carts of trays can be moved through a heated tunnel, so a continuous belt may be driven through a tunnel or oven enclosure. This approach is used in a continuous belt or conveyor drier, and a great number of designs are possible. Some of the more common features are uniform automatic feeding of product to the belt in a controlled thin layer, zoned heat and airflow control in different sections, tumbling over of product onto a second strand of belt, automatic collection of dried product, and, of course, continuous operation. The drying capacity of such driers generally is stated in terms of weight of product dried from one moisture level to another per square meter of belt surface per hour. This also can be expressed in terms of kilograms of water removed per square meter of belt surface per hour under defined operating conditions.

Belt Trough Drier. A special kind of air convection belt drier is the belt trough drier in which the belt forms a trough. The belt is usually of metal mesh, and heated air is blown up through the mesh. The belt moves continuously, keeping the food pieces in the trough in constant motion to continuously expose new surface. This speeds drying, and with air of about 135°C, vegetable pieces may be dried to 7–5% moisture in about 1 h.

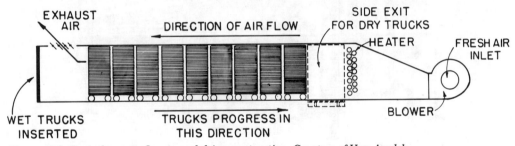

Figure 10.8. Typical counterflow tunnel drier construction. *Courtesy of Van Arsdel.*

But not all products may be dried this way since certain sizes and shapes do not readily tumble. Fragile apple wedges may break. Onion slices tend to separate and become entangled. Fruit pieces that exude sugar on drying tend to stick together and clump with the tumbling motion. These are but a few additional factors that must be considered in selecting a drier for a particular food.

Air Lift Drier.   Several types of pneumatic conveyor driers go a step beyond tumbling to expose more surface area of food particles. These generally are used to finish-dry materials that have been partially dried by other methods, usually to about 25% moisture, or at least sufficiently low so that the material becomes granular rather than having a tendency to clump and mat. One type of air lift drier is illustrated in Fig. 10.9. This might be used to finish-dry semimoist granules coming from a drum

Figure 10.9. One type of air lift drier. *Courtesy of Reitz Manufacturing Co.*

drier. Such granules at about 25% moisture can be brought to about 6% moisture more efficiently in a heated airstream than on the drum. This is because the more difficult moisture to remove in this falling rate period of dehydration is more easily evaporated from suspended particles in intimate contact with the heating medium. The suspended particles when dry are separated from the air and collected in a cyclone-type separator, which is described in the subsection on spray driers.

Fluidized-Bed Drier.   Another type of pneumatic conveyor drier is the fluidized-bed drier. This is similar in principle and construction to the fluidized-bed freezer described in chapter 9. In fluidized-bed drying (Fig. 10.10), heated air is blown up through the food particles with just enough force to suspend the particles in a gentle boiling motion. Semidry particles such as potato granules enter at the left and gradually migrate to the right, where they are discharged dry. Heated air is introduced through a porous plate that supports the bed of granules. The moist air is exhausted at the top. The process is continuous and the length of time particles remain in the drier can be regulated by the depth of the bed and other means. This type of drying can be used to dehydrate grains, peas, and other particulates.

Spray Driers.   By far the most important kind of air convection drier is the spray drier. Spray driers turn out a greater tonnage of dehydrated food products than all other kinds of driers combined. There are various types of spray driers designed for specific food products. Spray driers are limited to foods that can be atomized, such as liquids and low-viscosity pastes and purees. Atomization into minute droplets results in drying in a matter of seconds with common inlet air temperatures of about 200°C. Since evaporative cooling seldom permits particles to get warmer than about 80°C and

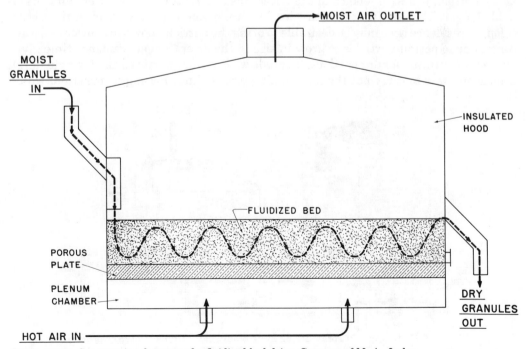

Figure 10.10. Construction features of a fluidized-bed drier. *Courtesy of M. A. Joslyn.*

properly designed systems quickly remove the dried particles from heated zones, this method of dehydration can produce exceptionally high quality with many highly heat-sensitive materials, including milk, eggs, and coffee.

In typical spray drying, the liquid food is introduced as a fine spray or mist into a tower or chamber along with heated air. As the small droplets make intimate contact with the heated air, they flash off their moisture, become small particles, and drop to the bottom of the tower from where they are removed. The heated air, which has now become moist, is withdrawn from the tower by a blower or fan. The process is continuous in that liquid food continues to be pumped into the chamber and atomized, along with dry heated air to replace the moist air that is withdrawn, and the dried product is removed from the chamber as it descends.

The principal components of a spray drying system differ in construction depending on the product to be dried. In the case of milk, the system includes tanks for holding the liquid, a high-pressure pump for introducing the liquid into the tower, spray nozzles or a similar device for atomizing the milk, a heated air source with blower, a secondary collection vessel for accumulating product drawn from the tower, and means for exhausting the moistened air (Fig. 10.11).

The main purpose of the drying tower or chamber is to provide intimate mixing of heated air with finely dispersed droplets. In the various spray driers shown in Fig. 10.12, the heated air and the atomized droplets may enter the tower together at the top or bottom or may enter separately, the particles may be made to descend straight down or take a spiral path, and the chamber may be vertical or horizontal.

As in tunnel driers, introduction of droplets and air in the same direction results in quick initial drying and slower final drying; countercurrent streams may be favored for highly hygroscopic materials. Further, if a liquid product is introduced at the top of the tower, it descends through and out of the tower in one pass; if product is introduced at the bottom, it first ascends and then descends and its time in the drier can thus be made longer. This also is true if the droplets are given a spiral motion in the tower. A longer residence time may be desirable to bring the particles down to a lower moisture content or to permit particles to grow in size in the drier (longer residence time gives greater opportunity for dry particles to collide with less dry particles and form clusters). This is one way to carry out the instantizing process known as agglomeration, which

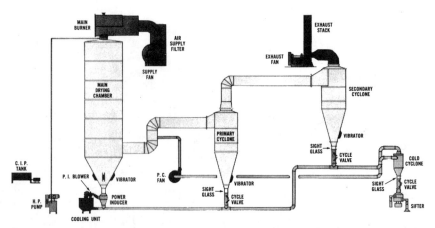

Figure 10.11. Diagram of spray drying system suitable for milk. *Courtesy of the De Laval Separator Co.*

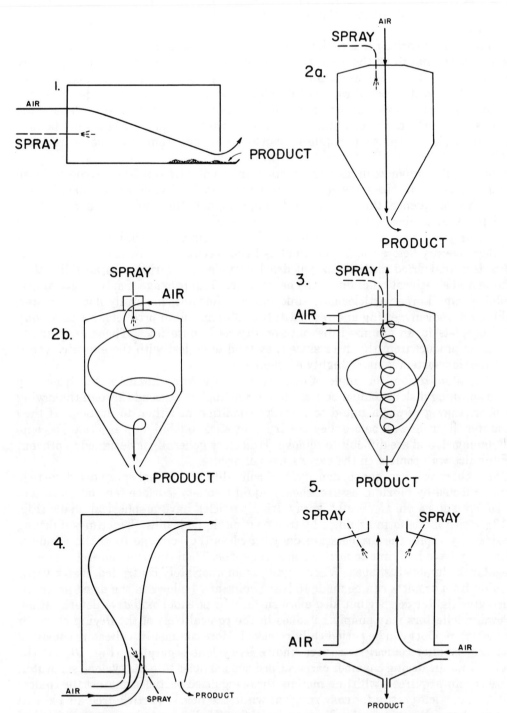

Figure 10.12. Some types of spray driers. Source: *Brown et al., Food Dehydration, 2nd ed., Vol. 1, W.B. Van Arsdel, M.J. Copley, and A.I. Morgan, Jr., (Editors). AVI Publishing Co., Westport, CT, 1973.*

yields clusters that have many voids, sink in water, and are therefore easier to dissolve than certain spray dried particles which are small in size, float on water, and are difficult to wet.

As important to dried product characteristics as the geometry and air pattern in the chamber is the nature of the atomization. Atomizers are of two main types: pressure spray nozzles and centrifugal spinning disks, or baskets. Spinning disks and baskets, from which deposited food throws out droplets, are favored where passage through a fine-hole pressure nozzle can damage the food, as might be the case in denaturing proteins of egg white. Viscous liquids and purees with fine pulp also may not be able to pass through a fine pressure nozzle but can be easily spun from a high-speed, rotating disk.

Small droplets promote quick drying, and uniform droplet size is necessary for even drying. Actually, the size and trajectory of the largest droplets determine drying time and, as a consequence, the size of the drying chamber. No atomizers have yet been developed that produce all droplets of the same size, but the object of their design is to make droplet size as uniform as possible. If not uniform, the small droplets dry first and then overdry before the larger droplets have become dry. The droplet size determines the final dried particle size; if dried particle size varies substantially, then settling and stratification of fines may occur in the final package. Particle size affects solubility rate. Large particles may sink and very fine ones generally float on water, making for uneven wetting and reconstitution of nonuniform products. Further, very small droplets in an atomized distribution dry as minute fines. These are hard to recover as product from the drier since they tend to be lost with the exit air even if the collection system is made highly efficient.

During atomization, the angle of departure from a spray nozzle or the trajectory from a spinning disk also must be considered. As droplets descend through the drying chamber, they go from a liquid to a sticky condition and then to dryness. If they encounter the drier wall before they are dry, they stick and build up as a cake, become heat damaged, and are difficult to remove. Trajectory generally is designed to prevent or minimize wall contact in the early stages of drying.

The appearance, size, shape, density, and solubility of the final spray dried particle can be affected by nozzle pressure, shear, liquid viscosity, surface tension, nature of the solids, and so on. Generally, spray dried particles have a spherical shape (Fig. 10.13), which is the form assumed by free-floating liquid bodies. Sometimes if drying is extremely rapid, the droplets are dehydrated as they emerge from the atomizer before they have had time to form a spherical shape. Then the dried particles may be irregular or dumbbell shaped. When drying is appropriately controlled, water vapor escaping from droplets can be made to leave voids and hollows in the dried particles, which give lighter density but also more surface for possible oxidative deterioration.

Powder collectors may simply be zones in the conical base of the drying chamber from which product can be periodically removed. More commonly, collectors consist of secondary smaller conical structures known as cyclone separators (Fig. 10.11). The exit air from the drying chamber carries the dried particles into the cyclone separator, where the air acquires a whirling motion, throwing dried particles against the conical wall. The particles settle for easy removal while the nearly particle-free air exits at the top. Since the exit air is never entirely free of fine particles, another kind of collector may be employed above the cyclone. This is a bag collector or filter just preceding air exhaust to the atmosphere. Product fines remaining in the bag collector

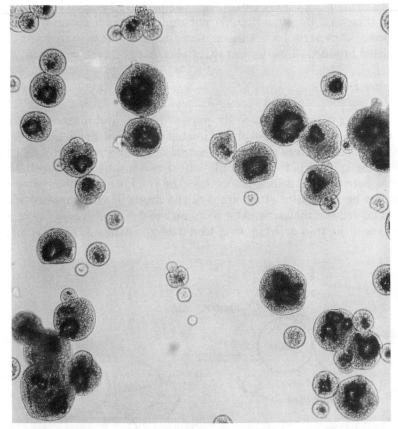

Figure 10.13. Photomicrograph of spray dried whole milk particles × 110. *Courtesy of Coulter and Jenness.*

for long periods exposed to heated exiting air generally become heat damaged and represent lower quality product.

One type of spray drying foams liquid food, such as milk or coffee, before spraying it into the drier. The result is a faster drying rate from the expanded foamed-droplet surface area and lighter-density dried product. This is known as foam-spraying drying.

It was stated that when particles are dry, they do not stick to the drier wall. An exception is thermoplastic substances such as juices high in sugar. Even when dry, these melt, stick, and build up on the wall. One kind of spray drier has a double wall and circulates cold water or cool air so as to chill the lower portion of the inner wall where dried juice particles would accumulate. Thus prevented from melting and fusing, these juices too may be spray dried and collected in particulate form.

Another type of spray drier has been developed especially to handle thermoplastic materials and other highly heat-sensitive foods. This is known as the BIRS spray drier. The BIRS drier uses countercurrent cool, dry air of about 30°C and 3% RH. To give the droplets sprayed in at the top of the tower sufficient time to dry at this relatively low temperature, the drying tower is built exceptionally tall. It may be 67 m high and 15 m in diameter. As droplets descend, they dry in about 90 sec. Products like orange, lemon, and tomato juices, otherwise difficult to spray dry because of thermoplasticity, can be dried this way. Because there is not rapid escape of steam from particles in this cool process, such particles are less puffed and more dense than many conventionally spray dried products. Low temperature also favors flavor retention.

Drum or Roller Driers

In drum or roller drying, liquid foods, purees, pastes, and mashes are applied in a thin layer onto the surface of a revolving heated drum. The drum generally is heated from within by steam. Driers may have a single drum or a pair of drums (Fig. 10.14). The food may be applied between the nip where two drums come together, and then the clearance between the drums determines the thickness of the applied food layer; or the food can be applied to other areas of the drum. Food is applied continuously and the thin layer loses moisture. At a point on the drum or drums a scraper blade is positioned to peel the thin dried layer of food from the drums. The speed of the drums

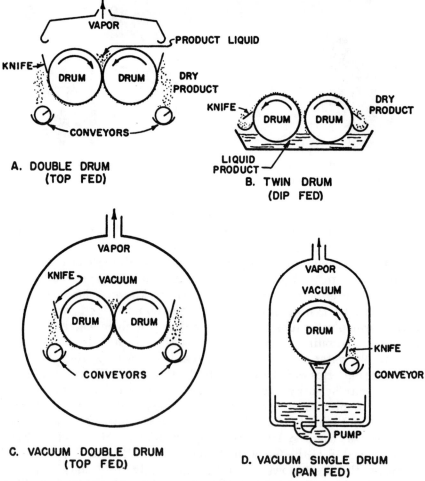

Figure 10.14. Several types of drum driers. Source: *Hall and Hendrick, Drying of Milk and Milk Products*, 2nd ed. AVI Publishing Co., Westport, CT, 1971.

is so regulated that the layer of food will be dry when it reaches the scraper blade, which also is referred to as a doctor blade. The layer of food is dried in one revolution of the drum and is scraped from the drum before that position of the drum returns to the point where more wet food is applied. Using steam under pressure in the drum, the temperature of the drum surface may be well above 100°C, and often is held at about 150°C. With a food layer thickness commonly less than 2 mm, drying can be completed in 1 min or less, depending on the food material. Other features of drum driers include hoods above drums to withdraw moisture vapor and conveyors in troughs to receive and move dried product.

Typical products dried on drums include milk, potato mash, heat-tolerant purees such as tomato paste, and animal feeds. But drum drying has some inherent limitations that restrict the kinds of foods to which it is applicable. To achieve rapid drying, drum surface temperature must be high, usually above 120°C. This gives products a more cooked flavor and color than when they are dried at a lower temperature. Drying temperature can, of course, be lowered by constructing the drums within a vacuum chamber (Fig. 10.14c,d), but this increases equipment and operating costs over atmospheric drum or spray drying.

A second limitation is the difficulty of providing zoned temperature control needed to vary the drying temperature profile. This is particularly important with thermoplastic food materials. Whereas dried milk and dried potato are easily scraped from the hot drum in brittle sheet form, this is not possible with many dried fruits, juices, and other products which tend to be sticky and semimolten when hot. Such products tend to crimp, roll up, and otherwise accumulate and stick to the doctor blade in a taffylike mass.

This condition can be substantially improved by a cold zone to make the tacky material brittle just prior to the doctor blade. But zone-controlled chilling is not as easy to accomplish on a drum of limited diameter, and therefore limited arc, as it would be in perhaps 6 m of length of a horizontal drying belt 45 m long. One means of chilling is by directing a stream of cool air onto a segment of the product on the drum prior to the doctor blade. A system for doing this and providing additional zoned temperature control around the drum is shown in Fig. 10.15.

For relatively heat-resistant food products, drum drying is one of the least expensive dehydration methods. Drum dried foods generally have a somewhat more "cooked" character than the same materials spray dried; thus drum dried milk is not up to beverage quality but is satisfactory as an ingredient in less delicately flavored manufactured foods. More gentle vacuum drum drying or zone-controlled drum drying increases dehydration costs.

Vacuum Driers

Vacuum dehydration methods are capable of producing the highest quality dried products, but costs of vacuum drying generally also are higher than other methods which do not employ vacuum. In vacuum drying, the temperature of the food and the rate of water removal are controlled by regulating the degree of vacuum and the intensity of heat input. Heat transfer to the food is largely by conduction and radiation. Vacuum drying methods usually can be controlled with a higher degree of accuracy than methods depending on air convection heating.

All vacuum drying systems have four essential elements: a vacuum chamber of heavy construction to withstand outside air pressures that may exceed internal pressure by

Figure 10.15. Double-drum drier with zoned temperature control. *Courtesy of Jones Div., Beloit Corp.*

as much as 9800 kg/m^2 (2000 lb/ft^2); a heat supply; a device for producing and maintaining the vacuum; and components to collect water vapor as it is evaporated from the food. Typical arrangements of these elements are shown in Fig. 10.16.

The vacuum chamber generally contains shelves or other supports to hold the food; these shelves may be heated electrically or by circulating a heated fluid through them. The heated shelves are called platens. The platens convey heat to the food in contact with them by conduction, but where several platens are arranged one above another, they also radiate heat to the food on the platen below. In addition, special radiant heat sources such as infrared elements can be focused on the food to supplement the heat conducted from platen contact.

The device for producing and maintaining vacuum is outside the vacuum chamber and may be a mechanical vacuum pump or a steam ejector. A steam ejector is a kind of aspirator in which high-velocity steam jetting past an opening draws air and water vapor from the vacuum chamber by the same principle that makes an insect spray gun draw fluid from the can.

The means of collecting water vapor may be a cold wall condenser. It may be inside the vacuum chamber or outside the chamber but must come ahead of the vacuum pump so as to prevent water vapor from entering and fouling the pump. When a steam ejector is used to produce the vacuum, the same steam ejector can condense water vapor as it is drawn along with the air from the vacuum chamber, and so a cold wall vapor condenser may not be needed except where a very high degree of efficiency is required. In Fig. 10.16, the system at the top employs steam ejectors connected to the vacuum chamber; the middle system uses a refrigerated condenser and vacuum pumps; the lower system employs a refrigerated condenser and steam ejectors.

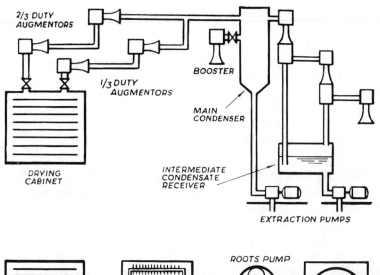

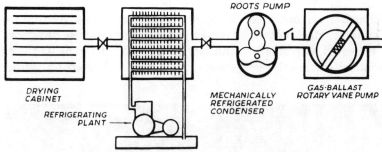

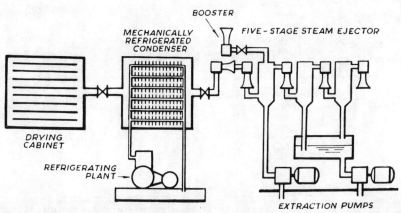

Figure 10.16. Elements of vacuum dehydration systems. *Courtesy of Columbine Press, U.K., A/S Atlas, Denmark.*

Atmospheric pressure at sea level is approximately 15 psi, or sufficient pressure to support a 30-in. column of mercury. This is also stated as 760 mm of mercury (Hg) or 760 torr. At 1 atmo, pure water boils at 100°C; at 250 mm Hg, pure water boils at 72°C; at 50 mm Hg, pure water boils at 38°C. High-vacuum dehydration operates at still lower pressures such as fractions of a millimeter of mercury. Freeze-drying generally will operate in the range of 2 mm to about 0.1 mm Hg.

Vacuum Shelf Driers.   One of the simplest kinds of vacuum driers is the batch-type vacuum shelf drier (Fig. 10.17). If liquids such as concentrated fruit juices are dried above about 5 mm Hg, the juice boils and splatters, but in the range of about 3 mm Hg and below, the concentrated juice puffs as it loses water vapor. The dehydrated juice then retains the puffed spongy structure seen in Fig. 10.17. Since temperatures well below 40°C can be used, in addition to quick solubility there is minimum flavor change or other kinds of heat damage. A vacuum shelf drier is also suitable for the dehydration of food pieces. In this case, the rigidity of the solid food prevents major puffing, although there also is a tendency to minimize shrinkage.

Continuous Vacuum Belt Drier.   Vacuum driers can be engineered for continuous operation. A diagram of a continuous vacuum belt drier is shown in Fig. 10.18. This drier is used commercially to dehydrate high quality citrus juice crystals, instant tea, and other delicate liquid foods.

The drier consists of a horizontal tanklike chamber connected to a vacuum-producing, moisture-condensing system. The chamber is about 17 m long and 3.7 m in diameter. Within the chamber are mounted two revolving hollow drums. Around the drums is connected a stainless steel belt which moves in a counterclockwise direction. The drum on the right is heated with steam confined within it. This drum heats the belt passing over it by conduction. As the belt moves, it is further heated by infrared radiant

Figure 10.17. Batch-type vacuum shelf drier. Source: *Ponting et al., Food Dehydration, 2nd ed., Vol. 2, W.B. Van Ardel, M.J. Copley, and A.I. Morgan, Jr. (Editors). AVI Publishing Co., Westport, CT, 1973.*

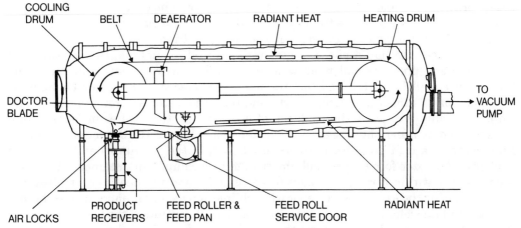

Figure 10.18. Continuous vacuum belt drier. *Courtesy of Votator Div., Chemetron Corp.*

elements. The drum to the left is cooled with cold water circulated within it and cools the belt passing over it. The liquid food in the form of a concentrate is pumped into a feed pan under the lower belt strand. An applicator roller dipping into the liquid continuously applies a thin coating of the food onto the lower surface of the moving belt. As the belt moves over the heating drum and past the radiant heaters, the food rapidly dries in the vacuum equivalent to about 2 mm Hg. When the food reaches the cooling drum, it is down to about 2% moisture. At the bottom of the cooling drum is a doctor blade which scrapes the cooled, embrittled product into the collection vessel. The belt scraped free of product receives additional liquid food as it passes the applicator roller and the process repeats in continuous fashion.

Products dried with this equipment have a slightly puffed structure. If desired, a greater degree of puffing can be achieved. This has been done in the case of milk by pumping nitrogen gas under pressure into the milk prior to drying. Some of the gas goes into solution in the milk. Upon entering the vacuum chamber this gas comes out of solution violently and further puffs the milk as it is being dried.

Freeze-Drying. Freeze-drying has been developed to a highly advanced state. Much of the development work has been aimed at optimizing the process and equipment to reduce drying costs, which still may be two to five times greater per weight of water removed than other common drying methods. Freeze-drying can be used to dehydrate sensitive, high-value liquid foods such as coffee and juices, but it is especially suited to drying solid foods of high value such as strawberries, whole shrimp, chicken dice, mushroom slices, and sometimes food pieces as large as steaks and chops. These types of food, in addition to having delicate flavors and colors, have textural and appearance attributes that cannot be well preserved by any current drying method except freeze-drying. A whole strawberry, for example, is soft, fragile, and almost all water. Any conventional drying method that employs heat would cause considerable shrinkage, distortion, and loss of natural strawberry texture. Upon reconstitution, such a dried strawberry would not have the natural color, flavor, or turgor and would be more like a strawberry preserve or jam. This can be largely prevented by drying from the solidly frozen state, so that in addition to low temperature, the frozen food has little chance to shrink or distort while giving up its moisture.

Food Science

The principle behind freeze-drying is that under certain conditions of low vapor pressure, water can evaporate from ice without the ice melting. When a material can exist as a solid, a liquid, and a gas but goes directly from a solid to a gas without passing through the liquid phase, the material is said to sublime. Dry ice sublimes at atmospheric pressure and room temperature. Frozen water will sublime if the temperature is 0°C or below and the frozen water is placed in a vacuum chamber at a pressure of 4.7 mm or less. Under such conditions the water will remain frozen, and water molecules will leave the ice block at a faster rate than water molecules from the surrounding atmosphere reenter the frozen block. Figure 10.19 is a diagrammatic illustration of a food piece being freeze-dried. Within the vacuum chamber, heat is applied to the frozen food to speed sublimation. If the vacuum is maintained sufficiently high, usually within a range of about 0.1–2 mm Hg, and the heat is controlled just short of melting the ice, moisture vapor will sublime at a near maximum rate. Sublimation takes place from the surface of the ice, and so as it continues, the ice front recedes toward the center of the food piece; that is, the food dries from the surface inward. Finally, the last of the ice sublimes and the food is below 5% moisture. Since the frozen food remains rigid during sublimation, escaping water molecules leave voids behind them, resulting in a porous spongelike dried structure. Thus, freeze-dried foods reconstitute rapidly but also must be protected from ready absorption of atmospheric moisture and oxygen by proper packaging.

In Fig. 10.19, a heating plate is positioned above and below the food to increase the heat transfer rate, but an open space is left with expanded metal so as not to seal off escape of sublimed water molecules. Nevertheless, as drying progresses and the ice front recedes, the drying rate drops off for several reasons. The porous dried layer ahead of the receding ice layer acts as an effective insulator against further heat transfer and slows the rate of escape of water molecules subliming from the ice surface.

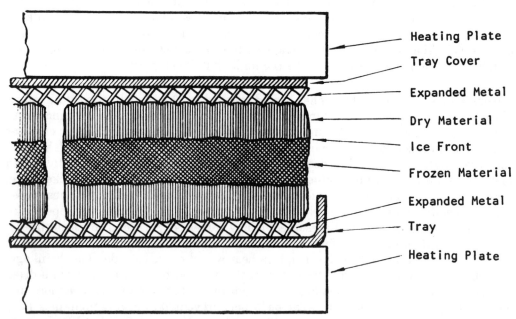

Figure 10.19. Schematic drawing of how a food piece is freeze-dried. *Courtesy of Columbine Press, U.K., and A/S Atlas, Denmark.*

But, in well-engineered freeze-drying systems, the growing porous dried layer generally interferes more with heat transfer than with water mass transfer. Some of the more practical means of increasing overall drying rates have therefore made use of energy sources with penetrating power, such as infrared and microwave radiations, to pass through dried food layers into the receding ice core.

A typical freeze-drying curve for asparagus is shown in Fig. 10.20, which also includes plots of the temperatures of the heating plates and the food surface during the drying run. At the start of the drying operation, no moisture has yet been removed, and the frozen product is below −30°C at its center and surface. The chamber is evacuated to a pressure of 150 µm and the heating platen is set at 120°C. As drying progresses and ice sublimes, the food surface temperature begins to rise from contact with the heated platen, but the receding ice core remains frozen, cooled by the latent heat of sublimation. The platen temperature now must be regulated to establish a delicate balance. Sufficient heat is needed to provide the driving force for rapid sublimation, but not so much as to melt the ice. If the ice was pure water, its melting point would be 0°C, but since some of the ice is frozen with solute as a eutectic, the maximum ice temperature that can be tolerated is usually somewhat below −4°C, depending on the particular food. As more and more ice sublimes, the dried shell surface temperature continues to rise, approaching the 120°C platen. It then becomes necessary to gradually decrease the platen temperature to about 65°C to prevent scorching of the dried food surface. As the platen temperature is decreased, heat transfer to the remaining ice core also drops off, augmenting the insulating effect of the growing dried layer. The result is a further decrease in the drying rate. Ultimately, all of the ice is sublimed, the entire dried mass reaches the 65°C temperature of the heating platen, and moisture is down to about 3%. This may require 8 h or longer. The dried product can now be removed from the vacuum chamber. However, the dried porous product is under high vacuum; if the vacuum is broken by admitting air, the product would instantaneously absorb this air into its pores, resulting in impaired storage stability. Therefore, it is common practice to break the vacuum with inert nitrogen gas. The nitrogen-impregnated product is then packaged, also under nitrogen.

Today, food companies wishing to install freeze-drying equipment on a major scale must consider the process from an overall systems approach. This includes material handling, the freezing operation, loading of drier trays, the drying operation, high vacuum and condenser requirements, unloading of trays, packaging requirements, and, of course, equipment, labor, and utility costs. Many equipment companies have designed total systems that can be custom engineered for a specific product and the needs of the manufacturer. It is common for such equipment companies, working with food manufacturers, to design and install entire freeze-drying plants. Seldom are two such plants quite the same. One type of plant layout is illustrated in Fig. 10.21. It also is sometimes advantageous to combine freeze-drying with air drying. Vegetable pieces may be air dried to about 50% moisture and then freeze-dried down to 2–3% moisture, as in the "Aire Freez" process of the California Vegetable Concentrates Company. This combination gives a high quality product at lower cost than with freeze-drying alone.

Atmospheric Drying of Foams

Vacuum drying methods, and freeze-drying in particular, can produce dehydrated foods of exceptional quality. With liquids and purees, nearly the same quality can be

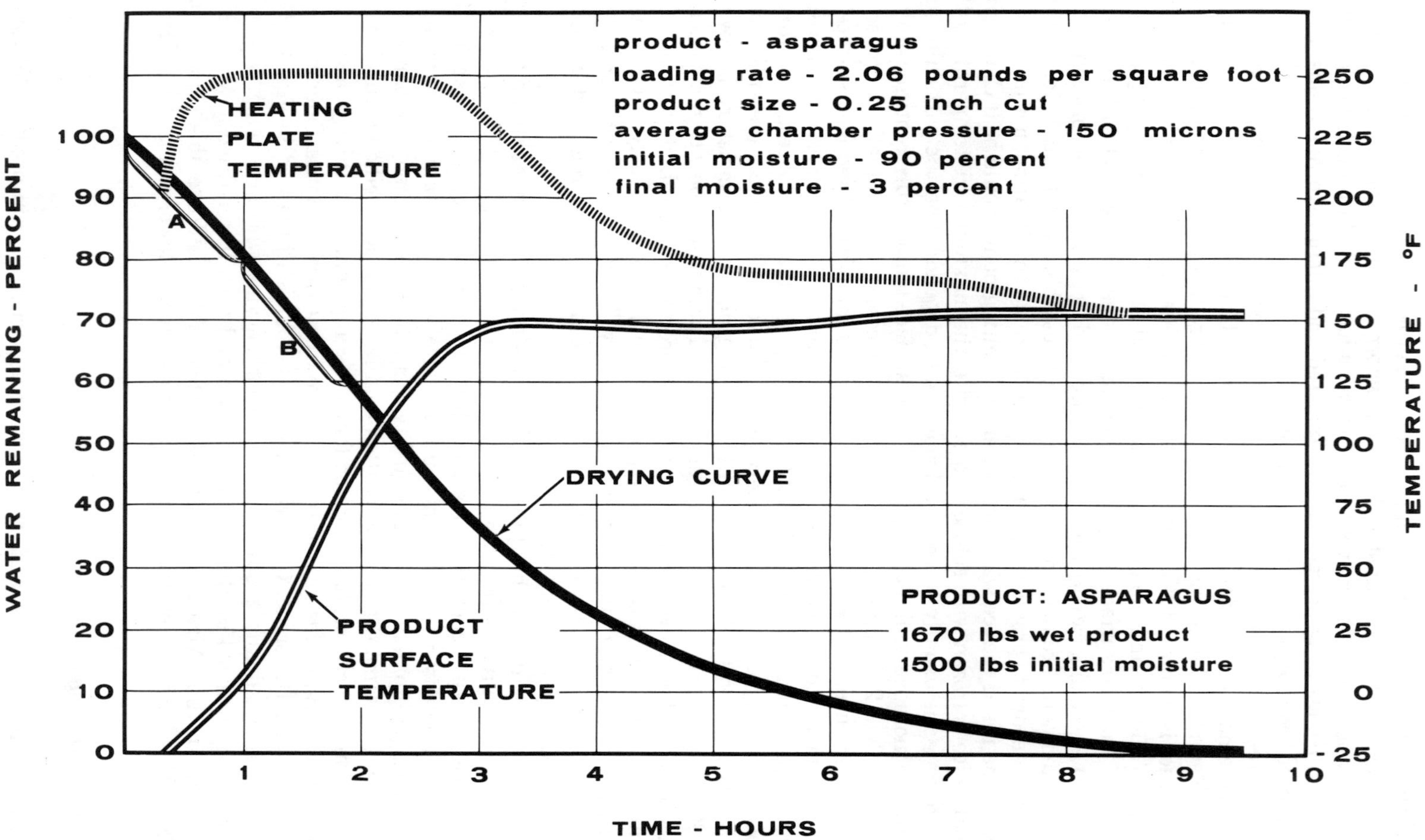
product - asparagus
loading rate - 2.06 pounds per square foot
product size - 0.25 inch cut
average chamber pressure - 150 microns
initial moisture - 90 percent
final moisture - 3 percent
HEATING PLATE TEMPERATURE
A
B
DRYING CURVE
PRODUCT SURFACE TEMPERATURE
PRODUCT: ASPARAGUS
1670 lbs wet product
1500 lbs initial moisture
WATER REMAINING - PERCENT
TEMPERATURE - °F
TIME - HOURS
0
10
20
30
40
50
60
70
80
90
100
-25
0
25
50
75
100
125
150
175
200
225
250
1
2
3
4
5
6
7
8
9
10

Figure 10.20. Drying curve and significant temperature changes during freeze dehydration of asparagus. *Courtesy of FMC Corp.*

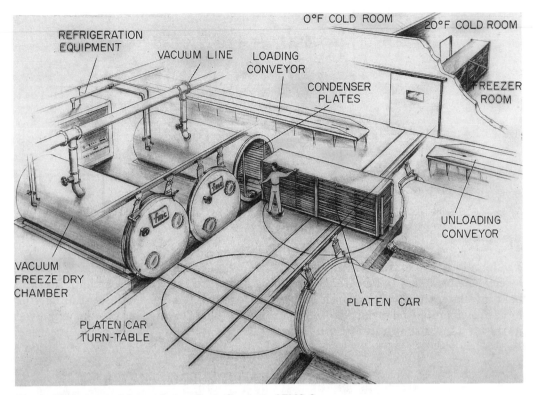

Figure 10.21. Atypical freeze-drying plant. *Courtesy of FMC Corp.*

obtained at atmospheric pressure with less expensive equipment and operating costs. This has been done in some instances by drying prefoamed liquid foods. As mentioned earlier, foaming is done to expose enormous surface area for quick moisture escape. This, in turn, can permit rapid atmospheric drying at somewhat reduced temperatures. In this type of drying, naturally foaming foods such as egg white are mechanically whipped to a foam density of about 0.3 g/cm^3. Foods that do not whip as readily, such as concentrated citrus juices, fruit purees, and tomato paste, are supplemented with low levels of an edible whipping agent belonging to such groups of materials as vegetable proteins, carbohydrate gums, or monoglyceride emulsifiers prior to being whipped. Stable foams are then cast in thin layers onto trays or belts and are dried by various heating schemes.

One such dehydration method is known as foam-mat drying (Fig. 10.22). In one particular type of foam-mat drier, the foam is deposited on a perforated tray or belt support as a uniform layer approximately 3 mm thick. Just before the perforated support enters the heated oven, it is given a mild air blast from below. This forms small craters in the stiff foam which further expands the foam surface and increases the drying rate. At oven temperatures of about 80°C, foam layers of many foods can be dried to about 2–3% moisture in approximately 12 min.

Another system casts similar stable foams on a nonperforated stainless steel belt at a uniform thickness of approximately 0.4 mm. The belt is heated from below by condensing steam and from above by high-velocity heated air. Product temperature is kept below 80°C, and drying time is 1 min or less. The exceptionally rapid drying rates are

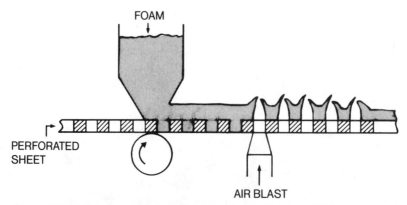

Figure 10.22. Cratering technique of foam-mat drying. *Courtesy of U.S. Department of Agriculture, Western Regional Research Laboratory.*

due largely to the extreme thinness of the foam layers and to the method of heating by condensing steam. When steam condenses under the belt, it gives up both sensible heat and latent heat of condensation, which together provide a substantial driving force to evaporate food moisture. An illustration of the integrated equipment employed in this system is shown in Fig. 10.23.

FOOD CONCENTRATION

Foods are concentrated for many of the same reasons that they are dehydrated. Concentration can be a form of preservation, but only for some foods. Concentration reduces weight and volume and results in immediate economic advantages. Nearly all liquid foods to be dehydrated are concentrated before they are dried because in the

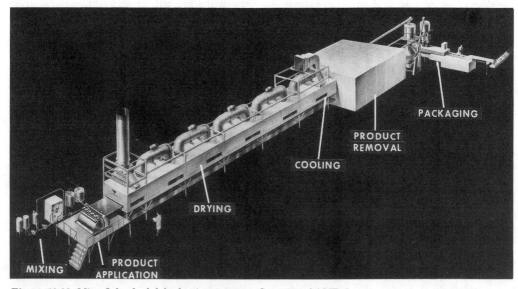

Figure 10.23. Microflake food dehydration systems. *Courtesy of AMF, Inc.*

early stages of water removal, moisture can be more economically removed in highly efficient evaporators than in dehydration equipment. Further, increased viscosity from concentration often is needed to prevent liquids from running off drying surfaces or to facilitate foaming or puffing. Also, some concentrated foods are desirable components of diet in their own right. For example, concentration of fruit juices plus sugar yields jelly. Many concentrated foods, such as frozen orange juice concentrate and canned soups, are easily recognized because of the need to add water before they are consumed. However, maple syrup and butter are somewhat less obvious concentrated foods. In the case of maple syrup, the dilute maple tree sap is concentrated from about 2% solids to 66% solids by boiling off water in open pans or kettles. In making butter, constituents of cream are concentrated from about 40% solids to 85% solids by breaking the fat emulsion and draining the buttermilk, which is largely water, from the churn. In these two cases, water removal is accompanied by other changes that we wish to achieve. As in dehydration, however, most food concentration aims at minimal alteration of food constituents.

The more common concentrated foods include evaporated and sweetened condensed milks, fruit and vegetable juices and nectars, sugar syrups and flavored syrups, jams and jellies, tomato paste, many types of fruit purees used by bakers, candy makers, other food manufacturers, and many more.

Preservative Effects

The levels of water in virtually all concentrated foods are in themselves more than enough to permit microbial growth. Yet although many concentrated foods such as nonacid fruit and vegetable purees may quickly undergo microbial spoilage unless additionally processed, such items as sugar syrups and jellies and jams are relatively immune to spoilage. The difference, of course, is in what is dissolved in the remaining water and what osmotic concentration is reached. Sugar and salt in concentrated solution have high osmotic pressures. When these are sufficient to draw water from microbial cells or to prevent normal diffusion of water into these cells, a preservative condition exists. Heavy syrups and similar products will keep indefinitely without refrigeration even if exposed to microbial contamination, provided they are not diluted above a critical concentration by moisture pickup.

The critical concentration of sugar in water to prevent microbial growth will vary depending on the type of microorganism and the presence of other food constituents, but usually 70% sucrose in solution will stop growth of all microorganisms in foods. Less than this concentration may be effective but for shorter periods of time, unless the foods contain acid or are refrigerated. Salt is highly preservative when its concentration is increased, and levels of 18–25% in solution generally will prevent all growth of microorganisms in foods. Except in the case of certain briny condiments, however, this level is rarely tolerated in foods. Removal of water by concentration also increases the level of food acids in solution. This is particularly significant in concentrated fruit juices.

In the sugar industry, juice squeezed from sugar cane contains approximately 15% sucrose and is highly perishable. Evaporators are an essential part of the equipment in sugar processing plants and are used to remove most of the water from cane or beet juice prior to subsequent crystallization steps in the production of dry granulated sugar. However, concentrated syrups with sugar levels of approximately 70% are sold also as important items of commerce. These syrups, with consistencies similar to honey, are pumped into tank cars and delivered to storage tanks of bakeries and confectionery manufacturers. Preservation is quite satisfactory provided there is no moisture condensation from the air onto interior tank surfaces. Sugar in this form, already in solution and easy to pump, is more convenient and economical than granulated sugar to use in many manufacturing operations.

Reduced Weight and Volume

Whereas the preservative effects of food concentration are important, the principal reason for most food concentration is to reduce food weight and bulk. Tomato pulp, which is ground tomato minus the skins and seeds, has a solids content of only 6%, and so a 3.78-liter can would contain only 231 g of tomato solids (Table 10.2). Concentrated to 32% solids, the same can would contain 1.38 kg of tomato solids, six times the value of the original product. For a manufacturer needing tomato solids, such as a producer of soups, canned spaghetti, or frozen pizzas, the savings from concentration are enormous in cans, transportation costs, warehousing costs, and handling costs throughout operations. This is especially so since much of the U.S. tomato crop is grown in the Sacramento Valley area of California and shipped to manufacturing plants in Chicago and eastern areas of the country. These savings can be even greater by eliminating small containers and shipping the aseptically packed concentrate in bulk (Fig. 10.24). For the same reasons, millions of tons of concentrated fruits, juices, vegetable products, milk products, and other commodities are used industrially, in restaurant operations, and in the home. Large quantities of concentrated buttermilk, whey, blood, yeast, and other food by-products, are used also in animal feeds by poultry and livestock growers.

Methods of Concentration

Solar Concentration

As in food dehydration, one of the simplest methods of evaporating water is with solar energy. This was done to derive salt from seawater from earliest times and

Table 10.2. Specific Gravity and Solids Relationship of Tomato Pulp and Commercial Tomato Concentrates

Total Solids (%)		Specific Gravity at 68°F (20°C)	Dry Tomato Solids	
			lb per gal at 68°F	g per liter at 20°C
6.0	Tomato pulp	1.025	0.51	61
10.8		1.045	0.94	113
12.0		1.050	1.05	126
14.2	Tomato puree	1.060	1.25	151
16.5		1.070	1.47	177
25.0		1.107	2.31	277
26.0		1.112	2.41	289
28.0	Tomato paste	1.120	2.61	314
30.0		1.129	2.82	339
32.0		1.138	3.03	364

SOURCE: Adapted from National Food Processors Assoc. data.

Figure 10.24. Aseptic rail car for tomato paste and other concentrates. *Courtesy of Fran Rica Mfg., Inc.*

is still practiced today in the United States in man-made lagoons. However, solar evaporation is very slow and is suitable only for concentrating salt solutions.

Open Kettles

Some foods can be satisfactorily concentrated in open kettles that are heated by steam. This is the case for some jellies and jams and for certain types of soups. However, high temperatures and long concentration times damage most foods. In addition, thickening and burn-on of product to the kettle wall gradually lower the efficiency of heat transfer and slow the concentration process. Kettles and pans are still widely used in the manufacture of maple syrup, but here high heat is desirable to produce color from caramelized sugar and to develop typical flavor.

Flash Evaporators

Subdividing the food material and bringing it into direct contact with the heating medium can markedly speed concentration. This is done in flash evaporators of the kind shown in Fig. 10.25. Clean steam superheated at about 150°C is injected into food which is pumped into an evaporation tube where boiling occurs. The boiling mixture then enters a separator vessel in which the concentrated food is drawn off at the bottom and the steam plus water vapor from the food is evacuated through a separate outlet. Because temperatures are high, foods that lose volatile flavor constituents will yield these to the exiting steam and water vapor. These can be separated from the vapor by essence-recovery equipment on the basis of different boiling points between the essences and water.

Thin-Film Evaporators

In thin-film evaporators (Fig. 10.26), food is pumped into a vertical cylinder which has a rotating element that spreads the food into a thin layer on the cylinder wall.

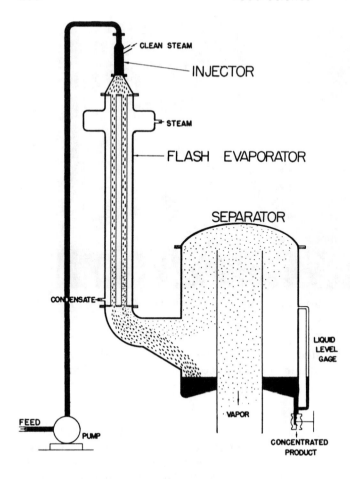

CLEAN STEAM

INJECTOR

STEAM

FLASH EVAPORATOR

SEPARATOR

CONDENSATE

LIQUID
LEVEL
GAGE

FEED

PUMP

VAPOR

CONCENTRATED
PRODUCT

Figure 10.25. Components of flash evaporator. *Courtesy of Oscar Krenz.*

The cylinder wall of double jacket construction usually is heated by steam. Water is quickly flashed from the thin food layer and the concentrated food is simultaneously wiped from the cylinder wall. The concentrated food and water vapor are continuously discharged to an external separator, from which product is removed at the bottom and water vapor passes to a condenser. In some systems the water vapor temperature is raised by mechanical vapor recompression to yield steam for reuse to save energy. Product temperature may reach 85°C or higher, but since residence time of the concentrating food in the heated cylinder may be less than a minute, heat damage is minimal.

Vacuum Evaporators

Heat-sensitive foods are most commonly concentrated in low-temperature vacuum evaporators. Thin-film evaporators frequently are operated under vacuum by connecting a vacuum pump or steam ejector to the condenser.

It is common to construct several vacuum vessels in series so that the food product moves from one vacuum chamber to the next and thereby becomes progressively more concentrated in stages. The successive stages are maintained at progressively higher degrees of vacuum, and the hot water vapor arising from the first stage is used to heat the second stage, the vapor from the second stage heats the third stage, and so on. In

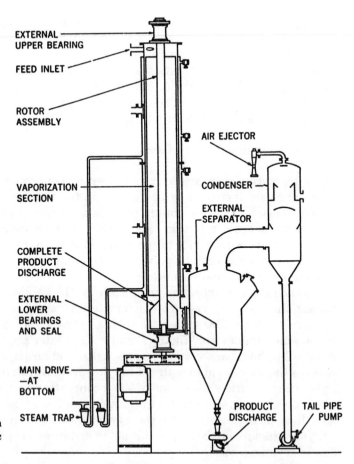

EXTERNAL UPPER BEARING

FEED INLET

ROTOR ASSEMBLY

VAPORIZATION SECTION

COMPLETE PRODUCT DISCHARGE

EXTERNAL LOWER BEARINGS AND SEAL

MAIN DRIVE —AT BOTTOM

STEAM TRAP

AIR EJECTOR

CONDENSER

EXTERNAL SEPARATOR

PRODUCT DISCHARGE

TAIL PIPE PUMP

Figure 10.26. Agitated thin-film evaporator. *Courtesy of Buflovak Equipment Div., Blaw-Knox Co.*

this way, maximum use of heat energy is obtained. Such a system, called a multiple-effect vacuum evaporator (See Fig. 5.9), may be sizable and expensive. Systems employed in the grape juice industry continuously concentrate juice from an initial solids content of 15% to a final solids concentration of 72% at rates of 4500 gal of single strength juice per hour. Similar systems concentrate tomato juice from 6% solids to 30% solids at rates of 15,000 gal or more of single strength juice per hour. Use of energy-saving mechanical vapor recompression is common.

Even with efficient vacuum evaporators where water may boil at 30°C or slightly lower, some volatile flavor compounds are lost with the evaporating water vapor. These volatile essences can be recovered, or "stripped," from the water vapor and returned to the cool concentrated food as has been mentioned earlier. However, it is possible to concentrate foods at still lower temperatures and further minimize heat damage and volatile flavor loss; one method of doing so is known as freeze concentration.

Freeze Concentration

As discussed previously, when a solid or liquid food is frozen, all of its components do not freeze at once. First to freeze is some of the water which forms ice crystals in the mixture. The remaining unfrozen food solution is now higher in solids concentration.

It is possible, before the entire mixture freezes, to separate the initially formed ice crystals. One way of doing this is to centrifuge the partially frozen slush through a fine-mesh screen. The concentrated unfrozen food solution passes through the screen while the frozen water crystals are retained and can be discarded. Repeating this process several times on the concentrated unfrozen food solution can increase its final concentration several-fold. Freeze concentration has been known for many years and has been applied commercially to orange juice.

Ultrafiltration and Reverse Osmosis

Low-temperature separation and concentration processes employing perm-selective membranes are increasingly being used in the food industry. These applications are largely dependent on membrane properties such as water permeability rate, solute and macromolecule rejection rates, and length of useful membrane life. Different membranes are required for different liquid foods. Synthetic membranes are manufactured from cellulose acetate, polyamide, and other materials, with considerable control over their physical and chemical properties.

Ultrafiltration membranes are generally "less tight" than reverse osmosis membranes; that is, they restrict macromolecules such as proteins but with moderate pressure allow smaller molecules such as sugars and salts to pass through. Reverse osmosis membranes are "tighter," and with greater pressure will permit the passage of water but hold back various sugars, salts, and larger molecules. In nature, osmosis involves the movement of water through a perm-selective membrane from a region of higher concentration to a region of lower concentration. The region of lower concentration generally contains solutes in solution and has associated with it an osmotic pressure. It is possible to reverse the normal flow of water through the membrane by applying pressure on the solute side of the membrane in excess of the osmotic pressure. This is reverse osmosis.

Applied to food concentration, ultrafiltration and reverse osmosis processes involve pumping liquid foods under pressure against perm-selective membranes in a suitable support. Equipment may be similar to pressure filters in design. Filtrates passing through one membrane may be further modified by passing through a second tighter membrane. This is done in the processing of cheese whey. One may force the whey through a reverse osmosis membrane and remove much of the water, thus concentrating virtually all of the whey solids. The whey also may be forced through an ultrafiltration membrane first, which would concentrate the lactalbumin protein above the membrane. The filtrate then may be forced through a reverse osmosis membrane selected to retain and concentrate lactose but allowing lower-molecular-weight salts to be removed with the water. Not only are valuable food constituents concentrated in this process, but the water discharged is very low in organic matter (low biological oxygen demand), which decreases its pollution load. Such a process is illustrated in Fig. 10.27. Besides various applications in the dairy industry, these membrane processes are being used to concentrate fruit juices, coffee and tea extracts, egg white and whole egg, soy proteins, enzymes, and other materials.

Changes During Concentration

Concentration processes that expose food to 100°C or higher temperatures for prolonged periods can cause major changes in organoleptic and nutritional properties.

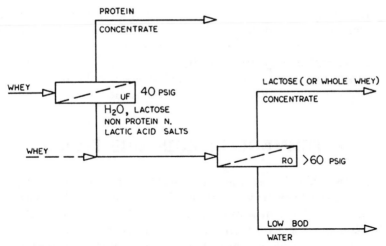

Figure 10.27. Schematic showing use of ultrafiltration and reverse osmosis to separate and concentrate components of cheese whey. *Courtesy of Abcor, Inc. (Mr. B. S. Horton).*

Cooked flavors and darkening of color are two of the more common results. In addition to the desirability of controlled amounts of these changes in maple syrup, heat-induced reactions also characterize certain candies such as caramel. In caramel production, sugar–milk mixtures are intentionally concentrated at high temperature. With most other foods the lower the concentration temperature the better, since the reconstituted concentrated food should resemble as closely as possible the natural product. Even at the lowest temperatures, however, concentration can cause other changes that are undesirable. Two such changes involve sugars and proteins.

All sugars have an upper limit of concentration in water beyond which they are not soluble. For example, at room temperature, sucrose is soluble to the extent of about 2 parts sugar in 1 part water. If water is removed beyond this concentration level, the sugar crystallizes out. This can result in gritty, sugary jellies or jams. It also results in a condition knows as "sandiness" in certain milk products when lactose crystallizes due to overconcentration. Since the amount of sugar that can be in solution decreases with decreasing temperature, a concentrated product may be smooth in texture at room temperature but become gritty or sandy when put into a refrigerator. This condition occurs in the manufacture of ice cream due to lactose crystallization during freezing if the concentration of lactose from concentrated milk ingredients is excessive.

As for effects on proteins, it has been pointed out that proteins can be easily denatured and precipitated from solution. One cause of denaturation can be high concentration of salts and minerals in solution with the protein. As protein-containing foods such as milk are concentrated, the levels of milk salts and minerals can become sufficiently high to partially denature the milk protein and cause it to slowly gel. The gelling may not show up immediately but only after weeks or months of storage, as frequently occurs in cans of evaporated and certain other condensed milks. The gelation of concentrated milk and other proteinaceous foods is an extremely complex phenomenon and is affected by many variables in addition to degree of concentration.

Microbial destruction, another type of change that may occur during concentration, will be largely dependent on temperature. Concentration at a temperature of 100°C or slightly above will kill many microorganisms but cannot be depended on to destroy bacterial spores. When the food contains acid, such as fruit juices, the kill will be greater, but again sterility is unlikely. On the other hand, when concentration is done under vacuum, many bacterial species not only survive the low temperatures but multiply in the concentrating equipment. It, therefore, is necessary to stop frequently and sanitize low-temperature evaporators, and where sterile concentrated foods are required, to resort to an additional preservation treatment.

INTERMEDIATE-MOISTURE FOODS

Water activity was briefly discussed in Chapter 7. In recent years, adjustment and control of water activity to preserve semimoist foods has attracted increasing attention. Intermediate-moisture foods or semimoist foods, in one form or another, have been important items of diet for a very long time. Generally, they contain moderate levels of moisture, of the order of 20–50% by weight, which is less than is normally present in natural fruits, vegetables, or meats but more than is left in conventionally dehydrated products. In addition, intermediate-moisture foods contain sufficient dissolved solutes to decrease water activity below that required to support microbial growth. As a consequence, intermediate-moisture foods do not require refrigeration to prevent microbial deterioration. In the past there have been various kinds of intermediate-moisture foods: natural products such as honey; manufactured confectionery products high in sugar, plus jellies, jams, and bakery items such as fruit cakes; and partially dried products including figs, dates, jerky, pemmican, pepperoni, and the like. Sweetened condensed milk with a sugar level of about 63% based on the water content also should be considered an intermediate-moisture food. In all of these products, preservation is partially from high osmotic pressure associated with the high concentration of solutes; in some, additional preservative effect is contributed by salt, acid, and other specific solutes.

Principles Underlying Technology

Essential to any discussion of intermediate-moisture foods is understanding water activity (A_w) and its relationship to food properties and stability. Water activity may be defined in a number of ways. Qualitatively, A_w is a measure of unbound, free water in a system available to support biological and chemical reactions. Water activity, not absolute water content, is what bacteria, enzymes, and chemical reactants encounter and are affected by at the microenvironmental level in food materials. Two foods with the same water content can have very different A_w values, depending on the degree to which the water is free or otherwise bound to food constituents.

Figure 10.28 is a representative water sorption isotherm for a given food at a given temperature. It shows what the final moisture content of the food will be when it reaches moisture equilibrium with atmospheres of different relative humidities. For example, at the temperature for which this sorption isotherm was established, this food will attain a moisture content of 20% at 75% RH. If this food were previously dehydrated to below 20% moisture and placed in an atmosphere of 75% RH, it would absorb moisture until it reached 20%. Conversely, if this food were moistened to greater than 20% water and then placed at 75% RH, it would lose moisture until it reached the equilibrium value of 20%. Under such conditions some foods may reach moisture

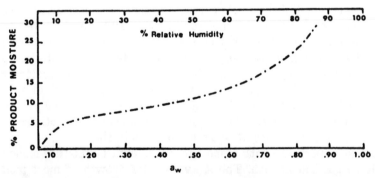

Figure 10.28. Generalized moisture sorption isotherm for a food product.

equilibrium in the very short time of a few hours, others may require days or even weeks. When a food is in moisture equilibrium with its environment, then the A_w of the food will be quantitatively equal to the RH divided by 100.

Water activity can be defined in still other terms in accordance with Raoult's law. Thus, A_w of a solution is quantitatively equal to the vapor pressure of the solution divided by the vapor pressure of pure water. This also is equal to the mole fraction of pure water in the solution, which is equivalent to the number of moles of water in the solution divided by the total number of moles present. Thus, a 1 molal solution of sucrose would contain 1 mole of sucrose and 55.5 moles of water (1000 g/18 g), and assuming it behaved as an "ideal" or "perfect" solution, would have an A_w value of 55.5/56.5 or 0.98. Such a solution would be quite dilute and if it constituted the water phase of a food would not of itself generally inhibit microbial growth.

Determining Water Activity

The foregoing relationships provide the means for measuring the A_w at various moisture contents and temperatures. One method involves placing small portions of the food in jars maintained at a fixed temperature and at different relative humidities with standard sulfuric acid or salt solutions. The samples are periodically weighed until they reach moisture equilibrium as indicated by no further gain or loss in weight. The equilibrium moisture content of each portion is next plotted at its corresponding RH. This plot yields a moisture sorption isotherm for the specific food (at the temperature chosen) of the kind indicated in Fig. 10.28. From the resulting moisture sorption isotherm curve, the RH corresponding to each moisture content divided by 100 is equal to the A_w at that moisture level.

In practice, the A_w of an experimental food formulation can be readily determined instrumentally. In this case, a sample of the food is placed in a vessel of limited headspace at a chosen temperature. The vessel is provided with a sensitive hygrometer sensor not in contact with the food but connected to a potentiometric recorder. As the food exchanges moisture with the headspace, a curve of RH is traced. The A_w then corresponds to the RH/100 at equilibrium. RH also can be measured with highly sensitive wet bulb and dry bulb temperature probes.

As stated earlier, A_w is a measure of free or available water, to be distinguished from unavailable or bound water. These states of water also bear a relationship to the characteristic sigmoid shapes of water sorption isotherm curves of various foods. Thus, according to theory, most of the water corresponding to the portion of the curve below its first inflection point (below 5% moisture in Fig. 10.28) is believed to be tightly bound water—often referred to as an adsorbed monomolecular layer of water. Moisture coresponding to the region above this point and up to the second inflection point (about 20% moisture in Fig. 10.28) is thought to exist largely as multimolecular layers of water less tightly held to food constituent surfaces. Beyond this second inflection point, moisture generally is considered to be largely free water condensed in capillaries and interstices within the food. In this latter portion of the sorption isotherm curve, small changes in moisture content result in great changes in the A_w.

Of the greatest importance with respect to intermediate-moisture foods is the effect of A_w on microorganism growth. The A_w values for growth of most food-associated bacteria, yeasts, and molds have received considerable study. The minimum A_w below which most important food bacteria will not grow is about 0.90, depending on the specific bacterium. Some halophilic bacteria may grow down to an A_w of 0.75, and certain osmophilic yeasts even lower, but these seldom are important causes of food spoilage. Molds are more resistant to dryness than most bacteria and frequently will grow well on foods having an A_w of about 0.80, and slow growth may appear after several months at room temperature on some foods even at an A_w as low as 0.70. At A_w values below 0.65, mold growth is completely inhibited, but such low A_w generally is not applicable in the fabrication of intermediate-moisture foods. This level would correspond to total moisture content well below 20% in many foods; such foods would lose chewiness and approach a truly dehydrated product. For most items, A_w values between 0.70 and 0.85 are required for semimoist texture. These levels are sufficiently low to inhibit common food-spoilage bacteria. Where they are not sufficiently low for long-term inhibition of mold growth, an antimycotic such as potassium sorbate is included in the food formulation to augment the preservative effect.

Although A_w values for microbial inhibition commonly are cited in the literature to two or three decimal places, this should not convey the impression that an A_w given as a minimum for growth of a particular microorganism is an absolute value. It can be influenced by such factors as the pH, temperature, nutritional status in terms of microbial requirements, and the nature of specific solutes in the water phase. Although these influences frequently are small, it is prudent to confirm the efficacy of a target A_w to prevent microbial spoilage of a new intermediate-moisture formulation by running appropriate bacterial plate counts. Bacteriological tests also are necessary from a public health standpoint.

When attempting to fabricate an intermediate-moisture food, one selects an appropriate target A_w and then selects ingredients to provide solute concentrations to yield the desired A_w. The total solute concentration corresponding to any A_w can be easily calculated from equations based on Raoult's law provided the food's water phase behaves as an ideal solution.

As solutions become more concentrated and more complex, however, they fail to behave in ideal fashion; then calculations relating solute concentrations and A_w become only approximations. For an A_w of 0.995, for example, theory calls for a total solute concentration of 0.281 molal. Sucrose and glycerol, which do not dissociate in solution, closely approach this ideal. Sodium chloride and calcium chloride, which dissociate to yield two and three ions, respectively, also approach ideal behavior in such dilute solutions when the sum of the concentrations of their ions is considered. However, in concentrated solutions, solutes become more effective in lowering A_w than would be predicted on the basis of ideal behavior. This is not due to suppression of ion dissociation, which in itself would be expected to contribute an opposite effect, but rather is thought to result from increased total hydration of large numbers of solute molecules. This also is so with solutes which do not dissociate such as sucrose and glycerol, favored ingredients in the formulation of intermediate-moisture foods. Such phenomena make it necessary to augment mathematical calculation of A_w by experimental measurement when attempting to establish intermediate-moisture food compositions.

Much of what has been said thus far with regard to A_w has had to do with microbial inhibition; however, A_w affects many other properties of foods, including chemical reactivity and equilibria, enzymatic activity, flavor, texture, color, and stability of nutrients.

Products and Technology

Aside from semimoist dog food, few intermediate moisture foods have been specifically developed for human consumption. However, several common foods meet the definition of intermediate moisture food. Jams, jellies, some processed/fermented meats, dried fruits, confections, bakery products, and snacks are common examples. The major concern with such foods is the control of moisture. Loss of moisture results in undesirable changes in texture, whereas pick up of moisture results in the potential for microbial growth. The major way in which changes in moisture are controlled is by packaging which inhibits moisture transfer.

References

Bhandari, B.R., Senoussi, A., Dumoulin, E.D., and Lebert, A. 1993. Spray drying of concentrated fruit juices. Drying Technol. *11*(5), 1081–1092.

Boersen, A.C. 1990. Spray drying technology. J. Soc. Dairy Technol. *43*(1), 5–7.

Bouman, S., Brinkman, D.W., de. Jong, P., and Wallewijn, R. 1987. Multistage evaporation in the dairy industry: Energy savings, product losses and cleaning. *In* Preconcentration and Drying of Food Materials: Thijssen Memorial Symposium: Proc. of the Int. Symp. on Preconcentration and Drying of Foods, Eindhoven, Netherlands, Nov. 5–6, 1987, S. Bruin (Editor), pp. 51–60.

Charm, S.E. 1978. Dehydration of foods. *In* The Fundamentals of Food Engineering, S.E. Charm (Editor). AVI Publishing Co., Westport, CT, pp. 298–408.

Chung, D.S. and Chang, D.I. 1982. Principles of food dehydration. J. Food Protect. *45*(5), 475–478.

Dalgleish, J. 1990. Freeze-drying for the Food Industries. Elsevier Science Publishers, London.

Deshpande, S.S., Cheryan, M., Sathe-Shridharn, K., Salunkhe, D.K. 1984. Freeze concentration of fruit juices. CRC Crit. Rev. Food Sci. Nutr. *20*(3), 173–248.

Erickson, L.E. 1982. Recent developments in intermediate moisture foods. J. Food Protect. *45*(5), 484–491.

Fast, R.B. and Caldwell, E.F. 1990. Breakfast Cereals and How They are Made. American Association of Cereal Chemists, St. Paul, MN.

Furuta, T., Hayashi, H., and Ohashis, T. 1994. Some criteria of spray dryer design for food liquid. Drying Technol. *12*(1/2), 151–177.

Hall, G.M. 1992. Fish Processing Technology. Blackie, London/VCH, New York.

Heid, J.L. and Joslyn, M.L. 1967. Food processing by drying and dehydration. In Fundamentals of Food Processing Operations: Ingredients, Methods, and Packaging, J.L. Heid and M.A. Joslyn (Editors). AVI Publishing Co., Westport, CT. pp. 501–540.

Iglesias, H.A. and Chirife, J. 1982. Handbook of Food Isotherms: Water Sorption Parameters for Food and Food Components. Academic Press, New York.

Jayaraman, K.S., Das-Gupta, D.K. 1992. Dehydration of fruits and vegetables recent developments in principles and techniques. Drying Technol. *10*(1), 1–50.

Jen, J.J. ed. 1989. Quality Factors of Fruits and Vegetables: Chemistry and Technology. ACS Symposium Series. No. 405, American Chemical Society, Washington, D.C.

Karel, M. 1982. Water activity and intermediate moisture foods. *In* Chemistry and World Food Supplies, The New Frontiers. Chemrawn II, W. Shemil, ed. Pergamon Press, Oxford, pp. 465–475.

Lee, D.S. and Pyun, Y.R. 1993. Optimization of operating conditions in tunnel drying of food. Drying Technol. *11*(5), 1025–1052.

Merlo, C.A., Pedersen, L.D., Rose, W.W. 1985. Hyperfiltration/Reverse Osmosis. A Handbook on Membrane Filtration for the Food Industry. Technical Information Center, Office of Scientific and Technical Information, U.S. Department of Energy, Washington, DC.

Owusa-Ansah, Y.J. 1991. Advances in microwave drying of foods and food ingredients. J. Inst. Can Sci. Technol. Aliment. *24*(3/4), 102–107.

Quarles, S.L. and Wengert, E.M. 1989. Applied drying technology. Forest Products J. *39*(6), 25–38.

Retsina, T. 1988. Agglomeration of Powders. British Food Manufacturing Industries Research Association, Leatherhead.

Rockland, L.B. and Stewart, G.F. 1981. Water Activity: Influences on Food Quality. Academic Press, New York.

Sano, Y. 1993. Gas flow behaviour in spray dryer. Drying Technol. *11*(4), 697–718.

Shi, X.Q. and Maupoey, P.F. 1993. Vacuum osmotic dehydration of fruits. Drying Technol. *11*(6), 1429–1442.

Shukla, T.P. 1991. Osmotic dehydration. Cereal Foods World *36*(8), 647.

Taoukis, P.S., Breene, W.M., and Labuza, T.P. 1987. Intermediate moisture foods. Adv. Cereal Sci. Technol. *9*, 91–128.

Van Arsdel, W.B., Copley, M.J., and Morgan, Jr., A.I. 1973. Food Dehydration. 2nd ed. Vols. 1 and 2. AVI Publishing Co., Westport, CT.

11

Irradiation, Microwave, and Ohmic Processing of Foods

Both irradiation and microwave heating employ radiant energies which affect foods when their energy is absorbed, whereas ohmic heating raises the temperature of foods by passing an electrical current through the food. Each requires special equipment to generate, control, and focus this energy. Each of these are relatively new technologies as applied to foods. Food irradiation is used primarily as a preservation method, but it also has potential as a more general unit operation to produce specific changes in food materials. Microwave energy, on the other hand, has been employed especially to produce rapid and unique heating effects, one application of which can be food preservation. Ohmic heating is the newest and least used of the three technologies. Like microwave heating, ohmic heating can preserve foods by the application of heat and has the ability to very rapidly heat foods with minimal destruction.

FOOD IRRADIATION

The discoveries of artificially produced radiations such as X-rays and radioactivity of natural materials date back to 1895–1896. Food irradiation studies are more recent, having begun in earnest shortly after World War II. The impetus for this research came largely from intensive investigations of nuclear energy, which led to developments in the economic production of radioactive isotopes and to evolution of high-energy accelerators. In the period since 1945, food irradiation has been investigated intensively. Much of this work has had to do with the safety and wholesomeness of irradiated products. In 1963, the U.S. Food and Drug Administration approved irradiation-sterilized bacon, the first in a growing list of proposed products. This approval, which limited the types of energy sources and doses that could be employed, was revoked in 1968. The safety of irradiated foods has continued to receive study since then, with several countries gradually adding to the list of approved irradiated foods. In 1983 the FDA approved irradiation as a means of controlling microorganisms on spices, and in 1985 the FDA widened the allowed uses of irradiation to additional foods such as strawberries, poultry, ground beef, and pork.

245

In its early development, irradiation was thought of as a process to preserve foods for extended periods by sterilization, much as thermal processing does. However, this has proven to be impractical for many products because the amount of irradiation required to commercially sterilize foods causes its own form of deterioration. Freezing prior to irradiation can reduce the damage, but this makes the process excessively expensive.

More recent developments have focused on the use of lower doses of irradiation which are less damaging to the food yet have desirable effects. As currently practiced, irradiation is used for three purposes. First, it can be used as an alternative to chemical fumigation to control insects in foods such as spices and fruits and vegetables. The second use is to inhibit sprouting or other self-generating mechanisms of deterioration. The third use is to destroy vegetative cells of microorganisms including those that might cause human disease. This results in an increase in safety and shelf life.

Forms of Energy

There are several forms of radiant energy emitted from different sources. These belong to the electromagnetic spectrum of radiations and differ in wavelength, frequency, penetrating power, and the effects they have on biological systems. Some of these forms of radiant energy and their bactericidal effects are indicated in Table 11.1

A light bulb emits visible energy. This is radiated from the bulb filament and travels in all directions; however, it can be focused and aimed at a target. Similarly, an infrared heat lamp contains a glowing element that radiates infrared energy. This can be directed at a steak, which absorbs the infrared energy and becomes warm or, indeed,

Table 11.1. Bactericidal Effects of Different Wavelengths of Radiant Energy

Classification	Wavelength (nm)	Germicidal Effects
Invisible (long)		
radio	Very long	None
infrared (heat)	800 and longer	Temperature may be raised
Visible		
red, orange, yellow, green, blue, violet	400–800	Little or none
Invisible (short)		
ultraviolet total range	13.6 to 400	
	320–400	Photographic and fluorescent range
	280–320	Human skin tanning, antirachitic-vitamin D
	200–280	Maximum germicidal power
	150–200	Shuman region
	100	Ozone forming, germicidal in proper concentration
X-rays	100–150	Marginal
Alpha, beta, and gamma rays	Less than 100	Germicidal
Cosmic rays	Very short	Probably germicidal

SOURCE: Weiser (1971)., Mountney, and Gould

cooked if the quantity of the absorbed energy is sufficiently high. There are other forms of energy which produce neither light nor heat and cannot readily be detected by the human eye or sense of touch, like radiowaves, ultraviolet light, or cosmic rays, but they are present nevertheless.

Certain types of energy radiations are emitted from the breakdown of atomic structure. Materials that undergo such changes are said to be radioactive. Some elements such as uranium are naturally radioactive; others can be made radioactive by high-energy bombardment of their atoms, as in the case of cobalt–60. Another form of energy is associated with the flow of electrons such as can be emitted from a cathode tube. These electrons or cathode rays can be given various degrees of acceleration and increased energy by passage through special electronic devices.

Some of the kinds of energy indicated above are used to a limited degree in food preservation. Ultraviolet light, especially within the wavelength range of 200–280 nm, is employed to inactivate microorganisms on the surface of foods. The severe limitation here is the low degree of penetration of ultraviolet light into foods, restricting its usefulness to outermost surface of food or liquids that can be exposed in thin layers. Treatment of equipment surfaces, water, and air used in food plants are additional current applications of ultraviolet light. X-rays have greater penetrating power than ultraviolet light and have received consideration as a means of preserving foods. However, X-rays cannot easily be focused, leading to low efficiency of use with current equipment. Thus, X-ray food applications to date have been experimental rather than commercial.

Currently, when the term *food irradiation* is used, it generally is understood to mean processing with a limited number of kinds of radiant energy that together are referred to as "ionizing" radiations. These are chosen because they have penetrating power but do not produce radioactivity in treated foods. They also do not produce significant heat in foods, and so the additional term *cold sterilization* has been applied to this kind of food preservation.

Ionizing Radiations and Sources

Natural radioactive elements and artificially induced radioactive isotopes, which can be produced in nuclear reactors, emit a variety of radiations and energy particles during radioactive decay. Among these are alpha particles, which are really helium atoms minus two outer electrons; beta particles or rays, which are high-energy electrons also referred to as cathode rays; gamma rays or photons, which are a type of X-ray; and neutrons. These radiations have different penetrating powers: Alpha particles will not even penetrate a sheet of paper, beta particles or electrons are more penetrating but can be stopped by a sheet of aluminum, and gamma rays are highly penetrating and will go through a block of lead if it is not too thick. Neutrons have great penetrating power and are of such high energy that they can alter atomic structure and make atoms that they strike radioactive. Such atoms, in turn, emit high-energy radiations.

The most suitable emissions for food irradiation have good penetrating power so that they will inactivate microorganisms and enzymes not only on the surface but deep within the food. On the other hand, high-energy emissions as neutrons would break down atomic structures in the food and make the food radioactive. Therefore, gamma rays and beta particles are those used most often.

Gamma and beta rays used in food irradiation may be derived from approved spent fuel elements after their use in a nuclear reactor. These fuel elements, which eventually develop fission fragments and other impurities, making them unsuitable for further use in a nuclear reactor, still possess intense radioactivity. Such spent fuel elements can be placed in an appropriately shielded and enclosed region, and the food brought into the path of their radiation. In early experimental food irradiation facilities, spent fuel elements were placed in shielded pits under about 5 m of water. Cans of food were lowered into vertical cylinders immersed in the water and surrounded by the fuel elements at the bottom of the pit. The containers were then held there for sufficient time to absorb an appropriate radiation dose. Current facilities are less cumbersome and make considerable use of artificially induced radioactive elements such as ^{60}Co for the radiating fuel. Where ^{60}Co is used, it is employed primarily as a gamma ray source, since beta particles may be more efficiently produced by electronic machines.

Units of Radiation

Various terms have been used to quantitatively express radiation intensity and radiation dosage:

• A roentgen of radiation is equivalent to the quantity of radiation received in 1 h from a 1-g source of radium at a distance of 1 yd. This is also the quantity of radiation that will produce 2.08×10^9 ion pairs per cubic centimeter of dry air, or one electrostatic unit of charge of either sign per cubic centimeter of air under standard conditions of temperature and pressure.
• The energy required to produce ion pairs in air can also be expressed in terms of electron volts. Approximately 32.5 ev are required to produce one ion pair in air. An electron volt is the energy equivalent of 1.6×10^{-19} J.
• A rad is a measure of ionizing energy absorbed. It has a quantitative equivalent of 10^{-5} J absorbed per gram of absorbing material.
• The rad and its multiples (krad, Mrad) are commonly used as units of absorbed radiation dose.
• Another unit, the Gray (Gy) equals 10^2 rads.

In irradiation processes the dose of radiation that a substrate receives is important. Different materials absorb radiation energy to different degrees.

A rad of radiation dosage represents the same amount of absorbed energy whether it comes from rays, particles, or a mixture of the two. The magnitude of a radioactive isotope power source is expressed in terms of curies, which is a measure of disintegrations per second. The strength of gamma radiations is expressed in terms of roentgens. The intensity or energy level of beta particles emitted from a linear electron accelerator is defined in terms of joules or electron volts. However, the length of time food is exposed to such sources, and the absorption properties of the food (and its container), will determine the number of rads received by the food, which is the effective dosage that produces changes in the food's microflora, enzymes, and other constituents.

Radiation Effects

Ionizing radiations penetrate food materials to varying degrees depending on the nature of the food and the characteristics of the radiations. Gamma rays have greater penetrating power than β particles. The efficacy of radiations in producing radiation effects, however, also is dependent upon their abilities to alter molecules and their ionization potential, that is, their abilities to knock electrons out of atoms of the materials through which they pass. Beta particles generally have greater ability to produce ionizations in matter through which they pass than gamma rays. Electron beams of higher energy levels penetrate more and produce more altered molecules and total ionization along their traveled paths than lower-energy electron beams.

Just as neutrons possessed of extremely high energy can alter atomic nuclei so as to make them radioactive, there are energy levels beyond which gamma rays and electron beams may induce radioactivity in foods. These energy levels are far in excess of what is needed to alter molecules, produce ionization, and inactivate microorganisms in foods. They also are far in excess of the energy levels of such isotope sources as ^{60}Co and ^{137}Cs, or 1.6×10^{-12} J electron beams, that in the past were considered safe by the FDA for irradiation processing of approved foods.

When ionizing radiations of moderate energy level pass through foods, there are collisions between the ionizing radiations and the food at molecular and atomic levels. Ion pair production results when the energy from these collisions is sufficient to dislodge an electron from an atomic orbit. Molecular changes occur when collisions provide sufficient energy to break chemical bonds between atoms; an important consequence of this is the formation of free radicals.

Free radicals are parts of molecules, groups of atoms, or single atoms that possess an unpaired electron. Stable molecules almost always possess an even number of electrons, and an unpaired electron configuration is an extremely unstable form. Free radicals, therefore, have a great tendency to react with one another and with other molecules to pair their odd electrons and attain stability.

The formation of ion pairs, free radicals, reaction of free radicals with other molecules, recombination of free radicals, and related physical and chemical phenomena provide the mechanisms by which microorganisms, enzymes, and food constituents are altered during irradiation.

Direct Effects

In the case of living cells and tissues, destructive effects and mutations from radiation were originally thought to be due primarily to direct contacts of high-energy rays and particles with vital centers of cells, much as a bullet hits a specific target. The same theory of action was extended to explain changes in nonliving materials and foods. Thus, a change in the color or texture of a food would be due to direct collision of a gamma ray or high-energy beta particle with a specific pigment or protein molecule. Such direct hits unquestionably do occur, but their frequency of occurrence at a given radiation dose probably is not sufficient to explain the major portion of radiation effects in a given substrate.

Indirect Effects

Direct hits need not occur for radiation to affect living or nonliving substrates. Just as radiations colliding with a cell or specific food molecule would produce ion pairs and free radicals, much the same occurs when high-energy radiations pass through water. In this case, water molecules are altered to yield highly reactive hydrogen and hydroxyl radicals. These radicals can react with each other, with dissolved oxygen in the water, and with many other organic and inorganic molecules and ions that may be dissolved or suspended in the water. Thus two hydroxyl radicals upon combining form hydrogen peroxide,

$$\cdot OH + \cdot OH \rightarrow H_2O_2;$$

two hydrogen radicals produce hydrogen gas,

$$\cdot H + \cdot H \rightarrow H_2;$$

a hydrogen radical plus dissolved oxygen yields a peroxide radical,

$$\cdot H + O_2 \rightarrow \cdot HO_2;$$

two peroxide radicals produce hydrogen peroxide and oxygen,

$$\cdot HO_2 + \cdot HO_2 \rightarrow H_2O_2 + O_2.$$

Hydrogen peroxide is a strong oxidizing agent and a biological poison. Hydroxyl and hydrogen radicals are strong oxidizing and reducing agents, respectively. They can enter also into reaction with organic materials and grossly alter molecular structure. Since living cells and food materials are mostly water, the activity imparted to this solvent by radiation constitutes a most important factor contributing to lethal and sublethal changes in living cells and to alteration of food constituents.

A substrate receiving ionizing radiation probably will experience some direct effects and will certainly be affected by indirect effects.

In food irradiation preservation, the primary goal is to inactivate undesirable microorganisms and enzymes while producing minimum changes in other food constituents. Microorganisms and enzymes can be inactivated by direct hits from radiations as well as by indirect effects. Other food constituents, largely in aqueous solution, are largely affected by indirect effects from free radicals produced during radiolysis of water. Therefore, attempts to minimize changes in foods during irradiation have been focused on limiting indirect effects.

Limiting Indirect Effects.   Efforts to limit the indirect effects of radiations have been largely directed at minimizing free radical formation from water and reaction of free radicals with food constituents. Three approaches that have had varying degrees of success, depending on the food material, illustrate this reasoning:

• Irradiation in the frozen state. Free radicals are produced even in frozen water, though possibly to a lesser extent. The frozen state also hinders free radical diffusion and migration to food constituents beyond the site of free radical production. Thus, freezing can limit undesirable reactions.
• Irradiation in a vacuum or under inert atmosphere. As indicated earlier, a hydrogen radical reacting with oxygen will produce a highly oxidative peroxide radical. Peroxide radicals produce hydrogen peroxide. By removing oxygen from the system, such reactions are minimized and food constituents are more protected. However, removal of oxygen and minimization of these reactions also has a protective effect on food microorganisms, limiting the benefits that can be obtained. There is also the problem of getting oxygen out of food systems.

• Addition of free radical scavengers. Ascorbic acid is an example of a compound that has a great affinity for free radicals. Addition of ascorbic acid and certain other materials to food systems results in consumption of free radicals through reaction with these and a sparing of other sensitive pigments, flavor compounds, and food constituents. But a problem exists in incorporating such scavengers throughout non-liquid foods.

Sensitivity to Ionizing Radiations

It is beyond the scope of this chapter to consider the many changes that occur within living systems and biological materials exposed to radiation. Ionizing radiations can alter the structures of organic and biochemical compounds essential to normal life when these radiations are received in high dosage. In foods, just as excessive amounts of heat deteriorate foods, excessive dosages of radiation adversely affect proteins, carbohydrates, fats, vitamins, pigments, flavors, enzymes, and so on. Excessive dosages also can change the protective properties of certain packaging materials such as plastic films and plastic or enamel interior can coatings. In this latter case, however, such dosages generally are in excess of sterilizing or pasteurizing requirements compatible with acceptable food quality.

Excessive dosage has meaning only in terms of specific substrates. As in the case of heat, different food materials vary greatly in sensitivity to ionizing radiations. Bacon, for example, can withstand a radiation dosage of 5.6 Mrad (56 kGy) and retain highly satisfactory organoleptic qualities. Such bacon is microbiologically sterile. Certain proteins, on the other hand, become highly disorganized at far lower doses, showing varying degrees of molecular uncoiling, unfolding, coagulation, molecular cleavage, and splitting out of amino acids, odorous compounds, and ammonia. Egg white is a particularly sensitive mixture of proteins and becomes thin and watery with a moderate radiation dose of 0.6 Mrad (6 kGy). This dosage is insufficient to produce sterility if the egg contains spores of certain bacteria. Therefore, a higher dosage to ensure sterility would be impractical since it would make fresh egg quite unacceptable for most food uses. Many foods cannot be irradiation-sterilized for similar reasons. Some of these, however, are given improved keeping quality by irradiation pasteurization at lower doses.

Some of the more important overall effects and uses of different irradiation doses are indicated in Fig. 11.1. About 10,000 rads will inhibit sprouting of potatoes and slightly more will destroy insects. Several hundred thousand rads will kill yeasts and molds, and this level will pasteurize many foods. Gross destruction of bacterial spores, producing food sterility, requires several million rads.

Dose-Determining Factors

When the purpose of irradiation is food preservation, the choice of dosage must take into account several factors. The more important of these include safety and wholesomeness of the treated food, resistance of the food to organoleptic quality damage, resistance of microorganisms, resistance of food enzymes, and cost. Safety and wholesomeness involve considerations beyond the absence of dangerous radioactivity and pathogens, which will be discussed later.

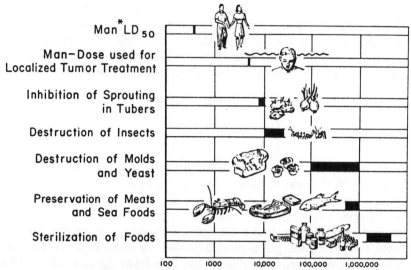

Figure 11.1 Approximate dosages in rads of ionizing radiation for specific effects. *LD_{50} is the whole-body radiation dose which is lethal to 50% of the people exposed. Source: *Goldblith, Food Processing Operations, Vol. 1, M.A. Joslyn and J.L. Heid* (Editors). AVI Publishing Co., Westport, CT, 1963.

Resistance of Food

The extent to which irradiation affects the organoleptic quality of foods varies widely and depends, in complex ways, on the chemical composition and physical structure of foods. These differences in foods set the upper limits of radiation dose that produce foods consumers will accept. Much of the acceptability data on specific foods has been obtained in studies conducted with volunteer troops of the U.S. Army. Pork loin, chicken, bacon, and shrimp have withstood sterilizing doses of 4.8 Mrad well. In some instances off-flavors detected in newly irradiated items largely disappear on storage. Some vegetables also have tolerated 4.8 Mrad. Various fruits have withstood sterilizing doses of 2.4 Mrad. More sensitive meats, fish, and fruits have been found quite acceptable at pasteurizing doses in the range of about 10^5–10^6 rads. These tolerances are reflected in the irradiation product and process specifications suggested in Table 11.2.

Resistance of Microorganisms

The most radiation-resistant microorganism of consequence in foods is *Clostridium botulinum*. Some viruses and microorganisms are yet more radiation resistant but are easily controlled by mild heating prior to irradiation. Many conditions in foods can prevent growth and toxin formation by *C. botulinum*. Among these are pH 4.6 and below, aerobic conditions, extreme dryness of certain foods, refrigeration temperatures below 3°C, and certain preservative chemicals. In foods where these conditions will not exist, *C. botulinum* must be assumed to be present and radiation dosage sufficient for its destruction employed.

As is the case for heat preservation, and based on similar logic, radiation dosages required to destroy spores of *C. botulinum* in various foods have been established. In the irradiation destruction curve shown in Fig. 11.2, D_M is the radiation dose giving

Table 11.2. Irradiation Doses Used for Treating Foods

Dose Range (kGy)	Objectives	Examples and Applications
0.05–0.15	Extension of storage life by inhibition of sprouting	Potatoes, onions, garlic, yams
0.1–0.3	Destruction of parasites to prevent transmission to man through food	Meat
0.1–0.5	Insect disinfestation	Grains, beans, rice, flour, dried fruits, dates, coffee beans
0.075–1.1	Quarantine control against insect pests and plant diseases	Mangoes, beans, fruit paw paws
0.5–1.5	Delay in maturation	Mushrooms, fruit
1.0–5.0	Extension of storage life at ambient temperatures by reducing numbers of bacteria, molds, yeasts	Fruit, vegetables, starch
0.5–10	Extension of refrigerated storage life	Meat, poultry, fish
2.5–10	Increased digestibility, reduction in cooking time	Soybeans, broad beans, lentils, dehydrated vegetables
3.0–13	Elimination of specific pathogens, e.g., salmonellae which cause food poisoning	Frozen meat, animal feeds, poultry, eggs, coconut, spices
35–60	Sterilization of foods to allow longterm storage without refrigeration	Meat

SOURCE: Australia, Parliament, Senate Standing Committee on Environment, Recreation and the Arts, 1988. Use of Ionizing Radiation, Australian Government Publishing Service. Canberra: Australia Govt. Pub. Service.

a 90%, reduction in population. In beef substrate (above pH 4.6), the value of D_M is 0.4 Mrad (4 kGy). It can be calculated that if l kg of beef contained a million botulinum spores and the food received a radiation dose of 12 D_M, then there would be only one chance in a billion that a l-kg can of such food would contain live spores (in calculating, one should not overlook the fact that a rad is an amount of energy per *gram* of material). A 12 D_M dosage (12 × 0.4 Mrad) is 4.8 Mrad. Such a dosage provides a wide margin of safety.

For foods with pH 4.6 and below, *C. botulinum* is not a problem, but other spoilage organisms must be inactivated. The most resistant of these has been found to have a D_M of about 0.2 Mrad. For sterilization with a substantial margin of safety, a 12 D_M dosage (equivalent to 2.4 Mrad) also may be employed.

Resistance of Enzymes

Most food enzymes are more resistant to ionizing radiations than even spores of *C. botulinum*. Enzyme destruction curves comparable to bacterial destruction curves have been established. It has been found that D_E values (radiation doses producing 90%

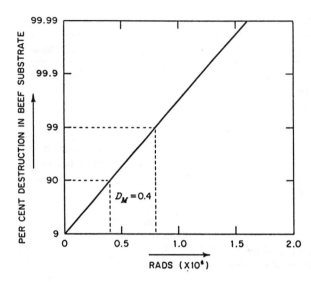

Figure 11.2 The D_M value for *Clostridium botulinum*. Source: *Desrosier and Rosenstock, Radiation Technology in Food, Agriculture, and Biology. AVI Publishing Co., Westport, CT, 1960.*

reduction in enzyme activity) are of the order of 5 Mrad. Four D_E values would produce nearly total enzyme destruction, but such a dosage of 20 Mrad would be highly destructive to food constituents. For these reasons, irradiation alone is not suitable where substantial enzyme destruction is required for storage stability.

This problem has been resolved by the use of various combination processes. Enzymes are readily inactivated by blanching and by certain chemicals. Temperatures of about 70°C for a few minutes are fairly effective. The combination of microorganism-destructive radiation doses plus such heat treatments are extremely effective.

Cost

Cost is another dose-determining factor. Higher doses are obtained by using stronger radiation sources or by exposing foods to less intense radiations for longer periods of time. Either practice increases processing costs. In the case of some foods, irradiation pasteurization may be economically feasible, whereas irradiation sterilization would not be. More broadly speaking, for some foods, preservation by irradiation can be more costly than preservation by heat, refrigeration, or freezing. Where these methods are applicable, there would be little incentive to use irradiation.

On the other hand, irradiation is uniquely suited to certain applications. Low-dose-irradiation pasteurization has extended the normal storage life of refrigerated marine products, meats, fruits, and vegetables from a few days up to several weeks. This can influence marketing practices and safety. The potential of irradiation stabilization is indicated by the irradiated moist shrimp shown in Fig. 11.3, which were stored for a year at room temperature before being prepared as illustrated. Irradiation is being considered as a method to reduce the risk of pathogenic microorganisms in foods that are commonly contaminated. Poulty products and Salmonella, or ground beef and certain types of *Escherichia coli* are examples. Low-dose irradiation reduces the numbers of these and other pathogens reducing the risk of food-borne disease. These processes are a form of pasteurization and not intended to make the products commercially sterile.

Figure 11.3. Radiation-sterilized shrimp stored 1 year at room temperature (4.8 Mrads). *Courtesy of U.S. Army Natick Research Laboratories.*

Controlling and Measuring Radiation

There are many similarities between radiation preservation and the principles discussed in Chapter 8. Like heat, radiations capable of cold sterilization can destroy microorganisms and inactivate many food enzymes, but they also can damage food constituents, and so radiation dose must be carefully controlled. As with heat, it is not just the intensity of the radiation source that is important but the amount of radiation that the food absorbs; thus, processing time is important. Radiation energy must be provided in such a manner that it reaches every particle of food within the mass or container. In the case of heat preservation, conduction and natural convection help distribute heat throughout the container; in the case of cold sterilization by irradiation, with the exception of limited diffusion of free radicals (indirect effect), these processes do not occur. An adequate killing dose must be obtained by uniformly irradiating throughout the entire food mass.

Safety and Wholesomeness of Irradiated Foods

The complex question of safety and wholesomeness of irradiated foods has been investigated in the United States by the Office of the Surgeon General of the Department of Defense, the Nuclear Regulatory Commission, the U.S. Department of Agriculture, the Food and Drug Administration, and others. Several international groups have likewise studied the safety of food irradiaton. In addition to safety from a microbiological standpoint, these studies have been concerned with (a) effects of irradiation treatments on the nutrient value of foods, (b) possible production of toxic substances from irradiation, (c) possible production of carcinogenic substances in irradiated foods, and (d) possible production of harmful radioactivity in irradiation treated foods. These studies have uniformly concluded that irradiation does not result in an unsafe product, particularly at the lower doses now being considered for pasteurization, insect control, and sprout inhibition.

Future for Food Irradiation

The use of irradiation for food in the United States must be specifically approved by the Food and Drug Administration on a food-by-food basis. No current approvals are in effect for food sterilization, but the FDA has approved several low-dose applications. For example, fruits and vegetables may receive low doses to kill insects, slow ripening, or inhibit sprouting. Irradiation of potatoes would be a specific example. Irradiation of poultry, pork, spices, herbs, and other seasonings to reduce microorganisms is also permitted. In approving these applications the FDA not only indicates which foods can be irradiated, but it also sets the amount of radiation that can be applied. Current rules also require that all foods be labeled in a specific way so that consumers know that the food has been irradiated.

Despite these controls, irradiation of food is controversial in the United States and other parts of the world. In some places there is much less controversy. How important food irradiation will eventually become is difficult to assess. Much will depend on future policies of the FDA and similar agencies abroad, relative to safety and approval of specific foods so processed. The improved keeping qualities and microbiological safety of irradiated foods could play a substantial role in international food exports and imports. To this end, international meetings have been held to consider the problems of drafting uniform guidelines and legislation pertaining to traffic in irradiated foods. These meetings have concluded that there is no toxicological hazard resulting from irradiating foods with a dose of up to 1 Mrad (10 kGy). This level of irradiation would help control several pathogens and extend the storage life of many foods. However, it is doubtful that irradation will become commonly used for producing shelf-stable commercially sterile foods in the near future. There was one commercial irradiation facility in operation in the United States in 1993.

MICROWAVE HEATING

Unlike ionizing radiations, microwave energy in food applications is used for its heating properties. Microwave energy is similar to the energy that carries radio and television programs and to the energy involved with radar.

Properties of Microwaves

Microwaves are electromagnetic waves of radiant energy, differing from such other electromagnetic radiations as light waves and radio waves only in wavelength and frequency. Microwaves fall between radio waves and infrared radiations, with wavelengths in the range of about 25 million to 0.75 billion nanometers, which is equivalent to about 0.025–0.75 m. The wavelengths of radio waves and infrared radiations, in comparison, are measured in kilometers and micrometers, respectively (Table 11.1). Wavelengths of electromagnetic radiations are inversely related to frequency, which is the rapidity with which the waveform occurs. Microwave wavelengths of about 0.025–0.75 m correspond to frequencies of about 20,000–400 MHz. (1 Hz = 1 cycle/ sec.). Because microwave frequencies are close to the frequencies of radio waves and overlap the radar range, they can interfere with communication processes, and so the use of specific microwave frequencies comes under the regulations of the Federal Communications Commission. For food applications the approved and most commonly used microwave frequencies are 2450 MHz and 915 MHz.

Microwaves, like light, travel in straight lines. They are reflected by metals, pass through air and many, but not all, types of glass, paper, and plastic materials, and are absorbed by several food constituents including water. When they are reflected or pass through a material without absorption, they do not impart heat to the object. To the extent that they are absorbed, they heat the absorbing material. In heating the material, they lose electromagnetic energy. The terms *loss factor* and *loss tangent* are used to indicate the microwave energy "lost" in passing through, or being entirely absorbed by, various materials under defined conditions. Materials that are highly absorbent of microwaves are said to be highly "lossy." Highly lossy materials are rapidly heated by microwaves. Loss factors for various substances are given in Table 11.3. Since foods differ in composition and in the physical distribution of components, foods vary in their heating patterns from microwave radiation.

The loss factor is also a measure of the degree of penetration of microwaves into materials. Since microwaves lose energy in the form of heat as they penetrate, the greater the loss factor; and the more heat that is produced, the shorter is the distance they can penetrate before all of their energy is consumed. It has been determined that 900-MHz microwaves lose more energy than 2450-MHz microwaves in certain materials, whereas the reverse is true in other materials; in some materials, loss is the same at both frequencies. Where depth of penetration is desired in a given material, one can choose the microwave frequency with the lower loss factor. It can be shown

Table 11.3. Dielectric Constant (ϵ'), Dielectric Loss Factor (ϵ'') and Half-Power Depth (HPD) of Various Materials at 2450 MHz

	ϵ'	ϵ''	HPD (cm)
Water (distilled at 25°C)	78	12.0	1.0
Water + 0.5M NaCl (25°C)		32	0.26
Ice (−12°C)	3.2	0.003	800
Beef (bottom round, cooked, 30°C)		12	0.7
Ham (precooked, 20°C)		23	0.4
Potato			
raw (25°C)		16	0.66
mashed (30°C)		24	0.48
Paper		0.15	14.8
Polyethylene	2.3	0.003	700

SOURCE: Adapted from Schiffman (1990).

that under similar conditions, by the time half of their incident energy is lost, 900-MHz microwaves will penetrate water to a depth of 76 mm, whereas 2450-MHz microwaves will penetrate to a depth of only about 10 mm.

Mechanism of Microwave Heating

Common alternating electric current reverses its direction 60 times a second. Microwaves do the same, but at frequencies corresponding to 915 or 2450 MHz. Food and certain other materials contain molecules that act as dipoles, that is, they exhibit positive and negative charges at opposite ends of the molecule. Such molecules also are said to be polar. Water molecules are polar with the negative charge centered near the oxygen atom and the positive charge nearer the hydrogen atoms.

When microwaves pass into foods, water molecules and other polar molecules tend to align themselves with the electric field. But the electric field reverses 915 or 2450 million times per second. The molecules attempting to oscillate at such frequencies generate intermolecular friction, which quickly causes the food to heat. Quite the same phenomenon occurs in dielectric heating, which is like microwave heating but employs radiations in the frequency range of about 1–150 MHz. Although microwaves generate heat within the food, components with different loss factors do not immediately heat up equally. However, as heat is generated, it also is conducted between food components, tending to equalize temperature. In liquid foods, the heat also is moved by convection. However, these secondary effects must not be confused with the prime mechanism of intermolecular friction, which occurs within the food at the sites of billions of molecules simultaneously.

Differences from Conventional Heating

In conventional heating, employing a direct flame, heated air, infrared elements, direct contact with a hot plate, and so on, the heat source causes food molecules to react largely from the surface inward, so that successive layers heat in turn. This produces a temperature gradient which can burn the outside of a piece of food long before the temperature within has risen appreciably. This is why steak can be crusted on the outside but still be rare on the inside.

In contrast, microwaves penetrate food pieces up to several centimeters of thickness uniformly, setting all water molecules and other polar molecules in motion at the same time. Heat is not passed by conduction from the surface inward, but instead it is generated quickly and quite uniformly throughout the mass. The result is an internal boiling away of moisture. The steam also heats adjacent food solids by conduction. Incidentally, as long as there is free water being converted to steam, the temperature of the food piece does not rise much above the boiling point of water, except as the steam within the food may be under some pressure as it attempts to escape. As a result there is virtually no surface browning or crusting from excessive surface heat. This is a limitation of microwaves in such operations as bread baking, meat cooking, and the like, where crusting or browned surfaces are desired. In such cases, if microwave heating is employed, it must be preceded, accompanied, or followed by a conventional kind of heating to produce such surfaces. On the other hand, low thermal gradient microwave heating lends itself to numerous special applications, as indicated in the list of food applications at the end of this section.

Microwave Generators and Equipment

The most commonly used type of microwave generator is an electronic device called a magnetron. The components of a magnetron are shown in Fig. 11.4. A magnetron is a kind of electron tube within a magnetic field which propagates high-frequency radiant energy. The power output of different size magnetrons is rated in kilowatts. A larger magnetron, or several smaller ones working together, will heat a given quantity of food to a given temperature in a shorter time than a smaller one. Also, it is well to recognize that since microwave energy heats only objects into which it is absorbed, there is a relationship between food load and heating time to a given temperature. Thus, 2 kg of water will take essentially twice the time to come to a boil as will 1 kg of water.

A simple microwave oven consists of a metal cabinet into which is inserted a magnetron (Fig. 11.5). The cabinet frequently is equipped with a metal "fan" to distribute

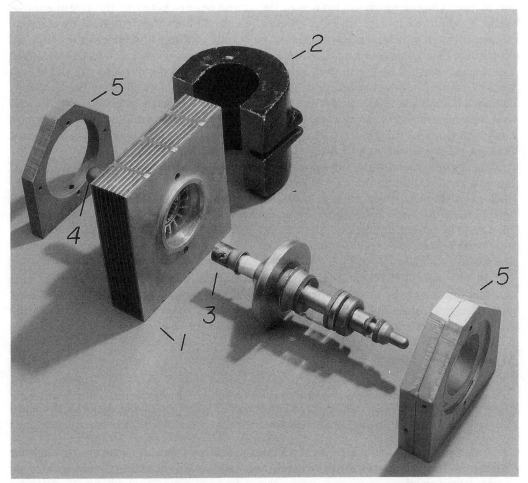

Figure 11.4. Components of a cavity magnetron. (1) Air-cooled anode showing multicavity arrangement; (2) permanent magnet; (3) cathode assembly; (4) antenna radiating elements; (5) mounting blocks. *Courtesy of Raytheon Co.*

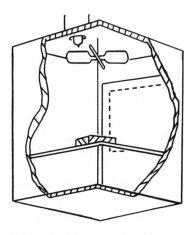

Figure 11.5. Simple microwave oven construction with magnetron at top. Source: *Copson, Microwave Ovens and Frozen Food Make Cents, Microwave Power Institute, 1976.*

the microwaves throughout the cabinet as they are reflected and bounce off the metal fan blades. The microwaves also are reflected and bounce between the metal cabinet walls. Food placed in such an oven (generally raised above the oven floor by a screen or rack through which microwaves can pass) is thus contacted by microwaves from all directions. This speeds heating time and facilitates steam escape. If the food is wrapped, the wrapper should be perforated, or otherwise allow for steam escape to prevent it from bursting. Microwave radiations are not dangerous when confined within properly designed equipment. Since microwaves can cause damage to the eyes and other tissues that may absorb them, safety engineering of microwave ovens and related equipment has evolved to a high degree. All microwave ovens have interlocks or equivalent devices that cut off the power supply when the oven door is opened.

More complex microwave tunnel ovens are equipped with an endless moving belt of low-loss material on which food is conveyed past magnetrons in continuous fashion. Such ovens generally are open at the inlet and outlet ends to receive and discharge the product. In such a case the microwaves are prevented from escaping through the open ends by providing trapping materials which absorb stray microwaves, by providing metal reflectors to turn would-be stray microwaves back into the oven chamber, and by other means.

It is also possible to heat liquid materials continuously with microwaves. In this case the liquid may be pumped through a coil of low-loss glass or like material placed within the microwave heating zone; or the magnetron(s) may be positioned surrounding a low-loss tube through which the liquid is pumped.

Microwave Food Applications

The current and potential uses of microwave heating in the food industry are many and are of growing importance. The following industrial applications listing by the Cryodry Corporation, a manufacturer of microwave heating systems, is highly illustrative.

1. **Baking.** Internal heating quickly achieves desired final temperature throughout the product. Microwaves can be combined with external heating by air or infrared to obtain crust.
2. **Concentrating.** Permits concentration of heat-sensitive solutions and slurries at relatively low temperatures in relatively short times.

3. **Cooking.** Microwaves cook relatively large pieces without high-temperature gradients between surface and interior. Well suited for continuous cooking of meals for large-volume institutional feeding.

4. **Curing.** Effective for glue-line curing of laminates (as in packaging) without direct heating of the laminates themselves.

5. **Drying.** Microwaves selectively heat water with little direct heating of most solids. Drying is uniform throughout the product; preexisting moisture gradients are evened out. Drying is at relatively low temperatures; no part of the product need be hotter than the vaporizing temperature.

6. **Enzyme Inactivation (Blanching).** Rapid uniform heating to the inactivating temperature can control and terminate enzymatic reactions. Microwaves are especially adaptable to blanching of fruits and vegetables without leaching losses associated with hot water or steam. Also, does not overcook the outside before core enzymes are inactivated.

7. **Finish-drying.** When most of the water has been removed by conventional heating methods, microwaves remove the last traces of moisture from the interior of the product quickly, and without overheating the already dried material.

8. **Freeze-drying.** The ability of microwave energy to selectively heat ice crystals in matter makes it attractive for accelerating the final stages of freeze-drying.

9. **Heating.** Almost any heat transfer problem can benefit from the use of microwaves because of their ability to heat in depth without high-temperature gradients.

10. **Pasteurizing.** Microwaves heat a product rapidly and uniformly without the overheating associated with external, high-temperature heating methods.

11. **Precooking.** Microwaves are well suited for precooking "heat and serve" items because there is no overcooking of the surface and cooking losses can be negligible. When the consumer reheats the food by conventional methods, the desired texture and appearance of conventionally cooked items can be imparted.

12. **Puffing and Foaming.** Rapid internal heating by microwaves causes puffing or foaming when the rate of heat transfer is made greater than the rate of vapor transfer out of the product interior. May be applied to the puffing of snack foods and other materials.

13. **Solvent Removal.** Many solvents other than water are efficiently vaporized by microwaves, permitting solvent removal at relatively low temperatures.

14. **Sterilizing.** Where adequate temperatures may be reached (acid foods), quick, uniform come-up time may permit high-temperature–short-time sterilization. Selective heating of moisture-containing microorganisms makes possible the sterilization of such materials as glass and plastic films, which are not themselves heated appreciably by microwaves. This application must be considered cautiously, since escaping steam temperatures generally are not sufficient to kill bacterial spores.

15. **Tempering.** Because the microwave heating effect is roughly proportional to moisture content, microwaves can equalize the moisture in a product that came from a process in a nonuniform condition.

16. **Thawing.** Controlled, rapid thawing of bulk items is possible due to substantial penetration of microwaves into frozen materials.

It must be recognized that several of the above applications may be achieved by other heating methods or combination processes. The choice of method must then depend on relative product quality and cost.

OHMIC HEATING

Ohmic heating was briefly mentioned in Chapter 5. Ohmic heating is one of the newest methods of heating foods. It is often desirable to heat foods in a continuous system such as a heat exchanger rather than in batches as in a kettle or after sealing in a can. Continuous systems have the advantages that they produce less heat damage in the product, are more efficient, and they can be coupled to aseptic packaging systems. Continuous heating systems for fluid foods that contain small particles have been available for many years. However, it is much more difficult to safely heat liquids containing larger particles of food. Beef stew would be one example. This is because it is very difficult to determine if a given particle of food has received sufficient heat to be commercially sterile. This is especially critical for low-acid foods such as beef stew which might cause fatal food poisoning if underheated. Products tend to become overprocessed if conventional heat exchangers are used to add sufficient heat to particulate foods. This concern has hindered the development of aseptic packaging for foods containing particulates. Ohmic heating may over come some of these difficulties and limitations.

Considerable heat is generated when an alternating electric current is passed through a conducting solution such as a salt brine. In ohmic heating a low-frequency alternating current of 50 or 60 Hz is combined with special electrodes. Products in a conducting solution (nearly all polar food liquids are good conductors) are continuously passed between these electrodes. In most cases the product is passed between several sets of electrodes, each of which raises the temperature. Figure 11.6 shows a diagram of such a system.

The major advantage of ohmic heating is that the food particles do not experience a significant temperature gradient from their outside to inside. This means that they can be heated without the usual heat damage associated with excessive surface heating. This is because there are no direct heat transfer surfaces to degrade product. The solid pieces and the liquid are heated nearly simultaneously.

After heating, products can be cooled in continuous heat exchangers and then aseptically filled into presterilized containers in a manner similar to conventional aseptic packaging. Both high- and low-acid products can be processed by this method.

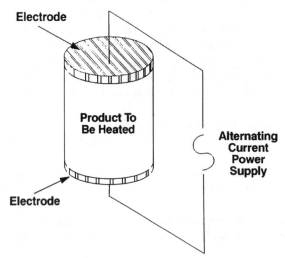

Figure 11.6. Principle of ohmic heating in which an alternating electrical current is passed through a food in a conducting fluid. Source: *D.L. Parrot Food Technology 46*(12)69, 1992.

References

Biss, C.H., Coombes, S.A., and Skudder, P.J. 1989. The development and application of ohmic heating for the continuous heating of particulate foodstuffs. *In* Process Engineering in the Food Industry: Development and Opportunities. R.W. Field and J.A. Howell (Editors). Elsevier Applied Science, London, pp. 17–25.

Buffler, C.R. 1993. Microwave Cooking and Processing: Engineering Fundamentals for the Food Scientist. Chapman & Hall, London, New York.

Decareau, R.V. 1992. Microwave Foods: New Product Development. Food & Nutrition Press, Trumbull, CT.

Derr, D.D. 1993. Food irradiation: What is it? Where is it going? Food Nutr. News *65*(1), 5–6.

Diehl, J.F. 1990. Safety of Irradiated Foods. Marcel Dekker, New York.

Diehl, J.F. 1992. Food irradiation: Is it an alternative to chemical preservatives? Food Addit. Contam. *9*(5), 409–416.

Diehl, J.F. 1993. Will irradiation enhance or reduce food safety? Food Policy *18*(2), 143–151.

Fryer, P.J., De Alwis, A.A.P., Koury, E., Stapley, A.G.F., and Zhang, L. 1993. Ohmic processing of solid-liquid mixtures: Heat generation and convection effects. J. Food Eng. *18*(2), 101–125.

Mason, J.O. 1992. Food irradiation promising technology for public health. Public health Rep. *107*(5), 489–490.

Mudgett, R.E. 1986. Microwave properties and heating characteristics of foods. Food Technol. *40*(6), 84–93.

Parrott, D.L. 1992. Use of ohmic heating for aseptic processing of food particulates. Food Technol. *46*(12), 68–72.

Pauli, G.H. and Takeguchi, C.A. 1986. Irradiation of foods: FDA perspective. Food Rev. Int. *2*(1), 79–107.

Rizvi, S.S.H. 1986. Engineering Properties of Foods. Marcel Dekker, New York.

Robins, D. 1991. The Preservation of Food by Irradiation: A Factual Guide to the Process and its Effect on Food. IBC Technical Services, London.

Rubbright, H.A. 1990. Packaging for microwavable foods. Cereal Foods World *35*(9), 927–930.

Ruley, J. 1989. The nutritional effects of microwave heating. Br. Nutr. Found. Nutr. Bull. *14*(1), 46–62.

Sastry, S.K. and Palaniappan, S. 1992. Ohmic heating of liquid particle mixtures. Food Technol. *46*(12), 64–67.

Satin, M. 1993. Food Irradiation: A Guidebook. Technomic Publ., Lancaster, PA.

Schiffmann, R.F. 1990. Microwave foods: basic design considerations. Tech. Assoc. Pulp Paper Ind. *73*(3), 209–212.

Shukla, T.P. 1992. Microwave ultrasonics in food processing. Cereal Foods World *37*(4), 332.

Thayer, D.W. 1990. Food irradiation: Benefits and concerns. J. Food Qual. *13*(3), 147–169.

Thomas, M.H. 1988. Use of ionizing radiation to preserve food. *In* Nutritional Evaluation of Food Processing. 3rd ed. E. Karmas and R.S. Harris (Editors). Chapman & Hall, London, New York.

Thorne, S. (Editor). 1991. Food Irradiation, Chapman & Hall, London, New York.

Truswell, A.S. 1987. Food irradiation. Br. Med J. *294*(6585), 1437–1438.

Weiser, H.H., Mountney, G.L., and Gould, W.A. 1971. Practical Food Microbiology and Technology. 2nd ed. AVI Publishing Co., Westport, CT.

World Health Organization. 1988. Food Irradiation: A Technique for Preserving and Improving the Safety of Food. World Health Organization, Geneva.

12

Fermentation and Other Uses of Microorganisms

FERMENTATIONS

Fermentations occur when microorganisms consume susceptible organic substrates as part of their own metabolic processes. Such interactions are fundamental to the decomposition of natural materials, and to the ultimate return of chemical elements to the soil and air without which life could not be sustained.

Natural fermentations have played a vital role in human development and are probably the oldest form of food preservation. Although the growth of microorganisms in many foods is undesirable and considered spoilage, some fermentations are highly desirable. Fruit and fruit juices left to the elements acquired an alcoholic flavor; milk on standing became mildly acidic and eventually became cheese; cabbage turned to sauerkraut. These changes tasted good and so early civilizations encouraged the conditions that permitted them to occur. Sometimes the desired results were obtained repeatedly, but this was not always so. It soon was also discovered that certain alcoholic fruit juices and sour milks would keep well, and so part of the food supply was converted into these forms as a means of preservation.

Today, other methods of food preservation are superior to fermentation as means of preserving many foods. In technically advanced societies the major importance of fermented foods has come to be the variety they add to diets. In many less developed areas of the world, however, fermentation and natural drying are still the major food preservation methods, and, as such, are vital to survival of much of the world's population.

The various preservation methods discussed thus far, based on the applications of heat, cold, removal of water, application of radiation, and other principles, all have the common objective of decreasing the numbers of living organisms in foods, or at least holding them in check against further multiplication. In contrast, fermentation, whether for preservation purposes or not, encourages the multiplication of microorganisms and their metabolic activities in foods. But only selected organisms are encouraged, and their metabolic activities and end products are highly desirable. A partial list of fermented foods from various parts of the world is given in Table 12.1. The increasing application of biotechnology and genetic engineering techniques to food production is bringing added importance to food fermentations.

Table 12.1. Some Industrial Fermentations in Food Industries

Lactic acid bacteria
  Vegetables and fruits
    cucumbers → dill pickles, sour pickles, salt stock
    olives → green olives, ripe olives
    cabbage → sauerkraut
    turnips → sauerrüben
    lettuce → lettuce kraut
    mixed vegetables, turnips, radish, cabbage → Paw Tsay
    mixed vegetables in Chinese cabbage → Kimchi
    vegetables and milk → Tarhana
    vegetables and rice → Sajur asin
    dough and milk → Kishk
    coffee cherries → coffee beans
    vanilla beans → vanilla
    taro → poi
  Meats → sausages such as salami, Thuringer, summer, pork roll, Lebanon bologna,
        cervelat
  Dairy products
    sour cream
    sour milk drinks—acidophilus, yoghurt, cultured buttermilk, Bulgarian, skyr, gioddu,
        leban, dadhi, taette, mazun
    butter—sour cream butter, cultured butter, ghee
    cheese—unripened → cottage, pot, schmierkase, cream
          whey → mysost, primost, ricotta, schottengsied
          ripened → Cheddar, American, Edam, Gouda, Cheshire, provolone
Lactic acid bacteria with other microorganisms
  Dairy products
    with propionic acid bacteria—Emmenthaler, Swiss, Samso, Gruyère cheeses
    with surface-ripening bacteria—Limburger, brick, Trappist, Münster, Port de Salut
    with yeasts—kefir, kumiss or kumys
    with molds—Roquefort, Camembert, Brie, hand, Gorgonzola, Stilton, Blue
  Vegetable products
    with yeasts—Nukamiso pickles
    with mold—tempeh, soya sauce
Acetic acid bacteria—wine, cider, malt, honey, or any alcoholic and sugary or starchy
products may be converted to vinegar
Yeasts
  malt → beer, ale, porter, stout, bock, Pilsner
  fruit → wine, vermouth
  wines → brandy
  molasses → rum
  grain mash → whiskey
  rice → saké, sonti
  agave → pulque
  bread doughs → bread
Yeasts with lactic acid bacteria
  cereal products → sour dough bread, sour dough pancakes, rye bread
  ginger plant → ginger beer
  beans → vermicelli
Yeasts with acetic acid bacteria
  cacao beans
  citron
Mold and other organisms
  soybeans—miso, chiang, su fu, tamari sauce, soy sauce
  fish and rice-lao, chao

Courtesy of C. S. Pederson.

Definitions

The term fermentation has come to have somewhat different meanings as its underlying causes have become better understood. The derivation of the word fermentation signifies a gentle bubbling condition. The term was first applied to the production of wine more than a thousand years ago. The bubbling action was due to the conversion of sugar to carbon dioxide gas. When the reaction was defined following the studies of Gay-Lussac, fermentation came to mean the breakdown of sugar into alcohol and carbon dioxide. Pasteur later demonstrated the relationship of yeast to this reaction, and the word fermentation became associated with microorganisms, and still later with enzymes. The early research on fermentation dealt mostly with carbohydrates and reactions that liberated carbon dioxide. It was soon recognized, however, that microorganisms or enzymes acting on sugars did not always evolve gas. Further, many of the microorganisms and enzymes studied also had the ability to break down noncarbohydrate materials such as proteins and fats, which yielded carbon dioxide, other gases, and a wide range of additional materials.

Currently, the term fermentation is used in various ways which require clarification. When chemical change is discussed at the molecular level, in the context of comparative physiology and biochemistry, the term fermentation is correctly employed to describe the breakdown of carbohydrate materials under *anaerobic* conditions. In a somewhat broader and less precise usage, where primary interest is in describing the end products rather than the mechanisms of biochemical reactions, the term fermentation refers to breakdown of carbohydrate and carbohydratelike materials under either *anaerobic* or *aerobic* conditions. Conversion of lactose to lactic acid by *Streptococcus lactis* bacteria is favored by anaerobic conditions and is true fermentation; conversion of ethyl alcohol to acetic acid by *Acetobacter aceti* bacteria is favored by aerobic conditions and is more correctly termed an oxidation rather than a fermentation. Common usage frequently overlooks this distinction and considers both types of reactions to be fermentations. In this and subsequent chapters the common usage of the term fermentation, referring to both the anaerobic and aerobic breakdown of carbohydrates, will be followed.

But the word fermentation also is used in a still broader and less precise manner. The term *fermented foods* is used to describe a special class of food products characterized by various kinds of carbohydrate breakdown; but seldom is carbohydrate the only constituent acted upon. Most fermented foods contain a complex mixture of carbohydrates, proteins, fats, and so on, undergoing modification simultaneously, or in some sequence, under the action of a variety of microorganisms and enzymes. This creates the need for additional terms to distinguish between major types of change. Those reactions involving carbohydrates and carbohydratelike materials (true fermentations) are referred to as "fermentative." Changes in proteinaceous materials are designated *proteolytic* or *putrefactive*. Breakdowns of fatty substances are described as *lipolytic*. When complex foods are "fermented" under natural conditions, they invariably undergo different degrees of each of these types of change. Whether fermentative, proteolytic, or lipolytic end products dominate will depend on the nature of the food, the types of microorganisms present, and environmental conditions affecting their growth and metabolic patterns. In specific food fermentations, control of the types of microorganisms and environmental conditions to produce desired product characteristics is necessary.

Benefits of Fermentation

In addition to the roles of fermentation in preservation and providing variety to the diet, there are further important consequences of fermentation. Several of the end products of food fermentation, particularly acids and alcohols, are inhibitory to the common pathogenic microorganisms that may find their way into foods. The inability of *Clostridium botulinum* to grow and produce toxin at pH values of 4.6 and below has already been cited. Increasing the acidity of foods by fermentation is very common. Foods as diverse as yogurt, hard sausages, and sauerkraut all contain acid as a result of fermentation.

When microorganisms ferment food constituents, they derive energy in the process and increase in numbers. To the extent that food constituents are oxidized, their remaining energy potential for humans is decreased. Compounds that are completely oxidized by fermentation to such end products as carbon dioxide and water retain no further energy value. Most controlled food fermentations yield such major end products as alcohols, organic acids, aldehydes, and ketones, which are only slightly more oxidized than their parent substrates, and so still retain much of the energy potential of the starting materials. Fermentation processes are attended by temperature increases. The energy dissipated as heat represents a fraction of the total energy potential of the original food material no longer recoverable for nutritional purposes.

1) Fermented foods can be *more* nutritious than their unfermented counterparts. This can come about in at least three different ways. Microorganisms not only are catabolic, breaking down more complex compounds, but they also are anabolic and synthesize several complex vitamins and other growth factors. Thus, the industrial production of such materials as riboflavin, vitamin B_{12}, and the precursor of vitamin C is largely by special fermentation processes.

2) The second important way in which fermented foods can be improved nutritionally has to do with the liberation of nutrients locked into plant structures and cells by indigestible materials. This is especially true in the case of certain grains and seeds. Milling processes do much to release nutrients from such items by physically rupturing cellulosic and hemicellulosic structures surrounding the endosperm, which is rich in digestible carbohydrates and proteins. Crude milling, however, practiced in many less developed regions, often is inadequate to release the full nutritional value of such plant products; even after cooking, some of the entrapped nutrients may remain unavailable to the digestive processes of humans. Fermentation, especially by certain molds, breaks down indigestible coatings and cell walls both chemically and physically. Molds are rich in cellulose-splitting enzymes; in addition, mold growth penetrates food structures by way of its mycelia. This alters texture and makes the structures more permeable to the cooking water as well as to human digestive juices. Similar phenomena result from the enzymatic actions of yeasts and bacteria.

3) A third mechanism by which fermentation can enhance nutritional value, especially of plant materials, involves enzymatic splitting of cellulose, hemicellulose, and related polymers that are not digestible by humans into simpler sugars and sugar derivatives. This goes on naturally in the rumen of the cow through the enzymatic action of protozoa and bacteria. It also occurs in the process of preparing silage for animal feeding. Cellulosic materials in fermented foods similarly can be nutritionally improved for humans by the action of microbial enzymes.

Of course, such changes are accompanied by gross changes in texture and appearance of the starting food materials, just as all fermented foods are markedly altered from their unfermented counterparts. Such changes are not looked upon as quality defects. Quite the contrary; particularly in areas of the world where most of human nutrients are derived from plant sources, food materials markedly altered by fermentation commonly are more frequent and relished items of diet than are the natural plant components.

Microbial Changes in Foods

The normal microbial flora associated with foods can produce a very wide range of breakdown products. Depending on the major food substrates attacked, these microorganisms are designated proteolytic, lipolytic, or fermentative. Because of their generally broad complement of enzymes, few types of microorganisms are exclusively proteolytic, lipolytic, or fermentative. Rather, most types exhibit varying degrees of each property, depending on environmental conditions and other factors. Nevertheless, many organisms are characteristically dominant in one or another of these three basic kinds of change produced in food.

Proteolytic organisms, which break down proteins and other nitrogenous compounds, give rise to putrid and rotten odors and flavors considered undesirable beyond certain rather low levels. Similarly, lipolytic organisms, which attack fats, phospholipids, and related materials give rise to rancid and fishy odors and flavors not desired in most foods beyond minor levels. On the other hand, fermentative organisms convert carbohydrates and carbohydrate derivatives largely to alcohols, acids, and carbon dioxide. These end products are not generally offensive to our tastes and add zest to many foods. Moreover, when produced in sufficient concentrations, the alcohols and acids resulting from fermentation inhibit many proteolytic and lipolytic organisms that are capable of food spoilage if not controlled. Herein lies the principle of preservation by fermentation: encourage the growth and metabolism of alcohol and/or acid-forming microorganisms and suppress or control the growth of proteolytic and lipolytic types. Once the fermentative organisms are heavily established, they limit growth of the other types, not only by virtue of their production of alcohol and acid but also because they compete for and consume certain constituents of the food that otherwise would be utilized by the proteolytic and lipolytic organisms.

Fermentation technology is not as simple as the above indicates. It is complex, due to the large number of microorganism types and enzymes on the one hand and the diversity of food systems on the other. Processors rarely deal with systems in which one or two organism types work on one or two food constituents; nor do they generally want only alcohol or acid production to the total exclusion of protein and fat breakdown. The clean, tart taste of fresh cottage cheese is largely due to the conversion by fermentation of lactose into lactic acid. On the other hand, the more complex flavors of Cheddar and Limburger cheeses are due to different degrees of protein and fat breakdown in addition to lactic acid fermentation. To obtain these balanced flavors in certain foods, the fermentation processes must be controlled to balance the microorganism types that may grow in the foods.

Some of the more common and significant types of microbial activity in foods are indicated below. The complex intermediate steps leading to the final results are omitted.

Sugar fermented by yeasts, such as *Saccharomyces cerevisiae* and *Saccharomyces ellipsoideus,* yields ethyl alcohol and carbon dioxide in accordance with the following overall reaction:

$$C_6H_{12}O_6 \xrightarrow{\text{yeast}} 2C_2H_5OH + 2CO_2.$$

This is the basis of wine and beer production and the leavening of bread.

Alcohol from yeast-fermented cider, in the presence of oxygen, will be further fermented by bacteria such as *Acetobacter aceti* to acetic acid as in the reaction

$$C_2H_5OH + O_2 \xrightarrow{\text{Acetobacter aceti}} CH_3COOH + H_2O.$$

This is the mechanism of vinegar production.

Lactose (milk sugar), fermented by *Streptococcus lactis* bacteria, gives lactic acid, which curdles the milk to yield cottage cheese and curd from which other cheeses can be made.

Acids produced from fermentation, in the presence of oxygen, can be further broken down by molds. When this happens, the preservative action of the acid against other microorganisms is lost.

Proteins broken down by proteolytic bacteria such as *Proteus vulgaris* and other organisms yield a wide range of nitrogen-containing compounds that give putrid, fishy, or decayed odors to food.

Lipids broken down by lipolytic bacteria such as *Alcaligenes lipolyticus* and other organisms yield fatty acids. These and their subsequent breakdown products contribute to rancid odors or the characteristic odors of some aged cheeses.

Low-acid foods supporting growth of *Clostridium botulinum* may contain toxins produced by this bacterium. This food-poisoning organism will not grow in fermented foods high in acid.

The types of activities indicated can lead to many interesting and highly significant sequences of reactions. These sequences are either prevented or encouraged, as discussed in the next section, depending on the type of fermented food being produced.

Controlling Fermentations in Various Foods

Among the many factors that influence microorganism growth and metabolism, the most common for controlling food fermentations include level of acid, level of alcohol, use of starters, temperature, level of oxygen, and amount of salt. These factors also determine the types of organism that may grow in a fermented food on later storage.

Acid

The inhibitory effects of acid are exerted whether acid is added directly to the food, is a natural constituent of the food, or is produced in the food by fermentative microorganisms. If not a natural constituent of the food (as it is in oranges or lemons), then acid must be added or formed by fermentation quickly, before spoilage or other harmful microorganisms have a chance to increase substantially in numbers and produce their effects.

Food containing acid may be in a state of preservation, but if oxygen is available and surface molds grow and further ferment the acid, its preservative power is lost. In this way, proteolytic and lipolytic activity may gradually develop on the surface of such food. This can occur during the ripening of Cheddar cheese and constitutes a defect. Acid level can be effectively decreased by neutralization also. Certain yeasts will tolerate moderately high-acid conditions and produce alkaline end products, such as ammonia, from the breakdown of protein. These neutralize previously formed acid and permit subsequent growth of proteolytic and lipolytic bacteria. This is desirable and is encouraged in the surface ripening of Limburger cheese.

These types of changes also occur when raw milk is allowed to ferment naturally (Fig. 12.1). Raw milk generally will be contaminated with a wide variety of microorganisms. After a short period during which freshly drawn raw milk fails to support microbial growth (period of germicidal action), *Streptococcus lactis* dominates the fermentation and produces lactic acid. Eventually, this organism is inhibited from further growth by its own acidity. Bacteria of the genus *Lactobacillus,* also common to milk, are still more acid tolerant than *Streptococcus lactis.* The lactobacilli now take over the fermentation and produce still more acid until the new level becomes inhibitory to their further growth. In the high-acid environment, these lactobacilli gradually die off and acid-tolerant yeasts and molds become established. The molds oxidize acid and the yeasts produce alkaline end products from proteolysis, both of which gradually decrease the acid level to the point where proteolytic and lipolytic spoilage bacteria find the medium satisfactory. The growth of these organisms, especially the increased proteolytic activity, decreases the milk's acidity to the point where it can become more alkaline than the original raw milk. During the period of *Streptococcus* and *Lactobacillus* growth, the milk clots and the curd becomes firm, with little evidence of gas accumulation or development of off-odors. Mold and yeast growth followed by proteolytic and lipolytic bacterial growth digest this curd, produce a gassy condition, and develop off-odors characteristic of putrefaction.

In bread-making, the sugars of dough are fermented with yeast, producing alcohol, carbon dioxide, and minor fermentation products. In typical white bread, the fermentation is not intended for preservation purposes and provides little protection of this kind. Here we are interested in the leavening power of the carbon dioxide gas and the flavors from fermentation. However, there are many varieties of sour breads where the yeast fermentation is accompanied by lactic acid fermentations from organisms of the *Lactobacillus* group. In addition to imparting characteristic flavor, the acid inhibits growth of spore-forming bacteria of the genus *Bacillus* in the dough and later in the bread. Spores of this genus if present in the dough survive the temperatures of baking.

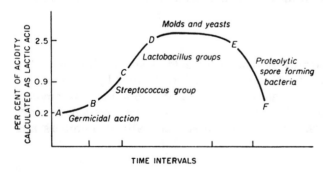

Figure 12.1. Sequence of changes in raw milk in relation to acid concentration. Source: *Weiser et al., Practical Food Microbiology and Technology, 2nd ed. AVI Publishing Co., Westport, CT, 1971.*

They then may produce a gummy condition known as "ropy" bread when nonacid bread is stored under damp conditions. This rarely occurs in sour breads.

Alcohol

Like acid, alcohol is a product of some fermentations and can be a preservative, depending on its concentration. The alcohol content of wines depends, in part, on the original sugar content of the grapes, the type of yeast, fermentation temperature, and level of oxygen. Just as with organisms producing acid, yeasts cannot tolerate their own alcohol and other fermentation products beyond certain levels. For many yeasts this occurs in the range of 12–15% alcohol by volume. Natural wines generally will contain 9–13% alcohol from fermentation. This is not sufficient in itself for complete preservation, and so such wines must receive, in addition, a mild pasteurization treatment. Fortified wines are natural wines to which additional alcohol is added to bring the final alcohol concentration up to about 20% by volume. Such wines may not require further pasteurization.

Use of Starters

When a particular type of microorganism is present in large numbers and is multiplying, it usually dominates its environment and keeps down the growth of other types of microorganisms. In early times, a winemaker or cheesemaker used this principle, without quite knowing why, when part of a previous batch of wine was poured back into fresh grape juice, or cheese milk into fresh milk for the next batch. Such practices continue today in many areas of the world. Fig. 12.2 illustrates one kind of primitive cheese-making currently practiced in Nepal in the Himalayas. Milk from the yak ox ferments under natural conditions until sufficient acid is produced to coagulate curd. The curd is squeezed through the fingers into noodlelike forms, which then are dried in the sun. The fermented milk from one day's operation is used as a starter to initiate fermentation of the next day's production.

In technologically advanced countries, starters of pure cultures obtained from commercial laboratories are used to help ensure controlled fermentation during cheese-making. These cultures, available in dehydrated and in concentrated frozen form, have been developed from selected strains of lactic acid organisms outstanding for their quick and dependable acid production under cheesemaking conditions. Such strains often are resistant to traces of antibiotics and pesticide residues, which may find their way into cheese milk from farm operations, and to bacterial viruses (phages), all of which could otherwise interfere with starter activity. Similarly, special cultures are available for the production of wine, beer, vinegar, pickles, sausage, bread, and other fermented foods. Frequently, the food is heated to inactivate detrimental types of contaminating organisms prior to starter addition.

Temperature

Various microorganisms may dominate a mixed fermentation depending on the fermentation temperature. The sauerkraut fermentation is particularly sensitive to temperature. The effects temperature can have in this fermentation on final acid concentration and time to reach various acidities are indicated in Table 12.2.

In sauerkraut production, three major types of organisms convert the sugar of cab-

Figure 12.2. A method of cheese-making still practiced in Nepal in the Himalayas. *Courtesy of Dr. E. Siegenthaler.*

bage juice to acetic acid, lactic acid, and other compounds. These bacteria include *Leuconostoc mesenteroides, Lactobacillus cucumeris,* and *Lactobacillus pentoaceticus. Leuconostoc mesenteroides* produces acetic acid, some lactic acid, alcohol, and carbon dioxide. The alcohol and acids also combine to form esters, which contribute to final flavor. *Lactobacillus cucumeris* produces additional lactic acid when *Leuconostoc mesenteroides* leaves off. *Lactobacillus pentoaceticus* produces still more lactic acid after *Lactobacillus cucumeris* ceases to be active. The desirable sequence of these fermentations is indicated in Fig. 12.3. *Leuconostoc mesenteroides* requires cool temperatures of about 21°C for optimum growth and fermentation in sauerkraut manufacture. The lactobacilli tolerate higher temperatures.

If temperatures much above 21°C are employed in the initial stages of the fermentation, the lactobacilli easily outgrow *L. mesenteroides* and then their high levels of acid production further prevent growth and fermentation of *L. mesenteroides*. Under these conditions, acetic acid, alcohol, and other desirable products of the *L. mesenteroides* fermentation are not formed. The sauerkraut fermentation, therefore, employs initial low temperatures, which then may be increased somewhat in the later stages of fermentation. This is but one example of manipulating temperature to favor the type of organism desired.

Table 12.2. Effect of Fermentation Temperature on Bacterial Growth and
Acid Production in Sauerkraut with 2.25% Salt Added

Temperature (°C)	Days	Total Acid (%)	Total Bacterial Count (× 100,000/ml)
7	1	0.04	40
	10	0.48	2,640
	20	0.70	2,105
18	1	0.16	2,150
	10	1.23	2,330
	20	1.71	560
32	1	0.71	6,400
	10	2.02	725
	20	—	—
37	1	0.72	15,600
	10	1.76	48
	20	—	—

SOURCE: Pederson and Albury (1954).

Oxygen

The aerobic nature of molds has been discussed. The acetobacter important in vinegar making also requires oxygen, but the yeast that produces alcohol from sugar does it better in the absence of oxygen. *Clostridium botulinum* is a strict anaerobe. Food processors provide or remove air or oxygen as required to encourage or inhibit particular microorganisms.

An organism may have different requirements with respect to oxygen for growth than it has for fermentation activity. Bakers' yeast *(Saccharomyces cerevisiae)* and wine yeast *(Saccharomyces ellipsoideus)* are good examples of this. Both grow better and produce greater cell masses under aerobic conditions, but they ferment sugars more rapidly under anaerobic conditions. Thus, in the commercial production of bakers' yeast, the yeast is grown under aerobic conditions by bubbling air through a yeast-inoculated molasses solution in large tanks. Fermentation is favored in the bread-

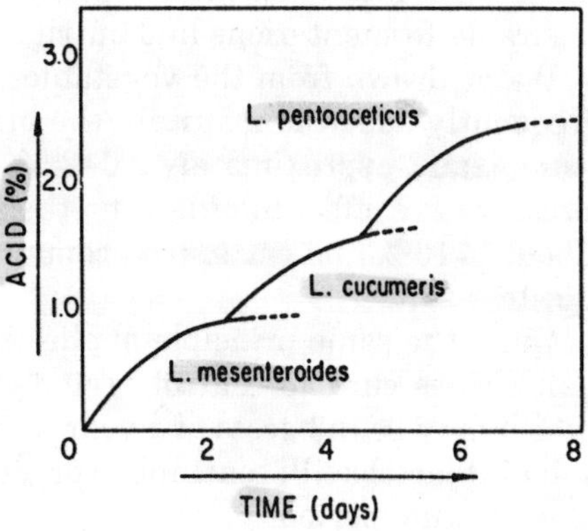

Figure 12.3. Sequence of acid fermentations in sauerkraut manufacture. Source: *Desrosier and Desrosier, Technology of Food Preservation, 4th ed. AVI Publishing Co., Westport, CT, 1977.*

making operation (after sufficient yeast population is established) by the relatively anaerobic conditions of large dough masses.

In traditional vinegar manufacture, the fermentations are separated principally on the basis of the relationships of the fermenting organisms to oxygen. In this two-step process, the first step—involving conversion of the sugar of apple juice to alcohol—may be started under aerobic conditions to stimulate yeast *growth* and increased cell mass. But conditions are soon made anaerobic to favor the *fermentation* of the sugar to alcohol. The second step involving the conversion of alcohol to acetic acid is promoted by highly aerobic conditions, since this transformation is really an oxidative fermentation. This conversion of alcohol to acetic acid is commonly carried out in a vinegar generator. Vinegar generators differ in design but generally consist of large tanks or vats packed with wood shavings to provide a large aerobic surface area. The alcoholic cider, after heavy inoculation with vinegar bacteria, is trickled through the wood shavings while air is blown up through the shavings. The vinegar is removed from the generator when its acetic acid concentration reaches 4% (or somewhat higher), since this is the minimum legal level for acetic acid in vinegar. Operation of a vinegar generator demands close control. Aerobic conditions can encourage mold development, and, as has been pointed out, molds can further break down acid. In addition, excessive aeration can itself oxidize acetic acid further to carbon dioxide and water.

Salt

Microorganisms can be separated on the basis of salt tolerance. The lactic acid-producing organisms used in fermenting olives, pickles, sauerkraut, certain meat sausages, and similar products generally are tolerant to moderate salt concentrations of the order of 10–18%. Many proteolytic and other spoilage organisms that can infect pickle and sauerkraut vats are not tolerant to salt above about 2.5%, and especially are not tolerant to a combination of salt and acid.

In these fermentations, added salt gives the lactic acid-producing organisms an advantage in getting under way even if proteolytic types are present on the cucumbers or cabbage. Once underway, the acid produced by the lactic acid organisms plus the salt strongly inhibits proteolytic and other spoilage types. The salt added to vegetable fermentations also draws water and sugar out of the vegetables. The sugar entering the salt brine provides readily available carbohydrate for continued fermentation in the brine, which complements fermentation within the vegetable tissue from inward diffusion of lactic acid microorganisms. In this way, salt makes the difference between desirable fermentations and outright spoilage.

Water drawn from the vegetables also tends to dilute the brine; thus salt must be frequently added to maintain the brine's preservative salt level. In the production of sauerkraut, approximately 2.0–2.5% salt generally is added to the cabbage, the major preservative effect coming from the acidity formed. Olives are placed in salt brines of about 7–10%, and cucumbers commonly are fermented in brines maintained at about 15–18% salt.

Quite the same principle applies in the making of cheese. It is common practice to salt cheese curd to control proteolytic organisms during the long ripening periods, which may be in excess of a year for certain types of cheese. In this case, various salt-tolerant lactobacilli continue to produce acid and further modify the cheese curd during the ripening period.

Many types of sausage and other fermented meats owe their unique flavors to fermentations by strains of *Leuconostoc, Lactobacillus,* and *Pediococcus* bacteria. Generally, fermentations by these organisms in meat products produce a less acid condition than is common in fermented vegetables. Such products as fermented sauerkraut and pickles have acidities in the range of about pH 2.5–3.5. Fermented meat sausages commonly have acidities in the range of pH 4.0–5.5. This degree of acidity would be marginal as a preservative were it not augmented by the presence of salt and other curing chemicals in the sausages, plus the effects from smoking, cooking, and partial drying of certain of these products.

The desire to reduce the salt content of some fermented products must be undertaken with caution, as this can encourage the growth of undesirable microorganisms including food-borne disease causing organisms. Reduction in salt content must often be accompanied by other methods of inhibiting undesirable organisms while still promoting desirable ones.

MICROORGANISMS AS DIRECT FOODS

Quite apart from the use of microorganisms to produce desirable changes in foods, microorganisms of various types are grown, harvested, and further processed to yield animal feed and human food. Strains have been selected for rapid growth on specific substrates, nutrient content including amount and quality of their protein, organoleptic properties, and other attributes. In some cases, the protein from microorganisms has been isolated and used in foods. The term *single-cell protein* (SCP) has been introduced to designate high-protein food from yeast and other microorganisms, although the practice of growing yeast for food goes back many years. The term single-cell protein can be misleading because it suggests a product that is all protein. Although food yeasts may contain at least one-third protein on a dry basis and it is possible to extract this protein and produce nearly pure protein isolates, this is not commonly done and the entire yeast cell is more often utilized as a food or feed supplement.

Brewers' yeast *(Saccharomyces cerevisiae* or *Saccharomyces uvarum)*, a by-product of beer-making, and bakers' yeast *(S. cerevisiae),* commonly grown on molasses and produced mainly for its leavening property, have long been used as sources of nutrients. These and other yeasts have different carbon and nitrogen assimilation patterns (Table 12.3) and, therefore, can be grown on a wide range of agricultural and industrial by-products, such as hydrolyzed plant tissues, cheese whey, ethanol, petroleum hydrocarbons, and other materials appropriately supplemented with nitrogen and mineral salts.

Yeast solids normally contain about 7–12% of nucleic acids, which can produce harmful effects when yeast is consumed in large amounts. Several methods involving extraction procedures and autolytic degradation by the yeast cell have been developed that can decrease nucleic acids to about 1% yet retain much of the protein. Such procedures need not be employed where the quantities of yeast consumed would contribute less than about 2 g of yeast nucleic acid per day to the adult diet.

GENETIC ENGINEERING

Humans have been breeding food animals, plants, and microorganisms in order to improve characteristics such as yield, disease resistance, appearance, processing attributes, and fermentation characteristics for centuries. Traditional breeding is ac-

Table 12.3. Assimilation Pattern of Commercially Grown Yeast Species

Nutrient	S. cerevisiae	S. uvarum	K. fragilis	C. utilis	C. tropicalis
Glucose	+	+	+	+	+
Galactose	+	+	+	−	+
Maltose	+	+	+	+	+
Sucrose	+	+	−	+	+
Lactose	−	−	+	−	−
Xylose	−	−	−	+	+
KNO$_3$	−	−	−	+	−
Ethanol	(+)	−	+	+	−

SOURCE: Reed and Peppler (1973).
+ means assimilation (growth)
− means no assimilation
(+) means a few strains assimilate

complished by mating a male and a female in hopes that the offspring will have the desired characteristics. In the case of plants and microorganisms this is sometimes accomplished by direct mutation of the genes. Genes contain all the inheritable traits of living organisms. Such breeding is, in reality, selecting and directing the genetic makeup of the animal, plant, or microorganism.

The problems with conventional breeding and mutation as methods of selecting for desirable traits are that they are not always predictable nor successful and can be time-consuming. It is also not possible to cross the species barrier with conventional breeding; that is, desirable traits of oranges, such as ability to produce high amounts of ascorbic acid, cannot be transferred to apples.

In recent years, techniques for more directly manipulating the genetic characteristics of organisms, commonly referred to as genetic engineering, have been developed, following major advances in molecular biology. Genetic engineering through the use of recombinant DNA techniques, cell hybridization, spheroplast or protoplast fusion, and other methods can now remove genes from cells of one organism and reinsert them into the cells of organisms and program them to do specific functions. Progress to date has been greatest with microorganisms, including yeasts, and plants, but progress has also been made with animals.

All of these processes are similar in that they identify the specific genes responsible for desirable traits in one species or type of organism and then transfer these specific genes to a different organism. In this way, the recipient organism acquires these traits. For example, in humans, pancreatic cells produce the protein insulin which is required to control blood sugar. People whose pancreas does not make insulin have diabetes and must take insulin. The genes from human cells which tell the pancreas to make insulin have been transferred to bacteria. The bacteria containing the genes for insulin are thus able to make insulin in culture. This insulin is collected, purified, and used to treat diabetics.

In the food industry, these new techniques are being used to improve yields of traditional fermentation products; convert underutilized raw materials into useful substrates; produce new and improved enzymes, flavoring agents, sweeteners, gums, and other food ingredients; and improve performance of cultures under economical processing conditions. For example, virus-resistant strains of important fermentation

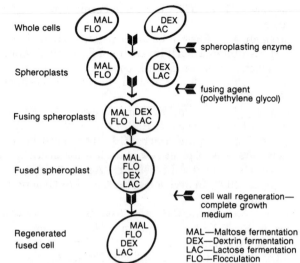

Figure 12.4. Production of improved yeast by spheroplast fusion. *Courtesy of Food Engineering and Labatt Brewing Co.*

MAL—Maltose fermentation
DEX—Dextrin fermentation
LAC—Lactose fermentation
FLO—Flocculation

microorganisms have been developed, as well as organisms which produce enzymes used to make foods such as cheese. In the brewing industry, cell hybridization has been used to produce improved yeast strains. As outlined in Fig. 12.4, the process involves removing cell walls from two yeast strains possessing different desirable attributes, promoting interchange of genetic material through fusion of their spheroplasts, and providing a medium and conditions for cell wall regeneration. The new yeast, which is capable of division and replication, can ferment maltose, dextrin, and lactose and thus has a wider range of utilizable substrates than either starting strain. Further, the property of flocculation facilitates removal of the yeast from the fermentation wort, making for better clarification of fermented beverages and more efficient reuse of yeast. Other important characteristics of the hybridized yeast resulting from initial strain selection include alcohol tolerance, production of desirable flavor compounds, and genetic stability.

Genetic engineering is also finding uses in agriculture. Genes from bacteria which can kill certain insects but are harmless to humans have been transferred to plants. The plants then produce the protein that is toxic to the insect, so that when the insect eats the plant, the insect dies. These proteins have no effect on humans because they are inactivated in the human stomach. This type of genetic engineering may lead to a large reduction in the use of synthetic pesticides.

References

Bacus, J. 1984. Update: Meat fermentation 1984. Food Technol. *38*(6), 59–63.

Anon. 1988. Food Biotechnology. Present and Future: A Study. The Administration, Washington, DC.

Glick, B.R. 1994. Molecular Biotechnology: Principles and Applications of Recombinant DNA. ASM Press, Ft. Washington, PA.

Harlander, S.K. and Labuza, T.P. (Editors). 1986. Biotechnology in Food Processing. Noyes Publications, Park Ridge, NJ.

Kosikowski, F.V. 1977. Cheese and Fermented Milk Foods. 2nd ed.. F.V. Kosikowski, Brookton-dale, NY.

Kurmann, J.A., Rasic, J.L., and Kroger, M. 1992. Encyclopedia of Fermented Fresh Milk Products. An International Inventory of Fermented Milk, Cream, Buttermilk, Whey, and Related Products. Chapman & Hall, London.

Montville, T.J. 1990. The evolving impact of biotechnology on food microbiology. J. Food Safety *10*(2), 87–97.

Mountney, G.J., Gould, W.A., and Weiser, H.H. 1988. Practical Food Microbiology and Technology. 3rd ed. Chapman & Hall, London, New York.

Paredes-Lopez, O. and Harry, G.I. 1988. Food biotechnology review: Traditional solid-state fermentations of plant raw materials application, nutritional significance, and future prospects. Crit. Rev. Food Sci. Nutr. *27*(3), 159–187.

Peters, P. 1993. Biotechnology: A Guide to Genetic Engineering. Wm. C. Brown Publishers, Dubuque, IA.

Prescott, S.C., Dunn, C.G., and Reed, G. 1982. Prescott & Dunn's Industrial Microbiology. 4th ed. G. Reed (Editor). Chapman & Hall, London, New York.

Pederson, C.S. and Albury, M.N. 1954. The influence of salt and temperature on the microflora of sauerkraut fermentation. Food Techol. *8*, 1–5.

Reed, G. and Nagodawithana, T.W. 1991. Yeast Technology. 2nd ed. Chapman & Hall, London, New York.

Reed, G. and Peppler, H.J. 1973. Yeast Technology. AVI Publishing Co., Westport, CT.

Smith, J.L. and Palumbo, S.A. 1981. Microorganisms as food additives. J. Food Prot. *44*(12), 936–955.

Steinkraus, K.H. 1983. Handbook of Indigenous Fermented Foods. Marcel Dekker, New York.

Steinkraus, K.H. 1989. Industrialization of Indigenous Fermented Foods. Marcel Dekker, New York.

Williams, J.G. 1993. Genetic Engineering. Bios Scientific Publishers, Oxford.

Wood-Brian, J.B. 1985. Microbiology of Fermented Foods. Chapman & Hall, London, New York.

13

Milk and Milk Products

Milk and milk products cover a very wide range of raw materials and manufactured products. No attempt is made in this chapter to deal with all of them. Rather, the properties and processing of fluid milk and some of the more common products manufactured from it, such as specialty milks, ice cream, and cheese, are discussed. Butter and margarine are considered in Chapter 16 on fats and oils.

Milk is a unique substance in that it is both consumed as fluid milk with minimal processing and it is the raw material used to manufacture a wide variety of products. Milk also has unique nutritional properties that make it an especially important food, particularly for the young. It is commonly pasteurized and homogenized and its composition is very close to what it was when taken from the cow. Milk can be separated into its principal components, cream and skim milk, which may be further separated into butterfat, casein and other milk proteins, and lactose. These are sold and used as products in their own right; or they may be further processed into butter, cheese, ice cream, and other well-known dairy products. Similarly, the milk may be modified by condensing, drying, flavoring, fortifying, demineralizing, and still other treatments. Whole milk or its components may be combined in various proportions for incorporation into numerous manufactured food products, such as milk chocolate, bread, cakes, sausage, confectionery items, soups, and many other food products not primarily of dairy origin.

FLUID MILK AND SOME OF ITS DERIVATIVES

Milk is the normal secretion of the mammary glands of all mammals. Its purpose is to nourish the young of the species. The nutritional needs of species vary and so it is not surprising that the milk from different mammals differs in composition.

Table 13.1 gives typical analyses of milks produced by various animals and used for human food. Although the cow is the principal source of milk for human consumption in the United States and many other parts of the world, in India most milk is obtained from the buffalo, in southern Europe the milk of goats and sheep predominates, and in Lapland the milk of the reindeer is consumed.

The principal constituents of milk—fat, protein (primarily casein), milk sugar or lactose, and the minerals of milk, which collectively are referred to as ash—vary not only in amounts among the different animal species but also, with the exception of lactose, somewhat in chemical, physical, and biological properties. Thus, the fatty

Table 13.1. Percentage Composition of Milks Used for Human Food

	Total Solids	Fat	Crude Protein	Casein	Lactose	Ash
Cow	12.60	3.80	3.35	2.78	4.75	0.70
Goat	13.18	4.24	3.70	2.80	4.51	0.78
Sheep	17.00	5.30	6.30	4.60	4.60	0.80
Water buffalo	16.77	7.45	3.78	3.00	4.88	0.78
Zebu	13.45	4.97	3.18	2.38	4.59	0.74
Woman	12.57	3.75	1.63	—	6.98	0.21

SOURCE: B. L. Herrington for all except human data, which is from Webb B.H., Johnson A.H., and Alford J.A. (1974).

acids of goat milk fat have different melting points, susceptibility to oxidation, and flavor characteristics than those of cow's milk. Similarly, milk protein of various species may differ with respect to heat sensitivity, nutritional properties, and ability to produce allergic reactions in other species.

This high degree of variability among the milks of different animals becomes especially important in processing operations. The conditions for condensing, drying, cheese-making, and so on that are optimum for cow's milk may not be satisfactory when applied to a dairy in India because the composition of the milk is different. In the remainder of this chapter, unless otherwise indicated, discussion will apply to the milk from cows.

Even the milk from cows will vary in composition, depending on many factors. These include the breed, animal-to-animal variability, age, stage of lactation, season of the year, the feed, time of milking, period of time between milkings, the physiological condition of the cow, including whether it is calm or excited, whether it is receiving drugs, and so on. All of these factors also affect the quality of the milk. Because of these sources of variation, seldom do literature values on the composition of milk agree exactly. Nevertheless, it is useful to remember the approximate composition of cow's milk, since most commercial milk supplies contain the mixed milk from several farms and variations tend to average out. The approximate composition of milk in Table 13.2 is on such a mixed milk. All of the solids in milk (total solids) amounts to approximately 13%. The terms *solids-nonfat* or *milk solids-nonfat* (MSNF) refer to total solids minus the fat, in this case 9%. Milk solids-nonfat also are referred to as *serum solids*. The

Table 13.2. Approximate
Composition of Cow's Milk

Constituents	%
Water	87.1
Fat	3.9
Protein	3.3
Lactose (milk sugar)	5.0
Ash (mineral)	0.7
	100.0
Solids-nonfat	9.0
Total solids	12.9

market price of milk purchased in bulk generally is based on its fat content and to a lesser extent on its solids-nonfat content. These solids of milk further determine the approximate yields of other dairy products that can be manufactured from the milk (Table 13.3).

The most important single factor governing the composition of cow's milk is the breed of the cow. The principal milk-producing breeds are the Ayrshire, Brown Swiss, Guernsey, Holstein, and Jersey. Holsteins generally produce the most milk, but Guernseys and Jerseys produce milk with the highest fat contents (around 5%).

Legal Standards

Milk is the most legally controlled of all food commodities in the United States and many other countries. The minimum standard for fat is regulated by law in each state in the United States with values ranging from 3.0% to 3.8%. Regulations also cover total solids, which for most states in the United States range from 11.2% to 12.25%. There also are standards of composition and regulations against conditions that would constitute adulteration for milk and all important milk products. In the case of fluid milk, federal standards include minima of 3.25% fat and 8.25% MSNF.

Each state and many cities in the United States regulate veterinary inspections on farms and specific sanitary requirements throughout the entire chain of milk handling and milk processing. This is essential to protect health since improperly handled milk can be a source of serious disease.

Milk has been referred to as a human's most nearly perfect food from a nutritional standpoint. Its wholesomeness and acceptability further depend on the strictest sanitary control, and the sanitary practices employed by the dairy industry have for many years been the guide to the entire food industry.

Milk is also highly regulated with respect to pricing structure and permissible marketing practices. An example of the former is that the same supply of milk often

Table 13.3. Approximate Milk Equivalents of Dairy Products

Product	kg Milk Required to Make 1 kg of Product
Butter	22.8
Cheese	10.0
Condensed milk—whole	2.3
Evaporated milk—whole	2.4
Powdered milk	7.6
Powdered cream	19.0
Ice cream—per 3.8 liters (1 gal)[a]	6.8
Ice cream—per 3.8 liters[b] (eliminating fat from butter and concentrated milk)	5.4
Cottage cheese	6.25 (skim milk)
Nonfat dry milk solids	11.0 (skim milk)

SOURCE: Milk Industry Foundation.
[a]The milk equivalent of ice cream per 3.8 liters (1 gal) is 6.8 kg (15 lb).
[b]Plant reports indicate that 81.24% of the butterfat in ice cream is from milk and cream. Thus the milk equivalent of the milk and cream is about 5.4 kg (12 lb).

will be priced according to the end use. Thus, a supplier generally must charge more for milk going into fluid whole milk channels than for milk to be used for manufactured products such as cheese or butter, even when the same milk source is used for both purposes. Control over marketing practices has included laws against standardizing high- and low-fat milks by the addition of butterfat or skim milk although the final blend may be well above the legal minimum fat content.

Many dairy pricing and marketing laws were originally established to protect the interests of producers and processors in a given region. Often, with time, such regulations tend to restrict rather than help a particular segment of the food industry. This has been particularly true in the dairy industry, especially as nondairy or partial-dairy substitute products such as margarine, certain coffee whiteners, and synthetic milks have grown in importance.

Milk Production Practices

The udder of the cow produces milk from components withdrawn from the animal's bloodstream. The milking operation stimulates release of hormones, which in turn act on muscles in the udder causing let-down of milk into the four teat canals. Hand milking in the United States is largely a thing of the past. Milking machines working on a vacuum principle squeeze and suck milk from the teat canals into receiving vessels; or the milk is drawn under vacuum from the milking machine cups through pipes leading to a bulk holding tank in another room.

This tank is provided with refrigeration to quickly cool the milk to 4.4°C or lower to control bacterial growth. Milk secreted by a healthy udder is sterile but quickly becomes contaminated with microorganisms from the external body of the cow and from milk-handling equipment. Milk should not be held in the cold tank on the farm more than 2 days before it is transported to a milk receiving station or milk processing plant. Often it is transported the same day it is produced.

Cooled milk is most commonly transported in bulk tank trucks from the farm to the processing plant. The milk is pumped into insulated stainless steel tanks holding up to 6600 gal. The tank truck driver records the volume of milk collected and removes a small sample for later analysis of fat and total solids on which to base price to the farmer, and for microbiological tests.

The milk, maintained cold in the tank trucks, may go directly to a milk processing plant or milk may be brought to a central receiving station where it is pooled. Here high-fat milk may be blended with low-fat milk before it is shipped to a processing plant. Such a blending of natural milks is legal even in states where standardization by the addition of butterfat or skim milk is not permitted.

Quality Control Tests

Upon receipt of milk at a processing plant, several inspections and tests may be run to control the quality of the incoming product. Some of these tests may have been done at an earlier receiving station. These tests commonly include determination of fat and total solids by chemical or physical analyses; estimation of sediment by forcing milk through filter pads and noting the residue left on the pad; determination of bacterial counts, especially total count, coliform count, and yeast and mold count; determination of freezing point as an index to possible water pickup; and evaluation of milk flavor.

Tests for cells from the cow are sometimes conducted as an indicator of possible infection in the cow's udder. Under special circumstances, tests for detection of antibiotic residues from treated cows and for pesticide residues that may get into the milk from the feed or from other farm use also may be made.

Bacterial counts play a major role in the sanitary quality of milk on which grades are largely based. Generally, fluid whole milk for consumer use, also referred to as market milk, has higher standards placed on it than milk which will be used for manufacturing purposes. The *Grade A Pasteurized Milk Ordinance Recommendations* of the U.S. Public Health Service/Food and Drug Administration provides an excellent guide to the setting of microbiological and sanitary standards, and many cities and states have adopted or patterned their milk regulations after this code. Among various milk products recognized by this ordinance are Grade A Raw Milk for pasteurization, which may not exceed a bacterial plate count of 100,000/ml on milk from individual producers or 300,000/ml on commingled (blended) milk, and Grade A Pasteurized Milk, which may not exceed a total bacterial count of 20,000/ml or a coliform count of 10/ml. These bacterial counts are among many other requirements that have gone into establishing grades.

As for flavor, much milk is received that is not of top quality. Milk may acquire off-flavors from cows eating unusual feeds, by absorption of odors and flavors from unclean barns and excessive bacterial multiplication, by the action of the natural milk lipase enzyme breaking down fat, and by oxidation, which often is caused by the milk coming into contact with traces of copper or iron in valves, pipes, or other milk-handling equipment. As little as 1 part of copper in 10 million parts of milk can cause oxidized flavors that vary in degree and are described as metallic, cardboard, oily, fishy, and so on. For this reason, iron and copper must be kept from coming into direct contact with product or with cleaning water that can contaminate equipment surfaces contacting product; the metal of choice in milk-handling operations is stainless steel.

The type of off-flavor in milk is generally a good clue to its cause, as are defects uncovered by other quality control tests. Defects are reported to farmers with suggested methods of correction. Acceptable milk is now ready for processing.

Milk Processing Sequence

The first step in processing milk may be a further blending of different batches to a specified fat content. All the while the milk is held cold, preferably at 4.4°C.

Clarification

The milk is next passed through a centrifugal clarifier (Fig. 13.1) to remove sediment, body cells from the udder, and some bacteria. Removal of these impurities in the clarifier is facilitated by distributing the milk in thin layers over conical disks which revolve at high speed. Since the milk is in thin layers, these impurities, which differ in density from the liquid milk, need travel only a very short distance under the influence of centrifugal force to be removed from the milk.

Clarification is by no means intended to rid the milk completely of bacteria, and the clarifier was not designed for this purpose. A special machine known as a Bactofuge, operating under much greater centrifugal force, has been designed for a high degree of bacterial removal. But even such machines fail to remove all bacteria from milk

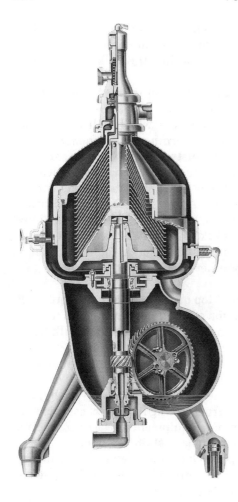

Figure 13.1. Centrifugal milk clarifier. *Courtesy of De La-val Separator Co.*

and could not be depended on to remove all pathogens. The clarified milk is now ready for pasteurization if it is to be processed as market milk.

Pasteurization

The aim of pasteurizing milk is to rid the milk of any disease-producing organisms it may contain and to reduce substantially the total bacterial count for improved keeping quality. Pasteurization also destroys lipase and other natural milk enzymes. Pasteurization temperatures and times for many years were selected to ensure destruction of *Mycobacterium tuberculosis,* the highly heat-resistant non-spore-forming bacterium that can transmit tuberculosis to humans. A treatment of 62°C (143°F) for 30 min, or its equivalent, was employed. In more recent years it was discovered that the organism causing Q fever, *Coxiella burnetii,* was slightly more resistant than the tuberculosis organism and required a treatment of 63°C (145°F) for 30 min, or its equivalent, to ensure its destruction.

The two accepted methods for milk pasteurization today are (1) the batch (holding) method of heating every particle of milk to not less than 63°C and holding at this temperature for not less than 30 min and (2) the high-temperature–short-time (HTST) method of heating every particle of milk to not less than 72°C (161°F) and holding for not less than 15 sec.

Pasteurized milk is not sterile and so it must be quickly cooled following pasteurization to prevent multiplication of surviving bacteria. Pasteurization at these temperatures does not produce an objectionable cooked flavor in milk and has no important effect on the nutritional value of milk. Although slight vitamin destruction may occur, this is easily made up by other foods in a normal diet.

Batch pasteurization is carried out in heated vats provided with an agitator to ensure uniform heating, a cover to prevent contamination during the holding period, and a recording thermometer to trace a permanent record of the time-temperature treatment. But batch pasteurization of milk has largely been replaced with HTST pasteurization.

HTST pasteurization requires the more complex system described in Chapter 8, with its heating plates, holding tube, flow diversion valve, and time–temperature recording charts. Such a HTST system is shown in Fig. 13.2, which also includes such additional equipment as a vacuum chamber to the left, to remove volatile off-flavors from the pasteurized milk.

All pasteurization equipment must be of approved design, and milk inspectors visit often to check on proper equipment operation.

Raw milk contains several enzymes. One of considerable importance in public health work is alkaline phosphatase. This enzyme has heat-destruction characteristics that closely approximate the time–temperature exposures of proper pasteurization. Therefore, if alkaline phosphatase activity beyond a certain level is found in pasteurized

Figure 13.2. Typical HTST pasteurization system for milk. *Courtesy of D. K. Bandler.*

milk, it is evidence of inadequate processing. This enzyme has the ability to liberate phenol from phenolphosphoric acid compounds. Free phenol gives a deep blue color with certain organic compounds. This is the basis for the phosphatase test. In one form of the phosphatase test, disodium phenyl phosphate is the source of phenol and 2,6,dichloroquinonechlorimide is the indicator reagent. Milk is incubated with the disodium phenyl phosphate and then the indicator reagent is added. A blue color indicates improper pasteurization or recontamination with unpasteurized product.

Homogenization

After pasteurization, milk may be homogenized, or homogenization may come just before the pasteurization step if the milk has first been warmed to melt the butterfat.

Milk and cream have countless fat globules that vary from about 0.1 to 20 μm in diameter. These fat globules have a tendency to gather into clumps and rise due to their lighter density than skim milk. Skim milk from which the cream has been removed has virtually no fat globules. The purpose of homogenization is to subdivide the fat globules and clumps to such small size that they will no longer rise to the top of the milk as a distinct layer in the time before the milk is normally consumed. This is an advantage since it makes the milk more uniform and prevents the cream from rising to the top of a container. In addition, subdivision and uniform dispersion of the fat gives homogenized milk a richer taste and a whiter-appearing color, as well as greater whitening power when added to coffee, than the same milk not homogenized.

In one type of homogenizer valve assembly (Fig. 13.3), large fat globules in milk entering at the bottom are sheared as they are pumped under pressure through a tortuous path. They emerge at the top about one-tenth of their original diameter.

Homogenization and cooling of the milk is followed by packaging in paper cartons, plastic jugs, or other types of packages. These are then delivered in refrigerated trucks to retail and other outlets.

Related Milk Products

The processing sequence just described is basic to the production of fluid milk (i.e., market milk). However, slight departures or additional steps in the sequence are employed to produce a number of closely related milk products.

Vitamin D Milk

Milk normally contains vitamin D, but the amount varies with the cow's diet, and with her exposure to sunlight. Since the diet of many children is deficient in vitamin D, it has become common practice to add the vitamin to milk. Milk can be increased in vitamin D activity by irradiating the milk with ultraviolet light, which, in effect, converts the milk sterol, 7-dehydrocholestrol, into vitamin D_3. But the vitamin D level that can be produced this way is somewhat limited. More practical has been the addition of a vitamin D concentrate to milk at a level to bring the potency up to 400 units of vitamin D per quart. Most of the milk consumed in the United States contains added vitamin D. It is generally added before pasteurization.

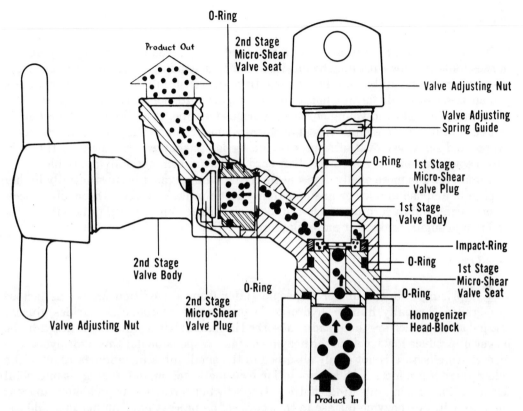

Figure 13.3. Diagram of two-stage homogenizer valve assembly. *Courtesy of Crepaco, Inc.*

Multivitamin Mineral Milk

Most nutritionists oppose the indiscriminate fortifying of general-purpose foods since normal dietary variety tends to supply all needed nutrients and excesses of certain nutrients can be harmful. Nevertheless, fortified milks have been available in some areas in past years.

Vitamins and minerals have been added to give each quart the formerly recommended minimum daily requirements of vitamin A, vitamin D, thiamin, riboflavin, niacin, iron, and iodine. Vitamin C was not commonly added since it is quickly destroyed during milk processing and normal storage. Such fortified milks represented but a small fraction of total milk production, and the new standards of identity for fluid milk products do not permit these multivitamin mineral milks.

Low-Sodium Milk

People with high blood pressure or edema may be on a restricted sodium diet. Low-sodium milk, available in a number of cities, is prepared by passing the milk through an ion exchange resin that replaces sodium with potassium. Low-sodium milk generally contains about 3–10 mg of sodium per 100 ml, whereas untreated milk contains about 50 mg/100 ml.

Soft-Curd Milk

The casein of milk coagulates and forms curd when acted on by enzymes and acid of the stomach. This curd may be harder or softer depending on the amount of casein and calcium in the milk and other factors. Human breast milk forms a soft fluffy curd; pasteurized cow's milk forms a harder more compact curd. Soft-curd milk may be easier to digest by infants and young children. Various treatments are known for producing soft curd milk. These include heat treatments comparable to those used in producing evaporated milk, removal of some calcium by ion exchange, treatment of the milk with enzymes, and other methods. Soft curd milks are commercially available.

Human milk is more easily digested than cow's milk, which is substantially higher in protein and ash and lower in sugar than human milk. To correct these differences, in preparing infant formulas it is common practice to dilute cow's milk with water and to add sugar, making it more like human milk.

Low-Lactose Milk

A surprisingly large number of people suffer from a condition known as lactose intolerance. Normally, humans hydrolyze lactose into its two monosaccharides, glucose and galactose, which are then readily absorbed in the small intestine. Some individuals, however, produce low levels of the enzyme lactase, responsible for this hydrolysis. The disaccharide lactose is not easily absorbed in the small intestine and passes on to the colon where it produces fluid accumulation by osmotic action and undergoes microbial fermentation causing intestinal distress. One way to overcome this problem is to treat the milk with the enzyme lactase in the course of its processing. Another is to add the enzyme to regular milk in the home and hold the milk in the refrigerator for a prescribed time to give the enzyme an opportunity to work. Manufacturers are providing lactase for this purpose.

Sterile Milk

Milk may be sterilized rather than pasteurized by using more severe heat treatments. If the temperature is sufficiently high, the time may be very short, preventing cooked flavor and color change. A typical heat treatment is of the order of 150°C for 2–3 sec. The milk is then quickly cooled and aseptically packaged in cans or appropriate cartons. Such milk is sometimes referred to as ultrahigh-temperature (UHT) processed milk. A system for heating and cooling such milk is shown in Fig. 13.4. In the past, sterile milk found its greatest use where refrigeration was not always available. The potential energy savings from elimination of refrigeration requirements is focusing greater attention on this type of product today.

Evaporated Milk

Evaporated milk is the most widely used form of concentrated milk. It is concentrated to approximately 2.25 times the solids of normal whole milk. To produce evaporated milk, raw whole milk generally is clarified, concentrated, fortified with vitamin D (to give 400 units per 0.946 liter when the evaporated milk is diluted with an equal volume of water), homogenized, filled into cans, sterilized in the cans in large continu-

Figure 13.4. System for ultrahigh temperature (UHT) processing of milk and other products. *Courtesy of Cherry-Burrell.*

ous pressure retorts at temperatures of about 118°C for 15 min, and cooled. This heat treatment gives evaporated milk its characteristic caramelized cooked color and flavor.

Whole milk also is being concentrated approximately 2 : 1 and 3 : 1 and sterilized outside the can by UHT techniques followed by aseptic canning. This gives sterile 2 : 1 evaporated milk and sterile 3 : 1 concentrate without cooked color and flavor. When concentrated milks are heated, the proteins have a tendency to gel and thicken on storage. Various prewarming treatments, addition of stabilizing phosphates and other salts permitted in specified low levels, and, more recently, membrane treatment of the milk to slightly modify its composition may minimize this problem, but its complete elimination is rarely achieved.

Sweetened Condensed Milk

Unlike evaporated milk, sweetened condensed milk is not sterilized, but multiplication of bacteria present in this product is prevented by the preservative action of sugar. The product is made from pasteurized milk that is concentrated and then supplemented with sucrose. Concentration and sugar addition are adjusted to give a sugar concentration of about 63% in the water of the final product. Preservation of milk with sugar has largely given way to milk preservation by heat. However, the combination of sugar and milk solids is convenient in food manufacture, and large amounts of sweetened condensed milk are today used by the baking, ice cream, and confectionery industries. The different steps involved in the manufacture of evaporated milk and sweetened condensed milk are outlined in Fig. 13.5.

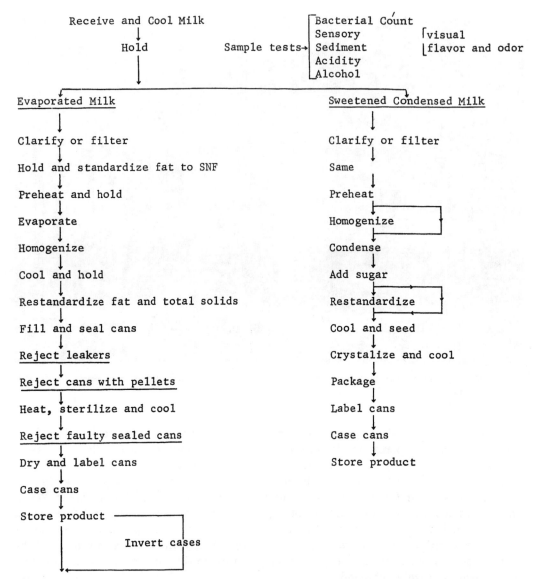

Figure 13.5. Steps involved in evaporated milk and sweetened condensed milk processing. Source: *Hall and Hendrick, Drying of Milk and Milk Products, 2nd ed. AVI Publishing Co., Westport, CT, 1971.*

Dried Whole Milk

Whole milk is dehydrated to about 97% solids principally by spray drying and vacuum drying. The drying operation is quite efficient, but on storage, the whole milk product soon acquires off-flavors, frequently of an oxidized character. How to completely prevent these off-flavors is still not known, and this is why dried whole milk of beverage quality is not yet as important a commercial product as is dried skim milk of beverage quality. However, large quantities of dried whole milk are used in the manufacture of other food products where storage flavor is not as apparent.

Separation of Milk

The products discussed to this point result largely from processing whole milk that has not been separated into its components. But milk can be readily separated into its two principal fractions, cream and skim milk. The separation is made in a centrifugal cream separator, which looks and functions quite like a milk clarifier (Fig. 13.1) but has separate discharge nozzles for cream and skim milk. The cream separator bowl rotates at a speed of several thousand revolutions per minute. Milk enters the top center of the bowl. The skim milk having a heavier density than the whole milk or cream is driven by centrifugal force to the outside of the bowl while the lighter cream moves toward the center of the bowl. The machine can be adjusted to separate cream over a wide range of fat contents for different uses.

Skim milk may be used directly as a beverage or it may be concentrated or dried for use in manufactured foods and animal feeds. Similarly, cream may be used directly, or it may be frozen, concentrated, dried, or further separated to produce butter oil and serum solids. All of these forms are used in manufactured foods.

Low-Fat Milks

Skim milk, which may contain 0.5% fat or less, has been consumed as a beverage for many years. Other low-fat milks may contain 1% or 2% fat. Because fat-soluble vitamins are removed when the fat is separated, the new federal standards for skim milk and low-fat milk require addition of vitamin A. Vitamin D also may be added, but it is optional.

Milk Substitutes

The prices of milk, cream, and other dairy products are largely determined by their milk fat (butterfat) contents. Butterfat for many years has sold for about five times the price of common vegetable fats and oils. As is discussed in Chapter 16, current technology can modify many fats and oils of vegetable, marine, or animal origin to perform nearly interchangeably in numerous food applications. This, of course, is the basis for the margarine industry.

Recently, various substitutes for dairy products other than butter have become commercially important. These have included vegetable fat frozen desserts (ice cream substitutes), vegetable fat coffee whiteners, and vegetable fat whipped toppings. Vegetable fat milk substitutes also have appeared. When these milk substitutes are made by combining nondairy fats or oils with certain classes of milk solids, the resulting products are referred to as filled milks. The term *imitation milk,* on the other hand, has been used to describe products that resemble milk but contain neither milk fat nor other important dairy ingredients. Use of the term *imitation,* with its negative connotation, is no longer a legal requirement for these types of products provided they meet certain nutritional requirements and are labeled so as not to be misleading. Their compositions from a nutritional standpoint are of the greatest significance in view of the important role milk has in the diets of persons of all ages.

Today it is possible to manufacture fluid milk substitutes of considerable quality. Generally, such products are made from skim milk or reconstituted skim milk powder plus coconut fat or some other vegetable fat. Additional ingredients such as monoglyceride and diglyceride emulsifiers, carotene (for color), vitamin D, and other substances commonly are added. The products are pasteurized, homogenized, packaged, and marketed quite like market milk. These filled milks may not be called milk but go under such names as Melloream (Mellorine is the name of a vegetable fat ice cream), protein drink, and other trade names.

Imitation milks and beverages made from casein derivatives (caseinates) or soybean protein and vegetable oils represent a still further departure from natural milk. Such products are being manufactured and sold in Asia and other regions. Carefully formulated, they can be of considerable importance nutritionally, since they generally can be produced at a lower cost than natural milk.

Milk components such as milk proteins, lactose, butterfat, and modified forms of these are increasingly being used as functional and nutritional ingredients in other manufactured foods. Milk components find their way into a variety of products ranging from confections to infant formula.

Growth Hormones and Milk Production

In the 1950s it was discovered that cows would produce more milk when injected with an extract of the pituitary gland from other slaughtered cows. Subsequently, it was discovered that a small protein was responsible for this increased milk production and that this protein also affected the growth rate of young animals. The protein is now called Bovine Growth Hormone (BGH) or more technically, Bovine Somatotropin (BST). In fact, all animals have some type of similar growth hormone, including humans. At one time it was hoped that BST could be used to treat human dwarfism which results from a genetic inability to produce human growth hormone. Unfortunately, growth hormones differ among species and so hormone from one species is not active in another.

By transferring the gene responsible for making BGH from a cow to a bacterium, scientists have learned how to make an inexpensive and readily available source of BGH. When given injections of BGH, lactating cows not only produce 10–15% more milk, but they do more efficiently with regard to the amount of feed required. The dairy industry around the world is interested, of course, in increased efficiency and has begun to use BGH. In 1993, the U.S. Food and Drug Administration approved the use of BGH (BST) to treat lactating cows in the United States.

This technology is not without controversy. Some worry about the safety of milk from BGH-treated cows. However, numerous scientific studies and several review panels have found no reason for concern. Others worry that BGH will cause economic disruption in the dairy industry and that it will speed the decline in the numbers of family farms in the United States.

ICE CREAM AND RELATED PRODUCTS

Ice cream was known in England in the early 1700s, but was still a rare item when served in the United States to White House guests by Dolly Madison in 1809. Today, multicylinder continuous freezers can turn out over 1000 gal of uniformly frozen ice cream per hour. In the United States about 1 billion gal of ice cream and related products are consumed annually.

Composition of Ice Cream

Dairy ingredients in many forms are used in the manufacture of ice cream and related products. These may include whole milk, skim milk, cream, frozen cream, butter, butter oil (which contains about 99% butterfat), condensed milk products, and dried milk products. Ice cream is composed of milk fat (butterfat) and milk solids-nonfat (MSNF) derived from these ingredients, plus sugar, stabilizer, emulsifier, flavoring materials, water, and air.

The mixture of these constituents, before the air is incorporated and the mixture frozen, is known as the ice cream mix. The mix composition may be made richer or leaner in fat, MSNF, and total solids, depending on market requirements; but in addition, a mix of chosen fat and MSNF composition can be formulated from various combinations of the basic dairy ingredients. In typical commercial operations the supply and cost of dairy ingredients varies throughout the year, and so the ice cream plant manager frequently adjusts mix formulas to keep the overall ice cream composition constant at the lowest possible cost.

Typical compositions of commercial ice creams and related products are listed in Table 13.4. A good average ice cream would contain about 12% milk fat, 11% MSNF, 15% sugar, 0.2% stabilizer, 0.2% emulsifier, and a trace of vanilla. This would give 38.4% total solids and the remainder would be water. To this might be added other ingredients such as nuts, fruit, chocolate, eggs, and additional flavorings. Deluxe and

Table 13.4. Approximate Percentage Composition of Commercial Ice Cream and Related Products

Milk Fat	MSNF[a]	Sugar	Stabilizer and Emulsifier	Approximate Total Solids
		Economy Ice Cream		
10	10–11	13–15	0.30–0.50	35.0–37.0
12	9–10	13–15	0.25–0.50	
		Good Average Ice Cream		
12	11	15	0.30	
14	8–9	13–16	0.20–0.40	37.5–39.0
		Deluxe Ice Cream		
16	7–8	13–16	0.20–0.40	
18	6–7	13–16	0.25	40.0–41.0
20	5–6	14–17	0.25	
		Ice Milk		
3	14	14	0.45	31.4
		Good Average Ice Milk (Soft Serve)		
4	12.0	13.5	0.40	
5	11.5	13.0	0.40	29.0–30.0
6	11.5	13.0	0.35	
		Sherbert		
1–3	1–3	26–35	0.40–0.50	28.0–36.0
		Ice		
—	—	26–35	0.40–0.50	26.0–35.0

SOURCE: Arbuckle (1986).
[a] Milk solids-nonfat.

French ice creams may have 18% fat; economy ice creams, 10% fat; and ice milk products only 4% fat. Fruit-flavored sherbets usually contain less than 2% fat, and fruit ices generally contain no fat.

Milk fat is the most expensive major ingredient of ice cream, and so the higher the fat content, generally the more expensive the product. State and federal regulations covering compositions of frozen desserts are largely based on milk fat and total milk solids contents. For example, according to federal standards, plain ice cream may contain no less than 10% milk fat and 20% total milk solids, whereas fruit, nut, or chocolate ice cream may contain no less than 8% milk fat and 16% total milk solids. There are also allowances for other ingredients. Products with leaner compositions may not be called ice cream. Federal standards also specify minimum compositions of ice milk and other frozen desserts. Proposals to change the standards have recently resulted in regulations permitting replacement of up to 25% of the MSNF in ice cream and related products with whey solids. Standards are subject to further changes, especially as the demand for lower-fat products increases.

These compositions are based on the ice cream exclusive of air; that is, percentages are based on weight of the ice cream. But ice cream is made to contain a great deal of air and is truly a whipped product. This air, uniformly whipped into the product as small air cells, is necessary to prevent ice cream from being too dense, too hard, and too cold in the mouth.

The increase in volume caused by whipping air into the mix during the freezing process is known as overrun. The usual range of overrun in ice cream is from 70% to 100%. If ice cream has 100% overrun, then it has a volume of air equal to the volume of mix that was frozen. In other words, 1 liter of mix makes 2 liters of frozen ice cream of 100% overrun. The overrun of any ice cream can be calculated from the formula:

$$\% \text{ overrun} = \frac{(\text{volume of ice cream} - \text{volume of mix}) \times 100}{\text{volume of mix}}$$

The maximum allowable overrun also is specified in federal and state standards by defining the minimum permissible weight per volume of product.

Functions of Ingredients

Each of the major ingredients in ice cream serves specific functions and contributes particular attributes to the final product. Milk fat gives the product a rich flavor and its smooth texture and body. The fat also is a concentrated source of calories and contributes heavily to the energy value of ice cream. Milk solids-nonfat contribute to the flavor and also give body and a desirable texture to ice cream. Higher levels of MSNF also permit higher overruns without textural breakdown. Sugar not only adds sweetness to the product but lowers the freezing point of the mix so that it does not freeze solid in the freezer. The sugar may be sucrose from cane or beet sources or it may be dextrose from corn syrup, or dextrose–fructose mixtures.

The stabilizers used in ice cream are generally gums such as gelatin, gum guar, gum karaya, seaweed gums, pectin, or manufactured gums of the carboxymethyl cellulose type, which are cellulose derivatives. The stabilizers form gels with the water in the formula and thereby improve body and texture. They also give a drier product, which does not melt as rapidly or leak water. The stabilizers by binding water also help to prevent large ice crystals from forming during freezing, which would give the product a coarse texture.

Egg yolk is a good natural emulsifier due to its content of lecithin. Commercial emulsifiers are numerous and generally contain monoglycerides and diglycerides. Emulsifiers help disperse the fat globules throughout the ice cream mix and prevent them from clumping together and churning out as butter granules during the freezing–mixing operation. Emulsifiers also improve whipping properties to reach desired over-run and further help to make ice cream dry and stiff.

Flavors give ice cream variety and consumer appeal. Vanilla is still the most popular flavor, followed by chocolate, strawberry, and a very large number of fruit, nut, and other combinations.

Manufacturing Procedure

The first step in preparing ice cream mix is to combine the liquid ingredients in a mixing vat and bring them to about 43°C. The sugar and dry ingredients are next added to the warm mix which helps dissolve them. Gross particulates such as nuts or fruits are not added at this time since they would be disintegrated during subsequent processing. Instead, they are added during the freezing step.

Pasteurization

The mix is now pasteurized by a batch or continuous heating process. Pasteurization temperatures are higher than for plain milk since the high fat and sugar contents tend to protect bacteria from heat destruction. Common temperatures for batch pasteurization are 71°C for 30 min, and for continuous HTST pasteurization, 82°C for 25 sec. Except for the higher temperature, pasteurization equipment is much the same as that used for milk.

Homogenization

The pasteurized mix is next homogenized at the temperature it comes from the pasteurizer. A two-stage homogenizer may be used (Fig. 13.6) with the mix pumped at a pressure of 1.7×10^7 Pa (2500 psi) through the first-stage valve and 4.1×10^6 Pa (600 psi) through the second-stage valve. Homogenization breaks up fat globules and fat globule clumps and, together with the added emulsifiers, prevents churning of fat into butter granules during the freezing operation. Homogenization also improves the overall body and texture of ice cream. After homogenization, the mix is cooled to 4.4°C.

Ageing the Mix

The mix is held anywhere from 3 to 24 h at a temperature of 4.4°C or lower in vats. During ageing the melted fat solidifies, the gelatin or other stabilizer swells and combines with water, the milk proteins also swell with water, and the viscosity of the mix is increased. These changes lead to quicker whipping to desired overrun in the freezer, smoother ice cream body and texture, and slower ice cream melt-down. Some manufacturers of stabilizers and emulsifiers claim that through the use of their prod-

Figure 13.6. Ice cream homogenizer.

ucts, ageing time may be drastically reduced or even eliminated, but ageing is still employed in many ice cream plants.

Freezing

The mix is now ready to be frozen. The cold, thoroughly blended mix is pumped to a batch or continuous freezer. Continuous freezers with multiple freezing chambers (Fig. 13.7) are more common in large manufacturing operations.

Mix and air enter the freezing cylinders, which are chilled by circulating refrigerant between double walls. The main purposes of the freezing operation are to freeze the mix to about −5.5°C and to beat in and subdivide air cells.

Freezing must be quick to prevent the growth of large ice crystals that would coarsen texture, and air cells must be small and evenly distributed to give a stable frozen foam. These are accomplished within the freezing chamber, which is of the scraped-surface type described in Chapter 9. The freezing chamber is provided with a special mixing element or dasher (see Fig. 5.6). The rotating dasher with its sharp scraper blades shaves the layers of frozen ice cream off the inner freezer wall as they are formed. This prevents buildup of an insulating layer, which would decrease freezing efficiency of the still colder freezer wall. The ice cream scrapings mixed into the remaining mix in the freezing cylinder also serve to seed the mix with small ice crystals, which speeds freezing of the mass. The dasher's rods and bars also beat air into the freezing mass, much as in the whipping of cream or egg white.

Mix passing through the freezer cylinder is frozen and whipped in about 30 sec or

Figure 13.7. Multiple-chamber ice cream freezer. *Courtesy of D. K. Bandler.*

less to a temperature of about −5.5°C. At this temperature not all of the water is frozen, and the ice cream is semisolid; in this condition it is easily pumped out of the cylinder as a continuous extrusion by the incoming unfrozen mix and the propelling action of the dasher.

The semisolid ice cream emerging from the freezer goes directly into packaging cartons or drums. The consistency of the ice cream entering the cartons is that of soft ice-cream-like products sold at roadside stands.

Various kinds of nozzles and filling attachments are available to make novelty ice cream as the product is extruded from the freezer cylinder. Thus, for example, three flavors can be pumped from three freezer cylinders through a three-compartment square nozzle to give the common vanilla, chocolate, strawberry block.

Ice Cream Hardening

Cartons of semisolid ice cream are placed in a hardening room where a temperature of about −34°C is maintained. Storage in the hardening room freezes most of the remaining water and makes the ice cream stiff. When stiff, the product is ready for sale.

Physical Structure of Ice Cream

The physical structure of ice cream should be understood, since changes in physical structure are the cause of several common defects in this product. As mentioned earlier, ice cream is a foam containing air cells which constitute overrun, and that the overrun will give ice cream with approximately twice the volume of the original mix. The dairy foam illustrated in Fig. 13.8 is similar to the foam in ice cream. In the frozen ice cream foam, the films of mix surround the air cells. The fat globules are dispersed within the films or layers of mix. Also within the films are the frozen ice crystals. As ice cream ages in storage, foams can shrink. In addition, weakened films of mix can collapse, causing the ice cream to lose volume. This can be excessive if the mix is low in solids, and represents a serious defect. Figure 13.9 is a photomicrograph of the internal structure of ice cream with more detail. The white areas marked *b* are air cells. All the rest are films of frozen mix surrounding the air cells. Within the films are ice crystals, solidified fat globules, and insoluble as well as dissolved sugars, salts, proteins, and other mix constituents. If the ice crystals marked *a* become too large, as occurs when fluctuating storage temperatures permit repeated partial thawing and refreezing, the ice cream becomes coarse and icy. If there is too much lactose from excessive milk solids and it should crystallize out, the ice cream becomes grainy or sandy. In addition to foam collapse and loss of overrun from formulas low in solids, excessive shrinkage can result from partial melting at too high a freezer storage temperature. Shrinkage due to mechanical compaction also occurs when ice cream is dipped from tubs to make cones; this is called dipping loss. Other textural defects make ice cream gummy, crumbly, curdy, watery, and so on, due largely to poor mix formulations. Ice cream also may have flavor defects common to other dairy products; these include cooked flavor, oxidized flavor, or even rancidity if made from off-flavor dairy ingredients. In addition, ice cream may have a host of unnatural flavors from poor quality flavoring ingredients.

Other Frozen Desserts

Many frozen desserts other than ice cream are available. They differ in their compositions (Table 13.4) and physical characteristics. Most are manufactured with much the same equipment and in accordance with the same principles used in making ice cream. Several of these frozen desserts are simply varieties of ice cream. Thus, we have plain ice cream, fruit ice cream, and nut ice cream with about 8–14% milk fat; deluxe ice creams with about 16–20% milk fat; French ice cream and frozen custard, which contain liberal quantities of egg yolk; parfait and spumoni, which are high in fat and generally also contain fruits and nuts; and other products.

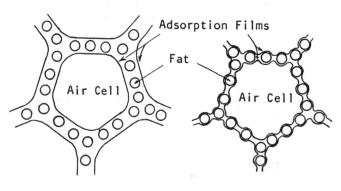

Fresh Foam   Old Foam

Figure 13.8. Diagram of three-phase dairy foam system. *Courtesy of H. H. Sommer.*

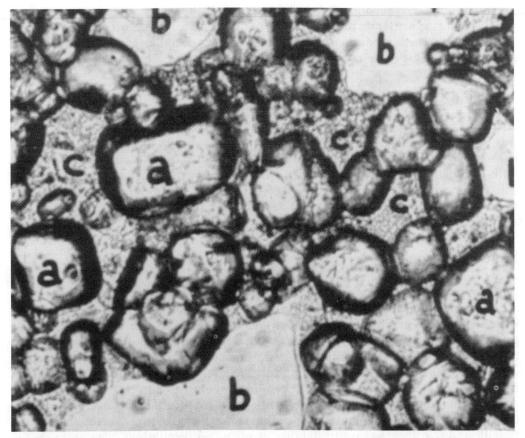

Figure 13.9. Photomicrograph of the internal structure of ice cream. Source: Arbuckle *Ice Cream, 4th ed. AVI Publishing Co., Westport, CT, 1986.*

Lower-fat products include "frozen shakes" and "soft ice milk" with milk fat contents from about 6% down to about 3%. These are the popular products served directly from the freezer at drive-in restaurants and other retail establishments. In most marketing areas, these products may not be called ice cream but instead are referred to as "soft serve" or by trade names.

Sherbets usually contain less than 2% milk fat and corresponding low levels of other milk solids. Sherbets, which are principally, water, sugar, and tart fruit flavorings, have low overruns of 30–40%. Ices are similar to sherbets but generally contain no dairy products and also have low overruns of 25–30%.

Many of these products and related items are loosely referred to by different names in different areas and parts of the world. Standards for several of these products, however, are quite specific. A newcomer among frozen desserts is frozen yogurt, which is made from milk or milk solids that have been fermented by lactic acid producing organisms.

Some ice cream and ice milk-type products are being manufactured with vegetable fat in place of milk fat with a cost advantage in manufacture. Federal law requires that such products be appropriately labeled if they move in interstate commerce. This is to protect dairy interests and to eliminate any question of intended deception. There are also a number of products in which all or part of the fat has been replaced with a fat substitute in an attempt to maintain the desirable texture of fat without the calories. The success of these products varies and no fat substitute has completely mimicked all properties of fat in ice cream and related products.

CHEESE

In addition to being delightful foods that contribute variety and interest to our diets, cheeses of various kinds always have been important sources of nutrients wherever milk-producing animals could be raised. Whereas today the gourmet may pay several dollars per pound for cheese, at the other extreme, in less developed regions where milk rapidly spoils because of lack of refrigeration, cheese may be a staple of diet sometimes made under the most primitive conditions.

Kinds of Cheese

Cheese may be defined as a product made from the curd of the milk of cows and other animals, the curd being obtained by the coagulation of the milk casein with an enzyme (usually rennin), an acid (usually lactic acid), and with or without further treatment of the curd by heat, pressure, salt, and ripening (fermentation) with selected microorganisms. This broad definition, however, does not cover all cheeses, since some are made from milk whey solids that remain after removal of coagulated casein. Further, vegetable fats and vegetable proteins are being used in cheese-like products.

Cheese-making is an old process and still retains aspects of an art even when practiced in the most modern plants. Part of this is due to the natural variation common to milk and the imperfect controllability of microbial populations. The basic cheese types evolved as products of different types of milk, regional environmental conditions, accidents, and gradual improvements by trial and error. There are over 800 names of cheeses, but many of the names describe similar products made in different localities or in different sizes and shapes. Of these, however, there are basically only about 18 distinct types of natural cheeses, reflecting the different processes by which they are made. These include brick, Camembert, Cheddar, cottage, cream, Edam, Gouda, hand, Limburger, Neufchatel, Parmesan, Provolone, Romano, Roquefort, sapsago, Swiss, Trappist, and whey cheeses.

The way some of the multiplicity of subtypes and different names has arisen can be illustrated by Cheddar cheese. In Fig. 13.10 are shown types of cheese hoops in which Cheddars may be pressed, giving rise to different sizes and shapes. The cheeses then get such names as Longhorns, Picnics, Daises, Twins, and so on. But they all are of the Cheddar type.

Classification by Texture and Kind of Ripening

A useful means of classifying the types and important varieties of cheeses is indicated in Table 13.5. It is based largely on the textural properties of the cheeses and the primary kind of ripening. Thus, there are hard cheeses, semihard cheeses, and soft cheeses, depending on their moisture content; and they may be ripened by bacteria or molds, or they may be unripened. The bacteria may produce gas, and so form eyes as

Figure 13.10. Various types of cheese hoops. *Courtesy of Damrow Co.*

Table 13.5 Classification of Cheeses

SOFT
  Unripened:
    Low Fat—cottage, pot, bakers'.
    High Fat—cream, Neufchatel (as made in United States).
  Ripened: Bel Paese, Brie, Camembert, cooked, hand, Neufchatel (as made in France).
SEMISOFT:
  Ripened principally by bacteria: brick, Munster.
  Ripened by bacteria and surface microorganisms; Limburger, Port du Salut, Trappist.
  Ripened principally by blue mold in interior: Roquefort, Gorgonzola, Blue, Stilton,
    Wensleydale.
HARD:
  Ripened by bacteria, without eyes: Cheddar, Granular, Caciocavallo.
  Ripened by bacteria, with eyes: Swiss, Emmentaler, Gruyere.
VERY HARD (grating):
  Ripened by bacteria: Asiago old, Parmesan, Romano, sapsago, Spalen.
PROCESS CHEESES:
  Pasteurized, cold-pack, related products.
WHEY CHEESES:
  Mysost, Primost, Ricotta.

SOURCE: Sanders (1953).

in the case of Swiss cheese, or they may not produce gas as in the case of Cheddar and so no eyes are formed. Among the soft and semisoft cheeses, Limburger is ripened primarily by bacteria, and Camembert by a mold; cottage cheese is not ripened.

The classification is extended to include "process" cheeses, which are essentially melted or blended forms of the above cheeses, and whey cheeses, which are made from the whey remaining after coagulation and removal of the casein. Whey cheeses are high in beta-lactoglobulin and alpha-lactalbumin, the second and third principal proteins in amount in milk. These are not coagulated by rennin or by the acid in most cheese-making processes and so they remain soluble in the whey. However, they can be easily coagulated from the whey as curds by heating.

All of the major types of cheese can fit into a classification such as this. The approximate percentage compositions of several of these cheeses are given in Table 13.6.

Cheddar Cheese—Curd-Making and Subsequent Operations

All cheese types begin with curd-making and then involve various manipulations of the curd or whey. Illustrative of the cheese-making process is the preparation of Cheddar, the most popular cheese in the United States, Canada, and England.

Milk contains fat, proteins (principally casein, less beta-lactoglobulin and still less alpha-lactalbumin), lactose, minerals, and water. When acid and/or the enzyme rennin are added to milk, the casein coagulates, trapping much of the fat, some of the lactose, and some of the water and minerals in the coagulant. This is the curd. The remaining liquid which contains dissolved lactose, proteins, minerals, and other minor constituents is the whey. In making Cheddar cheese, curd is formed under controlled conditions of temperature, acidity, and rennin concentration. This gives curd of the desired moisture content and texture for subsequent processing.

Cheese curd can be made from raw or pasteurized milk. When it is made from raw milk, the FDA requires that the finished cheese be ripened for 60 days or more as a

Table 13.6. Approximate Percentage Composition of Some Varieties of Cheese

Variety	Moisture	Fat	Protein	Ash (Salt-free)	Salt	Calcium	Phosphorus
Brick	41.3	31.0	22.1	1.2	1.8	—	—
Brie	51.3	26.1	19.6	1.5	1.5	—	—
Camembert	50.3	26.0	19.8	1.2	2.5	0.69	0.50
Cheddar	37.5	32.8	24.2	1.9	1.5	0.86	0.6
Cottage							
uncreamed	79.5	0.3	15.0	0.8	1.0	0.10	0.15
creamed	79.2	4.3	13.2	0.8	1.0	0.12	0.15
Cream	54.0	35.0	7.6	0.5	1.0	0.3	0.2
Edam	39.5	23.8	30.6	2.3	2.8	0.85	0.55
Gorgonzola	35.8	32.0	26.0	2.6	2.4	—	—
Limburger	45.5	28.0	22.0	2.0	2.1	0.5	0.4
Neufchatel	55.0	25.0	16.0	1.3	1.0	—	—
Parmesan	31.0	27.5	37.5	3.0	1.8	1.2	1.0
Roquefort	39.5	33.0	22.0	2.3	4.2	0.65	0.45
Swiss	39.0	28.0	27.0	2.0	1.2	0.9	0.75

safeguard against pathogens; such storage under the acid conditions of the cheese inhibits the common disease-producing organisms that could be present in the milk. But most Cheddar cheese is made from pasteurized milk, since pasteurization also destroys most spoilage types of organisms and undesirable milk enzymes and gives better control over subsequent fermentation of the curd.

Setting the Milk

The pasteurized whole milk is added to a vat and brought to about 31°C; a lactic acid-producing starter culture of *Streptococcus lactis* is added at a level of about 1.0% based on the milk. At this point, natural color may be added to the stirred milk if the Cheddar is to be of the orange-colored type. After about 30 min, a mildly acidic condition of about 0.2% acidity (calculated as lactic acid) will exist and the rennin enzyme in the form of a dilute solution is added. The commercial rennin preparation is known as rennet. Whereas the name rennin (also called chymosin) refers to the pure enzyme, commercial rennet, obtained from the fourth stomach of the calf, contains rennin and small amounts of other materials. Renninlike enzymes also are produced by selected microorganisms and are available to the cheese maker as commercial microbial rennets. Recently, rennin has been produced from microorganisms which have been genetically altered to produce the animal enzyme. The mild acidity improves the coagulating ability of the rennin.

Stirring is now stopped and the milk is allowed to set. In about 30 more minutes a uniform custardlike curd forms throughout the vat. Acid continues to be formed, as it will throughout the curd-making operation. The combination of rennin and acid forms a curd with a desirable elastic texture which, when subsequently heated or pressed, will shrink and squeeze out much of the trapped whey.

Cutting the Curd

The next step after setting is cutting the curd. This is done with curd knives that are made up of wires strung across a frame. One knife has the wires going vertically and the other horizontally. By drawing the knives through the length of the vat and then back and forth with the width of the vat, the curd is cut into small cubes. In the case of Cheddar, these are 0.25–0.5 in. on a side. The smaller the cube is, the greater the surface area, and so the quicker and more complete is the removal of whey from the cubes, which can lead to a drier cheese. The curd for different types of cheese, therefore, is cut into different size cubes.

Cooking

After cutting, which may take only 5–10 min, the cubes are gently agitated and the jacketed vat is heated with steam to raise the temperature of the curds and whey. The temperature is brought to about 38°C over a 30-min period and held at 38°C for about 45 min longer. This is known as "cooking."

Cooking at 38°C further helps to squeeze the whey from the curd cubes. Heat increases the rate of acid production and makes the curd cubes shrink. Both help expel the whey and toughen the curd cubes, which now take on a more rounded cottage-cheese-like form. During cooking, the curds continue to be gently agitated.

Draining Whey and Matting Curd

Agitation of the curds is stopped and they are permitted to settle. The whey is drained from the cheese vat and the curds are trenched along the sides of the vat to further facilitate whey drainage. After all of the whey has been drained, the curds are allowed to mat for about 15 min. During matting, the individual curd pieces fuse together to form a continuous rubbery slab (Fig. 13.11).

The process of matting and subsequent handling of matted curd is known as cheddaring and is unique to the production of the Cheddar type of cheese. Cheddaring involves cutting the matted curd into blocks, turning the blocks at 15-min intervals, and then piling the blocks on one another, two or three deep. The purposes of cheddaring are to allow acid formation to continue and to squeeze whey from the curd. The weight of the blocks on one another is a mild form of pressure. During cheddaring, the vat is maintained warm. The cheddaring operation of stacking and turning the blocks goes on for about 2 h or until the whey coming from the blocks reaches 0.5–0.6% acid.

Figure 13.11. Cutting matted curd in production of cheddar cheese. *Courtesy of F. V. Kosikowski.*

Milling and Salting

The curd blocks or slabs are now ready for the milling and salting operation. The rubbery slabs of cheddared curd are passed through a mill that cuts the blocks into small pieces. The milled pieces are spread out over the floor of the vat and sprinkled with salt. The amount of salt is about 2.5% of curd weight. The salt and curd are stirred to uniformly distribute the salt. The purposes of the salt are threefold: to further draw whey out of the curd by osmosis; to inhibit proteolytic and other types of spoilage organisms that might otherwise grow in later stages of the cheesemaking operation; and to add flavor to the final cheese.

Pressing

The milled and salted curd pieces are placed in hoops fitted with cheesecloth and the hoops are placed in a hydraulic press. Pressing at about 1.4×10^5 Pa (20 psi) continues overnight. Pressing determines the final moisture that the finished cheese will have. The more moisture or whey retained in the cheese from the press, the more acidity that can be fermented from it. This, in turn, affects the final texture of the cheese and what microorganisms can grow during the subsequent ripening period. And, of course, pressing determines the final shape of the cheese.

Curing or Ripening

After overnight pressing, the cheese is removed from the hoop and placed in a cool room at about 16°C and 60% RH for 3 or 4 days. This causes mild surface drying and forms a slight rind. To prevent mold from growing on the surface of the cheese, the cheese block or wheel is vacuum packaged in flexible film or dipped in hot paraffin. The film or coating, in addition to preventing surface molding, prevents the cheese from excessive drying out during the long ripening or ageing period. The cheese is boxed and placed in the curing room for ripening. The curing or ripening room generally will be at about 2°C and 85% RH.

Ripening is continued for at least 60 days whether the cheese milk was raw or pasteurized. For peak flavor, ripening may be continued for 12 months or longer. During this period, bacteria in the cheese and enzymes in the rennet preparation modify the cheese texture, flavor, and color by continuing to ferment residual lactose and other organic compounds into acids and aroma compounds, by partial hydrolysis of the milk fat and further breakdown of fatty acids, and by mild proteolysis of the protein. In the case of Cheddar, these changes are comparatively mild because of the types of organisms present (primarily lactic acid types) and the relatively low moisture content. The flavor is correspondingly mild when compared, for example, to Roquefort or Limburger cheese.

Advanced Processes

As pointed out, there is much hand labor in conventional Cheddar-making, and so considerable research has been directed to the development of mechanized and continuous cheese-making processes. Several mechanized schemes have been devised over the past 30 or so years, including the Ched-O-Matic and the Curd-A-Matic processes in the United States and various other processes in Europe and Australia. Most of these processes retain the classical steps of conventional Cheddar production and replace hand operations with mechanical equivalents.

An important departure from conventional cheese-making has involved cold milk renneting. In conventional cheese-making, rennet is added to warm milk and the system is permitted to set and form the curd. If rennet is added to cold milk, the casein is altered but the milk remains a liquid. If such milk is then heated to 32°C, instantaneous gelling occurs. It is thus possible to work with a liquid system, which is more easily pumped, metered, and so on, and then generate coagulated curd continuously by passing the liquid through a heat exchanger.

More recent advances utilize the treatment of milk by reverse osmosis and ultrafiltration. Not only do these treatments concentrate milk solids for efficient further processing but by careful choice of membranes the ratios of retained milk solids to whey can be altered. Thus, lactoglobulin and lactalbumin can be retained with the cheese solids rather than being lost to the whey, improving cheese yield and nutritional values; likewise, lactose levels can be reduced when less acidity is desired. The automated ultrafiltration system shown in Fig. 13.12 contains numerous membrane cartridges connected in series and provided with recirculation loops for progressive separation and concentration of milk solids. Such practices are increasingly influencing cheese-making methods and cheese properties.

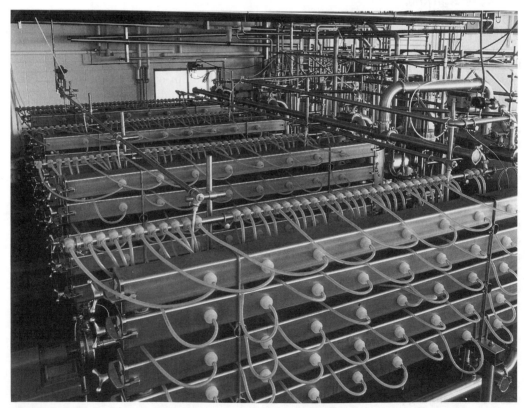

Figure 13.12. Ultrafiltration system for separating and concentrating milk solids. *Courtesy of Dorr-Oliver.*

Cottage Cheese

Cottage cheese is an example of a low-fat, soft cheese, generally coagulated with lactic acid rather than rennin. The curd is left in particulate form, that is, is not pressed. Further, it is not aged or ripened. Starting with pasteurized skim milk rather than whole milk, the curd-forming operations for low-fat cottage cheese have similarities to the early stages of Cheddar-making. The steps are as follows:

1. Pasteurized skim milk is warmed in a vat to 22°C.
2. A lactic starter (at about 1% level) is added to produce acid. In addition to *S. lactis,* the starter usually contains *Leuconostoc citrovorum,* a flavor-producing bacterium.
3. The vat is set and fermented for about 14 h (long-set method).
4. The coagulated milk is cut into small cubes.
5. The curd cubes are cooked for about 90 min with stirring and the temperature is gradually increased to 50°C.
6. After cooking, the whey is drained and the curd is washed with cold water to remove excess whey and limit acidity.
7. The curd is trenched to drain all water.
8. The curd may now be mildly salted, as a mild preservative measure and for flavor.
9. The curd also may be blended with sweet or soured cream to give 2% or 4% fat. The product is then called creamed cottage cheese.

Cottage cheese is packaged as loose curd particles and undergoes no further processing. It is highly perishable and must be kept refrigerated.

The principal variations in cottage-cheese-making have to do with the length of fermentation time in the vat. The 14-h holding time at 22°C is known as the *long-set* method. By using a larger amount of lactic starter (about 6%) and a temperature of 32°C, the proper degree of acidity for coagulation and curd cutting can be reached in 5 h; this is known as the *short-set* method. Another variation employs low levels of rennet plus the starter for milk coagulation.

Swiss Cheese

Like Cheddar, Swiss is a hard-type cheese. However, it is characterized by the formation of large holes, or eyes, and a sweet nutty flavor, which result from the activities of an organism known as *Propionibacterium shermanii.* This organism follows the lactic acid organisms and further ferments lactic acid (now in the form of lactate) to propionic acid and carbon dioxide. The propionic acid contributes to the nutty flavor, and the carbon dioxide gas collects in pockets within the ripening curd and forms holes or eyes.

Swiss cheese, also known as Emmental cheese, generally is made from raw milk. A multiple-organism starter containing lactic acid organisms, including *Lactobacillus bulgaricus* and heat-tolerant *Streptococcus thermophilus,* which produces lactic acid through the rather high cooking temperatures (about 53°C) that the curd reaches during processing, is added. The starter also may contain the eye-forming *Propionibacterium* or this may come in with the raw milk. Following an initial period of lactic acid fermentation, rennet is added to the kettle to coagulate the milk. The curd is cut with a harplike wire knife into rice-size particles. The curds and whey are now heated and cooked at about 53°C for about 1 h.

Unlike Cheddar-making, at this point the stirred heated curd is allowed to settle, a cloth with a fitted steel strip edge is slid under the curd, and the entire curd mass is hoisted from the kettle to drain (Fig. 13.13).

The entire curd from a kettle is placed in a single large hoop in which it is pressed for 1 day to begin forming a rind. The cheese wheel, which may weigh more than 90 kg, is removed from the hoop and placed in a large brine tank at about 10°C. It floats in this brine for about 3 days and its top is periodically salted. The salt removes still more water from the cheese surfaces than could be removed by pressing and thus produces the heavy protective rind.

The cheese is next removed from the brine to a warm ripening room maintained at about 21°C and 85% RH. The cheese remains here for about 5 weeks during which time the eyes are formed by the growth of the *Propionibacterium*. As the eyes are formed, the cheese becomes somewhat rounded (Fig. 13.14). The opening of the eyes also changes the sound of the cheese when it is thumped with the finger. After about 5 weeks, the cheese is moved to a colder curing room at about 7°C. Here it remains from 4 to 12 months to develop the full sweet nutty flavor associated with Swiss cheese.

In judging the quality of Swiss cheese much emphasis is given to the size, shape,

Figure 13.13. Draining Swiss cheese curd prior to placing it in the hoop. *Courtesy of Valio Finnish Coop. Dairies Assoc., Helsinki, Finland.*

Figure 13.14. Swiss cheese in curing room. *Courtesy of Swiss Cheese Union Inc., Bern, Switzerland.*

and gloss of the eyes (Fig. 13.15). This is not for appearance alone. Proper eye formation is an index of several other quality factors. For example, if the acidity is not properly controlled to produce a chewy elastic texture, then the curd would not be able to stretch and form the eye under the pressure of the generated carbon dioxide. Thus, excessive acid gives a brittle curd which forms cracks rather than eyes. So the eyes also are an index of texture. Similarly, the same organism that forms eyes produces the propionic acid necessary for the sweet nutty flavor. Good eye formation indicates active fermentation by this organism and well-developed flavor.

Blue-Veined Cheeses

Blue-veined cheeses are characterized by a semisoft texture and blue mold growing throughout the curd. There are four well-known varieties of blue-veined cheeses. Three are made from cow's milk: blue cheese, made in Denmark, the United States, and other countries; Stilton, made in England; and Gorgonzola, made in Italy. The fourth and perhaps the most famous blue-veined cheese is Roquefort, which is made from sheep's milk and is made in the Roquefort region of France.

All of the blue-veined cheeses acquire the characteristic blue marbling by having their curd inoculated with the blue-green mold *Penicillium roqueforti* prior to being hooped and pressed. Mold growth is encouraged during the ripening period which can be from 3 to 10 months at cool, moist, cavelike conditions of about 4°C and 90% RH.

Figure 13.15. Typical eye formation in quality Swiss cheese. *Courtesy of Swiss Cheese Union Inc., Bern, Switzerland.*

Molds are aerobic and so generally grow on the surface of cheeses and other foods. To permit the mold to grow throughout the cheese mass, it is common practice to pierce the pressed cheese when it is placed in the curing room. This allows air to penetrate the cheese and support mold growth throughout the mass. The blue-green color is from spores of the mold; in Fig. 13.16 the darkened lines where mold growth is heavy are visible along the pierced air channels. *Penicillium roqueforti* not only produces the mottled blue color but is an active splitter of milk fat. This gives rise to free fatty acids and ketones, which contribute to the sharp, peppery flavor of blue-veined cheeses.

Camembert

Another mold-ripened cheese is Camembert, which, like Roquefort, originated in France. However, this cheese is characterized by a soft cream-colored curd and a white feltlike mold growth that covers its entire surface (Fig. 13.17). The mold is *Penicillium camemberti,* and it is inoculated onto the pressed cheese curd after removal from the

Figure 13.16. Danish blue cheese showing heavy mold growth along air channels. *Courtesy of Danish Dairy Assoc., Aarhus, Denmark.*

hoop by spraying a mist of mold spores onto the cheese surfaces. Ripening, as in the case of Roquefort, is under damp conditions at about 7°C and 95% RH. But ripening time is only about 3 weeks. *Penicillium camemberti* is highly proteolytic and breaks down the curd protein, from the surface inward, to the texture of soft butter. If proteolysis goes too far because of prolonged storage in the supermarket or the home, the cheese develops a strong ammoniacal odor.

Limburger

Limburger is a semisoft cheese which, like Camembert, is ripened from the surface inward with a characteristic proteolytic decomposition. However, the major ripening agent is a surface bacterium called *Brevibacterium linens.*

Process Cheese

All of the cheeses described thus far are referred to as natural cheeses; that is, they are produced through a series of natural curd-making and ripening operations (cottage cheese is unripened). Process cheese is the name given to cheese made by mixing or grinding different lots of natural cheeses together and then melting them into a uniform mass.

This is done in part because different lots of natural cheese vary in moisture, acidity, texture, flavor, age, and other characteristics. A highly acidic cheese, for example, can be blended with a bland cheese to yield a more acceptable product. However, process cheeses have become so popular in their own right that they have necessitated plants for the production of natural cheese solely intended for conversion to process cheese.

In making process cheese, the mixed lots are melted together by heating to about

Figure 13.17. Camembert cheese showing white surface mold during curing. *Courtesy of Borden Co.*

71°C. This also pasteurizes the cheese. Emulsifiers such as sodium citrate and disodium phosphate are added to prevent fat separation and to add smoothness to the texture. The hot melted cheese is then filled into cartons and allowed to cool and solidify. The best known process cheese is the popular American cheese made from blended and melted Cheddar.

Process cheese may be used to prepare process cheese foods and spreads by mixing in additional dairy ingredients, fruits, vegetables, meats, and so on. The designations "Process Cheese Food" and "Process Cheese Spread" may be used only when the final products meet minimal federal standards with respect to fat and solids. Thus, "Process Cheese Food" must contain no less than 23% fat and no more than 44% moisture.

Cheese Substitutes

Various cheese substitutes, also referred to as cheese analogs, imitation cheese, and so on, are increasingly entering the marketplace (Fig. 13.18). They commonly have all or some of the milk fat replaced with vegetable fat and vegetable protein. The incentives for developing such products are lower cost, ready availability of substitute ingredients, changing consumer tastes, and real or perceived health benefits. Newer

Figure 13.18. Substitute American cheese being extruded and sliced. *Courtesy of Cheese Foods International, Ltd.*

products for which there is demand include cheese substitutes with reduced levels of fat, cholesterol, and sodium.

Other Dairy Products

There are several other related dairy foods that enjoy popularity. Junket dessert is sweet milk plus flavor that is coagulated with rennet to a custardlike consistency. Cultured buttermilk is pasteurized skim milk (or partially-skimmed milk) mildly coagulated with lactic acid culture (containing *Leuconostoc* bacteria for flavor) and consumed as a beverage. At one time, sour buttermilk was the liquid drained from the butter churn and allowed to ferment naturally, but today's cultured buttermilk is the result of a well-controlled process. Sour cream is fresh pasteurized cream mildly coagulated with lactic acid culture plus *Leuconostoc* flavor bacteria. It is higher in fat and heavier in consistency than cultured buttermilk.

Acidophilus milk is pasteurized milk or low-fat milk inoculated with *Lactobacillus acidophilus,* which is believed by some to provide health benefits by favorably altering the microflora of the intestinal tract. In the past the popularity of this product was limited by the flavor developed during fermentation. A more recent product has overcome this by adding the live organisms to pasteurized milk and refrigeration to prevent subsequent fermentation and flavor development.

Yogurt is pasteurized milk or low-fat milk coagulated to a custardlike consistency with a mixed lactic acid culture containing *Lactobacillus bulgaricus and Streptococcus thermophilus*. It is most often flavored with fruit preserves or other ingredients but is also consumed unflavored.

REDUCED FAT DAIRY PRODUCTS

Although dairy products are nutritionally important in the diet, especially of the young, there has been a desire to reduce not only calories but also the saturated fat and cholesterol in the diet. Therefore, a number of reduced-fat and low-fat dairylike products have appeared in recent years. Most of these products substitute nonfat food ingredients for all or a portion of the animal fat in dairy products. This is different than products in which vegetable fat replaces animal fat. Often the fat substitutes have calories themselves, but because the caloric density of fat is more than twice that of proteins or carbohydrates, replacement on an equal-weight basis results in an overall reduction in calories. One popular category of product is low-fat ice cream which has the eating characteristics of full-fat ice cream with less fat.

A wide variety of food ingredients have been used as fat replacers but all have the common property that they are designed to replace the functional characteristics of fat. The most important characteristic of fat is its texture or mouthfeel. As pointed out earlier, fat gives lubricity to foods, including dairy products such as ice cream. One fat substitute is made of proteins which have been processed into extremely small particles. When these particles are suspended in water (or ice milk) they impart a creamy texture to the fluid which closely mimics suspended fat particles. This creaminess is perceived in the mouth as similar to fat in such products as ice cream.

Other fat replacers are carbohydrate based. These food ingredients often bind large amounts of water and act as thickening agents. Like the particles of protein, this thickening can be perceived by the mouth as resulting from the inclusion of fat.

Further improvements in fat replacers can be expected. One product that has been waiting for FDA approval is a fat which is nonabsorbable. This food ingredient has the functional properties of fat in foods but is not digestible or absorbable by humans and, thus, does not contribute to calories or fat intake. Whether or not this material will be approved remains to be seen.

References

Arbuckle, W.S. 1986, Ice Cream. 4th ed. Chapman & Hall, London, New York.

Burton, H. 1988. Ultra-High-Temperature Processing of Milk and Milk Products. Chapman & Hall, London, New York.

Carci, M. 1994. Concentrated and Dried Dairy Products. VCH, New York.

DiLiello, L.R. 1982. Methods in Food and Dairy Microbiology. AVI, Westport, CT.

Food and Agricultural Organization. 1990. The Technology of Traditional Milk Products in Developing Countries. Food and Agriculture Organization of the United Nations, Rome.

Fox, P.F. 1987. Cheese. Chemistry, Physics, and Microbiology. Chapman & Hall, London, New York.

Fox, P.F. 1989. Functional Milk Proteins. Chapman & Hall, London, New York.

Fox, P.F. 1992. Advanced Dairy Chemistry. Chapman & Hall, London, New York.

Kosikowski, F.V. 1977. Cheese and Fermented Milk Foods. 2nd ed. F.V. Kosikowski, Brookton-dale, NY.

Kurmann, J.A., Rasic, J.L., and Kroger, M. 1992. Encyclopedia of Fermented Fresh Milk Products: An International Inventory of Fermented Milk, Cream, Buttermilk, Whey, and Related Products. Chapman & Hall, London, New York.

Lablee, J. 1987. Cheese manufacture. *In* Cheesemaking: Science and Technology, A. Eck (Editor). Lavoisier Publishing, New York, pp. 406–412.

Renner, E. 1983. Milk and Dairy Products in Human Nutrition. Volkswirtschaftlicher Verlag, Munchen.

Robinson, R.K. 1990. Dairy Microbiology. 2nd ed. Chapman & Hall, London, New York.

Robinson, R.K. 1994. Modern Dairy Technology. 2nd ed. Chapman & Hall, London, New York.

Saunders, G.P. 1953. Cheese varieties and descriptions. *In* Agriculture Handbook 54. U.S. Department of Agriculture, Washington, DC.

Scott, R. 1986. Cheesemaking Practice. 2nd ed. Elsevier Applied Science Publishers, New York.

Varnam, A.H. 1994. Milk and Milk Products: Technology, Chemistry and Microbiology. Chapman & Hall, London, New York.

Webb, B.H., Johnson, A.H., and Alford, J.A. 1974. Fundamentals of Dairy Chemistry. 2nd ed. AVI Publishing Co., Westport, CT.

Wong, N.P. 1988. Fundamentals of Dairy Chemistry. 3rd Ed. Chapman & Hall, London, New York.

14

Meat, Poultry, and Eggs

Humans are omnivorous and have consumed both animals and plants as foods throughout recorded history. However, before animals, birds, and fish can provide meat, eggs, or milk, their own physiological requirements for energy and synthesis must be satisfied. These requirements are met largely through the consumption of plant materials, which, if consumed directly by humans, could support a greater population than can the animal products derived from them. This is true with respect to total available calories, protein, and other nutrients needed to sustain life. In most cases, the amount of animal products consumed by a society is positively correlated with the affluence of the society.

Most human societies have preferred animal foods and have been willing to expend the greater effort generally required to satisfy this appetite when possible. In agriculturally advanced societies, it is possible to convert grain into meat (liveweight basis) at rates of about 2 kg/kg chicken, 4 kg/kg pork, and 8 kg/kg beef, although grass and forage crops also go into the feeding of beef. These conversion ratios are in part responsible for the relative prices of food.

Foods from animal products (including fish, which is discussed in Chapter 15) represent concentrated sources of many of the nutrients required by humans. This is to be expected since our tissues and body fluids are similar to their counterparts in other animals with respect to the elements and compounds they contain. Although it is probably true that we could supply all our nutritional needs directly from plant sources, this would require the consumption of a sizable number of plant types and a sophisticated knowledge of nutrition if no animal products such as dairy foods and eggs were included in the diet. This would be especially so with respect to meeting the requirements for essential amino acids, vitamins, and minerals. Moreover, farm animals convert large quantities of plant roughage materials unsuited for human consumption into human food. For example, humans cannot digest cellulose and, therefore, can derive no energy directly from it. However, ruminants such as cows can digest cellulose and turn it into foods (i.e., milk) which are useful to humans.

MEAT AND MEAT PRODUCTS

Meat and meat products generally are understood to include the skeletal muscles of animals; also included are the glands and organs of these animals (tongue, liver, heart, kidneys, brain, and so on). In a broader sense, meat also includes the flesh of poultry and fish, but these are generally considered separate from the red meats of terrestrial animals. In the United States, the principal sources of meat are cattle (beef), calves (veal), hogs (hams, pork, and bacon), sheep (mutton), and young sheep (lamb). Other societies consume different animals, including dogs, kangaroos, reindeer, and reptiles.

Meat products also include many by-products from animal slaughter: animal intestine for sausage casings; fat which is rendered into tallow and lard; hides and wool; animal scrap, bone, and blood used in poultry and other feeds; and gelatin, enzymes, and hormones used by the food, pharmaceutical, and other industries. For this reason the major meat processing companies are seldom in a single business but usually produce a wide variety of products.

Government Surveillance

Essential to the meat industry are two kinds of government surveillance: grading and meat inspection. It is important to understand that meat and poultry inspection in the United States is a mandatory program for all products in interstate shipment as well as products used within the states. Inspection is primarily concerned with health and safety matters. Grading is voluntary and is undertaken to inform consumers about the quality of meats and poultry.

Grading

The need for grading is clear. Like all natural products, meat is very heterogenous and varies. Animal carcasses are of all sizes, from many breeds, of varying ages, and have been fed on many different kinds of feeds. These factors result in cuts of meat varying in yield, tenderness, flavor, cookout losses, and general overall quality. A system of grading is essential to ensure that the wholesale buyer and ultimately the retail customer get what they pay for.

Quality grades are based on subjective evaluations of three main factors: carcass maturity, degree of fat marbling, and muscle firmness. Color is also considered. Maturity relates to tenderness; younger carcasses typically are more tender than those from older animals. Marbling is the deposition of fat within the lean muscle. Such intramuscular fat increases tenderness and palatability. In addition, lean meat should have a certain degree of firmness. Overly soft muscle is downgraded. Meat which is excessively dark when cut indicates possible stress on the animal prior to slaughter. Beef grades, in order of decreasing quality, are Prime, Choice, Select, Standard, Commercial, Utility, Cutter, and Canner. The grades, however, have very little relationship to the nutritional value of the cuts, except where one may wish to limit fat intake.

Of course, grades cannot be assigned until the animal has been slaughtered. A recently developed technique that may influence future grading and the purchase price paid for animals involves the use of ultrasonic energy to reveal the gross structure of meat prior to animal slaughter. Meat, fat, and bone reflect ultrasonic energy differently. By radiating such energy over the body of a live animal and recording the reflected energy pattern, it is possible to develop an X-ray-like cross-sectional view of portions of the animal carcass. In this way, purchases of live meat-yielding animals can be made more efficient in terms of intended end use than is now possible.

Meat Inspection for Wholesomeness

Federal employees inspect all meat going into interstate commerce, in accordance with the Federal Meat Inspection Act of 1906, to ensure a clean, wholesome, disease-free meat supply that is without adulteration. Unlike USDA grading practices, which are optional, inspection for wholesomeness is mandatory and is administered by US-DA's Food Safety and Inspection Service (FSIS).

If animals are diseased, the meat can carry a wide variety of organisms pathogenic to humans. These may include organisms capable of causing tuberculosis, brucellosis, anthrax, trichinosis and salmonellosis. There are some 70 such diseases that animals can transmit to man. For this reason, inspections are made by veterinarians or persons under their supervision at places of animal slaughter and at meat processing facilities. A federal law enacted in 1967 requires that all states adopt and enforce meat inspection practices at least comparable in thoroughness to the federal meat inspection laws without regard to where the meat will be shipped.

Slaughtering and Related Practices

There has been a law in the United States since 1958 that all animals coming under federal purchase must be rendered insensible to pain before being hoisted by their hind legs and bled. This practice has since been widely adopted. An exception exists in slaughtering according to religious ritual.

One common humane method of rendering an animal insensible is by striking it on the head with an air- or gunpowder-driven blunt or penetrating device. This has largely replaced stunning with a sledge hammer. Another method employs electric shock, and a third uses a tunnel filled with carbon dioxide through which the animal passes. Each of these various methods can differently affect blood hormone levels, muscle chemistry, and meat properties.

After stunning, hoisting, and bleeding, a modern slaughterhouse is an efficient continuous disassembly line. Virtually every component of the animal body is utilized, including the hide, viscera, blood, and carcass. The skinned, washed, and eviscerated carcass is then moved by monorail into a chill room, where the deepest part of the meat is chilled to 2°C in about 36 h. This prevents rapid bacterial spoilage.

The practice of resting animals before slaughter can help delay bacterial spoilage of meat. Animals store glycogen in their muscles as a source of reserve energy. After an animal is killed, this glycogen is converted under the anaerobic conditions in the muscles into lactic acid, which lowers the pH and acts as a mild preservative. But if animals are excited or exercised before slaughter, the glycogen is largely consumed and there is very little left to be converted to lactic acid in the postmortem tissues. Such meat can spoil more quickly. Additional research has shown that antemortem stress also can affect other carcass characteristics such as the defects of dark-cutting beef, and pale, soft, watery pork.

Structure and Composition of Meat

The gross structure of a cut of meat can be seen in Fig. 14.1. The dark areas are principal muscles and the white areas are fat; however, microscopic observation is required to see the fine structure of the muscles. Figure 14.2 is a diagram of a longitudi-

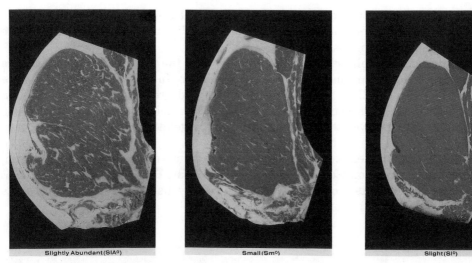

Slightly Abundant(SIA⁰)   Small (Sm⁰)   Slight(Sl⁰)

Figure 14.1. Different levels of fat marbling in beef according to U.S. Department of Agriculture standards: slightly abundant, small, and slight amounts of fat marbling. More marbling results in a higher grade. *Courtesy of National Livestock & Meat Board.*

nal section of lean muscle showing that the muscle is composed of bundles of hairlike muscle fibers. These protein muscle fibers are held together by proteinaceous connective tissue which merges to form a tendon which, in turn, connects the muscle to a bone. The muscle fibers themselves are elongated cells that contain many smaller highly oriented fibrils. A major protein of muscle fiber is myosin. The connective tissue contains two proteins called collagen and elastin. Collagen on heating in the presence of moisture dissolves and yields gelatin. Elastin is tougher and is a constituent of the

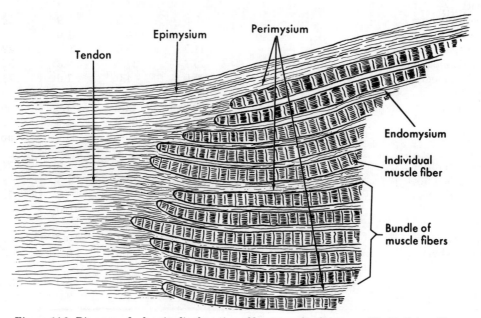

Figure 14.2. Diagram of a longitudinal section of lean muscle. *Courtesy of R. M. Griswold.*

ligaments. A cooked chicken leg nicely reveals the bundles of muscle fibers, the connective tissue between the bundles of muscle fibers, and the gelatinous substance in the connective tissue which is dissolved collagen.

When an animal is well fed, fat penetrates between the muscle fiber bundles; this is called fat marbling and makes muscle more tender. In addition, thinner muscle fibers are more tender than thicker muscle fibers and are more common in young animals. On cooking, muscle fibers contract and may become tougher, but cooking also melts the fat and dissolves the collagen into soluble gelatin, so the overall effect is increased tenderness.

The compositions of meat cuts will vary with the relative amounts of fat and lean, but a typical cut of beef may contain 60% water, 21% fat, 18% protein, and 1.0% ash. Compositions of the meats of other food animals, poultry, fish, and some milk products are given in Table 14.1 for comparison.

Ageing of Meat

Within a few hours after an animal is killed, rigor mortis sets in with a contraction of muscle fibers and an increasing toughness of the meat. This is correlated with the loss of glycogen and disappearance of ATP from the muscles of newly killed animals. If the meat is held cool, rigor mortis subsides in about 2 days, the muscles become soft again, and there is a progressive tenderization of the meat over the next several weeks. The tenderization is believed to be due principally to natural proteolytic enzymes in the meat, which slowly break down the connective tissue between the muscle fibers

Table 14.1. Typical Percentage Composition of Foods of Animal Origin (Edible Portion)

Food	Carbohydrate	Protein	Fat	Ash	Water
Meat					
beef, medium fat	—	17.5	22.0	0.9	60.0
veal, medium fat	—	18.8	14.0	1.0	66.0
pork, medium fat	—	11.9	45.0	0.6	42.0
lamb, medium fat	—	15.7	27.7	0.8	56.0
horse, medium fat	1.0	20.0	4.0	1.0	74.0
Poultry					
chicken	—	20.2	12.6	1.0	66.0
duck	—	16.2	30.0	1.0	52.8
turkey	—	20.1	20.2	1.0	58.3
Fish					
nonfatty fillet	—	16.4	0.5	1.3	81.8
fatty fish fillet	—	20.0	10.0	1.4	68.6
crustaceans	2.6	14.6	1.7	1.8	79.3
dried fish	—	60.0	21.0	15.0	4.0
Milk					
cow, whole	5.0	3.5	3.5	0.7	87.3
goat, whole	4.5	3.8	4.5	0.8	86.4
Cheese					
hard, whole milk	2.0	25.0	31.0	5.0	37.0
soft, partly whole milk	5.0	15.0	7.0	3.0	70.0

SOURCE: Food and Agriculture Organization.

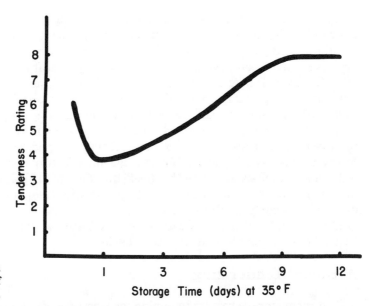

Figure 14.3. Effect of ageing on tenderness of beef. Courtesy of G. E. Brissey and P. A. Goeser.

as well as the muscle fibers themselves. The typical time course of meat tenderness during ageing is shown in Fig. 14.3.

Figure 14.4 compares photomicrographs of raw beef freshly slaughtered and the same beef after cold storage for 6 days. The freshly slaughtered beef is characterized by compactness of muscle fibers; the separation between muscle fibers and breaks in the fibers of the stored beef is evident.

Ageing or ripening of meat generally is done at 2°C by hanging the carcass in a cold

Figure 14.4. Photomicrographs of raw beef: (top) freshly slaughtered; (bottom) stored cold for 6 days. *Courtesy of Dr. Pauline Paul.*

room anywhere from 1 to 4 weeks. The best flavor and the tenderness develop in about 2–4 weeks. Humidity must be controlled and the meat may be covered with wrappings to minimize drying and weight loss. As pointed out in a previous chapter, ageing processes have been developed using higher temperatures for shorter times such as 20°C for 48 h. Tenderness results, but bacterial slime also develops quickly on the meat at this high temperature. In commercial practice, ultraviolet light may be used to keep down bacterial surface growth during quick ageing at high temperatures.

Because of the costs involved with ageing, not all beef is deliberately aged for increased tenderness and flavor before being shipped by meat packing companies. Further, some beef that is used for sausage manufacture may not be aged or even cooled following slaughter. So-called "hot beef" that has not yet passed through complete rigor mortis has superior water-holding characteristics compared with cold stored beef, a desirable property in the making of sausage meat emulsions. The superior water-holding capacity of hot beef can be retained also for later use if the beef is rapidly frozen before rigor has had time to subside.

Artificial Tenderizing

Cold room storage results in ageing, or ripening, of the meat with tenderizing from the meat's natural enzymes. There are several artificial means of tenderizing meat to various degrees.

Meat may be tenderized by mechanical means. During cold room storage the carcass can be hung in a manner to stretch the muscles and thereby encourage elongated, thinner muscle fibers. Further tenderizing of meat cuts can be obtained by pounding, cutting, or separating and breaking meat fibers with ultrasonic vibrations.

Meat may be tenderized somewhat by the use of low levels of salt, which solubilizes meat proteins. Salt is hygroscopic. Therefore, if salt is placed within the meat (e.g., ground hamburger), it holds water within the mass; if it is placed on the surface of the meat, it draws moisture out of the mass to the surface. Phosphate salts may be even more effective than common table salt in tenderizing meat, and either may be blended into ground meat or diffused into the flesh of fish, poultry, or meats to help retain juices and minimize bleeding or drip losses.

Another artificial tenderizing method involves the addition of proteolytic enzymes to the meat, such as bromelin from pineapple, ficin from figs, trypsin from pancreas, or papain from papaya. The native practice in tropical countries of wrapping meat in papaya leaves before cooking results in this kind of tenderization. Enzymes may be applied to meat surfaces, but penetration is slow; injection into the meat or into the bloodstream of the living animal before slaughter is more effective for large cuts. If this is done, then cold room ageing time is markedly reduced. Tenderizing enzymes function before cooking and during the cooking operation until the meat temperature reaches about 82°C; then they become heat inactivated.

Electrical stimulation of carcasses following slaughter is the newest commercial method of meat tenderization, although tenderizing effects on poultry killed electrically were noted by such early observers as Benjamin Franklin. Electrical tenderization involves application of sufficient voltage to cause rapid muscle contractions, which produce both physical and biochemical effects in the muscle tissue. These are associated with changes in levels of glycogen, ATP, lactic acid, pH, and enzyme activity. Through mechanisms not yet well understood, impulses of about 100–600 V over 1–2 min, given within about 45 min of slaughter, not only increase beef tenderness but are reported to improve lean meat color, texture, flavor, and to accelerate subsequent ageing. Both manual and automatic continuous electrical stimulation equipment have come into commercial use within the past decade.

Curing of Meat

Whereas ageing or ripening by cold room storage and tenderizing by artificial methods have as their prime objective increased tenderness, the curing of meat is a different process and has additional objectives. Curing refers to modifications of the meat that affect preservation, flavor, color, and tenderness due to added curing ingredients. Proper ageing still leaves the meat recognizable as a fresh cut, but curing is designed to grossly alter the nature of the meat and produce distinct products such as smoked and salted bacon, ham, corned beef, and highly flavored sausages including bologna and frankfurters.

Originally, curing treatments were practiced as a means of preserving meat before the days of refrigeration, and curing goes back at least to 1500 B.C. In less developed areas without modern preservation facilities, the prime objective of curing is still preservation. But where more effective preservation methods are available, the prime purpose of curing is to produce unique-flavored meat products; a secondary purpose is to preserve the red color of meat after cooking. Thus, cured corned beef when cooked remains red, whereas beef that is not cured turns brown on cooking. Similarly, cured ham retains its red color through cooking, but uncured pork becomes brown.

The principal ingredients used for curing or pickling meat are (1) sodium chloride, which is a mild preservative and adds flavor, (2) sodium nitrate and/or sodium nitrite, which help cured meats develop their unique flavor, act as preservatives and have antibotulinum activity, and fix the red color of cured meats; (3) sugar, which helps stabilize color and also adds flavor; and (4) spices, mainly for flavor. Sodium nitrate and nitrite have received considerable attention with respect to their safety in this application.

These ingredients are available in commercial mixtures or may be formulated by the meat processor and may be applied to meat in dry form by rubbing on surfaces or mixing directly into ground meats. In the case of hams or corned beef, they may be applied as a wet cure or pickle, by soaking in vats. When the meat cut is large and penetration of the cure is slow, the cure may be pumped directly into the meat via an artery, as is common with large hams; or the cure may be injected with multiple needles into slabs of bacon bellies, as shown in Fig. 14.5.

Meat Pigments and Color Changes

An understanding of meat pigments and the changes they undergo is important in meat processing. These changes are chemically complex, and only a few general principles are discussed here and outlined in Fig. 14.6. The chief muscle pigment is a protein called myoglobin. The physiological function of myoglobin is to store oxygen in the live animal's muscle. It has a purplish color when not carrying an oxygen molecule, but when exposed to oxygen, it becomes oxymyoglobin, which has a bright cherry-red color. Thus, when fresh meat is first cut, it is purple in color, but its surface quickly

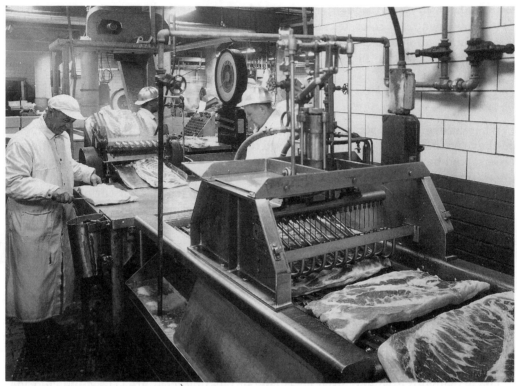

Figure 14.5. Automatic multiple needle injection of curing solution into bacon. *Courtesy of Swift and Co. (Mr. Ed Hois).*

becomes bright red upon exposure to air. Large cuts may be bright red on the surface but more purplish in the interior due to less oxygen within. The desirable bright red of oxymyoglobin is not entirely stable; on prolonged exposure to air and excessive oxidation it can shift to metmyoglobin, which has a brown color.

When fresh meat is cooked, these protein pigments are denatured and also produce a brown color. Steak cooked to a rare condition has less of the oxymyoglobin denatured and is more red. Well-done meat is more denatured and is more brown. Meats cured with nitrites are red and remain red through cooking. Nitrite combines with myoglobin to produce nitric oxide myoglobin, which is pink, in cured meats. Nitric oxide myoglobin on cooking is converted to nitrosohemochrome, which is pink or red as in cooked ham and bacon and quite stabile.

These pigment shifts, some of which are reversible, are affected by oxygen, acidity of the meat, and exposure to light; and the combination determines which pigments will dominate. Within the normal pigment shifts, the color of meat does not indicate wholesomeness or nutritional value; however, red color is often used by consumers to judge the freshness of meats. For this reason, packaging films are designed to protect meat color, largely by controlling diffusion of oxygen.

In the case of fresh meat cuts, films are used that allow oxygen to penetrate and keep myoglobin in the bright red oxymyoglobin form. However, cured meats are affected differently by oxygen—the pink nitric oxide myoglobin can be oxidized to the brown metmyoglobin. Thus, cured meats are generally vacuum packed to exclude air and

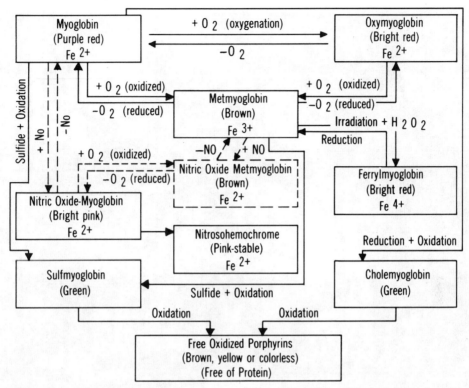

Figure 14.6. Pigment changes during the processing and handling of fresh and cured meats. *Courtesy of W.E. Kramlich.*

wrapped in films which are a high barrier to oxygen. Fresh or cured meats also can develop brown, yellow, and green discolorations from bacterial growth.

Smoking of Meats

Following curing, processed meats are often smoked. Smoking also was originally employed as a mild preservative, but today smoking is used mostly for its flavor contribution. At one time, smoking was done primarily in large smokehouses by hanging the meat over burning hardwood logs or wood chips; hickory smoke was preferred for flavor. If a smoke room is used, it should be at about 57°C to give the meat an internal temperature of about 52°C; smoking may take from 18 to 24 h. This is satisfactory in the case of pork products if the meat is cooked before smoking or will be cooked afterward. If, however, the meat is to be a ready-to-eat product without additional heat, then smoking must bring pork products to an internal temperature of 58°C or higher to ensure destruction of the trichinosis parasite; this procedure is required by the federal meat inspection laws. In Fig. 14.7, hams are being removed from a typical cabinet-type smokehouse.

Today there are several ways to generate smoke remotely and then circulate it into a smoke room or smoke tunnel (Fig. 14.8). In addition smoke can be generated in a special device without fire by high-speed frictional contact with the wood. The smoke can be given an electric charge and electrostatically deposited onto the meat surface.

Figure 14.7. Hams being removed from a cabinet-type smokehouse. *Courtesy of Swift and Co. (Mr. Ed Hois).*

There also are solutions of the flavor chemicals (i.e., liquid smoke) which have been extracted from smoke and are applied without the direct use of smoke.

Sausages and Table-Ready Meats

Cured meats especially, and uncured meats to a lesser extent, find their way into enormous quantities of sausage products. There are over 200 kinds of sausage products sold in the United States, the most popular of which are frankfurters. Most have their origins in countries outside the United States.

Figure 14.8. Mechanical sawdust burner and smokehouse controllers. *Courtesy of Swift and Co. (Mr. Ed Hois).*

Classification of sausage types is confusing but generally takes into account whether the ground meat is fresh or cured and whether the sausage is cooked or uncooked, smoked or unsmoked, dried or not, and fermented. Examples would be frankfurters— which generally are mildly cured, cooked, and smoked; fresh pork sausage—which is not cooked, smoked, or cured in manufacture; and Italian salami—which is cured, fermented, and dried.

Many sausages are prepared within a casing. Natural casings are made from cleaned animal intestines, the different sizes being used for different types of sausages. But natural casings are expensive and nonuniform and so artificial casings are more important. These casings are extruded tubes of regenerated collagen, cellulosic materials, or plastic films. The casings hold the ground meat together and prevent excessive moisture and fat losses during cooking and smoking operations. Large sausages such as bologna may have the casing removed after cooking and smoking, and then be sliced and packaged. Such products are known as table-ready meats.

Frankfurter Manufacture

The most important sausage product in the United States is frankfurters, and except for size, its production method is the same as for bologna. Generally, franks are made from finely ground cured beef, which is referred to as a frankfurter meat emulsion. The emulsion is pumped into great lengths of artificial casing which is automatically twisted every 6 in. to form links (Fig. 14.9). Following this procedure, the links are cooked by passing through hot water or steam and then hung for smoking, or smoking may precede the final cook.

"Skinless" franks are made in casings which are designed to be removed after cooking and smoking. Then after smoking, the casing is mechanically peeled from the now congealed form, which had its shape set within the casing during cooking.

These mechanical operations, which today dominate the frankfurter industry, are somewhat cumbersome, and so new continuous frankfurter processes have been developed. One is the Tenderfrank Process of Swift and Company in which no casings are used. In a continuous fashion the meat emulsion is injected into frank-shaped molds where it is coagulated with electronically generated heat. The franks are then conveyed through a tunnel in which they are smoked and then cooled, and from which they emerge for packaging. Another process that recently has been commercialized in Europe utilizes coextrusion (Fig. 14.10) to continuously form casing from a premixed collagen dough as meat emulsion simultaneously is extruded into the casing.

Freezing of Meat

Meat may be frozen and held in frozen storage for months in the case of pork and fatty meats, and for years in the case of beef. Storage time for pork and fatty meats is limited by gradual development of oxidized fat flavors. As with other frozen foods, to assure quality demands that the meat be quick frozen and then not thawed and refrozen to avoid excessive bleeding and drip when the product is finally thawed

Figure 14.9. Frankfurter emulsion in continuous length of casing being linked and loaded for subsequent processing. *Courtesy of Swift and Co. (Mr. Ed Hois).*

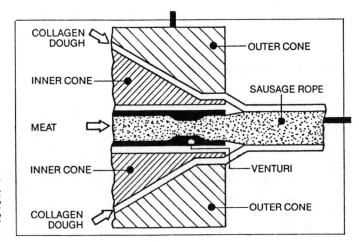

Figure 14.10. Special co-extrusion die for forming and filling sausage casing. *Courtesy of Food Engineering International and Unilever.*

and cooked. Few cured meats or sausages are commercially frozen since salt in their formulations increases the rate of fat oxidation and development of rancid flavors. Further, frozen storage tends to alter the flavor of the seasoning spices used in many sausage products.

Properly packaged fresh meat cuts store well in frozen condition but are not popular in supermarkets because customers like to see and feel frost-free meats and associate quality with unfrozen bright red cuts. They then take the meat home and often store it in a freezer. Restaurant operators use a great deal of frozen meat, which the customer does not see, and the military buys great quantities, some of which is imported from Australia and New Zealand.

Freezing temperatures can be used to destroy the trichinosis parasite in pork products. As was stated earlier, smoking or cooking to a uniform internal temperature of 58°C ensures destruction of the larvae of this organism. The USDA recommends that previously unheated pork products should be cooked by consumers to a uniform internal temperature of 77°C, followed by a short dwell time. This procedure provides a margin of safety, especially with fast cooking methods which may not provide uniform heating and the lethality of prolonged temperature rise. Frozen storage of pork products in accordance with the standards of temperature and time indicated in Table 14.2 also destroys the trichinosis parasite and is another recommended treatment by the USDA to render pork products safe. Concern over treatments for the safety of pork products

Table 14.2. U.S. Department of Agriculture
Standards for Pork Freezing

		Time (days)	
°C	°F	15-cm Diam	15- to 68-cm Diam
−15	+5	20	30
−23	−10	10	20
−29	−20	6	12

SOURCE: U.S. Department of Agriculture (1973).

is due in part to the fact that normal meat inspection practices are unable to detect the presence of trichina organisms with certainty.

Storage of Fresh Meat

Meat cutting in supermarkets is labor consuming, and as the cost of labor continues to increase, ways to reduce these costs are sought. Deboned and trimmed meat, in the unfrozen state, offers cost savings in labor and transportation and requires less energy for storage than does frozen meat. This has resulted in a shift to more centralized meat cutting at packing plants. These central processing lines reduce the size of the carcass to smaller sections called primal and subprimal cuts which may represent one-fourth to one-eighth of a whole carcass. The cuts are vacuum packaged in high barrier bags and paperboard boxes for shipment to retailers, where they are further reduced to retail sized cuts. They are then repackaged in the familiar tray and overwrap packages.

Beef and pork may be vacuum packed in a high barrier film and stored at 0°C (unfrozen) up to about 3 weeks. Such meat acquires the purplish color of myoglobin at its cut surfaces. When the film is removed, the myoglobin reoxygenates to oxymyoglobin and the meat reddens. Such meat is further cut into retail portions and wrapped in an oxygen-permeable film for consumer purchase. It also has been proposed that consumer cuts be centrally prepared and wrapped in oxygen-permeable film and then vacuum packed as a group in an oxygen-impermeable overwrap. When the latter is removed, the individual units would redden and be ready for direct sale.

Most recently, new technologies have been developed which are designed to allow centralized processing and packaging of retail cuts, thus eliminating entirely the need for further processing in the retail store. The most successful technology is called Modified Atmosphere Packaging (MAP). MAP utilizes sealed high barrier packages in which the air has been replaced with a mixture of gases which will reduce the rate of deterioration of the meat. Most often these gases include 10–50% carbon dioxide which inhibits the growth of many microorganisms which cause spoilage of refrigerated meats. For fresh red meats the gas mixture often contains 20–50% oxygen so that the myoglobin will be in the oxygenated cherry-red form. The meats must be sealed in high barrier films which will keep the air out and prevent the modified atmosphere from escaping.

Cooking of Meat

Cooking can make meat more or less tender than the original raw cut. When meat is cooked, there are three tenderizing influences: fat melts and contributes to tenderness; connective collagen dissolves in the hot liquids and becomes soft gelatin; and muscle fibers separate and the tissue becomes more tender. There also are two toughening influences: overheating can cause the muscle fibers to contract and the meat to shrink and become tougher; and moisture evaporates and the dried out tissue becomes tougher. Generally, lower cooking temperatures for a longer period of time produce more tender meat than higher temperatures for short periods of time to any given degree of doneness. But this also depends on the meat cut, which can vary considerably, as seen for beef in Fig. 14.11. Depending on the cooking method, the relationship between cooking temperature and tenderness can become quite complicated, especially with the newer

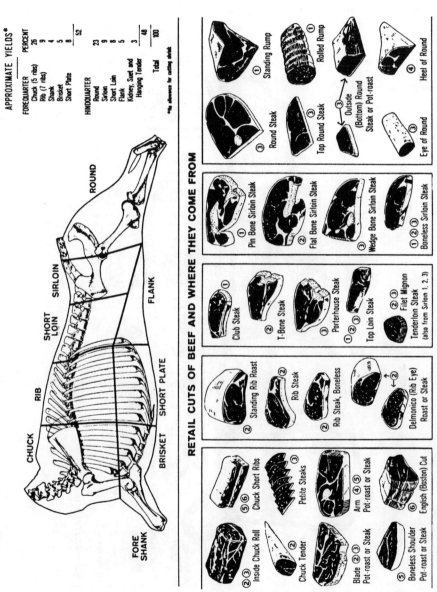

Figure 14.11. Wholesale and retail cuts of beef. *Courtesy of National Life Stock and Meat Board.*

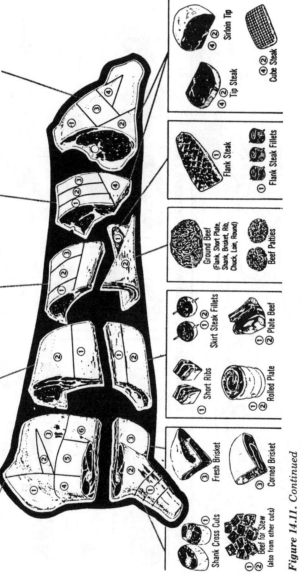

Figure 14.11. Continued

methods of microwave, dielectric, and infrared heating, and much research remains to be done.

The nutritional value of cooked meat generally remains very high. Normal cooking procedures do little to change the high value of the meat proteins, and minerals are heat resistant. Some minerals are lost in meat drippings, but, on the other hand, cooking dissolves some calcium from bone and so enriches the meat in this mineral. The B vitamins are heat sensitive and so cooking to well-doneness will destroy about 10% more of these vitamins than cooking to the rare condition. Even to the well-done stage, most meats retain about 70% of the B vitamins present in the uncooked meat. Advances have been made in the production of beef and pork of lower fat content via improved breeding and feeding of livestock in response to the growing demand for lower-fat products.

POULTRY

In the United States, the principal types of poultry are chickens, turkeys, ducks, and geese, and the amounts consumed are in that order. In other parts of the world, different types of fowl are consumed, such as emu and ostriches. Poultry is raised for meat and for eggs. We will first consider poultry for meat and confine discussion to chicken since much of the same technology applies to the other types of birds.

Production Considerations

In the past, most chicken meat came from egg-laying birds whose ability to lay eggs had declined, but, today, to satisfy the demand for meat, special genetic strains are produced that exhibit rapid growth, disease resistance, and good meat qualities such as tenderness and flavor. Chicken breeds include types with white feathers and types with brown and black feathers. The white-feathered types are favored for meat broilers because of the absence of dark pinfeathers.

Broiler farms are often quite large and raise several million birds per year. Chicks are highly susceptible to many diseases and so broiler producers must practice rigid husbandry with respect to bird housing, temperature and humidity control, sanitation, and feeding practices.

A remarkable achievement of breeders, poultry nutritionists, and feed manufacturers is that, with advanced technology, it is common to raise a 2.3-kg (5-lb) broiler in just 6 weeks with a feed conversion of 1.8 kg of feed per kg of bird. In other words, a 2.3-kg broiler is raised from a chick on just about 4.1 kg of feed. This is one reason why chicken may be purchased at a lower price on an edible weight basis than beef, which has a less efficient feed conversion ratio.

The market classification of poultry is generally based on age and liveweight. Going from smaller and younger to larger and older birds, the following designations are given to chickens: broiler or fryer, roaster, capon, stag, stewing chicken, and old rooster. The tenderness of the flesh generally decreases in the same order. Broilers or fryers are generally preferred for processing into fresh or frozen chicken where tenderness is essential. Processors of canned chicken or chicken soup are able to use older and tougher birds because the sterilization heat of canning usually tenderizes the meat.

Within these marketing classes there also are U.S. grade standards for quality of individual birds based on feathering, shape, fleshing, fat, and freedom from defects. There are three quality grades: A Quality or No. 1, B Quality or No. 2, and C Quality or No. 3. Birds are purchased by the processor from the grower depending on the type of products to be manufactured and competitive price.

Processing Plant Operations

Plants for dressing poultry vary in size up to the largest that can process more than 10,000 birds per hour. These modern plants are efficient continuous-line facilities in which the birds are moved from operation to operation via monorail. Live birds are shackled, electrically stunned, bled, scalded to facilitate feather removal, plucked of feathers, eviscerated, government inspected, washed and chilled, dried, packaged, and frozen if production calls for freezing. The operations can be partially mechanized and highly efficient in large plants, since the birds purchased are remarkably uniform with respect to size, shape, weight, and other characteristics. To ensure high quality uniform birds, large processors generally contract with growers in advance and set up rigid specifications on the kind of bird they want.

Slaughter and Bleeding

Birds generally are not fed for 12 h before slaughter to ensure that their crops are empty, which makes for cleaner operations. Bleeding time depends on efficiency of the cut, type of bird, and whether the bird was electrically stunned or not before cutting. Bleeding may take anywhere from 1 to 3 min depending on these factors. But bleeding must be quite complete in order to produce the desirable white or yellow skin color in the final dressed bird.

Scalding

After bleeding, the birds are conveyed through a scalding tank. Scalding loosens the feathers and makes for easier plucking and pinfeather removal. The higher the temperature, the shorter the time required, but careful time and temperature control is very important because at higher temperatures there is a greater danger of removal of portions of skin in the defeathering machines. Scalding may be done at 60°C in about 45 sec, or more safely with less chance of skin removal at 52°C in about 2 min. Optimum conditions must be established for the kind of bird being dressed.

Defeathering

Defeathering is commonly done mechanically by a device that has many rotating rubber fingers. This removes all but a few pinfeathers, which may be removed by dipping the bird in melted wax, chilling to harden the wax, and then peeling.

Eviscerating

Evisceration is generally done in a separate cool room (Fig. 14.12). Evisceration includes inspection of the viscera by a veterinarian or someone under such supervision.

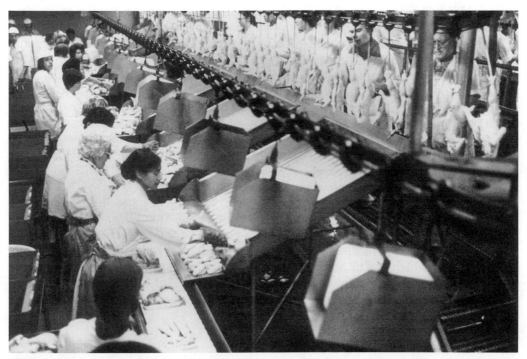

Figure 14.12. Poultry is efficiently eviscerated, processed, inspected, and packaged in an assembly line fashion. *Courtesy of J. M. Regenstein.*

The lungs and other organs that are difficult to dislodge may be removed by suction tubes. Birds passing inspection are thoroughly washed.

Chilling

The washed birds are rapidly chilled from about 32 to 2°C to prevent bacterial spoilage and to preserve quality. Chilling is done with ice slush, and the birds absorb a small amount of moisture from the slush, which makes them more succulent after packaging. But the maximum allowable water pickup is fixed by law. After chilling, birds are drained of excess moisture and are sized and graded for quality.

A few years ago, the majority of broilers were distributed as whole carcasses which were further cut by consumers. Now consumers prefer precut carcasses or, in many cases, they prefer to buy only selected portions of the carcass, such as breasts or thighs. Thus, partially automated processing lines which reduce whole carcasses into smaller portions have become common. Whole carcass sales now represent only about one-fifth of fresh poultry sales.

Packaging

The graded poultry or poultry parts may now be packaged as fresh poultry in boxes surrounded by crushed ice. If so, birds must be kept below 4°C and moved to retail channels rapidly, since shelf life may be only a few days. Shelf life depends on the bacterial load. Should this be about 10,000 organisms per square centimeter of surface, which is not uncommon, odor and slime will develop even at 4°C in about 6 days.

In order to overcome the drawbacks of shipping the added weight of ice and problems of its disposal, the poultry industry has rapidly moved to other technologies which eliminate the need for ice. For example, in the deep-chilling process, parts are sealed in consumer packages and then sent through a low-temperature blast tunnel and frozen to a depth of a few millimeters. This frozen crust quickly equilibrates with the rest of the carcass, which is then maintained at a temperature of −1 to −2°C which is above the freezing point of poultry meat. Modified atmosphere packaging (MAP) as discussed above for red meats is also used to ship prepackaged poultry products. Both of these technologies with controlled refrigeration can produce a product which has a shelf life of 18–21 days post-slaughter, which is sufficient time for distribution in most cases.

To prolong storage life, poultry can also be individually wrapped in low-moisture, low-oxygen transmission films or bags and frozen. When this is done, the bags are made to fit snugly, and the birds are vacuum packed in the bags to remove most of the air since the fat of chicken is highly susceptible to oxidation. The bags are made of plastic films which shrink when heated. Individual packages are passed thorough a hot air tunnel to tighten the bag around the carcass. This is particularly popular for turkey, for which the consumption is seasonal.

Government Inspection

In the United States all poultry sold in interstate commerce must be inspected for wholesomeness and, therefore, must be processed in plants having full-time government inspection service. Poultry is inspected live prior to slaughter, during evisceration, and during or after packaging. This inspection to protect the public health is mandatory. Inspection for quality grading by USDA representatives is optional, as in the case of meat. Poultry is graded as A or B grade. The two types of marks that may appear on poultry and represent these different forms of inspection are shown in Fig. 14.13.

Tenderness, Flavor, and Color

The color of the flesh of poultry and other birds is commonly described as dark or light. Dark muscle contains more myoglobin and is used by the bird for sustained activity. Light muscle has less myoglobin. Ducks or other birds which fly distances have dark breast muscle, whereas birds which do not fly distances, such as chickens and turkeys, have light colored breast muscle. The color of the skin of most birds reflects the amount of plant pigments consumed in the diet. Chickens raised on corn or other grains have very light colored skin. However, if highly pigmented plant material such as yellow flower petals are mixed into the diet, the skin will take on a

Figure 14.13. The U.S. Department of Agriculture grade shield of quality, and inspection mark of wholesomeness for poultry. *Courtesy of U.S. Department of Agriculture.*

yellow color. This does not directly affect the quality of the meat but, in some areas is taken as a measure of quality.

In general, the same factors favoring tenderness in red meat do so in poultry as well. Thus, meat from young birds is more tender than that from older ones, as is meat with less connective tissue (breast meat versus thigh meat) and more fat. In addition, birds grown in confined quarters without exercise yield more tender meat than do birds grown on open range.

Like meat and fish, poultry enters into a state of rigor mortis soon after being killed. Rigor mortis is associated with a conversion of glycogen to lactic acid, which has a mild preservative effect on the flesh, and with a contraction of the muscles and a stiffening of the tissues. Rigor mortis naturally subsides in poultry with a relaxation of the muscles after about 10 h or less. If poultry is cooked or frozen while the meat is in a state of rigor mortis, the meat may be excessively tough; this is avoided in good processing schedules.

Very fresh high quality chicken meat has little flavor. In fact scientific tests have shown that meat which has aged for a few days is preferred over meat that is only 1 day old. Chicken meat flavor also is affected by the feed received during growing. Excessive amounts of fish meal can give poultry a fishy flavor.

Contributing to tenderness and flavor, while at the same time providing added convenience during cooking, has been the development of the self-basting bird intended for roasting. This type of product is presently more common with turkey but can also be produced with chicken. To prepare self-basting birds, the manufacturer injects basting liquids at several points under the skin of the bird before packaging and freezing. The basting liquid generally contains vegetable oils, water, salt, emulsifiers, and artificial flavor and color. During roasting, the basting liquid moistens the skin and flesh and contributes succulence.

Nutritive Value

The composition of the edible parts of chicken depends on the cut and the method of cooking. Roasted white meat without the skin contains about 64% water, 32% protein, and 3.5% fat. Roasted dark meat without the skin contains about 65% water, 28% protein, and 6% fat. The skin is higher in fat. Chicken flesh contains more protein, less fat, and less cholesterol than red meat. The protein is of excellent quality and contains all of the essential amino acids needed by humans. The fat is more unsaturated than the fat of red meat and this can provide further nutritional advantages. Like other animal tissue, poultry flesh is a good source of B vitamins and minerals. Because of its high protein-to-fat ratio, chicken is a favored food of people who want to restrict fat intake and patients with vascular sclerotic tendencies.

Poultry Meat Products

The per capita consumption of poultry in the United States has increased over the last decade not only because people are eating more poultry directly, but also because the poultry industry has provided a large variety of products made from poultry meat. In many cases, these products are similar to products that might conventionally be made from red meats, including poultry based frankfurters, hams, sausages, bologna, salami, pastrami, ham, and other lunchmeats. Many of the newer products utilize mechanically separated poultry meat which is further ground to a fine emulsion. The emulsion may be cured, seasoned, smoked, and processed. Such products closely resemble their red meat counterparts, usually are lower in cost, may offer nutritional benefits, and must be appropriately labeled.

EGGS

Special strains of chickens are bred for large-scale egg production. Today, on the average, a hen often lays more than 260 eggs per year; in the United States about 70 billion eggs are produced each year. About 90% of these are consumed directly in the form of shell eggs. The remainder, broken out of the shell, is frozen, dried, or otherwise processed for use in the bakery, confectionery, and noodle industries, although there are also minor nonfood uses.

Egg Formation and Structure

Egg production is an integral part of the reproductive cycle in poultry (Fig. 14.14). Yolks containing the female germ cell are formed in the ovaries. These yolks drop into the mouth of the oviduct and then slowly pass down the oviduct. They are covered with layers of egg white from albumen-secreting cells, then with membranous tissue from other protein-secreting cells, and finally with calcium and other minerals from mineral-secreting cells near the bottom of the oviduct. This results in the egg shell. This process occurs whether the egg is fertilized or not. If fertilization is to take place, the sperm must travel up the oviduct and reach the yolk before the albumen and shell are deposited.

This sequence of events helps to explain several defects possible in shell eggs for human food: fertilized egg yolks produce embryos; ruptures in the ovary or oviduct can produce blood spots and sometimes meat specks; diseases of the ovary or oviduct can produce eggs infected with bacteria or parasites inside a sound shell. However, the contents of eggs from a healthy bird in unbroken shells are sterile when freshly laid.

The structure of the egg is diagrammed in Fig. 14.15. The central yolk is surrounded by a membrane called the vitelline membrane. Immediately beyond this is another membranous layer known as the chalaziferous layer. The yolk is connected to thick or firm albumen by two extensions of the chalaziferous layer called chalazaes. The yolk further is surrounded by layers of thin and thick albumen, and outside of these by another layer of thin albumen. This is surrounded by the shell, which has two inside shell membranes and an outer protective layer known as cuticle or "bloom."

The shell is porous and allows gases to pass in and out of the egg; these gases are used by the developing embryo in the case of a fertilized egg. On ageing, as air enters through the shell, the air cell at the blunt end between the shell membrane and the shell enlarges; a large air cell is an indication of storage and less fresh quality. When eggs are washed, the cuticle or bloom on the outside is removed, exposing the open pores of the egg shell. Under these conditions, bacteria can more easily enter the egg contents through the shell pores.

Composition

Eggs contain about two parts white to one part yolk by weight. The whole mixed egg contains about 65% water, 12% protein, and 11% fat. But the compositions of the

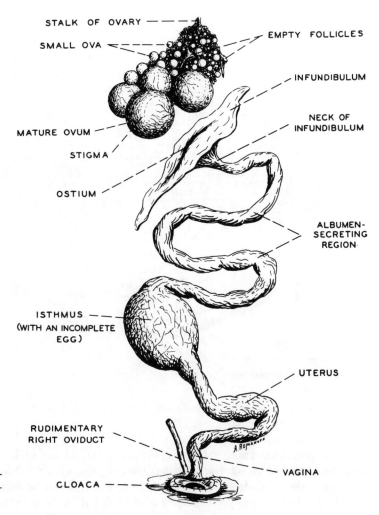

Figure 14.14. Reproductive organs of the hen. Source: *Romanoff and Romanoff.*

white and the yolk differ considerably. Virtually all of the fat is in the yolk, and when eggs are separated into white and yolk, we want to keep it this way since small amounts of fat adversely affect the whipping property of egg white. The 12% solids of egg white are virtually all protein (Table 14.3). The yolk is rich in fat-soluble vitamins A, D, E, and K and in phospholipids including the emulsifier lecithin. Nutritionally, eggs are a good source of fat, protein, vitamins, and minerals, especially iron.

Eggs also contain about 240 mg of cholesterol which is all contained in the yolk. For this reason, people who must restrict their cholesterol intake usually consume fewer whole eggs. This concern has led to a decline in the per capita consumption of eggs in recent years.

Quality Factors

Eggs may range in size from Peewee to Jumbo classifications, but quality grades are independent of size. The most common method of grading eggs is by candling, in which the egg is held up to a light source. Candling will reveal many defects—a cracked

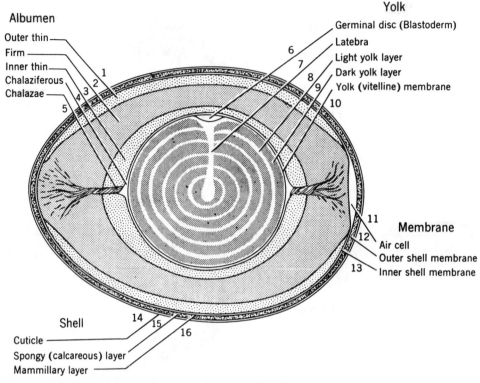

Figure 14.15. Structure of the hen's egg. *Courtesy of U.S. Department of Agriculture.*

shell, a fertilized yolk, a blood spot, an enlarged air cell, firmness of white which becomes thinner on ageing, and position of yolk which tends to drift off-center when the egg becomes stale. The extent of staleness also can be seen in broken-out eggs (see Fig. 6.13). Fresh eggs have a thick and high yolk rather than a flat yolk and a larger amount of thick white relative to runny thin white than do stale eggs, which spread out over a larger area than a fresh egg. Egg quality grades are based largely on these measures of freshness, since fresher eggs taste better, are easier to separate into whites

Table 14.3. Composition of the Hen's Egg

Fraction	%	% of Constituents			
		Water	Protein	Fat	Ash
Whole egg	100	65.5	11.8	11.0	11.7
White	58	88.0	11.0	0.2	0.8
Yolk	31	48.0	17.5	32.5	2.0
		Calcium carbonate	Magnesium carbonate	Calcium phosphate	Organic matter
Shell	11	94.0	1.0	1.0	4.0

SOURCE : U.S. Department of Agriculture.

and yolks for manufacturing purposes, and perform better in whipping and baking applications.

The shell color of eggs depends on the breed of chicken, but the yolk color depends largely on feed. Feeds high in carotenoids produce darker yolks, which are favored in some markets and in food manufacture to give a golden color to baked goods and products like noodles and mayonnaise. Despite common belief, shell color does not influence nutritional or other quality parameters.

Egg Storage

Because an abundance of eggs is produced in the spring of the year, eggs must be stored for use at other times. Storage is best at a temperature slightly above the freezing point of the egg. A temperature of $-1°C$ in warehouses is ideal; to minimize moisture loss from eggs, the relative humidity may be as high as 80%. In proper cold storage, Grade A quality can be maintained for as long as 6 months. After laying, eggs lose carbon dioxide through the porous shell, making the eggs more alkaline. The loss of carbon dioxide is associated with the staling process also, and some lengthening of storage stability can be achieved by storing eggs under carbon dioxide to minimize carbon dioxide loss.

More common, however, is the practice of spraying eggs to be stored with a light mineral oil. This closes the pores of the egg shell and retards both carbon dioxide and moisture loss. In another method of prolonging storage life, known as thermostabilization, eggs are dipped in hot water or hot oil for a brief period to coagulate a thin layer of albumen around the inside of the shell and thus further seal it. The heat also kills some of the surface bacteria.

Bacterial Infection and Pasteurization

As stated earlier, the contents of freshly laid eggs generally are sterile. However, the shell surface contains many bacteria, especially if the shell is soiled with chicken droppings. Even if the shell is not cracked, bacteria can enter through the natural shell pores. When eggs are washed, the shell cuticle is easily removed. If washing is not complete and the eggs are not dried, bacteria are especially likely to pass via the water through the shell. If the eggs are washed with warm water, increased temperature can make gases within the shell expand and escape through the pores. Then when the egg cools, a reduced pressure can result within the shell. This tends to draw bacteria and moisture from a wet shell into the egg through the pores. Cracked shells are obviously worse.

A particular group of bacteria belonging to the genus *Salmonella* are pathogenic to humans and commonly found in the poultry digestive tract. It is difficult to keep *Salmonella* organisms out of egg products and *Salmonella*-infected eggs have caused numerous outbreaks of disease. Because of the prevalence of *Salmonella* infections, food laws of the United States and several other countries require that all commercial eggs broken out of the shell for manufacturing use be pasteurized. This is a relatively new law in the United States, having been extended to egg white in 1966. Pasteurization of whole egg and egg yolk had been practiced for many years, but this was not the case with egg white.

Egg white is very sensitive to heat, being easily coagulated very near efficient pasteurization temperatures. For this reason an effective pasteurization treatment with minimal damage to the egg white was slow in being developed. The current pasteurization conditions for egg white or whole egg in the United States involve heating to the range of 60–62°C and holding for periods of 3.5–4.0 min. Egg white also may be pasteurized at lower temperatures of 52–53°C when combined with hydrogen peroxide. In one method the liquid whites are heated to this temperature and held for 1.5 min. Hydrogen peroxide at a level of 0.075–0.10% is metered into the egg white which is held at 52–53°C for 2 min more. The hydrogen peroxide is then broken down to water and oxygen by the addition of the enzyme catalase. Pasteurization processes may vary, but the treated eggs must be *Salmonella*-negative and meet other bacteriological standards.

More recently, concern has been over the occurrence of *Salmonella enteritidis* infection from diseased flocks. This human pathogen can enter eggs during their formation in the chicken before being laid. Contaminated eggs which have been eaten without sufficient cooking have been responsible for human disease including a number of deaths.

Freezing Eggs

Large quantities of eggs for use in food manufacturing are preserved by freezing. This is not done in the shell but rather with the liquid contents of the egg, which may be frozen as the whole egg or separated into yolk and white or various mixtures of yolk and white for special food uses.

Freezing plants generally are combined with egg-breaking facilities. The egg-breaking section of the plant receives eggs, may wash and dry them, and then break the egg contents from the shell. This used to be a hand operation but is now highly automated with egg-breaking machines. Where one operator can break and separate 60–90 dozen eggs per hour by hand, an operator–inspector can break and separate 600 dozen eggs per hour with these automatic machines. The operator–inspector also performs the very important function of rejecting spoiled eggs, one of which can ruin substantial amounts of good product. The common bacteria that are found in bad eggs generally fluoresce under ultraviolet light. This property has been used to help identify eggs to be rejected.

The whole or separated eggs are mixed for uniformity, screened to remove chalazae, membranes, or bits of shell, pasteurized, and placed in 13.6-kg (30-lb) cans or other suitable containers for freezing. Freezing generally is done in a sharp freezer room with circulating air at −30°C. Freezing may take from about 48–72 h.

Egg white and whole egg may be frozen as such, but egg yolk may not be frozen without additives since by itself it becomes gummy and thick, a condition known as gelation. Gelation of egg yolk on freezing is prevented by the addition of 10% sugar or salt or the addition of 5% glycerin. Sugar yolk is the product that goes to bakers, confectioners, and other users that can tolerate sugar in their end products, whereas mayonnaise manufacturers may use salt yolk. These ingredients may be dissolved in the yolk during mixing and prior to screening.

Drying Eggs

The whites, yolks, or whole eggs after pasteurization may be dried by any of several methods, including spray drying, tray drying, foam drying, or freeze-drying. Egg white contains traces of glucose. Whichever dehydration method is used, on drying or during subsequent storage at temperatures much above freezing, the glucose combines with egg proteins and the Maillard browning reaction occurs. This discolors the dried egg white. It has been found possible to prevent this browning reaction by removing glucose through fermentation by yeasts or with commercial enzymes. This is known as desugaring and is a step practiced prior to the drying of all egg white.

Egg Substitutes

The high level of cholesterol (about 240 mg per yolk) in egg yolk has caused many consumers to cut down on their consumption of eggs. Different approaches to reduce the cholesterol level have involved physically separating the yolk into high- and low-cholesterol fractions, decreasing the amount of yolk relative to albumen in the egg blend, and replacing yolk with a vegetable-oil-based yolk analog. One supplier formulates the "yolk" from corn oil, milk solids, emulsifiers, appropriate vitamins, and other additives, blends it with albumen, and then pasteurizes and freezes the product, which is sold for home, restaurant, and institutional use. Natural egg yolk, in addition to fat and protein, contains lecithin, which contributes to its emulsifying properties and the fat-soluble vitamins plus other nutrients. The importance of each of these constituents must not be overlooked in formulating an egg substitute.

References

Bechtel, P.J. 1986. Muscle as Food. Academic Press, Orlando, FL.

Beitz, D.C. and Hansen, R.G. 1982. Animal Products in Human Nutrition. Academic Press, New York.

Church, P.N. and Wood, J.M. 1992. The Manual of Manufacturing Meat Quality. Chapman & Hall, London, New York.

Cunningham, N.A. and Cox, N.A. 1987. The Microbiology of Poultry Meat Products. Academic Press, Orlando, FL.

Gracey, J.F. and Collins, D.S. 1992. Meat Hygiene. 9th ed. Bailliere Tindall, London.

Kinsman, D.M., Kotula, A.W., and Breidenstein, B.C. (Editors). 1994. Muscle Foods: Meat, Poultry and Seafood Technology. Chapman & Hall, London, New York.

Lawrie, R.A. 1985. Meat Science. 4th ed. Pergammon Press, Oxford.

Lawrie, R.A. 1992. Current concerns in the science and technology of red meat. Br. Nutr. Found. Nutr. Bull. *17*, 16–30.

McCoy, J.H. and Sarhan, M.E. 1988. Livestock and Meat Marketing. 3rd ed. Chapman & Hall, London, New York.

Mead, G.C. 1989. Processing of Poultry. Chapman & Hall, London, New York.

National Research Council. 1985. Meat and Poultry Inspection. The Scientific Basis of the Nation's Program. National Academy Press, Washington, DC. 209 pp.

Pearson, A.M. and Dutson, T.R. 1987. Restructured Meat and Poultry Products. Van Nostrand Reinhold, New York.

Pearson, A.M., Tauber, F.W., and Kramlich, W.E. 1984. Processed Meats. 2nd ed. Chapman & Hall, London, New York.

Price, J.F. and Schweigert, B.S. 1987. The Science of Meat and Meat Products. 3rd ed. Food & Nutrition Press, Westport, CT.

Schlenkrich, H. and Schiffner, E. 1988. Meat Microbiology. Deutscher Fachverlag, Frankfurt.

Smith, G.C. 1988. Laboratory Manual for Meat Science. 4th ed. American Press, Boston.

Stadelman, W.J. 1988. Egg and Poultry Meat Processing. VCH, New York.

Stadelman, W.J. and Cotterill, O.J. 1990. Egg Science and Technology. 3rd ed. Food Products Press, Binghamton, NY.

USDA. 1973. Treatment of pork products to destroy trichinae. Meat and Poultry Insp. Reg., Part 318, *10*, 125–131.

Weiner, P.D. 1987. Formulations for restructured poultry products. Adv. Meat Res. *3*, 405–431.

Woolcock, J.B. 1991. Microbiology of Animals and Animal Products. Elsevier, Amsterdam.

15

Seafoods

Foods derived from salt water are considered "seafoods," whereas all food derived from water environments, be they fresh or salt, are considered marine foods. The principal marine foods are saltwater fish, crustaceans and shellfish such as shrimp, lobster, crab, clams, and oysters, and certain freshwater fish and crustaceans. Seafood also includes other sea animals and plants such as seaweed and sea cucumber. Seafoods are also converted into large quantities of manufactured or processed foods, most of which are frozen or canned. Examples include precooked, battered, breaded, and frozen fillets, fish sticks, and shrimp, as well as canned tuna, salmon, and sardines. In addition, fish are salted, smoked, pickled, or dried. Currently, Americans consume about 16 lb of seafood per capita per year, compared to 112 lb of red meats and 64 lb of chicken. In other parts of the world, seafood makes up the major source of protein in the diet. Although the United States is a major exporter of fish, over 60% of the seafoods consumed in this country are from imports, reflecting market demand for selected species.

There are about 50 major species of fish, crustaceans, and shellfish available in the United States, varying with the season. In 1989, the U.S. Food and Drug Administration published its *Guide To Acceptable Market Names for Food Fish Sold in Interstate Commerce* and now requires that fish be appropriately named in accordance with this list to facilitate informed choice by consumers. For the last 30 years the three most important seafoods on an economic basis have been shrimp, salmon, and tuna. The trend is toward greater consumption of fresh or frozen fish, shrimp, and shellfish, whereas consumption of canned, smoked, salted, and pickled fish is decreasing.

Americans consume only certain parts of most fish, principally the muscles. Remaining parts plus large amounts of other fish such as menhaden, a fish not commonly used for human food, go into animal feeds. This represents about 50% of the weight of the U.S. fish catch, some of which returns in the forms of red meat, poultry, and eggs.

FISH PROCUREMENT

The usual methods of procuring fish for food are quite different from other food-gathering methods and significantly affect the quality of fish foods. Modern farmers, for example, have considerable control over plants and animals; they harvest and slaughter at times and in places largely of their own choosing. Hogs and cattle are fed under strict control and then brought to the place of slaughter according to schedule so that subsequent processing and refrigeration may follow immediately under optimized conditions. Food animals are also genetically bred for specific attributes. Not so with most seafood animals. Most fish are pursued and hunted and may be caught in the wild at great distances from processing facilities. These fish may lie poorly iced in the fishing vessel for a week or two, which generally is insufficient refrigeration, before being returned to port. Such "fresh fish" is not fresh at all; its quality is quite poor and becomes much worse before it reaches the consumer. This is aggravated by the fact that fish tissue is generally more perishable than most animal tissue under any circumstances, for reasons which will be indicated later. This practice is, however, changing to some degree. Oysters, lobsters, trout, catfish, salmon, and a few other species are now commonly "farmed" and this practice is on the increase worldwide. "Farming" fish is called aquaculture. Breeding programs and husbandry practices are being utilized in many cases. In the past, the waters of coastal nations were lightly fished by small boats. Fish were plentiful and boats returned to port in a day or so. Under such conditions, iced fish could be of high quality. In more recent years, with increased fishing pressure, improved methods of locating fish such as sonar and use of helicopters, and large trawler operations, vessels must travel farther to find fish. Under such conditions, since fish would have to remain in the ship's hold for several weeks or more, the need to prevent spoilage by processing on board has become essential .

Such processing is now being done on factory ships (Fig. 15.1) that accompany fleets of smaller fishing vessels. These have automatic equipment for eviscerating and cleaning fish and for converting livers to fish oil. They also have filleting machines,

Figure 15.1. Factory ship receiving fish from trawlers for processing at sea. *Courtesy of S. J. Holt.*

quick-freezing facilities, and fish meal processing plants to convert less edible portions of the fish to by-products. This is the current trend in the fishing industry of major fishing countries and will improve fish quality in the years ahead. Unfortunately, large numbers of vessels are not yet so equipped, and one of the major problems faced by food scientists is related to maintaining fresh quality.

MARINE FISH

One useful division of saltwater fish separates them into two groups depending on the depth of water in which they are found, which is correlated with great differences in the fish's fat content. Fish found in the middle and surface water layers of the sea are called pelagic fish and include herring, mackerel, salmon, tuna, sardines, and anchovies. This group includes many of the fatty fish, which in some instances have muscle containing 20% fat. The other group, called demersal fish, is found at or near the bottom of the sea, ordinarily on the continental shelves, and includes cod, haddock, whiting, flat fish such as flounder and halibut, ocean perch, shrimp, oysters, clams, and crab. Demersal fish usually have less than 5% fat and sometimes less than 1% fat in the muscle.

Composition and Nutrition

The composition and nutritional properties of the edible muscle of fish of a given species are quite variable, depending on season of the year, degree of maturity, and other factors. The herring, for example, may vary in muscle fat from about 8% to 20% with changes in the season and available food supply. The composition of most fish falls in the ranges of about 18–35% total solids, 14–20% protein, 0.2–20% fat, and 1.0–1.8% ash (see Table 14.1).

Nutritionally, fish proteins are highly digestible and at least as good as red meat with respect to content of essential amino acids. Consequently, the most important function of fish in all major fish-eating countries of the world is to provide high quality protein.

Because the fats of fish also are readily digestible and rich in unsaturated fatty acids, nutritionists frequently emphasize the importance of fish in the diet. But like all unsaturated fats, those in fish are highly susceptible to oxidation and the development of off-flavors and rancidity.

Fish are rich in vitamins. The fat of fish is an excellent source of vitamins A and D, and this was the reason for giving cod liver oil to children before multivitamin tablets were common. Fish muscle is a fair to good source of the B vitamins. Generally, shellfish and crustaceans are still richer in B vitamins than finfish.

Seafoods are a good source of important minerals and an excellent source of iodine in particular. Fish are lower in iron than most meats. Canned fish with the bones, such as salmon and sardines, are excellent sources of calcium and phosphorus.

Spoilage Factors

Fish tissue generally is more perishable than animal tissue, even under conditions of refrigerated or frozen storage. Beyond this generalization, it is difficult to make broad statements about the storage life of freshly caught fish because of the many variables that are encountered. Differences in tissue compositions of species, influence of season of the year on composition, differences between freshwater and saltwater fish and the effects of salt on the normal microflora of these fish, and varying procurement and holding practices on board fishing vessels are among these variables. Further, there presently are no rigid criteria that adequately differentiate such terms as truly "fresh," "good," or "acceptable," although grading systems based on taste panel results and selected chemical analyses can make useful distinctions. Certainly, the quality of a fish begins to change as soon as it is taken from the water, and "fresh" fish that is quite acceptable commercially is not fully the equivalent of the product when caught.

The stability data in Table 15.1 reflect commercial acceptability. Fresh fish held at a moderate temperature of 16°C remain good for only about 1 day or less. On ice at 0°C, finfish may remain good for periods up to about 14 days, but this is not true of all species. In contrast, beef may be aged at 2°C for several weeks to improve its texture and flavor. Even with mild salting and smoking, as in the case of finnan haddie and kippers, fish may remain good at 0°C for only a few weeks. Heavy salting and drying will, of course, preserve fish for long periods. There are several important reasons why fresh fish spoil rapidly, and these are microbiological, physiological, and chemical in origin.

Microbiological

Although the flesh of healthy live fish is bacteriologically sterile, there are large numbers of many types of bacteria in the surface slime and digestive tracts of living fish. When a fish is killed, these bacteria rapidly attack all constituents of the tissues. Furthermore, since these bacteria live on the cold-blooded fish at rather low ocean temperatures, they are well adapted to cold and continue to grow even under common refrigeration conditions.

Physiological

Fish struggle when caught and use up virtually all of the glycogen in their muscles, so little glycogen is left to be converted to lactic acid after death. Thus, the preservative

Table 15.1. Stability of Some Fish Products

Product	Approximate Number of Days Remaining in Good Condition	
	At 0°C[a]	At 16°C
Fresh cod	14	1
Fresh salmon	12	1
Fresh halibut	14	1
Finnan haddie	28	2
Kippers	28	2
Salt herring	1 yr	3 to 4 mo
Dried salt cod	1 yr	4 to 6 mo

[a]Assuming fish are immediately iced and never allowed to warm up.

effect of muscle lactic acid to slow bacterial growth is limited. This is in contrast to animal meat where animals are rested before slaughter to build up glycogen reserves.

Chemical

Associated with the fat of fish are phospholipids rich in trimethylamine. Trimethylamine split from phospholipids by bacteria and natural fish enzymes has a strong characteristic fishy odor. It is interesting to note that fish as taken from the water have little or no odor. Yet virtually all fish products that consumers encounter have a fishy odor and this is evidence of some deterioration. The fishy odor from liberated trimethylamine is further augmented by odorous products of fat degradation. The fats of fish are highly unsaturated and become easily oxidized, resulting in additional oxidized and rancid off-odors and off-flavors.

Preservation Methods

Because of the great tendency of fish to spoil, a number of methods of preservation have been developed over the years. The most basic methods are smoking and/or salting with subsequent drying. This is effective, but such preserved fish are not accepted in all cultures. Other societies find such preserved seafoods highly desirable.

Chemical preservatives such as sodium benzoate or sorbic acid can prolong storage life. In the United States, it used to be permissible to incorporate the antibiotic aureomycin in low levels into the ice used for packing fish on shipboard and in transit; the melting ice then held down bacterial growth. But this application is no longer allowed. Sodium nitrate and nitrite are permitted in a limited number of applications, as are salts of sorbic acid.

In recent years, irradiation with gamma rays to pasteurization doses has received much interest as an effective means of prolonging storage life of fresh refrigerated fish by 2–3 weeks. Experimental irradiation facilities have been tested to determine commercial feasibility and to establish data to support petition to the FDA for this irradiation application. But the current methods of greatest importance by far in preserving quality fish products still are refrigeration, freezing, and canning.

Shipboard Operations

Freezing can give excellent or poor quality results depending on how quickly the freezing is done after the fish is caught and on the freezing and storage temperatures. Ideally, fish should be gutted and frozen to −30°C within 2 h of being caught and held at this temperature. Unfortunately, this is costly and often unfeasible, and even supermarket frozen food cabinets generally are not kept below the −18° to −15°C range.

When fish are not processed and hard-frozen on board ship, and they generally are not except on the newer factory-type vessels, they commonly are stored with layers of ice at 0°C or in refrigerated chilled seawater at about −1°C. When the catch is held this way for a week or longer, it is no longer truly fresh on arrival at port. In Fig. 15.2 halibut is seen being unloaded from the cold hold of a fishing vessel. Not all fish held on ice or in chilled seawater is gutted on board ship, frequently this is done when the fish reaches port. Ungutted tuna often are chilled in ship brine wells down to

Figure 15.2. Unloading halibut from the cold hold of a fishing vessel. *Courtesy of National Marine Fisheries Service.*

−12°C, which partially freezes the fish. The tuna are then thawed and gutted weeks later in processing plants.

Processing Plant Operations

Fish that are not processed at dockside facilities commonly are packed with ice for transport to a processing plant. Here the fish are washed, gutted if not previously done, and then scaled, skinned, or filleted. There are machines for gutting, skinning, and filleting, but much of this work is still done by hand since machines need to be set for a given size of fish and size varies considerably.

Freezing

Fillets may be wrapped a few to a package for retail trade or packaged in sizable boxes and frozen. Small packages may be frozen in contact freezers of the type used to freeze retail portions of vegetables. Large boxes are frozen in room and tunnel blast freezers, preferably to a temperature of −30°C or below.

Cleaned whole fish also are frozen in the round. This often is done with larger fish, which after freezing are sawed perpendicular to their length into fish steaks; salmon and halibut are often handled this way. When individual fish are frozen, they are sometimes glazed with layers of ice by dipping in cold water. Dipping and freezing may be repeated several times to build up a thick glaze to protect the frozen fish from surface oxidation and from freezer burn (drying out) during frozen storage. Glazing also is practiced with frozen shrimp. When steaks are cut from large frozen fish, the newly cut surfaces are sometimes glazed to improve their keeping quality. It is common to add an antioxidant to the glazing water. Glazed frozen fish still require protective packaging in wrapping materials that are air-tight and moisture-tight.

Much fish goes into the preparation of prebreaded, precooked, and frozen fish sticks and individual fish portions. Generally, fish fillets that have been block-frozen in large boxes are used for this; they are cut to fish stick or portion size on band saws from the frozen blocks. The frozen portions are then battered and breaded and automatically conveyed to deep-fat fryers (Fig. 15.3). The fried pieces are then cooled, packaged, and refrozen.

The storage life of quality frozen fish further depends on its fat content. High quality, low-fat fish stored in the frozen state at $-21°$ to $-23°C$ may retain its quality for as long as 2 years.

Canning

Whereas high-fat fish do not store as well as low-fat fish in the frozen state, fish which contain more oil are more suitable for canning. Important examples are salmon, tuna, sardines, and herring. In the case of salmon, tuna, and sardines, additional fish oil, vegetable oil, or water commonly is added to the fish prior to can closure. Vegetable extracts are often added to help improve flavor. Canned fish products generally have a shelf life of several years.

Briefly, canning tuna involves the following steps:

1. Thaw the partially frozen tuna received from the fishing vessel.
2. Eviscerate, clean, and sort tuna for size.
3. Precook whole tuna in steam ovens (Fig. 15.4) to soften flesh for easy separation.
4. Let cool overnight, and then separate the light and the dark meat. The white meat gets the premium price, and the separation is still largely a hand operation which may involve 100 workers in a large plant.
5. Compact the tuna meat by machine into a cylindrical shape, cut off portions, and automatically fill into cans.
6. Add salt and vegetable oil or water to the cans.
7. Vacuum-seal the cans and sterilize them in a retort. The color of the flesh may darken somewhat due to browning reactions on heating.

Inspection and Grading

In the United States there is a voluntary fish inspection service and there are grade standards for fish and fishery products much like those of the USDA for agricultural products.

These and other standards—covering sanitation, quality, identity of fishery products, and processing plant operations—are administered by the National Marine Fisheries Service of the Department of Commerce. This agency also cooperates with the FDA, USDA, and state agencies, especially in matters relating to food safety and honest representation.

SHELLFISH

The term *shellfish* as generally used refers to the true shellfish (e.g., oysters and clams) and the crustaceans (e.g., lobsters and crabs). The high degree of perishability of finfish is shared by shellfish, except that most shellfish are even more perishable.

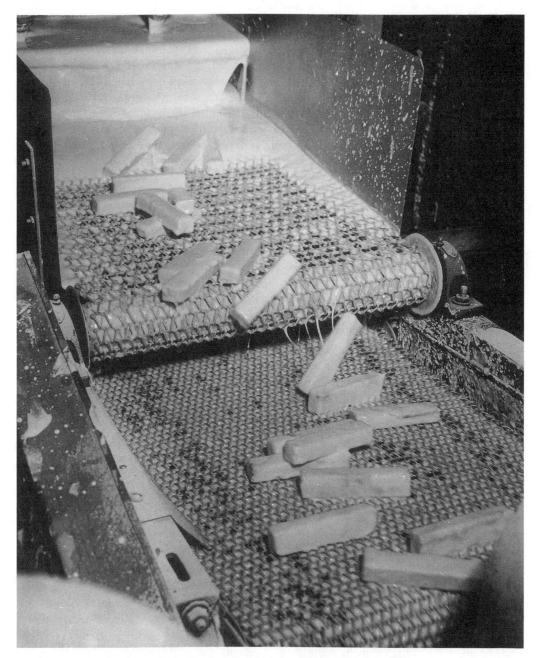

Figure 15.3. Fish portions being battered prior to frying. *Courtesy of J. W. Greer.*

Lobsters and crabs, for example, are best kept alive up to the point of their cooking or freezing, otherwise they deteriorate in quality in a matter of a day or less.

Shrimp

A favorite U.S. seafood, shrimp are caught in large trawling nets near U.S. coastal waters of the South Atlantic, Pacific, and the Gulf of Mexico, with additional quantities

Figure 15.4. Raw tuna being moved into precooker. *Courtesy of Starkist Foods, Inc.*

being imported from Central and South America where they are often farmed in aquaculture. Many are frozen in the raw state and in the breaded and precooked condition. Many are canned and some are freeze-dried.

After capture, heads are removed, the sooner the better for quality. This often is done on the shrimp boats. The landed iced shrimp are then unloaded at the processing plant where they are washed and sorted according to size. They then may be inspected, packed, and frozen in the shell without deveining. On the other hand, shrimp for breading and precooking prior to freezing are automatically peeled of shell and mechanically deveined by splitting and washing out the veinlike intestine.

Shrimp should be consumed or processed within 5 days of being caught even when they are iced. In addition to continued bacterial activity, iced shrimp will darken in color and may become black due to natural polyphenol oxidase enzymes contained in the shrimp (Fig. 15.5). This discoloration can be inhibited by addition of ascorbic acid or citric acid which can be added to the ice or used as a dip or spray. Shrimp quality is especially favored by very rapid freezing, such as is achieved with liquid nitrogen. Texture, color, and flavor are superior and drip loss is minimized when cryogenic freezing is used. However, these advantages are largely lost when shrimp are frozen by slower methods at sea and held on ice prior to arrival at the processing plant. To avoid this, some processors are supplying liquid nitrogen to shrimp boats. The pH of shrimp tissue is a fairly good index of shrimp quality. Freshly caught shrimp have a flesh pH of about 7.2, which increases even in ice storage. Quality remains generally good up to a pH of about 7.7. Above a pH of 7.9, shrimp become progressively spoiled.

Oysters and Clams

Oysters and clams remain essentially fixed in their environments and are harvested by raking the bottom or digging the mud close to shore. The U.S. oyster industry is concentrated in the Chesapeake Bay area and Long Island Sound, with lesser amounts of oysters coming from coastal waters of the South Atlantic and the Pacific Northwest states. The clam industry is more widely scattered along the entire seacoast.

Live oysters and clams are removed from the shell by hand, washed, sorted for size, and then further processed. Processing may simply involve packing in cans or jars and shipping in ice to market. Oysters and clams also are canned and sterilized in retorts

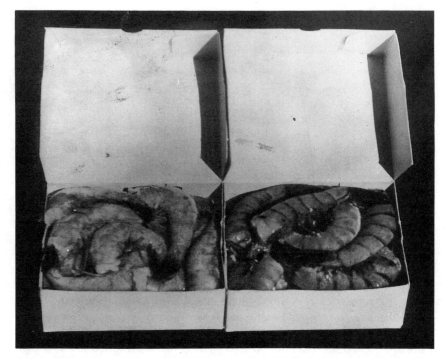

Figure 15.5. Shrimp darken if held before freezing. (Left) Frozen immediately after catching; (right) iced 5 days before freezing. *Courtesy of C. W. DuBois.*

and to a lesser extent are frozen. The frozen products may later be used in the manufacture of stews and chowders.

The condition of the water from which the clams and oysters are harvested is important. The oysters and clams can become infected with bacteria or virus from water which has become polluted from sewage. This is particularly serious because oysters and clams are frequently eaten raw. Outbreaks of infectious hepatitis and gastroenteritis have been traced to uncooked clams and oysters. For these reasons, commercial plants handling oysters and clams operate under regulations of the U.S. Public Health Service–FDA and usually also under state regulations.

Crabs

Crabs of major importance in the United States are the blue crab, Alaskan king crab, Dungeness crab, and tanner crab. For meat removal, crabs are cooked prior to the flesh being hand-picked. Meat is removed from the larger king crab with water jets and by passing legs through rubber rollers to squeeze out the meat. Crabmeat is canned or pressed into blocks and frozen. Whole cooked crabs and king crab legs with the shell are frozen also.

The yield of meat from crabs is low, averaging about 12% (20% for king crab). Disposal of the remaining shellfish "wastes" has become a difficult problem since recent environmental regulations prohibit the dumping of untreated shellfish wastes into coastal waters and treatment can be costly. This has focused research efforts on conversion of "wastes" into useful products such as chitin and chitin derivatives from the shells (also from the shells of shrimp) and enzymes from other tissues.

FISH BY-PRODUCTS

Parts of fish not used for human food such as the intestines, heads, and gills, as well as whole fish of less favored species, have been ground up, dehydrated, and converted to fish meal for animal and poultry feed. Such fish meal generally had a fishy odor and a high bacterial content and was not considered suitable for human food. Fish meal has also been used as fertilizer.

More recently, ways have been found to extract oils and fatty substances from ground fish tissue to an extent that little fish odor or flavor remains. This extracted tissue can be heated, stripped of solvent, and then dehydrated and milled to a bland, highly nutritious powder rich in high quality protein and minerals. When produced under proper bacteriological and sanitary control from selected species of fish, the product is known as fish protein concentrate (FPC) and can be used as human food.

Fish protein concentrate when properly manufactured and packaged can be readily incorporated into a wide variety of basic foods as an enrichment without adverse effects on acceptability of these items. Fish protein concentrate can contain 85–92% of high quality protein. It has been estimated that for a cost of approximately a cent per day per person, FPC could balance protein-deficient diets around the world. However, as yet, this product has acquired but limited commercial importance.

CONTAMINANTS IN FISH

Many foods taken from marine environments are near the top of the food chain and can concentrate undesirable pollutants from that environment. Unfortunately, human activity has resulted in pollution of many marine environments with biological and chemical pollutants, so it is not surprising that seafoods harvested from polluted water contain pathogenic microorganisms and toxicants. There are also naturally occurring toxicants which can be concentrated in seafoods.

Mercury is an example of a contaminant which has concentrated in seafoods. Mercury occurs naturally but usually not in toxic forms or amounts. It is also used in several industrial processes and may be unnaturally high due to industrial pollution. Mercury accumulation in fish depends not only on the level of mercury in the water but also on the type of fish, its natural food chain, and the age of the fish. Older and larger fish of a given species tend to have higher mercury levels, believed due to the longer time they have had to concentrate it in their tissues from their food supply.

Because mercury in sufficient concentration is toxic, regulatory agencies of various countries have set upper limits of mercury permitted in foods. Before 1979, the U.S. Food and Drug Administration's recommended maximum level for mercury in fish was 0.5 mg/kg; in some other countries this level was higher. Such levels are established on the basis of toxicological data plus an appropriate safety factor. At the present time, however, no one knows the absolute threshold level that is toxic in humans. In 1979 the U.S. action level was increased from 0.5 to 1.0 mg/kg of mercury in seafoods, and in 1984 this action level was made specific for methyl mercury rather than total mercury. The changes reflect newer information on toxicology of the compounds of mercury and economic considerations.

In addition to mercury, other pollutants reaching lakes, rivers, and the ocean can accumulate in fish. For example, PCBs (polychlorinated biphenyls) have been implicated as a human health risk, especially from lakes and rivers near sites where PCBs were used in various manufacturing processes. Other chemicals such as dioxin, pesticides such as DDT, endrin and dieldrin that may leach from soils, and heavy metals such as cadmium and lead can become pollutants. The situation is aggravated when the pollutant is not readily biodegradable, as is the case for most of the above materials.

It is not possible to immediately remove such pollutants from marine environments so regulatory agencies usually set limits on the amount of pollutant that seafood can contain and often advise people to limit the amount of a given fish they should consume.

Seafoods can also be contaminated by toxins and organisms of natural origin. Red tide which occurs without predictability is an example. The red color is due to immense numbers of the protozoa *Gymnocardium brevis,* which can infect fish and shellfish. Consumption of infected seafood can cause respiratory irritation in man, and such seafood has had to be withheld from the market on occasion.

Some fish themselves contain toxins which can cause death. Puffer fish, considered a delicacy in some Asian cultures, contain a potent toxin. The toxin is contained in the gonads of the fish and can be removed, but great care must be practiced.

Yet another problem is known as scromboid fish poisoning and is due to the formation of histamine resulting from microbiological decay when fish is held under poor storage conditions (above 0 to 6°C). Cooking does not provide protection against such toxins if already formed in the fish.

Also, fish may harbor parasitic tapeworms and roundworms that can infect humans. Such infections may result from the consumption of raw fish. These parasites are readily destroyed by common cooking and freezing procedures and by salting and/or smoking. Eating raw seafood which has not been frozen or cooked can be risky.

NEWER PRODUCTS FROM SEAFOOD

Seafoods, with only limited exceptions, are unique in that they are almost exclusively harvested from the wild. Humans have learned through agriculture to domesticate most other food sources, but seafoods depend to a large extent on nature. As such, it is recognized that the food from the sea is not an inexhaustible resource. Several new sources of seafood and new processes to better utilize existing seafoods have been developed in recent years. Oysters, salmon, catfish, and others are commonly farmed. Species which return to their spawning grounds after roaming free are sometimes "ranched." Ranching fish means hatching fry and turning them lose from a designated place. After growing in the wild, the fish will return to spawn and can be harvested.

Machines similar to those used to separate meat from bone have been successfully used to obtain minced fish flesh from filleting wastes and from underutilized species. This minced fish flesh is now being used in the processed seafoods of several countries. One newer process which provides high quality food from minced fish yields surimi. Minced fish is highly washed to remove solubles including pigments and flavors, leaving an odorless and flavorless high-protein flesh. This is combined with other ingredients to give the fish muscle emulsion texture and frozen stability. Flavors and color are added and the product extruded into shapes resembling other products such as crabmeat or lobster meat. Surimi production is a highly technical process requiring considerable skill. In the United States this process is mostly used to make artificial crab products. Surimi is especially popular in parts of Asia including Japan.

In order to provide a more constant and reliable source of certain seafoods, there has been a growth in various forms of aquaculture, involving both freshwater and saltwater species. Presently, about 15% of the world's seafood and 5% of U.S. seafood

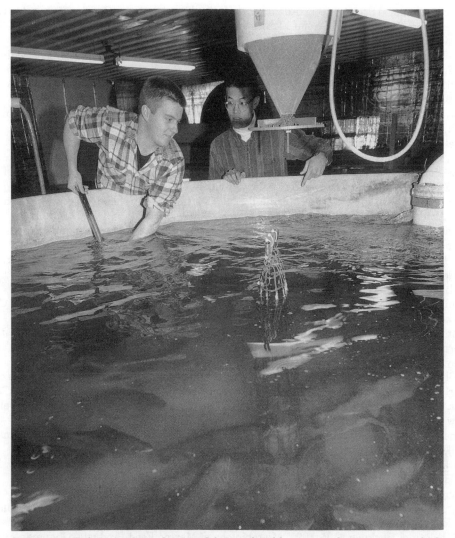

Figure 15.6. Fish (trout) being harvested from a closed loop aquaculture system in which water is purified and recycled. *Courtesy of Cornell University Photo Services.*

is produced under varying degrees of control, utilizing ponds, tanks, cages, nets and other forms of confinement (Fig. 15.6). In the southern part of the United States, for example, growing catfish in freshwater ponds has become successful. In northern Europe, raising salmon represents a large and growing business. Research has been concerned with fish genetics, nutrition, disease control, yields, fish qualities for food use, and development of markets. In aquaculture as well as ocean fishing there is no doubt that an enormous potential for food remains to be developed.

References

Ahmed, F.E. 1991. Seafood Safety. National Academy Press, Washington, DC.
Aitken, A. et al. 1982. Fish Handling & Processing. 2nd ed. Aberdeen Ministry of Agriculture, Edinburgh.

American Chemical Society. 1992. Advances in Seafood Biochemistry: Composition and Quality: Papers from the American Chemical Society Annual Meeting, New Orleans, Louisiana. Technomic Publ. Co., Lancaster, PA.

Bligh, E.G. and Ackman, R.G. (Editors). 1991. Seafood Science and Technology. Proceedings of the International Conference Seafood 2000 Celebrating the Tenth Anniversary of the Canadian Institute of Fisheries Technology of the Technical University of Nova Scotia, 13–16 May, 1990, Halifax, Canada. Fishing News Books, Oxford.

Bonnell, A.D. 1994. Quality Assurance in Seafood Processing: A Practical Guide. Chapman & Hall, London, New York.

Connell, J.J. 1990. Control of Fish Quality. 3rd ed. Fishing News Books, Oxford.

Gorga, C. and Ronsivalli, L.J. 1988. Quality Assurance of Seafood. Van Nostrand Reinhold, New York.

Hall, G.M. 1992. Fish Processing Technology. Blackie, New York.

Hall, S. and Strichartz, G.R. 1990. Marine Toxins. Origin, Structure, and Molecular Pharmacology. American Chemiscal Society, Washington, DC.

Karmas, E. 1982. Meat, Poultry, and Seafood Technology. Recent Developments. Noyes Data Corp., Park Ridge, NJ.

Kinsella, J.E. 1987. Seafoods and Fish Oils in Human Health and Disease. Marcel Dekker, New York.

Lovell, R.T. 1991. Foods from aquaculture. Food Technol. *45*(9), 87–92.

Martin, R.E. and Collette, R.L. 1990. Engineered Seafood Including Surimi. Noyes Data Corp., Park Ridge, NJ.

Martin, R.E., Flick, G.J., Hebard, C.E., and Ward, D.R. 1982. Chemistry and Biochemistry of Marine Food Products. AVI Publishing Co., Westport, CT.

McVey, J.P. 1983. Handbook of Mariculture. Vol. 1. CRC Press, Boca Raton, FL.

Negedly, R. 1990. Elsevier's Dictionary of Fishery, Processing, Fish, and Shellfish Names of the World in Five Languages, English, French, Spanish, German, and Latin. Elsevier, New York.

Pigott, G.M. and Tucker, B.W. 1990. Seafood. Effects of Technology on Nutrition. Marcel Dekker, New York.

Regenstein, J.M. and Regenstein C. 1991. An Introduction to Fish Technology. Chapman & Hall, London, New York.

Sikorski, Z.E., Pan, B.S., and Shahidi, F. (Editors). 1994. Seafood Proteins. Chapman & Hall, London, New York.

Stansby, M.E. 1990. Fish Oils in Nutrition. Chapman & Hall, London, New York.

Straus, K. 1991. The Seafood Handbook: Seafood Standards: Establishing Guidelines for Quality. Seafood Business, Rockland, ME.

Ward, D.R. and Hackney, C.R. 1991. Microbiology of Marine Food Products. Chapman & Hall, London, New York.

16

Fats, Oils, and Related Products

Fats and oils have been discussed in several previous chapters. The major portions of fats and oils are made up of fatty acid esters of glycerol. Edible fats and oils come from both plant and animal sources and have important functional and nutritional properties in foods. Certain components of fats are required nutrients as well as carriers of the fat-soluble vitamins. Fats and oils can deteriorate in foods and are susceptible to oxidation and rancidity. They also have shortening, lubricating, emulsifying, and whipping properties, and high caloric value. The terms *fat* and *oil* only indicate whether the material is liquid or solid: fats that are liquid at room temperature are called oils.

EFFECT OF COMPOSITION ON FAT PROPERTIES

The structural formula of a typical triglyceride molecule of a fat is shown in Fig. 16.1. In this case, three different fatty acids are esterified or connected to glycerol. There are numerous fatty acids. The structure of the fatty acids that are esterified to glycerol largely determine the properties of fats, including whether they are solid or liquid at room temperature. It is well to review some of the more important properties these different fatty acids contribute to fats before considering the processing and utilization of fats and oils.

Short-chain fatty acids give softer fats of lower melting points than do long-chain fatty acids. Fatty acids can have areas of unsaturation within their molecules due to the absence of hydrogen atoms at certain points. This is where double bonds occur in the fatty acids. In the triglyceride molecule of Fig. 16.1, all three fatty acids are of the same length—each contains 18 carbon atoms—but the degree of unsaturation of each is different. The top fatty acid, stearic acid, is fully saturated (i.e., it has no place where additional hydrogen atoms could be placed); the middle one, oleic acid, has one double bond and is missing two hydrogen atoms; the third fatty acid, linoleic acid, has two double bonds and is missing four hydrogen atoms, so it is the most unsaturated.

The greater the degree of unsaturation in the fatty acids of fat molecules, the softer the fat is at a given temperature and the lower its melting point. When there is a considerable degree of unsaturation, the fat will be liquid at room temperature and will be called an oil.

By chemical means, hydrogen can be added to an oil to saturate its fatty acids, thereby converting it to a solid. This process is termed *hydrogenation* and commonly converts a vegetable oil to a solid shortening. Partial hydrogenation produces an

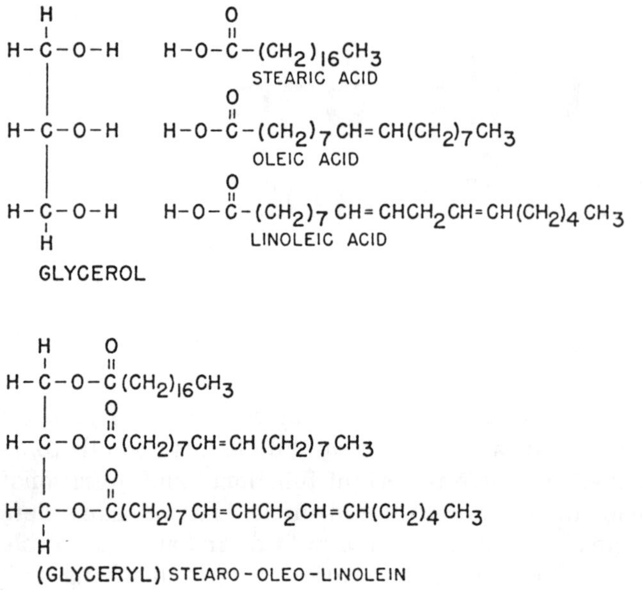

$$
\begin{array}{ll}
\text{H} & \text{O} \\
\text{|} & \text{||} \\
\text{H–C–O–H} & \text{H–O–C–(CH}_2)_{16}\text{CH}_3 \\
\text{|} & \quad\quad\quad \text{STEARIC ACID} \\
& \text{O} \\
& \text{||} \\
\text{H–C–O–H} & \text{H–O–C–(CH}_2)_7\text{CH=CH(CH}_2)_7\text{CH}_3 \\
\text{|} & \quad\quad\quad \text{OLEIC ACID} \\
& \text{O} \\
& \text{||} \\
\text{H–C–O–H} & \text{H–O–C–(CH}_2)_7\text{CH=CHCH}_2\text{CH=CH(CH}_2)_4\text{CH}_3 \\
\text{|} & \quad\quad\quad \text{LINOLEIC ACID} \\
\text{H} & \\
\text{GLYCEROL} &
\end{array}
$$

$$
\begin{array}{l}
\text{H} \quad \text{O} \\
\text{|} \quad \text{||} \\
\text{H–C–O–C (CH}_2)_{16}\text{CH}_3 \\
\text{|} \quad\quad \text{O} \\
\quad\quad\quad \text{||} \\
\text{H–C–O–C (CH}_2)_7\text{CH=CH (CH}_2)_7\text{CH}_3 \\
\text{|} \quad\quad \text{O} \\
\quad\quad\quad \text{||} \\
\text{H–C–O–C(CH}_2)_7\text{CH=CHCH}_2\text{CH=CH(CH}_2)_4\text{CH}_3 \\
\text{|} \\
\text{H} \\
\text{(GLYCERYL) STEARO–OLEO–LINOLEIN}
\end{array}
$$

Figure 16.1. Structural formula of a typical triglyceride molecule and its components.

intermediate degree of solidification. The degree of hydrogenation is important in determining the solid properties of foods containing fats.

Unsaturated fatty acids are highly reactive with oxygen at the points of unsaturation. Therefore, hydrogenation, which saturates fats, makes them more resistant to oxidation and more stable against oxidized flavor development.

Fatty acids, like many organic compounds, exhibit isomerism. Fatty acid isomers have the same numbers of carbon, hydrogen, and oxygen atoms but in different geometrical arrangements, which result in different chemical and physical properties. Fatty acids with the same empirical formula may have straight chains or branched chains, as with *n*-butyric acid and isobutyric acid. Fatty acids with one or more double bonds can show two types of isomerism, namely, positional and geometric isomerism. Positional isomerism has to do with the position of the double bond or bonds along the carbon chain. Geometric isomerism is due to restricted rotation of two carbon atoms connected by a double bond. In this case, hydrogen atoms (or other groups) attached to the carbons of the double bond can be on the same side of the double bond (cis isomer) or on opposite sides of the double bond (trans isomer). Thus, oleic acid (cis form) can be converted to elaidic acid (trans form):

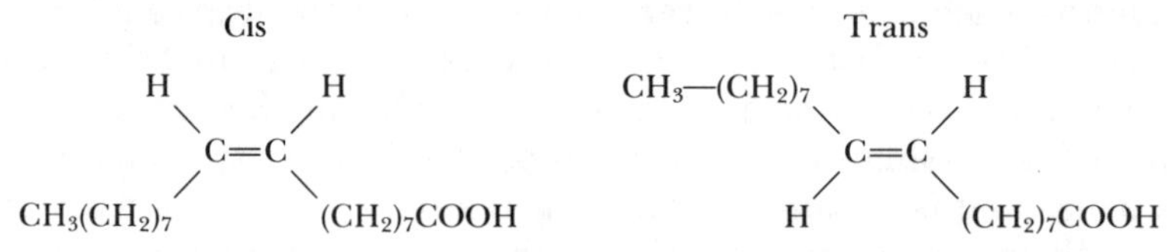

$$
\begin{array}{cc}
\text{Cis} & \text{Trans} \\
\\
\text{H} \quad\quad \text{H} & \text{CH}_3\text{—(CH}_2)_7 \quad\quad \text{H} \\
\diagdown \quad \diagup & \diagdown \quad\quad\quad \diagup \\
\text{C=C} & \text{C=C} \\
\diagup \quad \diagdown & \diagup \quad\quad\quad \diagdown \\
\text{CH}_3\text{(CH}_2)_7 \quad (\text{CH}_2)_7\text{COOH} & \text{H} \quad\quad (\text{CH}_2)_7\text{COOH} \\
\\
\text{Oleic acid (mp 14°C)} & \text{Elaidic acid (mp 44°C)}
\end{array}
$$

Most naturally occurring unsaturated fatty acids are in the cis form but may be changed to the trans form under certain conditions of processing. Isomers of a given molecular weight generally differ in melting point, solubility, stability, biological and nutritional properties, and in other ways, and these differences are imparted to the fats containing them.

Natural fats contain more than one kind of triglyceride molecule. A given fat will generally contain a mixture of triglyceride molecules differing in the lengths and in the degrees of unsaturation of their fatty acids (Tables 16.1 and 16.2). Because of this, some molecules in the fat will be softer and some harder. The overall fat may be a liquid at room temperature but actually contain some solid fat molecules suspended in the liquid oil. Should the liquid fat be cooled, more of the fat molecules will solidify and they may form fat crystals, which separate from the liquid oil portion. This is one property that is used in separating fats into liquid and solid fractions. The liquid fraction will have a lower melting point and the solid fraction a higher melting point than the original mixture, and both fractions will be suitable for different food uses.

Figure 16.2 is a photomicrograph of a vegetable shortening with solidified fat crystals suspended in the liquid oil portion. The proportions of fat crystals and liquid oil depend on the melting points of the crystals, which will be low when the fatty acids in the crystals are of short length or are highly unsaturated. When this shortening is chilled, more crystals form and the shortening stiffens; when the shortening is heated, the crystals melt and the shortening becomes a liquid oil.

The fat and oil chemist and processor have as their raw material a versatile group of natural fats with different properties (Table 16.3). For example, the fat from cocoa, known as cocoa butter, is solid at temperatures below 30°C; on the other hand, the fat pressed from peanuts is liquid at room temperature. But the chemist and processor can also modify fats and oils of a given kind very readily by hydrogenation to make them stiffer, by temperature-controlled crystallization followed by separation into solid and liquid portions, and by still other means. The processor can then further blend various natural, hydrogenated, or crystallized fats into an endless number of mixtures and thereby further tailor special fats for the widest range of uses (Table 16.4).

SOURCES OF FATS AND OILS

Fats and oils may be of vegetable, animal, and marine origin. Examples of vegetable fats include the solid fat cocoa butter, and the liquid oils include corn oil, sunflower oil, soybean oil, cottonseed oil, peanut oil, olive oil, canola oil, and many more. Animal fats include lard from hogs, tallow from beef, and butterfat from milk. Fish oils include cod liver oil, oil from menhaden, and whale oil, although the whale is really a mammal.

Some of these fats and oils are chosen for food uses because of special flavor or physical attributes. For example, lard contributes a meaty flavor, olive oil has a distinctive flavor on salads, and butterfat has a buttery aroma and flavor. But in the production of shortenings, margarines, and frying fats, chemical and physical methods such as hydrogenation and crystallization are used to make many of the natural fats interchangeable with respect to texture and physical properties. Where permitted by law, flavors may be added to duplicate or even improve on the taste of a specific natural fat. Other fats and oils are chosen because they may have nutritional benefits. For example, oils from certain plants are highly unsaturated, which may have positive health benefits.

In the past, there has been spirited debate about the relative merits of different types of fat, and fierce competition among their makers. Prime examples of such controversies are butter versus margarine and lard shortening versus vegetable shortening. But in today's market, economics and availability clearly dictate what fats will be used when the fats can be made interchangeable for a specific application by modern technology.

Table 16.1. Vegetable Fats and Oils

Typical Percentage Composition and Iodine Value

Fatty Acid	Carbon Atoms	Cocoa Butter	Coconut	Corn	Cotton-seed	Olive	Palm	Palm Kernel	Peanut	Rape-seed	Saf-flower	Sesame	Soy-bean	Sun-flower
Caprylic	8	—	6	—	—	—	—	3	—	—	—	—	—	—
Capric	10	—	6	—	—	—	—	4	—	—	—	—	—	—
Lauric	12	—	44	—	—	—	—	51	—	—	—	—	—	—
Myristic	14	—	18	—	1	—	1	17	—	—	—	—	—	—
Palmitic	16	24	11	13	24	13	48	8	6	4	8	10	12	8
Palmitoleic	16	—	—	—	1	1	—	—	—	—	—	—	—	—
Stearic	18	35	6	4	3	2	4	2	5	2	3	5	2	5
Oleic	18	39	7	29	18	75	38	13	61	19	13	40	24	21
Linoleic	18	2	2	54	53	9	9	2	22	14	75	43	54	66
Linolenic	18	—	—	—	—	—	—	—	—	8	1	2	8	—
Arachidic	20	—	—	—	—	—	—	—	2	—	—	—	—	—
Gadoleic	20	—	—	—	—	—	—	—	—	13	—	—	—	—
Behenic	22	—	—	—	—	—	—	—	3	—	—	—	—	—
Erucic	22	—	—	—	—	—	—	—	—	40	—	—	—	—
Lignoceric	24	—	—	—	—	—	—	—	1	—	—	—	—	—
Iodinevalue		37	9	127	109	84	51	16	101	104	146	114	134	134

SOURCE: T. J. Weiss (1983).

Table 16.2. Animal Fats and Oils

| Fatty Acid | Carbon Atoms | Typical Percentage Composition and Iodine Value | | |
		Butter	Lard	Tallow
Butyric	4	3	—	—
Caproic	6	3	—	—
Caprylic	8	2	—	—
Capric	10	3	—	—
Lauric	12	3	—	—
Myristic	14	10	1	3
Myristoleic	14	1	—	—
Palmitic	16	26	25	28
Palmitoleic	16	1	2	3
Stearic	18	15	13	23
Oleic	18	29	47	40
Linoleic	18	2	12	2
Linolenic	18	2	—	—
Iodine value	—	35	63	41

SOURCE: T. J. Weiss (1983).

Because fats vary widely in price, and often can be made interchangeable or nearly so, it is important that deception be prevented and consumers indeed get what they choose to purchase. Thus, there are federal standards of identity for foods, labeling requirements for nonstandardized foods, and analytical tests to distinguish between certain fats to prevent adulteration. Butter cannot be called butter if it contains any fat other than butterfat. This is also true of olive oil. Substitute cocoa and chocolate

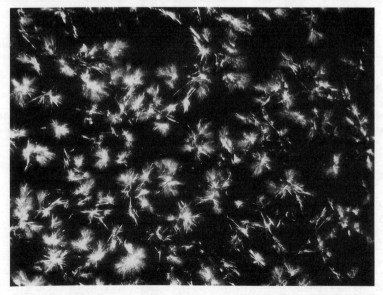

Figure 16.2. Photomicrograph of a shortening showing solidified fat crystals suspended in liquid oil. *Courtesy of V. C. Mehlenbacher, Garrard Press.*

Table 16.3. Physical Characteristics of Food Fats—Raw Materials

Fat	Solid Fat Index[a] at °C					Melting Point (°C) (Capillary)
	10	21	27	33	38	
Butter	32	12	9	3	0	36
Cocoa butter	62	48	8	0	0	29
Coconut oil	55	27	0	0	0	26
Lard	25	20	12	4	2	43
Palm oil	34	12	9	6	4	39
Palm kernel oil	49	33	13	0	0	29
Tallow	39	30	28	23	18	48

SOURCE: T. J. Weiss (1983).
[a]Solid Fat Index is related to the percentage of fat that exists in crystalline form at a given temperature.

products must be appropriately labeled when they contain vegetable fats other than cocoa butter. But today, margarine, shortenings, frying fats and many other products are made from a wide variety of fat sources mostly of plant origin in accordance with availability and price.

FUNCTIONAL PROPERTIES OF FATS

Apart from flavor differences, when fats are used as shortenings, tenderizers, lubricants, frying media, whipping agents, and for other purposes, there are special require-

Table 16.4. Physical Characteristics of Typical Fat Products

Product	Solid Fat Index[a] at °C					Melting Point (°C) (Capillary)	Consistency (Bloom) at 24°C
	10	21	27	33	38		
Shortenings							
cake and icing	28	23	22	18	15	51	40
cake mix	40	31	29	21	15	48	75
coating fat, winter	65	55	45	19	1	39	Hard & brittle
coating fat, summer	67	58	51	31	18	48	Hard & brittle
frying	44	28	22	11	5	43	70
pie crust	33	28	22	10	8	48	70
yeast dough	26	20	12	6	3	46	50
Margarines							
table (premium)	24	12	8	2	0	37	15
table (regular)	28	16	12	3	0	38	25
cake	29	19	17	11	7	46	40
pastry, roll-in	25	21	20	18	15	50	80
puff paste	28	25	24	22	19	51	110

SOURCE: T. J. Weiss (1983).
[a]Solid Fat Index is related to the percentage of a fat that exists in crystalline form at a given temperature.

ments in each of these capacities. In the case of butter or table margarine, a plastic texture is required such that the butter or margarine will not become too hard in the refrigerator to spread, or too soft on a summer day so that it will run.

Oils used in salad oils should be clear and pourable. They should not contain high-melting-point molecules that will solidify and crystallize when the salad oil is placed under refrigeration. Likewise, oils used in the preparation of mayonnaise should not form crystals when the mayonnaise is refrigerated, since these fat crystals could disrupt the state of emulsion and cause the mayonnaise to separate into fat and liquid phases.

Chocolate products should not melt at room temperature. They should be brittle and snap when bitten into, yet melt quickly in the mouth. Cocoa butter melts sharply above 30–36°C and possesses these properties. Chocolate substitutes contain fats other than cocoa butter that may have been tailored to closely match these properties of cocoa butter. In the case of chocolate-type coatings for certain confections and crackers, however, it may not be the objective to match chocolate but rather to produce a coating with a higher melting point to prevent melting when handled.

Where the fat is to be whipped into a fat–air emulsion, as in the case of buttercream-type icings, partially solidified shortenings generally produce a firmer whip than liquid oils. Also, where shortening is to be used in baked goods, a partially solidified fat generally functions better than most liquid oils, since the oils tend to separate and collect in pockets in the baked item.

Optimum fats for some of these and other applications cannot always be obtained from nature. More generally, they must be fabricated by blending a variety of fats and oils that have undergone various degrees of modification. This is particularly true for fats and oils used in manufactured foods produced with automatic equipment. Thus, the requirements of a shortening used in the high-speed continuous bread-making process are different from those of a shortening for bread made by the batch method.

PRODUCTION AND PROCESSING METHODS

There are only a few basic production methods to obtain fats and oils from animal, marine, and vegetable sources. These include rendering, pressure expelling, and solvent extraction, which are followed by various refining and modifying procedures.

Rendering

In the process of rendering, meat scraps are heated in steam or water to cause the fat to melt. The melted fat then rises and water and remaining tissue settle below. The melted fat is then separated by skimming or centrifugation. Dry-heat rendering cooks the tissue under vacuum to remove moisture; wet rendering utilizes water and steam; and low-temperature rendering uses just enough heat to melt the fat. Low-temperature rendering can produce a fat of lighter color, but where more meaty flavor is desired, higher-temperature rendering is used. Rendering also is used to obtain oil from whale blubber or fish tissue. In its simplest form, rendering can be carried out in a heated kettle; but large-capacity modern rendering plants are sophisticated and employ continuous rendering methods.

Pressing or Expelling

Various types of mechanical presses and expellers are used to squeeze oil from oilseeds. Seeds are usually first cooked slightly to partially break down the cell structure and to melt the fat for easier release of oil. The seeds also may be ground or cracked for the same purpose. The heat from cooking or grinding should not be excessive or it may darken the color of the oil. With some seeds (e.g., corn) only the germ portion of the seed is pressed to obtain oil, whereas with others, the entire seed is pressed. The expelled oil is usually further clarified of seed residues by being pumped through multiple filter cloths or by centrifugal clarification.

Solvent Extraction

It is common in large-scale operations to remove the oil from cracked seeds at low temperatures with a nontoxic fat solvent such as hexane. The solvent is percolated through the seeds, and after the oil is extracted, the solvent is distilled from the oil and recovered for reuse. Solvent extraction frequently will get more oil out of seeds than is possible by pressing. Combined processes employ pressing to remove most of the oil followed by solvent extraction to recover final traces. The oil-free residual seed meal is then ground for animal feed. The stages in solvent extraction of soybeans are shown in Fig. 16.3

Degumming

Vegetable oils obtained by pressing or solvent extraction always contain fatlike substances, such as phospholipids or fat–protein complexes, which are gummy. When wetted with water, these materials become insoluble in the oil and settle out. This is one way to obtain the phospholipid lecithin.

Refining

Whereas water will settle much of the gummy material, use of alkali solution will settle additional minor impurities from the oil. These include free fatty acids, which combine with alkali to form soaps. These soaps can be removed by filtration or centrifugation. Treatment with alkali is known as refining.

Bleaching

Even after degumming and refining, the oil from seeds contains various plant pigments such as chlorophyll and carotene. These can be removed by passing heated oil over charcoal or various adsorbent clays and earths. Animal fats generally can be bleached by heat alone.

Deodorization

Natural fats and oils from seeds, meat, and fish can contain low-molecular-weight odorous compounds. These are desirable in some products (e.g., olive oil, cocoa butter, lard, fresh butterfat, and chicken fat) and are not deliberately removed. But many other oils such as fish oils and several seed oils have disagreeable odors. These odors are removed by heat and vacuum or sometimes by adsorption onto activated charcoal. The heat is often supplied by injecting steam into the fat in low-pressure evaporators.

Figure 16.3. Stages in production of soybean oil. (Upper left) Soybeans; (upper right) flaked beans for solvent extraction; (lower left) Soybean oil; (lower right) residual ground meal. *Courtesy of Procter and Gamble Co.*

Hydrogenation

Hydrogenation to saturate fatty acid double bonds and thus change the viscosity of the fat is carried out by whipping deaerated hot oil with hydrogen gas plus a nickel catalyst in a closed vessel, known as a converter. When the desired degree of hardening of the fat is reached, the unreacted hydrogen gas is removed from the vessel by vacuum and the nickel catalyst is removed by filtration (Fig. 16.4).

Hydrogenation not only saturates many double bonds but also produces trans isomers of various unsaturated fatty acids. When oleic acid (cis form) is converted to its trans isomer, elaidic acid, its melting point of about 14°C is increased to about 44°C. This increase in melting point also occurs when other cis isomers are changed to the trans form. Thus, hydrogenation increases the hardness of fats by this mechanism as well as by saturating double bonds; the extent to which each mechanism occurs can be influenced by temperature, pressure, time, and other hydrogenation variables. Hydrogenation also changes the nutritional properties of some fats. Polyunsaturated fatty acids not only become more saturated but the essential linoleic acid, and possibly other unsaturated fatty acids, lose biological activity when they are converted to the trans form. This is one of the reasons that diets should acquire part of their fat from unhydrogenated sources.

Figure 16.4. Commercial-scale hydrogenation of fats. *Courtesy of Procter and Gamble Co.*

Winterizing and Fractionation

As pointed out above, fats and oils are made up principally of a mixture of various triglycerides. The triglycerides containing more saturated fatty acids, and fatty acids of longer chain lengths, tend to crystallize out and settle from the mixture when an oil is chilled. Crystallization and settling in a refrigerated product such as salad oil can be prevented by cooling and removing the crystals before the final product is bottled. The precooling treatment to remove fat crystals is known as winterizing. It may be done by simply setting drums of oil in a cold room at a selected temperature that is lower than the product will later experience, or by continuous processing through precisely controlled heat exchangers. One recent advance in fractionating the components of fats utilizes winterizing and the solubilities of fats in solvents such as acetone. Here the fat is dissolved in the solvent and chilled to produce a crystalline fraction. The filtrate is recovered and then chilled to a lower temperature to crystallize a second fraction, and so on.

Plasticizing and Tempering

The consistency and functional properties of more solid fats also are largely influenced by the state of crystallization. A given fat or oil can be modified by chilling and by agitation, both of which influence crystallization rate and crystal form. If a heated fat is allowed to cool slowly to a solidification temperature, it will have a different crystalline structure than if it is rapidly chilled to the same temperature; rapid chilling with agitation causes further differences. Controlled chilling with or without agitation to influence consistency and functional properties is referred to as plasticizing. It generally is done by pumping the melted fat or oil through a tubular scraped-surface heat exchanger for supercooling, and then through a second chilled cylinder provided with a high-speed shaft containing rows of pins that alternate with pins on the cylinder wall to provide intensive agitation. For some applications, the crystallized fat also will be given a controlled degree of aeration by introducing measured amounts of air or nitrogen prior to chilling. A freshly plasticized fat undergoes further changes in consistency and functional properties upon standing. After a period of about 2–4 days at a temperature of about 27°C, these changes essentially cease. Holding a newly plasticized fat at a controlled temperature until its properties become stabilized is known as tempering. The chemical or physical basis for tempering is poorly understood, but the improvements in tempered shortenings, such as an increased ability to emulsify water or air, are easily demonstrated. The ability of fat crystals to exist in different crystal forms is known as polymorphism.

Monoglyceride and Diglyceride Preparation

Glycerol esters containing only one or two fatty acid residues rather than the full three can be prepared from triglycerides plus about one-fifth their weight of additional glycerol. If the mixture is heated to about 200°C in the presence of a sodium hydroxide catalyst, some of the fatty acid molecules disassociate from the triglycerides and react with free hydroxyl groups of the added glycerol. Since there is an excess of glycerol relative to total available fatty acid molecules, some of the hydroxyl groups of the glycerol must remain unesterified. The reaction, known as glycerolysis, is carried out under an inert gas or under vacuum to prevent oxidation.

Monoglycerides and diglycerides are both hydrophilic because of their free hydroxyl groups and hydrophobic due to their fatty acid residues. They are, thus, partially soluble in water and partially soluble in fat, which makes them excellent emulsifying agents. Monoglycerides and diglycerides are commonly added to shortenings and many other food products for their emulsifying properties.

PRODUCTS MADE FROM FATS AND OILS

Food manufacturers use the many forms of fats and oils as ingredients in a wide variety of food products, or further process the fats and oils into manufactured products, a few of which are described in this section.

Butter

The raw material for butter is milk fat, usually in the form of cream, which is separated from the milk to contain about 30–35% fat. The cream is pasteurized at a somewhat higher temperature than for pasteurizing milk since the high fat content has a slight protective effect on bacteria. Sometimes the cream is slightly acidic due to lactic acid from fermentation of the cream when it is separated on the farms and goes through several days of handling. When this is so, the cream is neutralized with a food-grade alkali prior to pasteurization.

Depending on the color of the cream, a vegetable coloring material (e.g., extract of annatto seed or carotene) may be added to deepen the yellow color. A measured amount of a lactic acid, diacetyl-producing bacterial culture also may be added to the cream to improve the butter flavor. The cream is now ready for churning.

Churning, whether done as a batch operation or in a continuous process, is a mechanical agitation designed to reverse the natural emulsion in cream. The fat globules of cream are suspended in water such that the water is the continuous phase and the fat globules are the dispersed or discontinuous phase. Each fat globule is surrounded by a kind of phospholipid membrane, which contains lecithin, that helps to keep the globule emulsified or suspended in the water phase. The mechanical agitation of churning breaks this membranelike surface and causes globules to collide with one another. As a result, the globules clump together and form small butter granules; these grow in size and separate from the water phase of the cream. The resulting water phase or serum is known as buttermilk.

With the breaking of the emulsion, the butter granules mat together into a large mass of solid fat since the churn is operated at about 10°C. At this point, the agitation or tumbling action of the churn is stopped and most of the buttermilk is drained from the churn. The state of the emulsion is now reversed. The mass of butterfat is the major component and it traps about 15% of buttermilk within it. The butterfat is now the continuous phase, and the remaining buttermilk, which is largely water with dissolved lactose, casein, and other milk solids, is suspended as droplets within the mass of fat. This condition results after about 40 min of batch churning.

The butter mass in a batch-type churn now appears as in Fig. 16.5. The mass next is washed with pure water by turning a hose into the churn to remove surface-adhering buttermilk. At this point, the wash water is drained and salt is added to the churn. A small amount of pure water also may be added to contribute the maximum amount of water permitted by law in the final butter. The churn again is closed and given further tumbling action to work the butter. The purpose of working is to uniformly disperse the salt and to subdivide the water droplets into smaller and smaller size. Subdivision of water droplets by proper working in the churn prevents these droplets from running together and causing leaky butter.

The salt, which is added at a level of about 2.5% of the final butter, contributes flavor and also acts as a preservative. All of the salt goes into solution in the water droplets; since the amount of water is only about 15%, the salt concentration in the water is actually about seven times the 2.5% salt added. At this concentration, the salt is a strong preservative within the water droplets and largely prevents the growth of spoilage bacteria in these droplets.

The butter may now be packaged in large cartons or in smaller units with a machine known as a butter printer. In the latter case, the butter is loaded into the hopper of the machine from where it is mechanically extruded in the shape desired, cut to size, and wrapped.

Butter also may be made by high-speed continuous processes, but the basic principles cited for the batch churning operation apply to all. Typically, cream is pumped into cylindrical chilled churns that look somewhat like an ice cream freezer (Fig. 16.6). High-speed mixing within the cylinder forms the butter granules in a matter of seconds. The butter granules may be mechanically forced through perforated plates while the

Figure 16.5. Butter mass in batch-type churn. *Courtesy of J. A. Gosselin Co., Ltd.*

buttermilk simultaneously is drained from the cylinder. A salt solution is injected into an extension of the cylinder where the butter is further worked as it is extruded. The continuously extruded butter goes directly to automatic packaging equipment. The entire process can be operated under computer control (Fig. 16.7)

In the United States the legal standard for butter requires that the fat be entirely from butterfat and that the finished product contain not less than 80% butterfat by weight. Color and salt are optional.

Margarine

The term *margarine* applies to certain types of shortenings as well as table spreads. The consumption of table spreads in the United States has remained rather constant over the last 30 years; however, butter has declined in popularity while margarine has increased. Much of this is due to the lower price of margarine and nutritional differences of the two products.

In the United States, margarine is made largely from vegetable oils that have been hydrogenated or crystallized for the proper spreading texture. The vegetable oils may also be blended with lesser quantities of animal fats. The choice and blending of oils depend on seasonal availability, price, and nutritional differences like degree of saturation or unsaturation. Like butter, legal table margarine must contain no less

Figure 16.6. Continuous butter-making churn. *Courtesy of Anderson Bros. Mfg. Co.*

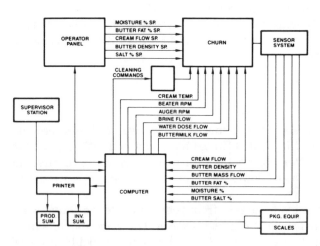

Figure 16.7. Computer controlled butter-making system. *Courtesy of Food Engineering and Zikonix, Inc.*

than 80% fat. Since the oils are naturally almost 100% fat, water is added to produce the desired water-in-oil emulsion, which is physically quite the same as in butter. To the oil and water phases are added emulsifiers, salt, butter flavor, color, and permissible chemical preservatives such as sodium benzoate. Vitamins A and D also may be added.

In the manufacture of margarine two mixtures usually are made: one of the oil and all other fat-soluble ingredients, and the other of water and all water-soluble ingredients. These two mixtures are then emulsified in a vat with vigorous agitation, which distributes the water phase as small droplets throughout the continuous oil phase. The emulsion, which would quickly separate if not stiffened by chilling, is now quickly cooled. In modern continuous systems (Fig. 16.8) this is done by pumping the emulsion through a series of heat exchangers which may have special agitators to further subdivide the water droplets throughout the fat as it stiffens. The emulsion is next passed through a chilled crystallizer to further solidify and plasticize the fat. Proper temperature control to develop optimum fat crystals is most important for producing the desired semiplastic consistency. The semisolid margarine is continuously extruded and packaged as in continuous butter-making.

Whereas butter may contain only butterfat, most margarines contain only vegetable fats. However, some vegetable fat margarines are made to contain about 5–40% butter to enhance flavor.

Butter and margarine are calorically concentrated and dense foods. Recently, there has been a move to reduce the caloric content of these spreads. Two technologies have been used. The first is to increase the amount of air incorporated into the spread by whipping. This does not reduce the caloric content on a weight basis but does on a volume basis because the volume is increased by about 50%. The second method is to

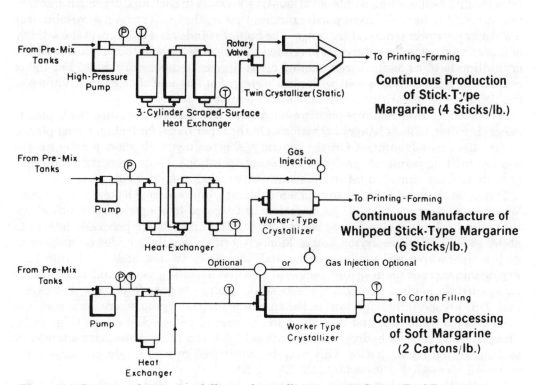

Figure 16.8. Diagrams of margarine chilling and crystallizing systems. Source: *Food Engineering*.

increase the amount of water incorporated in the spread by adding more plus better emulsifiers. By adding water, the caloric content of the spread can be reduced by 33% or more. Many of these products have become successful in the marketplace in recent years.

Shortenings and Frying Oils

Margarine is often used in baking when its flavor, which results from the milk ingredients in it, is desired. Other bakery shortenings are made entirely of vegetable oils and are flavorless or may have butter flavors added to them. Others, particularly those favored for pie crusts, are made of lard. Many contain blends of vegetable and animal fats. Emulsified shortenings in addition contain monoglycerides, diglycerides, and related compounds. These permit cakes to be made with higher levels of water and sugar, and therefore to be more moist and tender, than would be possible with unemulsified shortenings.

Shortenings may be prepared to all degrees of stiffness including pourability. Pourable shortenings are easily pumped and metered but are generally not equal to the plastic shortenings in bakery performance. One important function of a shortening is to hold air, whether beaten in a cake batter or creamed with other icing ingredients. This ability to hold air generally is increased by a plastic consistency of the shortening. Further, following baking, plastic shortenings remain dispersed within baked goods, whereas liquid shortenings have a tendency to leak and collect in pockets.

Two shortenings with the same initial hardness can have different softening properties as they are subjected to the same conditions of elevated temperature. This results from the different melting points of various triglycerides in each mixture. Such shortenings are said to have different plastic ranges; that is, they will remain semisolid over a wider or narrower temperature range. The Solid Fat Index is a measure of the solidity of fats at various temperatures; it is related to the percentage of the fat that exists in crystalline form as distinct from melted oil at a given temperature. Solid Fat Index curves therefore correlate well with plastic ranges of fats and shortenings. Values of the Solid Fat Index are given in Tables 16.3 and 16.4.

A shortening that remains plastic over a wide range of temperature (long plastic range) is suited to most bakery operations. On the other hand, for frying, a long plastic range offers no advantages. On the contrary, shortenings with short plastic ranges and low melting points are preferred for frying operations. These properties minimize greasiness from unmelted fat in the mouth when fried items are eaten.

Frying fats have additional requirements that are different from bakery shortenings. Whereas shortenings in baked goods seldom are exposed to temperatures much above the boiling point of water during baking, frying fats and oils are generally heated to about 160–190°C in the frying kettle. Monoglycerides and diglycerides decompose at such temperatures and produce smoke and, therefore, are not added to frying fats. Frying fats exposed to these high temperatures also must be given considerable stability against darkening, pyrolytic decomposition leading to gum formation, and oxidation. This is especially important in the case of industrial frying where fat is continuously filtered, reheated, and reused, as in the case of potato chip frying (Fig. 16.9). This generally calls for hydrogenation. Fats at high temperatures also have a tendency to foam in the frying kettle. This may be minimized by the incorporation of such materials as methyl silicones into the frying fats.

Frequently, somewhat conflicting attributes may be desired in fats and oils. For

Figure 16.9. Industrial potato chip fryer. *Courtesy of J. D. Ferry Co., Inc.*

example, users want frying fats with maximum stability but also with low melting points to minimize greasiness in the mouth. Hydrogenation can contribute to the former but may aggravate the latter. On the other hand, fats with saturated shorter chain fatty acids can have both stability and relatively low melting points without hydrogenation. Although there is often more than one approach to solve such problems, the solution frequently must involve compromise influenced by the specific food application.

Mayonnaise and Salad Dressings

Standards of identity for mayonnaise in the United States require that it be made from at least 65% vegetable oil, 2.5% acetic or citric acid, and egg yolk. It may contain salt, natural sweeteners, spices, and various flavoring ingredients from natural sources. The acid is a microbial preservative. The egg yolk provides emulsification and a pale yellow color which may not be imitated or intensified. Commercial mayonnaise generally contains 77–82% of a winterized salad oil, 5.3–5.8% liquid egg yolk, 2.8–4.5% of a 10% acetic acid vinegar, small amounts of salt, sugar, and spices, and additional water to make 100%. In mayonnaise the oil phase is present in greater quantity than the water phase. Generally, the phase in greater quantity becomes the external or continuous phase when emulsions are made. In the case of mayonnaise, however, this is reversed to give an oil-in-water emulsion that has the characteristic viscosity, mouthfeel, and taste. Such an "unnatural" emulsion is difficult to prepare and tends to be relatively unstable. If the oil used in mayonnaise preparation is not properly winterized, then fat crystallization in a refrigerator will break the emulsion. Even when the oil is winterized, the mayonnaise emulsion quickly breaks upon freezing.

Mayonnaise is prepared commercially by both batch and continuous methods, and there are many variations with respect to order and rates of ingredient additions to the mixers. Two-stage mixing is commonly employed with high-speed turbine blades in the first stage followed by yet more severe shearing of the oil into fine droplets in the second stage. The second-stage mixer may possess close clearance whirling teeth as in a colloid mill. Mayonnaise may be whipped with small quantities of inert gas such as nitrogen or carbon dioxide to produce a finished product with a specific gravity of 0.88–0.92. The gas commonly is pumped along with the emulsion into the second-stage mixer of enclosed design where the high shear mixing under pressure subdivides the gas into minute bubbles, which further contribute to "body" or firmness and texture. Mayonnaises with the same ingredient composition can vary greatly in their firmness, smoothness, sheen, spooning characteristics, and taste, and these properties are greatly influenced by the conditions of mixing. Mayonnaise is preserved against microbial spoilage by its acid content, but it is very sensitive to oxidative deterioration of flavor and should be refrigerated after jars are opened.

Salad dressings may be very similar to mayonnaise but generally contain less oil (35–50%) and contain a starch paste as a thickener. The egg yolk or other emulsifier, vinegar, and seasonings perform the same functions as in mayonnaise, and principles with respect to mixing and emulsion stability are similar except that special care must be given to starch cooking to develop the desired degree of thickening. For example, if the starch–water suspension is cooked together with the vinegar, acid hydrolysis tends to thin the starch paste. Therefore, it is preferred to add vinegar to the previously cooked starch paste, which is then blended with the oil, egg yolk, and other ingredients prior to final mixing–emulsification.

Pourable salad dressings, such as French dressing contain oil, vinegar, spices, and other ingredients. Pourable dressings may be fully emulsified or readily separate into oil and aqueous layers which are commonly shaken before use. Emulsifiers vary and include numerous gums as well as egg yolk. In the United States, French dressing composition is covered by a standard of identity that requires a minimum oil level of 35%, although higher levels of oil are common. Other pourable salad dressings can be quite variable in composition and flavoring ingredients. The separating types are well mixed but need not be emulsified prior to bottling.

FAT SUBSTITUTES

Because of the desire to reduce the caloric and fat content of the diet and to change the types of fat in the diet, many new fat substitutes have been or are being developed. These substitutes generally are designed to imitate the functionality of a particular fat in a food system at a reduced caloric content. They can generally be divided into two types: those which reduce caloric content by substituting a less calorically dense substance for the fat and those that behave like a fat but are not readily absorbed by the body and, thus, do not provide the same calories.

As pointed out earlier, fats in foods have several functions and some of those functions can be imitated by nonfat substances. For example, fats impart a smooth mouthfeel to many foods like chocolate and ice cream. Protein particles can be made very small, round, and hard. When suspended in water, these small particles thicken and provide a smooth mouthfeel similar to that of fat. Because protein contains only about 40% of the caloric content of fat, the net effect is a reduction in calories. Such "microparticulated" proteins now serve as a popular fat substitute in such products as ice cream. Carbohydrates which can provide increased viscosity to water and thus mimic oils in such products as pourable salad dressings also reduce caloric content. One drawback to using proteins and carbohydrates as fat substitutes is that they do not withstand the high temperatures of frying or cooking.

The second type of fat substitutes, the sugar esters, are chemically similar to natural fats but are not absorbed and metabolized in the body. These substances will withstand the high temperatures of frying but not contribute to caloric content of fried foods. Few of this type of fat substitute have been approved for foods at this time, but it is likely that they will be. The most well-known substance is Olestra® which is a chemical derivative of the common table sugar sucrose.

TESTS ON FATS AND OILS

Generally, fats and oils are tested to gain information related to performance in specific food applications, to measure degree of deterioration (such as oxidation or rancidity) as well as stability of the fat against such change, to check fat properties against purchase specifications, and to identify fats and oils against possible misrepresentation or adulteration. The important physical and chemical properties of fats and oils have been measured in various ways that have created a number of terms commonly associated with properties of fats.

Chemical Tests

The degree of unsaturation of the fatty acids in a fat or oil can be quantitatively expressed by the Iodine Value of the fat. Iodine Value refers to the number of grams of iodine absorbed by 100 g of fat. Since the iodine reacts at the sites of unsaturation much as would hydrogen in hydrogenation, the higher the Iodine Value the greater the degree of unsaturation in the fat.

The degree of oxidation that has taken place in a fat or oil can be expressed in terms of Peroxide Value. When the double bonds of unsaturated fats become oxidized, peroxides are among the oxidation products formed. Under standard conditions these peroxides can liberate iodine from potassium iodide added to the system. The amount of iodine liberated is then a measure of peroxide content, which correlates with degree of oxidation already experienced by the fat and probable tendency of the fat to subsequent oxidative rancidity. Oxidative rancidity results from the liberation of odorous products during breakdown of unsaturated fatty acids. These commonly include such compounds as aldehydes, ketones, and shorter-chain fatty acids. This is the type of fat deterioration that can often be prevented or minimized by the addition of chemical antioxidants, such as butylated hydroxyanisole (BHA) and butylated hydroxytoluene (BHT).

Fats also are degraded by the process of hydrolysis, which in the presence of moisture splits triglycerides into their basic components of glycerol and free fatty acids. The free fatty acids, especially if they are of short-chain length, cause off-odors and rancid flavors in fats and oils. This type of deterioration, referred to as hydrolytic rancidity, is to be distinguished from oxidative rancidity. Hydrolytic rancidity does not require oxygen to occur but is favored by the presence of moisture, high temperatures, and natural lipolytic enzymes. The term *Acid Value* refers to a measure of free fatty acids present in a fat. Acid Value is defined as the number of milligrams of potassium hydroxide necessary to neutralize 1 g of the fat or oil.

The average molecular weight of the fatty acids in a fat, which influences firmness of the fat as well as flavor and odor properties (low-molecular-weight fatty acids are more odorous), is another important property of fats and oils. The average molecular weight of the fatty acids in a fat is indicated by Saponification Value, which is the number of milligrams of potassium hydroxide required to saponify (convert to soap) 1 g of fat. Since 1 g of fat must contain more fatty acids if they are of short-chain length or fewer fatty acids if they are of long-chain length, and the fatty acids react with the alkali to give the soap, it follows the Saponification Value increases and decreases inversely with average molecular weight.

These are but a few of the chemical tests that have been applied to fats and oils. Much that they reveal can be learned today more quickly by instrumental analytical methods such as gas chromatography and infrared absorption analysis, and many of the classical chemical tests have been largely replaced by these newer methods.

Physical Tests

The most important physical characteristic of fats is their consistency under different temperature conditions. Most fats and oils do not melt or solidify sharply at a given temperature. Rather, because fats and oils are mixtures of triglyceride molecules, each with its own melting point, fats melt or solidify gradually over a temperature range.

There are various tests to indicate the beginning of melting of a fat or oil previously chilled to a specified temperature. In one type, the temperature at which the chilled cloudy fat within a capillary tube loses its cloudiness from melting of its solidified crystals is taken as the melting point. In another, the temperature at which chilled fat in a capillary tube softens just enough to slide within the tube is considered the melting point. A related type of test measures the temperature for a melted fat to go to the crystalline state by observing the point at which cloudiness of the fat is complete. This test can then be extended to determine the temperature at which the cloudy fat congeals.

The Solid Fat Index, mentioned earlier, is a measure of solidity of fats and is related to the percentage of the fat that is crystalline at specific temperatures. Crystallinity is measured by changes in volume that occur when fat crystals melt. The experimental method is referred to as dilatometry.

More sophisticated instrumental methods can be used to gain information on the state of crystallinity in a fat. One of these is X-ray diffraction and is based on the ability of crystals to deflect an X-ray beam according to the spacing between molecules within the crystal. Another is differential scanning calorimetry, which measures transition energies as fat crystals form or melt and precise temperatures at which these events occur.

Additional Tests

The consistency of semisolid fats can easily be measured by their resistance to penetration of a needle, ring, or cone. The response of fats and oils to frying tempera-

Table 16.5. Refined Corn Oil Analytical Data

Property	Value
Acidity (free fatty acid as oleic)	0.020 to 0.050
Acid value	0.04 to 0.10
Color (Lovibond)	20 to 25 yellow
	2.5 to 5 red
Cold test	Clear
Saponification value	189 to 191
Iodine value	125 to 128
Hehner value	93 to 96
Titer	18° to 20°C
Melting point	−16° to −11°C
Smoke point	221° to 260°C
Solidifying point	−20° to −10°C
Flash point	302° to 338°C
Fire point	310° to 371°C
Specific gravity	0.918 to 0.925
kg per liter	0.920 at 21°C

SOURCE: Corn Industries Research Foundation, Inc.

tures can be indicated by such measurements as smoke point, flash point, and fire point, which correspond to the temperatures at which these occurrences begin.

Many chemical and physical tests are highly useful in the identification and quality control of fats and oils as food ingredients (Table 16.5). Frequently, however, they may not correlate closely with optimum performance of a given fat in a specific application. For this reason, actual performance tests with the fat may be indispensable. Performance tests are essentially scaled-down versions of the actual application in which the fat will be used. Sometimes they impose still more severe conditions than will be encountered in the corresponding commercial operation to add conservatism to the evaluation.

Effective performance tests on fats and oils (or any other ingredients) will differ among plants manufacturing the same type products when there are differences in recipe, manufacturing procedures, or even stability requirements under varying distribution and marketing conditions. Thus, performance tests on shortening for cakes prepared in a small retail bakery would be different from the tests to be performed in a large wholesale bakery employing automated high-speed operations and exposing its finished products to the variables associated with national distribution.

References

Anon. 1988. Food Fats and Oils. 6th ed. The Institute, Washington, DC.

Bailey, A.E., Swern, D., Formo, M.W., and Applewhite, T.H. 1985. Bailey's Industrial Oil and Fat Products. 4th ed. John Wiley & Sons, New York.

Coenen, J.W.E. 1985. Hydrogenation of edible oils and fats. Dev. Food Sci. *11*(Pt. A), 369–391.

Erickson, D.R. et al. 1980. Handbook of Soy Oil Processing and Utilization. American Soybean Association and the American Oil Chemists' Society, St. Louis, MO.

Gurr, M.I. 1992. Role of Fats in Food and Nutrition. Chapman & Hall, London, New York.

Hoffmann, G. 1989. The Chemistry and Technology of Edible Oils and Fats and Their High Fat Products. Academic Press, San Diego, CA.

Kinsella, J.E. 1987. Seafoods and Fish Oils in Human Health and Disease. Marcel Dekker, New York.

Lawson, H.W. 1994. Food Oils and Fats: Technology, Utilization and Nutrition. Chapman & Hall, London, New York.

Nettleton, J. 1995. Omega–3 Fatty Acids and Human Health. Chapman & Hall, London, New York.

Phelan, J.A. 1986. Dairy spreads. J. Soc. Dairy Technol. *39*(4), 110–115.

Rossell, J.B. and Pritchard, J.L.R. 1991. Analysis of Oilseeds, Fats, and Fatty Foods. Elsevier Applied Science, New York.

Salunkhe, D.K. 1992. World Oilseeds: Chemistry, Technology, and Utilization. Chapman & Hall, London, New York.

Stansby, M.E. (Editor). 1990. Fish Oils in Nutrition. Chapman & Hall, London, New York.

Swern, D. 1982. Bailey's Industrial Oil and Fat Products. Vol. 2. 4th ed. John Wiley & Sons, New York.

Varela, M.G., Bender, A.E., and Morton, I.D. 1988. Frying of Food. Principles, Changes, New Approaches. VCH Publishers, New York.

Vergroesen, A.J. and Crawford, M. 1989. The Role of Fats in Human Nutrition: Introduction. 2nd Ed. Academic Press, San Diego, CA.

Weiss, T.J. 1983. Food Oils and Their Uses. 3nd ed. AVI Publishing Co., Westport, CT.

17

Cereal Grains, Legumes, and Oilseeds

Cereals are plants which yield edible grains such as wheat, rye, rice, or corn. Cereal grains provide the world with a majority of its food calories and about half of its protein. These grains are consumed directly or in modified form as major items of diet (flour, starch, oil, bran, sugar syrups, and numerous additional ingredients used in the manufacture of other foods), and they are fed to livestock and thereby converted into meat, milk, and eggs.

On a worldwide basis, rice is probably the single most important human food, with wheat not far behind. Nearly all rice grown goes directly to human food. Similar amounts of corn and wheat are grown, but much of the corn is used for feeding livestock, whereas only a small portion of wheat is used in animal feed. Although wheat is produced in many temperate-zone countries, over 90% of the rice is grown in Asia, where most of it is consumed. Much of the world's corn is grown in the United States. In recent years annual world production of wheat, rice, and corn has been about 560, 530, and 470 million metric tons, respectively.

The principal cereal grains grown in the United States are corn, wheat, oats, sorghum, barley, rye, rice, and buckwheat. In the United States, corn is by far the largest cereal crop; in recent years corn production has averaged about 200 million metric tons, but most of it is used for animal feeding. Wheat—with an annual production of about 66 million metric tons—is the largest U.S. cereal crop used primarily for direct human food.

Legumes are flowering plants having pods which contain beans or peas. Oilseeds are seeds which contain a high oil content and are widely grown as a source of oil. Both are considerably higher in protein than are cereal grains (Table 17.1). Legumes include the various peas and beans, most of which are low in fat, but a notable exception is the soybean. The term oilseed is applied to those seeds, including the soybean, which are processed for their oil. Other oilseeds include the peanut seed, cottonseed, sunflower seed, rapeseed, flaxseed, linseed, and sesame seed. The coconut also is an important oilseed. Cereal grains not only are comparatively low in protein but the proteins have deficiencies in certain essential amino acids, especially lysine. Legumes as well as many oilseeds are rich in lysine, though relatively poor in methionine.

Some oilseeds, such as the soybean, peanut, and coconut, are important foods in addition to being sources of oil. Oilseeds also yield great quantities of oilseed meals; for many years these were used principally to fatten livestock. Modern technology has

Table 17.1. Protein Content of Vegetable and Animal Products

Vegetable	Protein (%)	Animal	Protein (%)
Cereals	7–15	Whole milk	3.5
Legumes	20–25	Eggs	13
Oilseeds (defatted)	45–55	Meat (red)	16–22
Concentrates (soy, cottonseed)	60–80	Fish	18–25
		Meat (poultry)	20–25
Isolates (soy, wheat)	90–95	Nonfat dry milk	36

SOURCE: Horan (1974).

made it possible to separate high quality proteins from these meals, and today, oilseed proteins in their many forms are used to improve the nutritional properties of cereal products, to extend the meat supply, and to generally increase available protein worldwide.

CEREAL GRAINS

General Composition and Structure

The major constituents of the principal cereal grains are listed in Table 17.2. These grains contain about 10–14% moisture, 58–72% carbohydrate, 8–13% protein, 2–5% fat, and 2–11% indigestible fiber. They also contain about 300—350 kcal/100 g of grain. Although these are typical values, compositions vary depending on varieties of the particular grain, geographical and weather conditions, and other factors.

A moisture content of 10–14% is typical of properly ripened and dried grains. When the moisture content of grains from the field is higher than this, they must be dried to this moisture range, otherwise they may mold and rot in storage before they are further processed. Some molds which grow on cereal grains containing excessive moisture produce toxic metabolites which can cause disease in humans and animals consuming the grain. Cereal grains contain about two-thirds carbohydrate, most of which is

Table 17.2. Typical Percentage Composition of Cereal Grains

Grain	Moisture	Carbohydrate	Protein	Fat	Indigestible Fiber	Kilocalories (per 100 g)
Corn	11	72	10	4	2	352
Wheat	11	69	13	2	3	340
Oats	13	58	10	5	10	317
Sorghum	11	70	12	4	2	348
Barley	14	63	12	2	6	320
Rye	11	71	12	2	2	321
Rice	11	65	8	2	9	310
Buckwheat	10	64	11	2	11	318

in the form of digestible starches and sugars. The operations of milling generally remove much of the indigestible fiber and fat from these grains when they are to be consumed for human food.

The nutritional quality of cereal proteins is not as high as that of most animal proteins. Table 17.3 lists the patterns of the essential amino acids lysine, methionine (plus cystine), threonine, and tryptophan of several cereals compared with whole egg and an FAO recommended standard mixture of these amino acids. Because the first limiting amino acid of these cereals is lysine, the ratio of the lysine concentration in a cereal grain protein to the concentration in whole egg, or the FAO standard mixture, can be used as an index of quality. This ratio times 100 gives the chemical score of a cereal, which can be improved by the addition of lysine. The lysine limitation also can be overcome by consuming cereals with other foods high in lysine.

There are a few important structural features that the cereal grains have in common and that form the basis for subsequent milling and other processing operations. All of the cereal grains are plant seeds and as such contain a large centrally located starchy endosperm, which also is rich in protein, protective outer layers such as hull and bran, and an embryo or germ usually located near the bottom of the seed. These portions are seen in the diagrams for wheat and corn in Figs. 17.1 and 17.2.

For most food uses, processors remove the hulls, which are largely indigestible by man; the dark-colored bran; and the germ, which is high in oil, is enzymatically active, and under certain conditions would be likely to produce a rancid condition in the grain. Thus, the component of primary interest is the starchy, proteinaceous endosperm. Since the bran is rich in B vitamins and minerals, it is common practice to add these back to processed grains from which bran has been removed; this is known as enrichment.

Besides indigestibility of hulls, bran color, and possible rancidity from the germ, a further reason for removing these components in many cases is to improve the functional properties of the endosperm in manufactured food use. For example, white bread made from wheat flour would have less acceptable color, flavor, and volume if the bran and germ were not removed before the flour was ground. However, there also are applications in which unmilled whole grain—containing hulls, bran, and germ—is used. Grain for animal feed is an example; sprouted barley, used for its malting effect in the brewing industry, is another example. Whole wheat bread, preferred by many, utilizes flour from which the bran and germ have not been removed during milling.

The processing and utilization of the major cereal grains are discussed in the following sections.

Wheat

As with all cereal grains, there are many varieties of wheat differing in yield, in resistance to weather, insects, and disease, and in composition. Wheats are classified into two types: hard and soft. In comparison with soft wheat, hard wheat is higher in protein, yields a stronger flour, which forms a more elastic dough, and is better for bread-making when a strong elastic dough is essential for high leavened volume. In contrast, soft wheat is lower in protein, yields a weaker flour, which forms weak doughs or batters, and is better for cake-making. Wherever wheat is used for human consumption, the majority of it is first converted to flour.

Table 17.3. Amino Acid Patterns of Cereals Compared with Whole Egg and FAO Patterns

Cereal	mg Amino Acid/g N				Limiting Amino Acid	Chemical Score (Egg)	Chemical Score[a]
	Lysine	Methionine and Cystine	Threonine	Tryptophan			
Barley	216	246	207	96	Lysine	50	64
Cornmeal	167	217	225	38	Lysine	38	49
Millet	214	302	241	106	Lysine	49	63
Oats	232	272	207	79	Lysine	53	68
Polished rice	226	229	207	84	Lysine	52	66
Ragi	181	357	263	105	Lysine	42	53
Rye	212	210	209	46	Lysine	49	62
Sorghum	126	181	189	63	Lysine	29	37
Teff	174	301	213	93	Lysine	40	51
Wheat bulgur	161	219	177	66	Lysine	37	47
Wheat flour (white)	130	250	168	67	Lysine	30	38
Standard							
Hen's Egg	436	362	320	93	—	—	—
FAO/WHO 1973	340	220	250	60	—	—	—

Source: Jansen (1977).
[a]Source: FAO/WHO (1973).

$$\text{Chemical score} = \frac{\text{limiting amino acid (mg/g N) in sample} \times 100}{\text{concentration of same amino acid (mg/g N) in egg (or in FAO mixture)}}$$

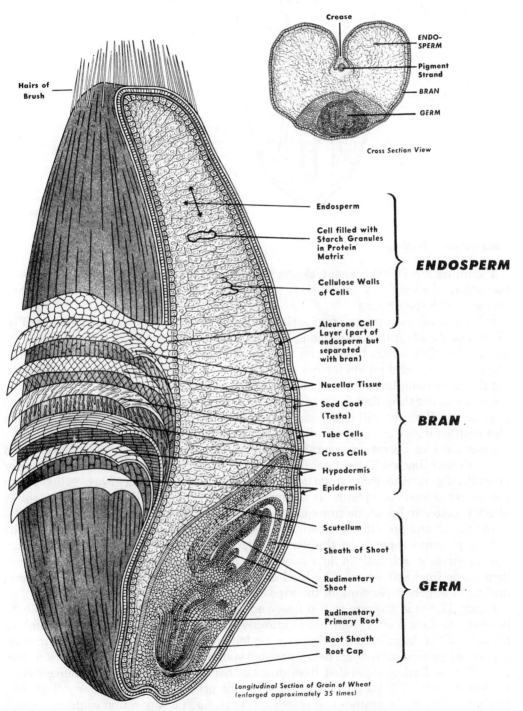

Crease

ENDO-SPERM

Pigment Strand

BRAN

GERM

Cross Section View

Hairs of Brush

Endosperm

Cell filled with Starch Granules in Protein Matrix

ENDOSPERM

Cellulose Walls of Cells

Aleurone Cell Layer (part of endosperm but separated with bran)

Nucellar Tissue

Seed Coat (Testa)

BRAN

Tube Cells

Cross Cells

Hypodermis

Epidermis

Scutellum

Sheath of Shoot

Rudimentary Shoot

GERM

Rudimentary Primary Root

Root Sheath

Root Cap

Longitudinal Section of Grain of Wheat (enlarged approximately 35 times)

Figure 17.1. Structure of a wheat kernel. *Courtesy of the Wheat Flour Institute.*

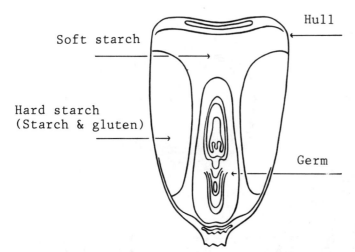

Soft starch

Hull

Hard starch
(Starch & gluten)

Germ

Figure 17.2. Diagram of a corn kernel. *Courtesy of J. T. Goodwin.*

Conventional Milling

The miller receives the wheat, cleans it of foreign seeds and soil, soaks or conditions the wheat to about 17% moisture to give it optimum milling properties, and then proceeds with the milling.

Milling involves a progressive series of disintegrations followed by sievings (Fig. 17.3). The disintegrations are made by rollers set progressively closer and closer together. The first rollers break open the bran and free the germ from the endosperm. The second and third rollers further pulverize the rather brittle endosperm and flatten out the more semiplastic germ. The flakes of bran and flattened germ are removed by the sieves under these first few sets of rollers. The pulverized endosperm is run through successive rollers set still closer together to grind it into finer and finer flour, which also is sifted under each set of rollers to remove the last traces of bran.

From such an operation several flour fractions having finer and finer endosperm particles are collected. These finer fractions also contain progressively lower and lower amounts of ground-up contaminating germ or bran, some of which always gets through the earlier sieves. As a result, as the flour is progressively milled, it becomes whiter in color, better in bread-making quality, but lower in vitamin and mineral content.

The starch and protein composition of flour—no matter how fine it is ground in the milling process—depends on the variety and kind of wheat that was ground. Thus, the protein-to-starch ratio of flour made from hard wheat will be greater than that of flour made from soft wheat. The kind of flour that is produced during conventional milling is largely dependent on the kind of wheat available.

Figure 17.4 is a diagram of two finely milled flours. The endosperm contains both protein (the dark matter) and starch granules (the white matter in this diagram). In addition to the large mixed endosperm agglomerates, there are smaller fragmented starch and protein particles. The fragmented starch and protein particles are too close in size to be further separated from one another by the sieves of the conventional milling operation. If they could be, it would be possible to separate any flour into fractions differing in protein and starch contents. Such a separation could yield both a hard and a soft flour from the same wheat. Further, a naturally hard wheat could be made to yield a soft flour plus a protein fraction, just as a naturally soft wheat could be made to yield a hard flour plus a starch fraction.

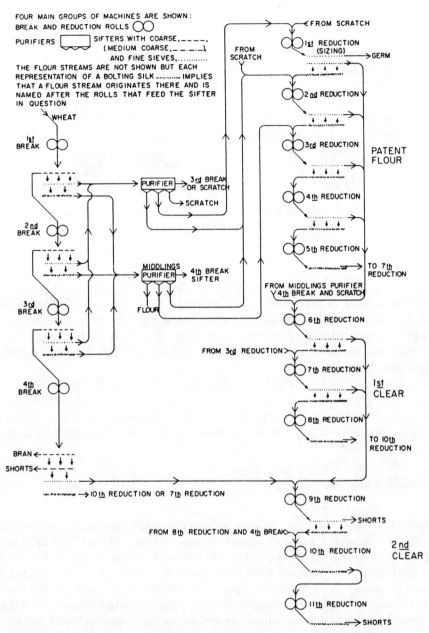

Figure 17.3. Flow diagram of typical wheat milling system. *Courtesy of R. A. Larsen.*

Turbomilling and Air Classification

Further processing can separate flour into higher protein or higher starch fractions in a process known as turbomilling. In turbomilling, flour from conventional milling is further reduced in particle size in special high-speed turbo grinders, which cause the endosperm agglomerates to abrade against each other in a high-speed air vortex. Although the resulting protein and starch particles are too close in size to be sepa-

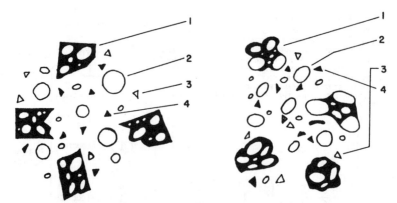

Figure 17.4. Particle types present in hard and soft wheat flour. (1) Endosperm agglomerates; (2) starch granules; (3) broken starch; (4) broken protein. *Courtesy of R. A. Larsen.*

rated by sieves, they do differ sufficiently in particle size, shape, and density to be separable in a stream of turbulent air. In this case, the slightly finer protein particles rise and the starch particles settle in the stream of air. The flour and air mixture is blown into a specially designed air classifier, which then may impose centrifugal force on the suspended particles, and two fractions of flour differing in protein and starch concentrations are recovered.

Turbomilling, developed in the late 1950s, probably is the greatest milling advance of the past century since it gives us the ability to separate flour into fractions and then blend the fractions in any desired ratio. Thus, turbomilling makes it feasible to custom-blend flours for bread-making, cake-making, cookie-making, and many other specific applications.

Uses of Wheat Flour and Granules

The uses of wheat flour in the baking industry include the making of breads, sweet doughs, cakes, biscuits, doughnuts, crackers, and the like. Wheat flour is also used in making breakfast cereals, gravies, soups, confections, and other articles. But a principal use of wheat flour, and coarser milled fractions of wheat, is in the preparation of alimentary pastes, such as macaroni, spaghetti, and other forms of noodles and pasta. Alimentary pastes like bakery doughs contain mostly milled wheat and water. The wheat, usually a hard durum wheat, is milled to yield coarse particles known as semolina, somewhat less coarse durum granulars, and finer durum flour. Alimentary pastes also may contain eggs, salt, and other minor ingredients. They differ from bakery doughs in that alimentary pastes are not leavened.

The unleavened dough is formed by mixing the ingredients in the ratio of about 100 parts of the wheat products to 30 parts of water. The dough then may be extruded in a thin sheet (Fig. 17.5), which is cut into flat noodles and dried in an oven to about 12% moisture; or the unleavened dough may be extruded in dozens of other shapes depending on the choice of dies. Figure 17.6 shows a die for extruding macaroni with a hole in the middle; this product also is oven-dried to about 12% moisture. Quick-cooking noodles, sometimes referred to as instant noodles, are made by steaming noodle

Figure 17.5. Extruded noodle dough being fed to cutter. *Courtesy of Braibanti Corp.*

dough and then frying it. Frying removes moisture and the noodles are not further oven-dried.

Rice

Rice is the staple food of billions of people worldwide. Whereas wheat for the most part is ground into flour, most of the world's rice is consumed as the intact grain, minus hull, bran, and germ. Therefore, the milling process must be designed not to disintegrate the endosperm core of the seed.

Milling

Rice milling begins with whole grains of rice being fed by machine between abrasive disks or moving rubber belts. These machines, known as shellers or hullers, do not crush the grains but instead rub the outer layer of hull from the underlying kernels. The hulls are separated from the kernels by jets of air, and the kernels, known as brown rice, move to another abrasive device called a rice-milling machine. Here, remaining inner layers of bran and germ are dislodged by the rubbing action of a ribbed rotor. The endosperms with bran and germ removed can now be further polished to a white, high glossy finish.

As in the case of wheat, the higher the degree of milling or polishing, the lower are

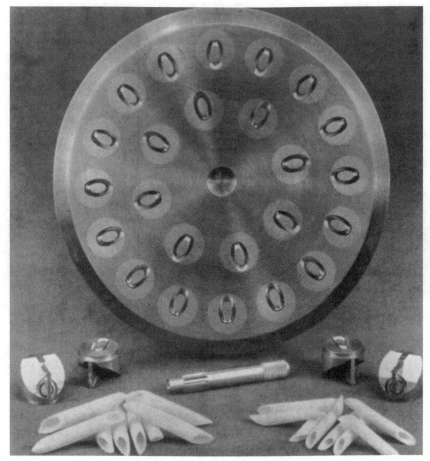

Figure 17.6. Die for extruding macaroni. *Courtesy of Glenn G. Hoskins Co.*

the remaining vitamin and mineral contents. This is particularly serious in the case of rice because entire populations depend on rice as the principal item of diet.

Enrichment

The two major ways to enrich rice differ from the simple admixture of vitamins and minerals in powder form that may be done in the case of flour. One method is to coat the polished rice with the enrichment mixture and then to further coat the grains with a waterproof edible film material. Upon hardening, the film material prevents the enrichment ingredients from dissolving away when the marketed rice is washed, as is common practice.

The second important method involves parboiling or steeping the whole rice grains in hot water before removal of hulls, bran, and germ in milling. Parboiling may be for about 10 h at 70°C, although several other time–temperature combinations can be used. This causes the B vitamins and minerals from the hulls, bran, and germ to leach into the endosperm. The rice is then dried, milled, and polished as before. Parboiled rice, processed for enrichment and other desirable changes in the rice kernels, has also been referred to as converted rice.

The principal nutrients used to enrich rice are thiamin, niacin, and iron; thiamin is particularly effective in reducing the incidence of beriberi where polished rice is a major item of diet. Legislation requires all rice sold in Puerto Rico to be enriched. Most of the rice sold in the United States is enriched. To be called enriched rice in the United States, the product must meet the standards indicated in Table 17.4.

Improved Varieties

Plant breeders are continuously at work improving the yields and properties of cereal grains. This includes considerations of soil types, weather conditions, response to fertilizer application, resistance to disease and insect attack, nutritional quality, storage stability, milling properties, cooking and processing characteristics, and other factors.

The development of a high-yielding strain of rice, designated IR–8, by the International Rice Research Institute in the Philippines, kindled hopes that the continued world shortage of this important food grain could be relieved. IR–8 has proven especially high yielding in the tropics. Consumer acceptability of this rice has not been universal, however, and other high-yield varieties with better milling and cooking characteristics have replaced much of the IR–8 in several countries. Meanwhile, it is of interest to note that the importance of rice as a dietary staple in some countries may be reduced as wheat becomes available in the form of bread and pasta. This tendency is now being seen in Japan and in parts of Indonesia.

Rice Products

Rice can be made quick-cooking or almost instant in terms of preparation time. This is done by precooking to gelatinize the starch, and then drying under conditions that will give the rice an expanded internal structure for quick absorption of water during subsequent preparation. Many patents exist.

Rice is also the basis of several prepared foods and dried mixes. Typically, these products contain quick-cooking rice and other ingredients such as spices, noodle products, and starch-based sauces.

Rice may be ground into flour and as such is used by people allergic to wheat flour. Rice is a source of starch. It is the grain that is used in preparing the Japanese fermented alcoholic beverage sake. Rice hulls, bran, and germ also are used as animal feed.

Table 17.4. Federal Standards for Rice Enrichment

	Minimum (mg/lb)	Maximum (mg/lb)
Thiamin	2.0	4.0
Riboflavin[a]	1.2	2.4
Niacin	16	32
Iron	13	26
Calcium[a]	500	1000
Vitamin D[a]	250	1000

[a]Optional ingredients.

Corn

Corn is consumed as human food in many forms. In its harvested wet form, it is consumed as a vegetable. The kernels of a special variety may be dried and consumed as popcorn. Popcorn pops because, on heating, moisture trapped in the center of kernels turns to steam and this escapes with force sufficient to explode the kernels. Popcorn might therefore be considered the original puffed cereal.

But the majority of corn consumed as human food has undergone milling and is consumed as a specific or modified fraction of the original cereal grain. Like the other cereal grains, corn is milled to remove hulls and germ. Both are fed to livestock. In addition, the germ is an important source of corn oil. Corn is milled in two basic ways; dry milling and wet milling.

Dry Milling

Corn kernels are first conditioned to about 21% moisture and then passed between special rotating cones that loosen the hulls and germ from the endosperm. The entire mixture is next dried to about 15% moisture to facilitate subsequent roller milling and sieving. The hulls may now be removed by jets of air. From here on, corn milling is much the same as wheat milling. The endosperm and loosened germs are passed through rollers that flatten the germ and crush the more brittle endosperm. Sieving now easily separates the flattened germ from the endosperm particles. The endosperm may be recovered in the form of coarse grits or corn meal, or it may be passed through finer rollers and reduced to corn flour.

Wet Milling

After the corn kernels are cleaned, the first step in wet milling is steeping the kernels in large tanks of warm water that generally contain acid and sulfur dioxide as a mild preservative. The softened kernels are next run through an attrition mill to break up the kernels. The pasty mass from this mill is then pumped to water-filled settling troughs. Here the lighter-density rubbery germ floats to the top and is skimmed off to be pressed for oil. The slurry now contains the hulls and the protein and starch fractions of the endosperm. The water slurry is passed through screens that remove the hulls.

The remaining water slurry containing the starch and protein fractions is now passed through high-speed centrifuges to separate the heavier starch from the lighter protein. The starch fraction is finally dried to yield the familiar corn starch. The protein fraction is also dried to yield corn gluten, which is rich in the corn protein known as zein. Corn gluten is commonly used in animal feeds. Separated zein has industrial uses including some as a food ingredient. Corn starch can be used as such in manufactured foods or be further converted into corn syrup by the hydrolytic action of acid or starch-splitting enzymes. The relationships between these various products from the wet milling of corn are shown in Fig. 17.7. Processes for the wet milling of wheat, rye, and oats also have been developed.

Corn Sugars

The corn syrup resulting from hydrolysis of starch may be used as a sweetener. It contains varying proportions of dextrins, maltose, and glucose, depending on the

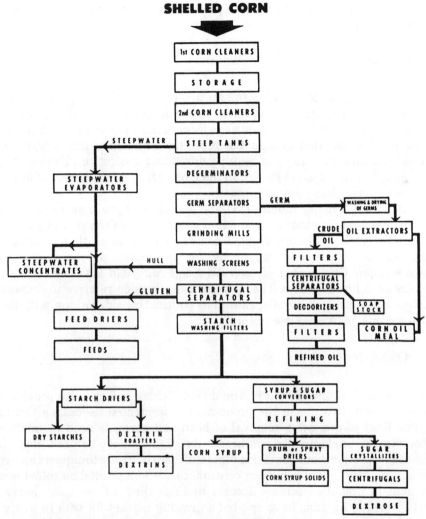

Figure 17.7. Flow Diagram of the wet-milling process. *Courtesy of Corn Refiners Association.*

method and degree of hydrolysis. More extensive hydrolysis yields a greater proportion of glucose, also known as dextrose. Since glucose is sweeter than dextrins or maltose, more extensive hydrolysis also yields a sweeter syrup. The glucose may further be enzymatically converted to fructose, which is still sweeter. Such a conversion is known as isomerization. Through hydrolysis and isomerization many sweeteners can be produced from corn starch; these include corn syrups, high-glucose (dextrose) syrups, glucose–fructose syrups, and high-fructose syrups. These syrups may be dehydrated to produce corn syrup solids, or they may be used to yield highly purified glucose or fructose by crystallization.

Blends of glucose and fructose can be made to equal the sweetness of cane or beet sugar (sucrose). Further, corn sugars and syrups in various proportions can yield a wider range of functional properties than is possessed by sucrose. The properties, ready availability, and favorable costs of corn-derived sweeteners in recent years have resulted in enormous quantities of these products replacing part or all of the sucrose in many food formulations.

Alcohol from Corn

Conversion of plant materials by fermentation to ethanol has long been known, and corn has long been an important ingredient in the manufacture of alcoholic beverages. In recent years, increased fuel oil costs have focused research on more efficient conversion of biomass into ethanol as a partial replacement for gasoline. A 90% gasoline–10% ethanol mixture (gasohol) has been produced and used in the United States and higher ethanol fuels are used in Brazil. In the United States, one source of fermentable sugar for fuel ethanol has been corn starch.

Since U.S. corn is a major human food and animal feed grain at home and abroad, its increased demand for ethanol production could increase the cost of feed and food throughout the world. This has created concern among some that a valuable food commodity should not be sacrificed to fuel for unessential purposes while millions experience hunger. One bushel of corn (25.4 kg) can yield 9.7 liters of anhydrous ethanol plus useful by-products. The cost of this conversion relative to conversions of other fuel-producing raw materials, as well as political considerations, will determine the future for this additional use of corn.

Barley, Oats, Rye

Barley, oats, and rye are used for animal feed. Barley and rye also provide sources of fermentable carbohydrate in the production of fermented beverages and distilled liquors. The flour of rye, after removal of bran and germ, is used in mixture with wheat flour in the production of rye bread. Rye flour cannot be used alone for this purpose since its protein would not form films sufficiently strong to support an expanded bread structure. Most oats for human consumption are marketed as rolled oats or as an ingredient in breakfast cereals. Recent findings that oat bran can lower serum cholesterol levels in humans have created a growing market for oats in many forms.

Barley is also used to produce barley malt. In this case, the whole barley seed is steeped in water to allow the live germ to sprout. The sprouted barley, having initiated growth, becomes much increased in enzymatic activity, especially starch-digesting amylase activity. The sprouted barley is next dried under mild heat so as not to inactivate its enzymes. The sprouted dried barley, now known as malt, is used in the brewing industry to help digest starchy material into sugars for rapid yeast fermentation. Malt also has a distinctive flavor which contributes to the flavor of brewed beverages such as beer. Malt further adds flavor to breakfast cereals and malted-milk concentrates. Malt syrups also find use in various bakery operations where amylase activity is desired.

Breakfast Cereals

The cereal grains find an important use in the manufacture of breakfast cereals. Most breakfast cereals are made from the endosperm of wheat, corn, rice, or oats. The endosperm may simply be broken or pressed, with or without toasting, to yield such uncooked cereals as farina and oatmeal.

But far more popular are the so-called ready-to-eat cereals. For these, the endosperm may be broken or ground into a mash, and then converted into flakes by squeezing the broken grits or mash between rollers. The mash also can be extruded into numerous shapes; or the endosperm may be kept intact as kernels to be puffed, as in the case of puffed rice. But in all cases, the flaked, formed, or puffed cereal must be oven-cooked and dried to develop toasted flavor and to obtain the crisp, brittle textures desired. This crispness requires that many ready-to-eat breakfast cereals be dried to about 3–5% moisture.

In Fig. 17.8 is a set of flaking rolls that may be used for producing corn flakes. The coarse pieces of corn endosperm or grits are cooked and then partially dried to a firm plastic consistency. The grits are then passed through these rolls, which squeeze them into individual flakes. The flakes are oven-toasted and dried to about 3% moisture.

Wheat or rice endosperms to be puffed are first cooked and then partially dried as individual kernels. These are next placed in a puffing gun which heats the kernels under pressure, converting moisture within the kernels into steam. When the gun is opened suddenly, the steam under pressure within the kernels expands explosively and puffs the kernels. In some cases, mashes of cereal doughs are extruded into moist pellets, which may be puffed in the same manner. The puffed cereals are toasted, often sugar coated, and dried.

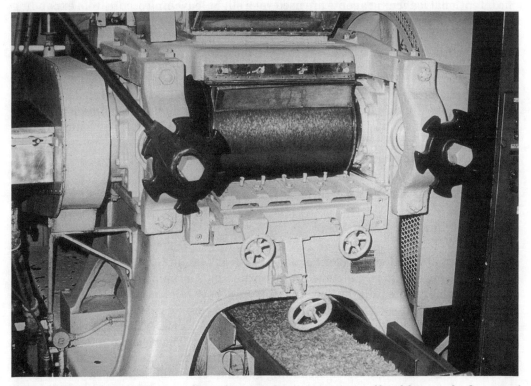

Figure 17.8. A set of cereal flaking rolls used to manufacture many types of breakfast cereals. *Courtesy of R. B. Gravani.*

SOME PRINCIPLES OF BAKING

Wheat flours find their principal applications in the production of bakery products. Most bakery products—unlike other wheat products such as alimentary pastes or noodles and unpuffed breakfast cereals—are leavened; that is, they are raised to yield baked goods of low density.

The term *baking* strictly refers only to the operation of heating dough products in an oven. But since there are many steps that must take place before the oven if baking is to be successful, baking has come to mean all of the science and technology that must precede the oven as well as the oven-heating step itself. We will consider baking in this broader sense.

Although there are a great many bakery products which grade one into another in terms of their formulas, methods of preparation, and product characteristics, it is possible to classify bakery products according to the way in which they are leavened. This classification, though not perfect, is useful. Four categories may be defined:

- *Yeast-raised goods* include breads and sweet doughs leavened by carbon dioxide from yeast fermentation
- *Chemically leavened goods* include layer cakes, doughnuts, and biscuits raised by carbon dioxide from baking powders and chemical agents
- *Air-leavened goods* include angel cakes and sponge cakes made without baking powder
- *Partially leavened goods* include pie crusts, certain crackers, and other items where no intentional leavening agents are used yet a slight leavening occurs from expanding steam and other gases during the oven-baking operation.

This classification refers to the intended source of the leavening gas; however, the intended source is not the only source of leavening, as will be seen shortly. Leavening gas can produce leavening only if it is trapped in a system that will hold the gas and expand along with the gas. Therefore, much of cereal science related to baking technology is really the engineering of food structures through the formation of correct doughs and batters to trap leavening gases, and then the coagulation or fixing of these structures by the application of heat. This brings in the need to understand further some of the properties of flour and certain other baking ingredients.

Major Baking Ingredients and Their Functions

Gluten and Starch of Wheat Flour

The principal functional protein of wheat flour is gluten. Gluten has the important property that when it is moistened and worked by mechanical action, it forms an elastic dough. It does this by forming linkages between protein molecules. These linkages form a three-dimensional structure which provides strength to the dough. The longer the dough is worked, the more linkages are formed. This is the reason that dough is kneaded when a strong structure is required. The resulting dough may be stretched in two directions and form sheets or films, or it may be stretched in all directions under the pressure of expanding gas and form bubbles as does bubble gum. However, gluten films weaken and then break down under excessive mechanical action such as over mixing of the dough. Additionally, on exposure to sufficient heat, the gluten coagulates and forms a semirigid structure. If the gluten has been expanded by gas prior to being heated, then this fairly rigid structure will be of a cellular character such as the inside of a loaf of bread.

The gluten of wheat flour has starch associated with it. Wheat starch does not form elastic films as does gluten; rather, the moistened starch, when heated, forms a paste and stiffens, or more correctly gelatinizes.

Thus, these two constituents of wheat flour together are capable of forming a batter or dough depending on the amount of water employed; and both the gluten and starch contribute to the semirigid structures resulting when such batters or doughs are heated.

The character of a dough or batter depends considerably on the type of flour used. As indicated earlier, strong flours containing more gluten, and gluten of a quality that will stretch farther before tearing, are the kind chosen for making bread because bread dough must be able to expand to a great degree and yield baked products of especially light density. Weaker flours generally contain less gluten and their films tear more readily; further, such films are less tough, and when baked, they yield structures that are less chewy and more tender. This is the kind of flour selected for making cakes and related products in which more tender and friable structures are desired. Figure 17.9 shows unbaked and baked doughs that were made from gluten separated from the same weight of cake flour, bread flour, and an intermediate flour known as all-purpose flour.

Leavening Agents

Yeast and baking powders are not the only effective leavening agents. Water in doughs or batters turns to steam in the oven, and the expanding steam contributes to leavening. Air in a dough or a batter similarly expands when heated in the oven and contributes to leavening. In yeast-leavened or chemically leavened goods, although carbon dioxide from fermentation or from baking powder is the major leavening gas, it is supplemented with expanding steam and expanding air from oven heat. Not only

Figure 17.9. Unbaked and baked gluten doughs from same weight of (left to right) cake flour, all-purpose flour, and bread flour. *Courtesy of the Wheat Flour Institute.*

are the amounts of gas leaveners produce important but also the rates of gas production and the time of gas production.

Yeast. Two forms of yeast are used in baking—moist pressed cakes and dehydrated granules. Both forms consist of billions of living cells of *Saccharomyces cerevisiae*. When rehydrated, the yeast begins metabolism and fermentation, of which carbon dioxide is a by-product. In the bread-making process and related sweet dough processes, yeast ferments simple sugars and produces carbon dioxide and alcohol. This fermentation is gradual, beginning slowly and increasing in rate with time. The increase in rate with time is due to two conditions in a dough: (1) yeast cells are multiplying and their enzymes are becoming more active while the dough is prepared and held and (2) sugar for fermentation is gradually being liberated from starch in the dough by the action of natural flour enzymes.

This gradual production of carbon dioxide is preferable to an immediate burst of gas because the film-forming property of gluten also develops gradually as the dough is being hydrated and mechanically kneaded. If the gas were to evolve before the film-forming property became developed, it would escape entrapment and there would be no leavening. Further, in the operations involved in converting the large dough mass into individual loaf-size pieces, there is considerable rough handling which tends to knock gas out of the rising dough pieces. Gradual and continued production of carbon dioxide up to the point of oven baking replenishes the lost gas and maximizes leavening.

The distribution of gas bubbles in the elastic dough prior to baking, the delicate structure of the leavened dough, and the origin of the cellular structure of the baked loaf can be seen in Fig. 17.10. The amounts of leavening gases and their rates of production must be balanced against the rate of development of the film-forming structure and its strength to hold the gas prior to and during baking.

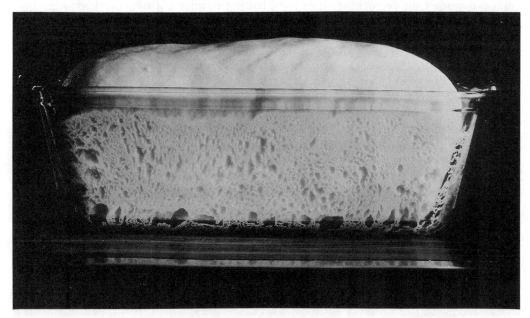

Figure 17.10. Gas bubbles formed in expanding dough prior to baking. *Courtesy of the Wheat Flour Institute.*

The heat of the baking operation kills the yeast and inactivates its enzymes; thus, fermentation and the release of carbon dioxide ceases. However, the bubbles already formed enlarge under the influence of heat due to expansion of carbon dioxide, expansion of entrapped air, and conversion of water into steam. As the temperature of the loaf rises, starch gelatinizes and gluten coagulates, resulting in a semirigid, less fragile structure.

Baking Powders. Baking powders used in making cakes and related goods contain particles of sodium bicarbonate as a source of carbon dioxide, and particles of an edible acid to generate the carbon dioxide when water and heat are supplied. The simplified overall reaction in the case of a baking powder containing sodium bicarbonate as the carbon dioxide source and monocalcium phosphate as the baking acid is as follows:

$$3CaH_4(PO_4)_2 + 8NaHCO_3 \rightarrow Ca_3(PO_4)_2 + 4Na_2HPO_4 + 8CO_2 + 8H_2O$$

Monocalcium phosphate	Sodium bicarbonate	Tricalcium phosphate	Disodium phosphate

Such a reaction takes place too rapidly and so its speed and time of occurrence must be controlled. Various baking powders differ in the times and rates of reactions, and baking powders are formulated to produce controlled release of gas for specific bakery product applications.

For example, in making cakes, all ingredients may be mixed together and then deposited as a fluid batter into pans from a large bakery hopper. The fluid batter with its weak cake flour has very little gluten development or other means to hold evolved carbon dioxide. Thus, if there is a major carbon dioxide evolution during mixing or holding of batters, the gas will largely escape from the batter, and leavening power will be lost. However, when the batter is placed in the oven, starch gelatinizes, gluten coagulates, and egg proteins, if present, coagulate. If gas is produced while this is taking place, the gas will be trapped and expand the solidifying mass, giving the desired volume increase and cellular structure.

On the other hand, too much gas may evolve in the oven due to an excessive amount of baking powder. This tends to overexpand the gas cells, which become weakened and collapse. The result is a coarse-grain structure with lowered volume. It also is possible to produce gas in the oven too slowly. When this happens, the gluten, starch, and eggs set the structure and the crust is formed before all the gas is released. The late gas can then rupture the crumb structure and produce cracks in the surface crust.

The times and rates of gas evolution from baking powders can be regulated by the selection of different baking acids that react faster or slower with sodium bicarbonate. These acids may be used also in different particle sizes or they may be coated with various materials to control their rates of solution, thereby further controlling their rates of reaction with sodium bicarbonate.

Baking powders are of two principal kinds: fast or slow acting. Some called double-acting powders, contain both a fast- and a slow-reacting acid in combination with sodium bicarbonate. Double-acting baking powders are compounded to give a quick burst of carbon dioxide in the batter stage to lighten the batter and make mixing easier, especially for the home baker who may mix by hand, and then to liberate additional carbon dioxide in the oven when the structure is being set.

Eggs

In addition to their nutrient, flavor, and color contributions, eggs can function as a principal structure builder in cakes. Like gluten, egg white is a mixture of proteins. It forms films and entraps air when it is whipped, and on heating, it coagulates to produce rigidity. The proteins of egg yolk have similar properties. This is particularly important when eggs are combined with relatively low levels of a weak flour, as is the case in the preparation of angel cakes and sponge cakes.

In these cakes the eggs are whipped and gently folded together with the other ingredients. The entrapped air in the egg foam is the primary leavening system since, generally, no baking powder is used. In the oven, the gluten, starch, and egg stiffen, and the subdivided air bubbles expand from heat. Steam generated from water enters the air bubbles and further serves to expand them. This is one reason why the whipping quality and foam stability of eggs are so important to the baker.

Shortening

Unlike flour and eggs, which are structure builders and tougheners, shortening is a tenderizer. But in many recipes, additionally, the beating of shortening is called for to entrap air prior to the incorporation of other ingredients to finish the batter. When the batter is baked in the oven, the shortening melts and releases the air bubbles which contribute to the leavening action of baking powder and expanding steam. The melted shortening then deposits around the cell walls of the coagulating structure to contribute a tenderizing effect and lubricate the texture.

The cellular structure of a cake (i.e., whether it is fine or coarse grain) and cake volume are affected by the number and size of air bubbles and water droplets trapped in the beaten shortening. These in turn are determined by the plasticity of the shortening and the use of emulsifiers. The state of emulsion is affected also by the other ingredients present and the sequence in which they are incorporated into the batter. The photomicrographs in Fig. 17.11 are of two layer cake batters with the shortening stained by a fat-soluble dye. The spheres are mostly air bubbles within fat globules.

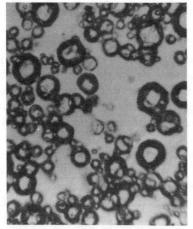 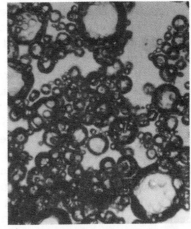

Figure 17.11. Photomicrographs of layer cake batters showing air bubbles within fat globules. *Courtesy of Dr. Andrea Mackey.*

Such differences between batters may be produced, for example, by creaming the shortening and sugar together prior to mixing in the remaining ingredients, in contrast to beating all of the ingredients together in a single step. Such modifications in mixing procedure can easily give differences in grain structure and cake volume from the same ingredient formulation.

Sugar

Sugar, like shortening, is a tenderizer in baked goods. It also adds sweetness and, in the form of sucrose, provides additional fermentable substrate in yeast-raised goods. Bakers' yeast cannot ferment sucrose directly, but hydrolyzes it first by means of the enzyme invertase into glucose and fructose. The yeast then immediately ferments the glucose; after the glucose is consumed, it proceeds to ferment the fructose. Sugar also has moisture-retaining properties in baked goods. In this respect the hydrolytic products of sucrose, namely, glucose and fructose which together are referred to as invert sugar, usually are superior to sucrose. This is one reason why invert sugar syrups are frequently used in addition to sucrose in various baked goods made without yeast. Corn syrups from the hydrolysis of starch, which contain glucose, maltose, and dextrins, also have this moisture-retaining property. Sucrose, fructose, glucose, maltose, and dextrins further contribute to the different kinds of browning that baked goods develop in the oven.

The Baking Step

Baking is a heating process in which many reactions occur at different rates. Some of these reactions include the following: (1) evolution and expansion of gases; (2) coagulation of gluten and eggs and gelatinization of starch; (3) partial dehydration from evaporation of water; (4) development of flavors; (5) changes of color due to Maillard browning reactions between milk, gluten, and egg proteins with reducing sugars, as well as other chemical color changes; (6) crust formation from surface dehydration; and (7) crust darkening from Maillard browning reactions and caramelization of sugars.

The rates of these different reactions and the order in which they occur depend to a large extent on the rate of heat transfer through the batter or dough. If the crust forms before the center of the mass is baked, because of too high top heat compared to bottom heat or too hot an oven, then the center of the baked item may remain soggy or late escaping gas may crack the crust. Quite apart from the temperature distribution in the oven, the rate of heat transfer is affected also by the nature of the baking pan.

Shiny pans reflect heat and slow heat transfer into the pan contents. Dull and dark-colored pans absorb heat more rapidly and speed heat transfer. The shape of pans also is obviously important: a shallow pan with the batter or dough in a thin layer will develop different thermal gradients and give different results than a smaller deeper pan containing the same weight of material to be baked.

Were all of the preceding factors not enough to provide causes for variations in baked goods, there also is the effect of altitude. Unless otherwise indicated, most bakery formulas were developed for use at altitudes near sea level. At elevations of about 900 m (3000 ft) and higher, excessive expansion of leavening gases under reduced atmospheric pressure causes stretching and weakening of the cellular structure being formed in the oven. The result can be collapsed items of coarse and irregular grain. Corrective measures at high altitudes, therefore, call for cake formulas with less baking powder, more or stronger flour as tougheners, or decreased levels of tenderizers such as shortening and sugars. Because of their tougher doughs, bread formulas are less sensitive to altitude than cake formulas.

Much of what has been said regarding the baking step applies principally to conventional baking in ovens where conduction, convection, and radiation heating contribute to the end result. Consumer products are being developed for baking in home microwave ovens. This involves special product formulation to overcome differences in water binding and browning, which can be achieved through the use of various starches, hydrocolloids, reducing sugars and other Maillard-producing ingredients.

The varieties of breads, cakes, and other bakery items can run into the thousands as ingredients, formulas, and preparation methods are changed. Today, the principles of cereal chemistry and baking technology are well understood and the many possible variables can be kept under fairly rigid control in large modern bakeries. This permits automated high-speed operations with uniform production rates of tens of thousands of units per hour. In smaller bakeries and in the home, however, baking remains more of an art than a science.

LEGUMES AND OILSEEDS

General Compositions

As stated earlier, legumes and oilseeds are considerably higher in protein than cereal grains, and oilseeds also are much higher in fat. Whereas different cereal grains may contain about 7–14% protein and about 2–5% fat, various mature dry legumes and oilseeds contain about 20–40% protein; fat levels in peas and beans are low but are 20–50% in oilseeds. These compositions are reflected in the meals and flours derived from legumes and oilseeds. Table 17.5 gives the compositions of some dehulled legume flours as well as data on protein yields that can be obtained from them. The high fat content of soybean is why this legume is also commonly listed among the oilseeds. If fat is removed from the dehulled soybean, the flour that can then be produced would be even more concentrated in protein than is indicated in Table 17.5.

Protein Supplementation and Complementation

Although cereal grains are relatively low in total protein and generally low in lysine and certain other amino acids, these shortcomings can be overcome by appropriate blending with legume or oilseed products.

The most obvious result of such blending is that the mixture is higher in protein than the cereal component alone. Beyond this, however, legumes and various oilseeds improve the quality of cereal proteins by supplementing them with limiting amino acids such as lysine (sometimes tryptophan or threonine). This is called protein supplementation. On the other hand, legumes and some oilseeds, which are deficient in methionine, can be supplemented by cereal grains, which are not deficient in this amino acid. Such mutual balancing of each other's amino acids is known as protein complementation. An example of this is shown in Fig. 17.12 where mixtures of corn

Table 17.5. Flour Composition and Yield of Protein Isolate from Grain Legumes, Dry Basis

Legume Flour	Protein (%) N × 5.7	Fat (%)	Fiber (%)	Ash (%)	Yield (g/100 g flour)	Protein Isolate			
						Nitrogen (%)	Yield (% of total protein)	Color	Whey N (% of total N)
Soybean	39.7	23.1	2.2	4.8	36.6	15.0	78.9	Cream	10.1
Lupine	40.8	7.9	1.5	3.1	30.8	15.2	65.6	White	21.5
Fababean	30.0	1.5	1.4	2.9	28.2	14.9	80.2	Tan	18.4
Pea bean	28.6	1.6	1.7	4.0	28.6	13.1	74.6	White	23.0
Mung bean	24.7	0.6	0.9	3.7	26.9	14.1	87.6	Yellow	11.4
Field pea	22.7	1.0	1.5	2.9	22.7	14.0	79.8	Cream	20.5
Lima bean	20.0	0.9	2.1	3.8	17.9	12.5	64.3	White	32.7
Lentil	19.8	1.1	1.1	3.4	19.0	13.3	72.8	Cream	18.2
Chickpea	19.2	5.6	1.3	2.6	18.5	13.6	74.6	Cream	17.9

SOURCE: Fan and Sosulski (1974).

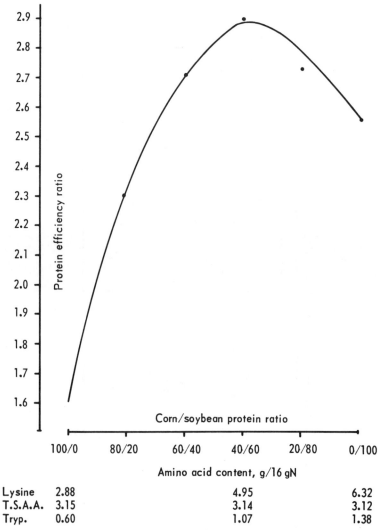

Figure 17.12. Complementation effects in rats fed combinations of soybean flour and whole corn flour at a constant level of dietary protein. Source: *Bressani et al.*, J. Food Sci. *39*, 577–580, (1974).

flour and soybean flour were fed to rats and their weight gains per gram of protein consumed (protein efficiency ratio) were measured. Optimum results were obtained with the 40% corn/60% soybean protein ratio. With less soybean, lysine became limiting; with more soybean, methionine was limiting. Much progress has been made over the past 30 years in using such mixtures of local crops to improve human nutrition.

Soybean Technology

Of the various legumes and oilseeds, the soybean is the outstanding source of protein due to its high protein content and the relative ease of its extractability. The soybean has been intensively studied and many processes have been developed to obtain and modify its protein for special food uses. Some of the more important processing operations and resulting products are indicated in Fig. 17.13. A food-grade flour of about 50% protein is obtained by dehulling and low-temperature extraction of the oil. Partially defatted flours also are available. The defatted flour can be further concentrated in protein by acid-washing starch and other components from the acid-precipitated protein, or it can be still further concentrated to the "isolate" stage by dissolving the defatted flour in alkali, filtering, reacidifying, and centrifuging the precipitated protein from the whey. The protein isolate can then be modified by enzymes and other treatments to affect its solubility, whipability, and other properties and then spray dried; or the isolate can be dissolved in alkali, forced through the holes of a spinnerette, and recoagulated into fibers in an acid bath. In this way, meatlike texture can be given to soy protein and used to manufacture meat analogs. The equipment for fiber production and the fibers being gathered and drawn from the bath are shown in Fig. 17.14. But soybean protein can be texturized in other less costly ways, including extrusion-cooking directly from soy flour. In this process, flour of about 50% protein is wetted to about 30–40% moisture and heat-coagulated under pressure. Further texturizing occurs as the dough expands and becomes oriented passing through the extrusion orifice. The dough is then cut into chunks and dried. Upon reconstitution and cooking, its texture and appearance are remarkably like meat (Fig. 17.15). Such products are less costly than meat and are increasingly being used as partial replacements for meat in meat-containing mixtures.

Peanuts

Like the soybean, the peanut, or groundnut, is both a legume and an oilseed. The shelled whole nuts contain about 25% protein and about 50% oil. Peanut flours, protein concentrates, and protein isolates can be produced, but their use as human food is limited. The protein of peanut is not as high in lysine as that of soybean.

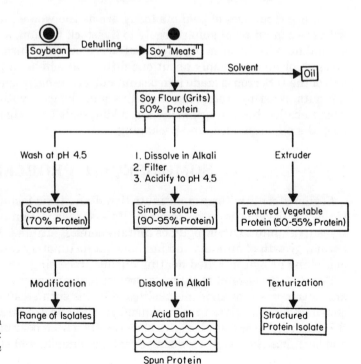

Figure 17.13. Types of protein products from soybean. Source: *McCleary et al., Food Canada* 33(11), 23–25 (1973).

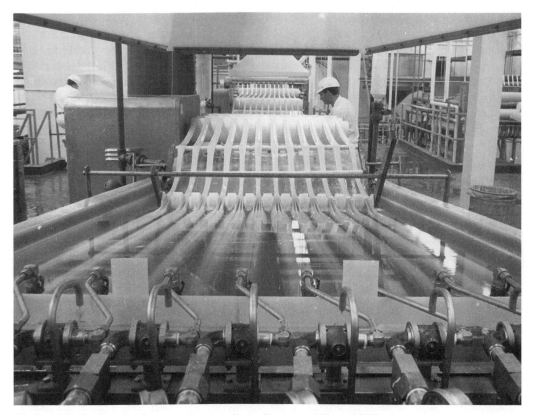

Figure 17.14. Production of soy protein spun fibers. *Courtesy of General Mills, Inc.*

The principal uses of peanuts today are as the whole nut, as a source of peanut oil with the peanut meal going largely to livestock feeding, and, in the United States, as ground nuts in the form of peanut butter. About two-thirds of the world's peanuts are pressed for oil and supply about one-fifth of all edible oil production. Somewhat over half of the U.S. crop is made into peanut butter. Its basic manufacture involves shelling the nuts, roasting, removing the skins and "hearts" with heat followed by rubbing (this is called blanching), grinding, adding salt and sugar for flavor, emulsifiers to keep the oil suspended, and packaging.

SOME SPECIAL PROBLEMS

Legumes contain certain antinutritional and toxic factors that must be inactivated if their full value is to be realized. Raw soybeans contain an antitrypsin factor or trypsin inhibitor. Other legumes contain hemagglutinins. These factors interfere with normal growth of animals and humans but fortunately can be inactivated by the heat of cooking or by controlled heating during processing.

Peanuts, because of their moisture content at harvest, may support mold growth and development of toxic metabolites of molds such as aflatoxins. Today, peanuts are stored under conditions to control mold growth and are carefully inspected to minimize this hazard. Aflatoxins also have been removed from peanut meal by solvent extraction and been inactivated by oxidizing agents, ammonia, and other treatments. Due to the

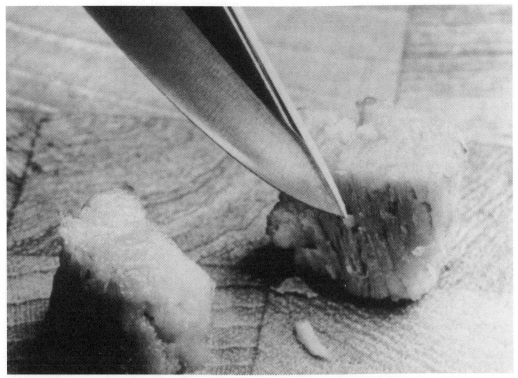

Figure 17.15. Textured soybean proteins made by the thermoplastic extrusion process. Source: Horan, *New Protein Foods, Vol. 1A, A. M. Altschul (Editor). Academic Press, New York, 1974.*

ubiquitous nature of molds, aflatoxins and other mycotoxins can never be completely eliminated from feeds and foods, although they can be decreased to insignificant levels. Dry and wet milling of corn, for example, removes a major portion of any aflatoxin which may have been present initially. Heat processing and cooking further reduce remaining aflatoxin. Currently, maximum permissible levels in the parts per billion range are enforced in many countries.

Cottonseed endosperm has pigment glands that contain the toxic pigment gossypol. Any gossypol that gets into the oil is largely removed during oil refining. The presence of gossypol in the meal has impeded acceptance of cottonseed flour and cottonseed protein for human food. It is possible to remove unruptured pigment glands by controlled disintegration of the seeds in hexane and centrifugal separation of the lighter glands from the rest of the endosperm. Glandless varieties of cottonseeds that are free of gossypol also have been developed by plant breeders.

Problems such as these, which generally yield to research and controlled processing, must always be considered when less common sources of food are proposed for use in technologically underdeveloped regions.

References

Anon. 1957. Rice. Fed. Reg. *22*, 6887–6888.
Anon. 1958. Rice. Fed. Reg. *23*, 1170–1171.

Cheng, L.M. 1992. Food Machinery. For the Production of Cereal Foods, Snack Foods, and Confectionary. Ellis Horwood, New York.

Christensen, C.M. 1982. Storage of Cereal Grains and Their Products. 3rd ed. American Association of Cereal Chemists, St. Paul, MN.

Dick, J.W. 1989. Pasta Science. Sci. Food Agric. *1*(2), 6.

Dupont, J. and Osman, E.M. (Editors). 1987. Cereals and Legumes in the Food Supply. Iowa State University Press, Ames, IA.

Fan, T.Y. and Sosulski, F.W. 1974. Dispersibility and isolation of proteins from legume flours. Can. Inst. Food Sci. Technol. J. *7*, 256–259.

FAO/WHO. 1973. Energy and Protein Requirements. Report of a Joint FAO/WHO Ad Hoc Expert Committee, Geneva. World Health Organization Tech. Rept. Ser. 522.

Fast, R.B. and Caldwell, E.F. 1990. Breakfast Cereals and How They Are Made. American Association of Cereal Chemists, St. Paul, MN.

Horan, F.E. 1974. Nurtition cereal blends—from conception to consumption. Cereal Sci. Today *19*, 112–117.

Hoseney, R.C and Rogers, D.E. 1990. The formation and properties of wheat flour doughs. Crit. Rev. Food. Nutr. *29*(2), 73–93.

Hoseney, R.C. 1986. Principles of Cereal Science and Technology. American Association of Cereal Chemists, St. Paul, MN.

Jansen, G.R. 1977. Amino acid fortification. *In* Evaluations of Proteins for Humans. C.E. Bodwell (Editor). AVI Publishing Co., Westport, CT.

Kent, N.L. 1983. Technology of Cereals. An Introduction for Students of Food Science and Agriculture. 3rd ed. Pergamon Press, New York.

Luh, B.S., (Editor). 1991. Rice. 2nd ed. Chapman & Hall, London.

Lusas, E.W, Erickson, D.R., and Wai, K.N. (Editors). 1989. Food Uses of Whole Oil and Protein Seeds. American Oil Chemists' Society, Champaign, IL.

Manley, D.J.R. 1991. Technology of Biscuits, Crackers, and Cookies, 2nd ed. Chapman & Hall, London.

Matthews, R.H. 1989. Legumes. Chemistry, Technology, and Human Nutrition. Marcel Dekker, New York.

Matz, S.A. 1992. Bakery, Technology and Engineering. Chapman & Hall, London. New York.

Matz, S.A. 1993. Snack Food Technology. 3rd ed. Chapman & Hall. London. New York.

Pomeranz, Y. and Munck, L. (Editors). 1981. Cereals, a Renewable Resource, Theory and Practice. American Association of Cereal Chemists, St. Paul, MN.

Salunkhe, D.K. and Kadam, S.S. 1989. CRC Handbook of World Food Legumes. Nutritional Chemistry, processing Technology, and utilization. CRC Press, Boca Raton, FL.

Stauffer, C.E. 1990. Functional additives for bakery foods. Chapman & Hall, London. New York.

Sultan, W.J. 1990. Practical Baking. 5th ed. Van Nostrand Reinhold, New York.

Welch, R. Oat Crop. 1995. Chapman & Hall, London.

Woodroff, J.G. 1983. Peanuts: Production, Processing, Products. 3rd ed. Chapman & Hall, London. New York.

18

Vegetables and Fruits

Vegetables and fruits have many similarities with respect to their compositions, methods of cultivation and harvesting, storage properties, and processing. In fact, many vegetables are considered fruits in the true botanical sense. Botanically, fruits are those portions of a plant that house seeds. Therefore, tomatoes, cucumbers, eggplant, peppers, okra, sweet corn, and other vegetables would be classified as fruits according to this definition. However, the important distinction between fruits and vegetables has come to be made on a usage basis: those plant items that are generally eaten with the main course of a meal are often considered to be vegetables; those that commonly are eaten alone or as a dessert are considered fruits. This is the distinction made by food processors, certain marketing laws, and the consuming public.

GENERAL PROPERTIES

Because vegetables are derived from various parts of plants, it is sometimes helpful to classify vegetables according to the plant part from which they are derived, such as roots, leaves, stems, buds, and so on (Table 18.1).

Fruits are the mature ovaries of plants with their seeds. The edible portion of most fruits is the fleshy part of the pericarp or vessel surrounding the seeds. Fruits, in general, are acidic and sugary. They commonly are grouped into several major divisions, depending principally on botanical structure, chemical composition, and climatic requirements. Thus, berries are generally small and quite fragile, although cranberries are rather tough. Grapes are also berries, which grow in clusters. Melons, on the other hand, are large and have a tough outer rind. Apricots, cherries, peaches, and plums contain single pits and are known as "drupes." "Pomes" contain many pits and are represented by apples, quince, and pears. Citrus fruits, characteristically high in citric acid, include oranges, grapefruit, and lemons. Tropical and subtropical fruits include bananas, dates, figs, pineapples, papayas, mangos, and others, but not the separate group of citrus fruits; these all require warm climates for growth.

GROSS COMPOSITION

The compositions of representative vegetables and fruits in comparison with a few of the cereal grains are shown in Table 18.2. The composition of vegetables and fruits depends not only on botanical variety, cultivation practices, and weather but also on

Table 18.1. Classification of Vegetables

	Examples
Earth vegetables	
roots	Sweet potatoes, carrots
modified stems	
corms	Taro
tubers	Potatoes
modified buds	
bulbs	Onions, garlic
Herbage vegetables	
leaves	Cabbage, spinach, lettuce
petioles (leaf stalk)	Celery, rhubarb
flower buds	Cauliflower, artichokes
sprouts, shoots (young stems)	Asparagus, bamboo shoots
Fruit vegetables	
legumes	Peas, green beans
cereal	Sweet corn
vine fruits	Squash, cucumber
berry fruits	Tomato, egg plant
tree fruits	Avocado, breadfruit

Courtesy of B. Feinberg.

Table 18.2. Typical Percentage Composition of Edible Portion of Foods of Plant Origin

Food	Constituent				
	Carbohydrate	Protein	Fat	Ash	Water
Cereals					
wheat flour, white	73.9	10.5	1.9	1.7	12
rice, milled, white	78.9	6.7	0.7	0.7	13
maize (corn) whole grain	72.9	9.5	4.3	1.3	12
Earth vegetables					
potatoes, white	18.9	2.0	0.1	1.0	78
sweet potatoes	27.3	1.3	0.4	1.0	70
Vegetables					
carrots	9.1	1.1	0.2	1.0	88.6
radishes	4.2	1.1	0.1	0.9	93.7
asparagus	4.1	2.1	0.2	0.7	92.9
beans, snap, green	7.6	2.4	0.2	0.7	89.1
peas, fresh	17.0	6.7	0.4	0.9	75.0
lettuce	2.8	1.3	0.2	0.9	94.8
Fruits					
banana	24.0	1.3	0.4	0.8	73.5
orange	11.3	0.9	0.2	0.5	87.1
apple	15.0	0.3	0.4	0.3	84.0
strawberries	8.3	0.8	0.5	0.5	89.9
melon	6.0	0.6	0.2	0.4	92.8

SOURCE: Food and Agriculture Organization (FAO).

the degree of maturity prior to harvest and the condition of ripeness, which continues after harvest and is influenced by storage conditions. Nevertheless, some generalizations can be made.

Most fresh vegetables and fruits are high in water, low in protein, and low in fat. The water content is generally greater than 70% and frequently greater than 85%. Interestingly, the water content of milk and apples is similar. Commonly, protein content is no greater than 3.5% and fat content no greater than 0.5%. Exceptions exist to these typical values: dates and raisins are substantially lower in moisture but cannot be considered fresh in the above sense; legumes such as peas and certain beans are higher in protein; a few vegetables such as sweet corn are slightly higher in fat; and avocados are substantially higher in fat. On the other hand, vegetables and fruits are important sources of both digestible and indigestible carbohydrates. The digestible carbohydrates are present largely as sugars and starches, and the indigestible cellulosic and pectic materials provide fiber, which is important to normal digestion. Fruits and vegetables are important sources of minerals and certain vitamins also, especially vitamins A and C. The precursors of vitamin A, including beta-carotene and certain other carotenoids, are present particularly in the yellow-orange fruits and vegetables and in the green, leafy vegetables. Citrus fruits are excellent sources of vitamin C, but green, leafy vegetables, and tomatoes are also good sources. Potatoes also are an important source of vitamin C in many countries, not so much because of the level of vitamin C in potatoes, which is not especially high, but rather because of the large quantities of potatoes consumed.

STRUCTURAL FEATURES

The structural unit of the edible portion of most fruits and vegetables is the parenchyma cell (Fig. 18.1). Although parenchyma cells of different fruits and vegetables differ

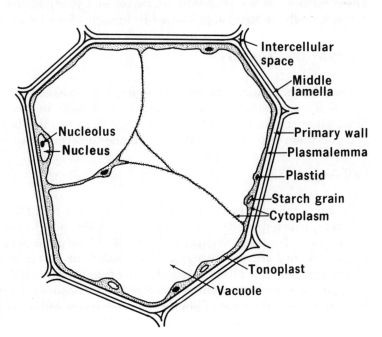

Figure 18.1. Diagram of a parenchyma cell. *Courtesy of B. Feinberg.*

somewhat in gross size and appearance, all have essentially the same fundamental structure. Parenchyma cells of plants differ from animal cells in that the actively metabolizing protoplast portion of plant cells represents only a small fraction (about 5%) of the total cell volume. This protoplast is rather filmlike and is pressed against the cell wall by the large water-filled central vacuole. The protoplast has inner and outer semipermeable membrane layers between which are confined the cytoplasm and its nucleus. The cytoplasm contains various inclusions, among them starch granules and plastids such as the chloroplasts and other pigment-containing chromoplasts. The cell wall, cellulosic in nature, contributes rigidity to the parenchyma cell and confines the outer protoplasmic membrane. It also is the structure against which other parenchyma cells are cemented to form extensive three-dimensional tissue masses. The layer between cell walls of adjacent parenchyma cells, referred to as the middle lamella, is composed largely of pectic and polysaccharide materials which act to cement cells together. Air spaces also exist, especially at the angles formed where several cells come together.

The relationships between these structures and their chemical compositions are further indicated in Table 18.3. Parenchyma cells vary in size from plant to plant but are quite large when compared to bacterial or yeast cells. The larger parenchyma cells may have volumes many thousand times greater than a typical bacterial cell.

Several types of cells other than parenchyma cells contribute to the familiar structures of fruits and vegetables. These include various types of tubelike conducting cells, which distribute water and salts throughout the plant. Such cells produce fibrous structures toughened by the presence of cellulose and the woodlike substance lignin. Cellulose, lignin, and pectic substances also occur in specialized supporting cells, which increase in importance as plants become older. An important structural feature of all plants, including fruits and vegetables, is protective tissue. This can take many forms but usually is made up of specialized parenchyma cells that are pressed compactly together to form a skin, peel, or rind. Surface cells of these protective structures on leaves, stems, or fruits secrete waxy cutin and form a water-impermeable cuticle. These surface tissues, especially on leaves and young stems, also contain numerous valvelike cellular structures (stomata) through which moisture and gases can pass.

Turgor and Texture

The range of textures encountered in fresh and cooked vegetables and fruits is great and, to a large extent, can be explained by changes in specific cellular components. Since plant tissues generally contain more than two-thirds water, the relationships between these components and water further determine textural differences.

Cell Turgor

Turgor is the rigidity of plant cells resulting from being filled with water. The state of turgor, which depends on osmotic forces, is the most important factor determining the texture of fruits and vegetables. The cell walls of plant tissues have varying degrees of elasticity and are largely permeable to water and ions as well as to small molecules. The membranes of the living protoplast are semipermeable; that is, they allow passage of water but selectively transfer dissolved and suspended materials. The cell vacuoles contain most of the water of plant cells; within this water are dissolved sugars, acids,

Table 18.3. Structural and Chemical Components of Plant Cells

Structure	Chemical Constituents
Vacuole	H$_2$O, inorganic salts, organic acids, oil droplets, sugars, water-soluble pigments, amino acids, vitamins
Protoplast	
membrane	
tonoplast (inner)	Protein, lipoprotein, phospholipids, phytic acid
plasmalemma (outer)	
nucleus	Nucleoprotein, nucleic acid, enzymes (protein)
cytoplasm	
active	
chloroplasts	Chlorophyll
mesoplasm (ground substance)	Enzymes, intermediary metabolites, nucleic acid
mitochondria	Enzymes (protein), Fe, Cu, Mo vitamin co-enzyme
microsomes	Nucleoproteins, enzymes (proteins), nucleic acid
inert	
starch grains	Reserve carbohydrate (starch), phosphorus
aleurone	Reserve protein
chromoplast	Pigments (carotenoids)
oil droplets	Triglycerides of fatty acids
crystals	Calcium oxalate, etc.
Cell wall	
primary wall	Cellulose, hemicellulose, pectic substances and noncellulose polysaccharide
middle lamella	Pectic substances and noncellulose polysacharides, Mg, Ca
plasmodesmata	Cytoplasmic strands interconnecting cytoplasm of cells through pores in the cell wall
surface materials (cutin or cuticle)	Esters of long chain fatty acids and long chain alcohols

Courtesy of B. Feinberg.

salts, amino acids, some water-soluble pigments and vitamins, and other low-molecular-weight constituents.

In the living plant, water taken up by the roots passes through the cell walls and membranes into the cytoplasm of the protoplasts and into the vacuoles to establish a state of osmotic equilibrium within the cells. The osmotic pressure within the cell vacuoles and within the protoplasts pushes the protoplasts against the cell walls and causes them to stretch slightly in accordance with their elastic properties. These processes result in the characteristic appearance of live plants and are responsible for the desired plumpness, succulence, and much of the crispness of harvested live fruits and vegetables.

When plant tissues are damaged or killed by storage, freezing, cooking, or other causes, denaturation of the proteins of the cell membranes occurs, resulting in the loss of perm-selectivity. Without perm-selectivity, osmotic pressure in cell vacuoles and protoplasts cannot be maintained, and water and dissolved substances are free to diffuse out of the cells and leave the remaining tissue in a soft and wilted condition.

Other Factors Affecting Texture

Whether a high degree of turgor exists in live fruits and vegetables or a relative state of softness develops from loss of osmotic pressure, final texture is further influenced by several cell constituents.

Cellulose, Hemicellulose, and Lignin.   Cell walls in young plants are very thin and are composed largely of cellulose. As the plant ages, cell walls tend to thicken and become higher in hemicellulose and in lignin. These materials are fibrous and tough and are not significantly softened by cooking.

Pectic Substances.   Pectin and related substances are complex polymers of sugar acid derivatives. The cementlike substance found especially in the middle lamella, which helps hold plant cells to one another, is a water-insoluble pectic substance. Upon mild hydrolysis, this substance yields water-soluble pectin, which can form gels or viscous colloidal suspensions with sugar and acid. Certain water-soluble pectic substances also react with metal ions, particularly calcium, to form water-insoluble salts such as calcium pectates. The various pectic substances may influence texture of vegetables and fruits in several ways. When vegetables or fruits are cooked, some of the water-insoluble pectic substance is hydrolyzed into water-soluble pectin. This results in a degree of cell separation in the tissues and contributes to tenderness. Since many fruits and vegetables are somewhat acidic and contain sugars, the soluble pectin also tends to form colloidal suspensions which thicken the juice or pulp of these products.

Fruits and vegetables also contain a natural enzyme that can further hydrolyze pectin to the extent that it loses much of its gel-forming property. This enzyme is known as pectin methyl esterase. Some products (e.g., tomato juice and tomato paste) contain both pectin and pectin methyl esterase. If freshly prepared tomato juice or paste is allowed to stand, the original viscosity gradually decreases due to the action of pectin methyl esterase on pectin gel. This can be prevented if the tomato products are quickly heated to a temperature of about 82°C to inactivate enzyme liberated from broken cells before the pectin is hydrolyzed. This treatment, known as the hot-break process, is commonly practiced in the manufacture of tomato paste and tomato juice products to yield products of high viscosity. In contrast, when low-viscosity products are desired, no heat is used and enzyme activity is allowed to proceed. This is the cold-break process. After the appropriate viscosity is achieved, the product can be heat treated, as in canning, to preserve it for long-term storage.

It often is desirable to firm the texture of fruits or vegetables, especially when products are normally softened by processing. In this case, advantage is taken of the reaction between soluble pectic substances and calcium ions to form calcium pectates. These calcium pectates are water insoluble; when they are produced within the tissues of fruits and vegetables, they increase structural rigidity. Thus, it is common commercial practice to add low levels of calcium salts to tomatoes, apples, and other vegetables and fruits prior to canning or freezing.

Starch.   The occurrence of starch within starch granules and the swelling and gelatinization of these granules in the presence of moisture and heat have previously been mentioned. When starch granules absorb water and gelatinize, they gradually lose their granular structure and produce a pasty, viscous colloidal suspension. The swelling of starch granules within the cells of plant tissues on heating causes a corresponding swelling of these cells and contributes to firm texture and plumpness.

On the other hand, starch swelling together with osmotic pressure can be so great as to cause plant cells to burst. When this happens, the viscous colloidal starch suspension oozes from the cells and imparts pastiness to the system. The same occurs when cells containing much starch are ruptured by processing conditions. This is particularly important in the case of potato products. The desirable texture of mashed potatoes and other potato products is a mealiness rather than a stickiness or pastiness. Therefore, in the production of dehydrated potato granules and flakes much of the technology of mixing and drying is aimed at minimizing both cell rupture and release of free starch. The same is true in the cooking and mashing of fresh potatoes, which, if excessive, can produce undesirable pastiness.

Color and Color Changes

Much of the appeal of fruits and vegetables in our diets is due to their desirable colors. The pigments and color precursors found in fruits and vegetables occur for the most part in the cellular plastid inclusions (e.g., chloroplasts and other chromoplasts) and to a lesser extent dissolved in fat droplets or water within the cell protoplast and vacuole. These pigments are classified into four major groups: chlorophylls, carotenoids, anthocyanins, and anthoxanthins. Pigments belonging to the latter two groups also are referred to as flavonoids, and include the tannins.

Chlorophylls

Chlorophylls are largely contained within the chloroplasts and have a primary role in the photosynthetic production of carbohydrates from carbon dioxide and water. The bright green color of leaves and other plant parts is due largely to oil-soluble chlorophylls, which in nature are bound to protein molecules in highly organized complexes. When plant cells are killed by ageing, processing, or cooking, the proteins are denatured and the magnesium bound in the chlorophyll may be released. This causes a chemical change of chlorophyll to pheophytin which is olive green or brown in color. Conversion to pheophytin is favored by acid pH and occurs less readily under alkaline conditions. For this reason, peas, beans, spinach, and other green vegetables, which tend to lose their bright green colors on heating, can be to some extent protected against such color changes by the addition of sodium bicarbonate or other alkali to the cooking or canning water. The addition of magnesium salts can help reduce the conversion of chlorophyll to pheophytin. However, this practice is not looked on favorably nor used commercially because alkaline pH tends to soften cellulose and vegetable texture and to increase the destruction of vitamin C and thiamin at cooking temperatures.

Carotenoids

Pigments belonging to the carotenoid group are fat soluble and range in color from yellow through orange to red. They often occur along with the chlorophylls in the chloroplasts but are present in other chromoplasts also and may occur free in fat droplets. Important carotenoids include the orange carotenes of carrot, corn, apricot, peach, citrus fruits, and squash; the red lycopene of tomato, watermelon, and apricot;

the yellow-orange xanthophyll of corn, peach, paprika, and squash; and the yellow-orange crocetin of the spice saffron. These and other carotenoids seldom occur singly within plant cells.

Of major importance is the relationship of some carotenoids to vitamin A. Some carotenoids serve as precursors to vitamin A. A molecule of orange beta-carotene is converted into two molecules of colorless vitamin A within the body. Some other carotenoids (e.g., alpha-carotene, gamma-carotene, and cryptoxanthin) are precursors of vitamin A also, but because of minor differences in chemical structure one molecule of each of these yields only one molecule of vitamin A.

In food processing, the carotenoids are fairly resistant to heat, changes in pH, and water leaching since they are fat soluble. However, they are very sensitive to oxidation, which results in both color loss and destruction of vitamin A activity.

Anthocyanins

Anthocyanin pigments belong to a group of plant chemicals known as flavonoids. They are water soluble and commonly are present in the juices of fruits and vegetables. The anthocyanins include the purple, blue, and red pigments of grapes, berries, plums, eggplant, and cherries. The color of anthocyanins depends on the pH. Thus, many of the anthocyanins that are violet or blue in alkaline media become red on addition of acid. The color of red fruits and vegetables shifts toward violet and gray-blue if the pH becomes basic. Red anthocyanins also tend to become more violet, blue, or colorless on reaction with metal ions, which is one reason for lacquering the inside of metal cans when the true color of anthocyanin-containing fruits and vegetables is to be preserved. The water solubility of anthocyanins also results in easy leaching of these pigments from cut fruits and vegetables during processing and cooking.   ·

Flavonoids

The yellow flavonoids are structurally related to anthocyanins and comprise a large group of chemicals which are widely found in plant foods. They also are pH sensitive, tending toward a deeper yellow in alkaline media. Thus, potatoes or apples become somewhat yellow when cooked in water with a pH of 8 or higher, which is common in many areas. Acidification of the water to pH 6 or lower favors a whiter color.

Tannins

Tannins are complex mixtures of phenolic compounds found in plants. Under most circumstances they are colorless, but on reaction with metal ions they form a range of dark-colored complexes which may be red, brown, green, gray, or black. They are responsible for the dark color found in the bark of oak, sumac, and myrobalen trees. The various shades of these colored complexes depend on the particular tannin, the specific metal ion, pH, concentration of the complex, and other factors not yet fully understood.

Water-soluble tannins appear in the juices squeezed from grapes, apples, and other fruits as well as in the brews extracted from tea and coffee. The color and clarity of tea are influenced by the hardness and pH of the brewing water. Alkaline waters that contain calcium and magnesium favor the formation of dark brown tannin complexes, which precipitate when the tea is cooled. If acid in the form of lemon juice is added to such tea, its color lightens and the precipitate tends to dissolve. Iron from equipment or from pitted cans has caused a number of unexpected colors to develop in tannin-containing products, such as coffee, cocoa, and foods flavored with these.

The tannins also are important because they possess astringency which influences flavor and contributes body to coffee, tea, wine, apple cider, beer, and other beverages. Excessive astringency causes a puckery sensation in the mouth, which is the condition produced when tea becomes high in tannins from overbrewing.

Betalains

Like anthocyanins, betalains are red water-soluble pigments but are chemically different and are less widely dispersed in the plant world. The primary food plant in which they occur is the red beet. They also occur in some cactus fruits and flowers. Betalains are degraded by thermal processing but occur in such high amounts that sufficient pigment remains for coloration. They are relatively stable compared to other natural red pigments, especially in the pH range of 4–6. They have been considered for use as naturally occurring food colorants.

ACTIVITIES OF LIVING SYSTEMS

Fruits and vegetables continue to respire after harvest and, thus, can be considered "alive." *Respiration* means the produce takes in oxygen and gives off carbon dioxide, moisture, and heat, which influence storage, packaging, and refrigeration requirements. Moisture and heat can build up in storage and packaging causing the growth of molds. Enough heat can be generated to damage the produce itself.

Fruits and vegetables, before and after harvest, also undergo changes in carbohydrates, pectins, and organic acids, which influence the various quality attributes of the products. Few generalizations can be given concerning changes in starches and sugars. In some plant products, sugars decrease quickly and starch increases soon after harvest. This is the case for ripe sweet corn, which can suffer flavor and texture quality losses in a very few hours after harvest. Unripe fruit, in contrast, frequently is high in starch and low in sugars. Continued ripening after harvest generally results in a decrease in starch and an increase in sugars as occurs in apples and pears. However, this does not necessarily mean that the starch is the source of the newly formed sugars. Further, changes in starch and sugars are markedly influenced by postharvest storage temperatures. For example, potatoes stored below about 10°C continue to build up high levels of sugars, whereas the same potatoes stored above 10°C do not. Thus, potatoes that are stored for dehydration are kept above 10°C in order to keep the level of reducing sugars low and to minimize Maillard browning reactions during drying and subsequent storage of the dried product.

After harvest, the changes in pectins of fruits and vegetables are more predictable. Generally, there is a decrease in water-insoluble pectic substance and a corresponding increase in water-soluble pectin. This contributes to the gradual softening of fruits and vegetables during storage and ripening. Further breakdown of water-soluble pectin by pectin methyl esterase also occurs.

The organic acids of fruits generally decrease during storage and ripening. This occurs in apples and pears and is especially important in oranges. Oranges have a long ripening period on the tree, and time of picking is largely determined by degree of acidity and sugar content, which have major effects on juice quality.

As acids disappear during ripening, more than just tartness of fruits is affected. Since many plant pigments are sensitive to acid, fruit color would be expected to change as the organic acid content changes. Additionally, the viscosity of pectin gel is affected by acid and sugar, both of which change in concentration with ripening. Further consequences of the live state of vegetables and fruits are indicated in the following sections on harvesting and processing.

The quality decline in stored respiring fruits and vegetables is termed senescence and results from the continued enzymic activity. A number of factors influence respiration rate and, hence, senescence rate. The two primary factors for a given produce item are temperature and the composition of the storage atmosphere. Reduced temperatures, lower oxygen levels, and raised carbon dioxide levels can be combined to greatly reduce the rate of senescence and increase storage time. This is not without limit. Many fruits and vegetables suffer "chill injury" from temperatures that are too low. If the carbon dioxide levels raise above 5–10%, damage can also occur. Storage of produce under such controlled conditions is termed *controlled atmosphere storage* and commonly practiced for some produce such as apples.

HARVESTING AND PROCESSING
OF VEGETABLES

Varietal Differences

The food scientist and vegetable processor must appreciate the substantial differences that cultivars of a given vegetable possess. In addition to differences in response to weather and in pest resistance, cultivars of a given vegetable differ in size, shape, time of maturity, and resistance to physical damage. These latter factors are of the greatest importance in the design and use of mechanical harvesting devices. Varietal difference is also important in the processing of fruits and vegetables. Certain types of apples become soft and loose texture more readily than others when heated. These apples might be more desirable for applesauce but less desirable for products such as pie fillings. Varietal differences in the effects of processing are important for most fruits and vegetables.

A varietal difference in resistance to tomato cracking is illustrated in Fig. 18.2. Varietal differences in warehouse storage stability and in suitability for different processing methods also exist. A cultivar of peas that is suitable for canning may be quite unsatisfactory for freezing, and cultivars of potatoes that are preferred for freezing may be less satisfactory for drying or potato chip manufacture. This should be expected since different cultivars of a given vegetable vary somewhat in chemical composition, cellular structure, and biological activity of their enzyme systems. Because of the importance of varietal differences, large food companies commonly provide special seed to farmers whose crops they contract to buy a year in advance. They also frequently manage their own vegetable farms to further guarantee a sufficient supply of high quality uniform raw materials.

Figure 18.2. Radial and concentric cracking varieties compared with resistant tomato type. *Courtesy of Campbell Soup Co.*

Harvesting and Preprocessing Considerations

When vegetables are maturing in the field, they are changing from day to day. There is a time when the vegetable will be at peak quality from the standpoint of color, texture, and flavor. Because this peak quality lasts only briefly, harvesting and processing of several vegetables, including tomatoes, corn, and peas, are rigidly scheduled to capture this peak quality.

After a vegetable is harvested, it may quickly pass beyond the peak quality condition. This is independent of microbial spoilage. One study on sweet corn showed that in just 24 h at room temperature 26% of the total sugars were lost with a comparable loss of sweetness in the corn. Even when stored just above freezing at 0°C, 8% of the sugar was lost in 24 h and 22% in 4 days. Some of this sugar was probably converted to starch; some was used during respiration. In similar fashion, peas and lima beans can lose over 50% of their sugar in just 1 day at room temperature; losses are slower under refrigeration but there is still a great change in vegetable sweetness and freshness of flavor within 2 or 3 days. Not all losses of sugar are due to respiration or conversion to starch. Some of the sugar in asparagus can be converted to fibrous tissue after harvest; this contributes to a more woody texture.

Along with the loss of sugar, the evolution of heat can be a serious problem when large stockpiles of vegetables are transported or held prior to processing. At room temperature, some vegetables will liberate heat at a rate of 60 Btu/lb/day. This is enough for each ton of vegetables to melt 800 lb of ice per day. Since the heat further deteriorates the vegetables and speeds growth of microorganisms, the harvested vegetables must be cooled if not processed immediately.

But cooling only slows down the rate of deterioration, it does not prevent it, and vegetables differ in their resistance to cold storage. As pointed out earlier, each type of vegetable has its optimum cold storage temperature which may be between about 0 and 10°C. Storage below 7°C in the case of cucumbers, for example, will result in pitting, soft spots, and decay. What actually happens is that at too low a temperature the normal metabolism of the living vegetable is altered and various abnormalities occur along with decreased resistance to invasion by microorganisms that are present and can grow at the low storage temperature.

The continual loss of water by harvested vegetables due to transpiration, respiration, and physical drying of cut surfaces results in wilting of leafy vegetables, loss of plumpness of fleshy vegetables, and loss of weight of both. Moisture loss cannot be completely and effectively prevented by hermetic packaging. When fresh vegetables are sealed in plastic bags, the bags become fogged with moisture, the carbon dioxide level increases, and oxygen decreases. Because these conditions accelerate the deterioration of certain vegetables, it is common to perforate such bags to prevent such deterioration and to minimize high humidity in the package, which encourages microbial growth.

Shippers of fresh vegetables and vegetable processors appreciate the perishability of vegetables and do everything they can to minimize delays in processing of the fresh product. In many processing plants it is common practice to process vegetables immediately from the fields. To ensure a steady supply of top-quality produce during the harvesting period, many large food processors employ trained field managers who can advise on growing practices so that vegetables will mature and can be harvested in rhythm with the processing plant capabilities. This minimizes pileup and need for storage.

Postharvest Practices

Cooling of harvested vegetables in the field is a common practice (see Fig. 9.2). Fresh produce is often transported in liquid-nitrogen-cooled trucks to processing plants or directly to market. At the processing plant, vegetables are cleaned, graded, peeled, cut, and so forth; some equipment is used for these operations, but use of hand labor is still common.

Washing

The choice of washing equipment and other equipment used in processing vegetables depends on the size, shape, and fragility of the particular kind of vegetable.

A flotation cleaner for peas and other small vegetables (Fig. 18.3) operates on the principle that sound peas will sink, whereas broken peas, weed seeds, and certain other kinds of contamination will float, provided a liquid of the proper density is employed. In this case, a mineral oil–water emulsion is used and its density can be further controlled by frothing the mixture with air. The sound peas that sink are moved on to further processing and the floating debris is pumped to waste. Another type of washer is the rotary washer in which vegetables are tumbled while they are sprayed with jets of water. This type of washer should not be used to clean fragile vegetables. Fragile vegetables such as asparagus are valued for their wholeness and cannot be washed in agitating equipment that would break them up. Asparagus may be washed by gentle spraying on a belt.

Vegetables are washed to remove not only field soil and surface microorganisms but also fungicides, insecticides, and other pesticides. There are laws specifying maximum levels of these contaminating materials that may be retained on vegetables. Modern instruments can detect many pesticide residues at levels as low as a few parts per billion. Wash water containing detergents and other sanitizers can reduce the level of many residues.

Figure 18.3. Flotation cleaner for shelled peas. *Courtesy of Key Equipment Co.*

Skin Removal

Several methods are used to remove skins from those vegetables requiring skin removal. Skins can be softened from the underlying tissue by submerging vegetables in hot alkali solution. Lye may be used at a concentration of about 1% and at about 93°C. The vegetables with loosened skins are then conveyed under high-velocity jets of water which wash away the skins and any residual lye (see Fig. 5.3). Since the cost of lye and of treating lye-containing waste waters can be appreciable, processors sometimes use less expensive hot-water scalding followed by a machine that slits the skin and gently squeezes the vegetable, such as tomatoes, through the slit skin.

Vegetables with a thick skin, such as beets and sweet potatoes, may be peeled with steam under pressure as they pass through cylindrical vessels. This softens the skin and the underlying tissue. When the pressure is suddenly released, steam under the skin expands and causes the skins to puff and crack. The skins are then washed away with jets of water.

Onions and peppers are best skinned by exposing them to direct flame or to hot gases in rotary tube flame peelers of the type shown in Fig. 18.4. Here, too, heat causes steam to develop under skins and puff them so that they can be washed away with water.

Cutting and Trimming

Many vegetables require various kinds of cutting, stemming, pitting, or coring. Asparagus spears are cut to precise length. The clippings from the base of the stalk,

Figure 18.4. Rotary tube flame peelers for onions and peppers. *Courtesy of Gentry International, Inc.*

which are more fibrous and tougher than the prized stalk, are used in soups and other heated products where heat tenderizes them. Brussels sprouts are trimmed largely by hand by pressing the base against a rapidly rotating knife. Green beans are cut by machine into several different shapes along the length of the vegetable or transverse to the length. Olives are pitted by aligning them in small cups and then mechanically pushing plungers through the olives (Fig. 18.5). Pimentos may then be mechanically stuffed into the holes.

Blanching

Most vegetables that do not receive a high-temperature heat treatment (as in normal canning) must be heated to a minimal temperature to inactivate natural enzymes before processing or storing (even when frozen). This special heat treatment to inactivate enzymes is known as blanching. Blanching is not indiscriminate heating. Too little is ineffective, and too much damages vegetables by excessive cooking, especially when the fresh character of the vegetable is to be preserved by freezing. Blanching is essential for vegetables that are to be frozen because freezing only slows enzyme action, it does not destroy or completely stop it. If blanching does not precede freezing, then the product, which is often held in the frozen state for many months, will slowly develop off-flavors and off-colors, and other kinds of enzymatic spoilage may result.

Two of the more heat-resistant enzymes in vegetables are catalase and peroxidase. If these are destroyed, then other enzymes that contribute to deterioration will be inactivated also. Effective heat treatments for destroying catalase and peroxidase in different vegetables are known, and sensitive chemical tests have been developed to detect the amounts of these enzymes that might survive blanching treatment.

Because various types of vegetables differ in size, shape, heat conductivity, and the natural levels of their enzymes, blanching treatments had to be established on an experimental basis. As with sterilization of food in cans, the larger the food item, the longer it takes for heat to reach the center. Peas are more rapidly blanched than corn on the cob. Small vegetables may be adequately blanched in boiling water in a minute

Figure 18.5. Olive orientater and pitter. *Courtesy of Atlas-Pacific Co.*

or two; large vegetables may require several minutes (Table 18.4). Blanching with steam under pressure at higher temperatures requires shorter times but runs a greater risk of heat damage to the vegetable.

Because much of the enzyme activity in sweet corn is within the cob, steam blanching (Fig. 18.6) to inactivate 100% of the enzyme activity requires excessive heat. Therefore, as a compromise, processors use blanching conditions that destroy only about 90% of the enzyme activity; this avoids excessive softening of the kernels, which would reduce the quality of the final, frozen product more than the slight residual enzyme activity does. Blanching with microwave energy, to rapidly heat the center of large items before the surfaces are overcooked, can be effective in applications such as this.

Canning

Large quantities of vegetable products are canned. A typical flow sheet for a vegetable canning operation (which also largely applies to fruits) is shown in Fig. 18.7. The unit

Table 18.4. Blanching Time for Vegetables for Freezing

Vegetable	Blanching Time in Water at 100°C (min)
Asparagus	
small (5/16 in. diam or less at butt)	2
medium (6/16 to 9/16 in. diam at butt)	3
large (10/16 in. diam and larger at butt)	4
Beans, green and wax	
small (less than 5/16 in. diam or Sieve No. 2 and smaller)	1–1½
medium (5/16–6/16 in. diam or No. 3 and 4)	2–3
large (6/16 in. diam and larger or No. 5 and larger)	3–4
Beets	
small, whole (1¼ in. diam or less)	3–5
diced (1¾ to 2¼ in. diam)[a]	3
Broccoli	
cut into pieces not more than 1 in. thick	2–3
Cabbage (summer)	
chop coarsely	1–1½
Cauliflower	
break into flowerettes or curds not over 2 in. in length by 1½ diam	3
Corn (cut or whole kernel)	
blanch on cob, cool, then cut	2–3
Corn-on-cob	
small (less than 1⅝ in. diam at butt)	7
medium (1⅝ to 2 in. diam at butt)	9
large (over 2 in. diam at butt)	11
Peas	1–1½
Spinach	1½
Swiss chard	2

SOURCE: Canadian Dept. of Agriculture.
[a]Cook in boiling water 2 min, peel, dice and blanch *or* cook through, then dice.

operations performed in sequence include harvesting, receiving, washing, grading, heat blanching, peeling and coring, can filling, exhausting to remove air, sealing, retorting, cooling, labeling, and packing. The vegetable may be canned whole, diced, pureed, as juice, and so on.

HARVESTING AND PROCESSING OF FRUITS

Varietal Differences

As with vegetables, the diversity of kinds of fruit is further enlarged by the numerous cultivars of a given fruit. There are, for example, about 1000 varieties of apples and about 3000 cultivars of pears, but of these only a few are commercially important. Although some fruit is marketed fresh, in many cases more is processed into a wide

Figure 18.6. Steam blanching corn on the cob. *Courtesy of Western Canner and Packer.*

range of products. Varietal differences are particularly important in selecting fruits for use in various products. Apples provide a good illustration. In addition to being eaten fresh, apples are used to make applesauce, canned apple slices, apple juice and cider, jellies, frozen slices, and dried slices. Apple cultivars differ in such properties as resistance to weather, insects, and disease; time of maturity and yield; storage stability; color of flesh; firmness when cooked; amount of juice; acidity; and solids content. For optimum results apple cultivars must be matched to particular end uses, and processing plants frequently are equipped to manufacture the products for which the local apple cultivars are best suited. This is also true of other fruits. Knowledge of varietal differences for the various fruits is highly specific; for this reason when a fruit processing use is contemplated, it is best to consult with state agricultural experiment stations or equivalent agencies.

Fruit Quality

Fruit quality depends on tree stock, growing practices, and weather conditions. More important, however, are the degree of maturity and ripeness when picked and the method of harvesting. There is a distinction between maturity and ripeness of a fruit. Maturity is the condition when the fruit is ready to eat or, if picked, will become ready to eat on further ripening. Ripeness is that optimum condition when color, flavor, and texture have developed to the peak. Some fruits are picked when they are mature but not yet ripe. This is especially true of very soft fruits like cherries and peaches; when fully ripe such fruits are so soft as to be damaged by the act of picking. Further, many

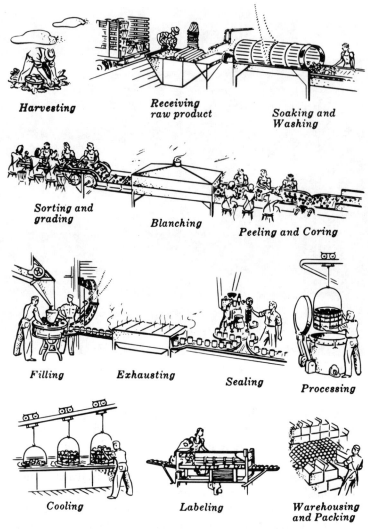

Figure 18.7. Typical vegetable canning operations. *Courtesy of American Can Co.*

fruits that continue to ripen off the tree are likely to become overripe before they can be utilized if picked at peak ripeness.

When to Pick

The proper time to pick fruit depends on several factors: the cultivar, location, weather, ease of removal from the tree which changes with time, and the purpose to which the fruit will be put. In oranges, for example, both the sugar and acid levels change as fruits ripen on the tree (sugars increase and acid decreases). The ratio of sugar to acid determines the taste and acceptability of the fruit and the juice. Since citrus fruits cease ripening once they are picked, the quality of citrus depends largely on harvesting at the proper time. In Florida, there are laws that prohibit picking citrus fruits until a sugar–acid ratio that assures good quality has been reached. In the case of many fruits to be canned, fruits are picked before they are fully ripe in terms of eating texture since canning will further soften the fruit.

Several ripeness classes of honey dew melons are recognized: unripe but mature, ripening initiated, ripe, early senescence, and senescence. Some of the changes in the fruit as it progresses through these stages of ripeness are indicated in Fig. 18.8. The unripe but mature stage is reached about 40 days after flowering, the earliest stage at which honey dews should be picked. If they are picked this early, however, gassing with the plant hormone ethylene is essential for proper subsequent ripening. With fruit picked at the class 2 stage of ripeness, ethylene gassing may be beneficial but is not essential for development of full ripeness. Beyond this stage, the natural production of ethylene by the fruit makes artificial gassing superfluous. Whether on the vine or picked, the fruit subsequently passes through early senescence, when it is still edible but past its prime, and senescence, when it is no longer of edible quality.

Quality Measurements

Many quality measurements can be made before a fruit crop is picked to determine if proper maturity or degree of ripeness has developed. Color may be measured with instruments of the kind discussed in previous chapters, or by comparing color of fruit on the tree with standard picture charts. Because shapes of fruits change as they mature, length and width measurements also can serve as a guide to correct picking time (Fig. 18.9).

Texture may be measured by a compression device such as the simple type of plunger shown in Fig. 18.10, which is pressed into the fruit and gives a reading as the spring contracts. Where individual units of a harvest may vary, fruit may be separated after picking based on texture. For example, firm cranberries bounce, whereas soft, overripe, or rotten ones do not bounce as high. In a cranberry separator (Fig. 18.11), berries are

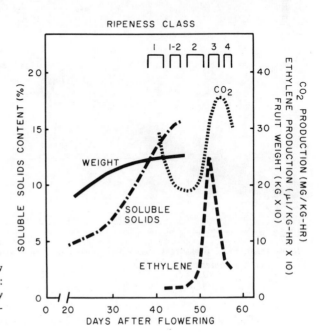

Figure 18.8. Some changes in honey dew melons with ripening at 20°C. Source: *Kasmire et al., Honey Dew Melon Maturity and Ripening Guide, University of California, Davis, 1970.*

Figure 18.9. Researchers study length–width index as guide to maturity of mangos.

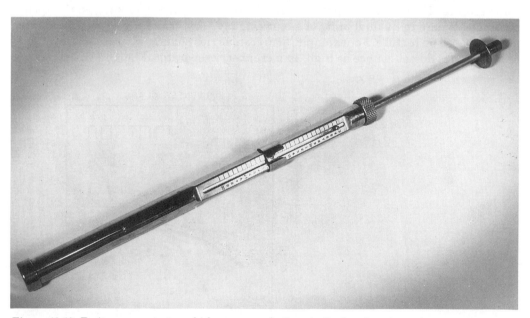

Figure 18.10. Fruit pressure tester which measures fruit maturity based on its resistance to puncture. *Courtesy of A. Kramer and B. E. Twigg.*

Figure 18.11. Separator for eliminating soft cranberries. *Courtesy of National Cranberry Association.*

given a chance to bounce over a wooden barrier: Those that make it are accepted; those that do not are automatically separated and used for less demanding purposes.

As fruits mature on the tree, their concentration of juice solids, which are mostly sugar, changes. The concentration of soluble solids in the juice can be estimated with a refractometer or a hydrometer. The former measures the ability of solutions to bend or refract a light beam, which is proportional to the solution's concentration; a hydrometer is a weighted spindle with a graduated neck that floats in the juice at a height related to the juice density.

The acid content of fruit, as already mentioned, changes with maturity and affects

Table 18.5. Seasonal Changes in Brix, Acid, and Ratios of Grapefruit Juice

	Texas			Florida		
	% Acid	Brix	Ratio	% Acid	Brix	Ratio
November	1.35	10.80	8.00	1.47	10.05	6.84
December	1.45	11.10	7.65	1.44	10.25	7.12
January	1.41	11.35	8.05	1.38	10.25	7.43
February	1.31	11.30	8.62	1.34	10.30	7.69
March	1.18	11.20	9.48	1.29	10.30	7.95
April	1.07	11.00	10.28	1.22	10.15	8.32
May	0.95	10.75	11.32	1.11	9.80	8.84
Average	1.245	11.07	9.05	1.32	10.15	7.79

Courtesy of E. M. Burdick

flavor. Acid concentration can be measured by a simple chemical titration on the fruit juice. But for many fruits the tartness and flavor are determined primarily by the ratio of sugar to acid. Percentage of soluble solids, which are largely sugars, is generally expressed in degrees Brix, which relates specific gravity of a solution to an equivalent concentration of pure sucrose. Therefore, in describing the taste or tartness of fruits and fruit juices, the terms *sugar-to-acid ratio* or *Brix-to-acid ratio* are commonly used. The higher the Brix, the greater the sugar concentration in the juice; the higher the Brix-to-acid ratio, the sweeter and less tart is the juice. The seasonal changes in degrees Brix, acid content, and Brix-to-acid ratio for grapefruit juice are listed in Table 18.5. The relationships of these quantities to quality standards of grapefruit juice are summarized in Table 18.6.

Harvesting and Processing

Much of the harvesting of most fruit crops is still done by hand. This labor may represent about half of the cost of growing the fruit. Therefore, development of mechani-

Table 18.6. Representative Standards for Grapefruit Juices[a]

	% Acid		Degrees Brix		Ratio	
	Min.	Max.	Min.	Max.	Min.	Max.
Fancy or Grade "A"						
grapefruit	0.90	2.00	9.5	None	7:1	14:1
grapefruit "sweet"	1.00	2.00	12.5	[c]	9:1[c]	14:1
blended with orange	0.80	1.70	10.0	None	8:1	17:1
blended "sweet"	0.80	1.70	12.5	[b]	10:1[b]	17:1
Standard or Grade "C"						
grapefruit	0.75	None	9.0	None	6.5:1	None
grapefruit "sweet"	0.85	None	12.5	[c]	9:1	None
blended with orange	0.65	1.80	9.5	None	7.5:1	None
blended "sweet"	0.65	1.80	12.5	[b]	10:1[b]	None

Courtesy of E. M. Burdick.
[a]These standards as established by the USDA are subject to frequent revision.
[b]When the Brix is above 16, the ratio may be less than 10:1.
[c]When the Brix is above 16, the ratio may be less than 9:1.

cal harvesters remains a top priority for agricultural engineers; also important is associated research to breed varieties that produce fruits of nearly equal size that mature uniformly and are resistant to mechanical damage. Mechanical damage can be subtle. For example, cherries that are not to be processed immediately frequently are picked with their stems attached to avoid the small break in the flesh that would allow microbial invasion. This is why fresh cherries with stems are sometimes seen in the supermarket.

Harvested fruit is washed to remove soil, microorganisms, and pesticide residues and then sorted according to size and quality. Sorting techniques have progressed from hand sorting to water sorting, which takes advantage of changes in density with ripening, to sophisticated automatic high-speed sorting in which compressed air jets separate fruit in response to differences in color and ripeness as measured by light reflectance or transmittance. Fresh fruit that is not marketed as such may be processed in many ways, one of the more important being freezing.

Freezing

Large amounts of high-quality fruit are frozen for home, restaurant, and manufacturing use by the baking and other food industries. Freezing is generally superior to canning for preserving the firmness of fruits. As with vegetables, fruit to be frozen must be stabilized against enzymatic changes during frozen storage and on thawing.

The principal enzymatic changes that are objectionable in the case of frozen fruits are oxidations, which cause darkening of color and alterations of flavor. A particularly important color change is enzymatic browning of lighter colored fruits such as apples, peaches, and bananas. This is due to oxidation of pigment precursors, often referred to as catechol–tannin substrates, by enzymes of the group known as phenol oxidases and polyphenol oxidases. Depending on the intended end use for the frozen fruit, various methods are employed to inactivate these enzymes or otherwise prevent oxidation.

Heat Blanching. Fruits generally are not heat-blanched because the heat causes loss of turgor, resulting in sogginess and juice drainage after thawing. Instead, chemicals are commonly used without heat to inactivate oxidative enzymes or to act as antioxidants; these chemicals are combined with other treatments as described in the following sections. An exception exists in the case of fruit slices to be frozen for later use in pies. Since the frozen fruit ultimately will receive heating during the baking operation, heat blanching before freezing is still sometimes practiced. In this case, calcium salts may be added to the blanching water, or added after blanching, to firm the fruit by forming calcium pectates. It also is not uncommon to add pectin, carboxymethyl cellulose, alginates, and other colloidal thickeners to such fruit prior to freezing.

Ascorbic Acid Dip. Ascorbic acid or vitamin C minimizes fruit oxidation primarily by acting as an antioxidant and itself becoming oxidized in preference to the catechol–tannin compounds. Ascorbic acid frequently is added to fruits dissolved in a sugar syrup. Levels of 0.05–0.2% ascorbic acid in an apple-syrup or peach-syrup mixture usually are effective, provided there is time for penetration prior to the freezing step. Peaches so treated may not darken in frozen storage at $-18°C$ in 2 years. Because increased acidity also helps retard oxidative color changes, ascorbic acid and citric acid may be used together. Citric acid further reacts with (chelates) metal ions and, thus, removes these catalysts of oxidation from the system.

Sulfur Dioxide Dip.   Sulfur dioxide (SO_2) is able to stabilize the color of fresh and processed fruits and vegetables. Sulfur dioxide also inhibits the activity of common oxidizing enzymes and has antioxidant properties; that is, it is an oxygen acceptor (as is ascorbic acid). Further, SO_2 reduces nonenzymatic Maillard-type browning by reacting with aldehyde groups of sugars so that they are no longer free to combine with amino acids. Sulfur dioxide also interferes with microbial growth. Inhibition of browning is especially important in dried fruits such as apples, apricots, and pears.

Sulfur dioxide dips have been used to prolong the fresh appearance of cut vegetables and lettuce and other uncooked vegetables in restaurants and salad bars. However, some people are severely allergic to SO_2; thus, the FDA has prohibited the use of SO_2 on fresh produce and has required that uses in processed products be limited to a residual of less than 10 parts per million or appropriately labeled so that sensitive individuals can avoid exposure.

Sugar Syrup.   The addition of sugar syrup, one of the oldest methods of minimizing oxidation, was used long before the browning reactions were understood and remains today a common practice. Sugar syrup minimizes oxidation by coating the fruit and thereby preventing contact with atmospheric oxygen. Sugar syrup also offers some protection against loss of volatile fruit esters, and it contributes sweet taste to otherwise tart fruit. Today it is common to dissolve ascorbic acid and citric acid in the sugar syrup for added effect or to include sugar syrup after an SO_2 treatment.

Vacuum Treatment.   When employed, vacuum treatments generally are used in combination with one of the chemical dips or with the addition of sugar syrup. The fruit submerged in the dip or in the syrup is placed in a closed vessel and vacuum is applied to draw air from the fruit tissue. When the vacuum is broken, the chemical dip or syrup enters the voids from which air was removed, effecting better penetration of the solution.

Concentration and Drying

Some high-moisture fruits may be pureed and concentrated to two or three times their natural solids content for more economical handling and shipping, or the fruits may be dried for various purposes to different moisture levels. Partially dried fruits (e.g., dried apricots, pears, prunes, figs, and raisins) are still largely prepared by sun drying in open wooden trays. When fruits are dried under temperature conditions that do not inactivate oxidative enzymes, SO_2 is commonly employed to minimize browning, as mentioned earlier. The SO_2 also keeps down microbial growth during the slow, low-temperature drying process.

FRUIT JUICES

Perhaps the most commonly processed juice is orange juice. However, several of the steps used in orange juice also are common in the manufacture of other juices, but equipment varies depending on the properties of the different fruits. The main steps in the production of most types of juice are extraction of the juice, clarification of the juice, juice deaeration, pasteurization, concentration (if solids are to be increased), essence add-back, canning or bottling, and freezing if the juice is to be marketed in this form.

Extraction

Juice extractors for oranges and grapefruit, whose peels contain bitter oils, are designed to cause the peel oil to run down the outside of the fruit and not enter the juice stream. Because bitter peel oil is not a problem in the case of apples, the whole apple is pressed after grinding.

Clarification

The juice pressed from most fruits contains small quantities of suspended pulp, which is often removed. This may be done with fine filters, but since these have a tendency to clog, it is common to use high-speed centrifuges, which separate the juice from the pulp according to their differences in density.

Many people prefer crystal clear apple juice. However, simple filtration of centrifugation may leave minute particles of pulp and colloidal materials suspended in the juice by the natural pectic substances of the fruit. Addition of commercial enzyme preparations that digest pectic substances causes the fine pulp to settle, which makes filtering or centrifuging more effective and produces clarified apple juice. Orange juice, on the other hand, is more acceptable if it retains a slight cloud of suspended pulp and so this is not removed.

Deaeration

Orange and other juices contain entrapped air and are deaerated by being sprayed into a vacuum deaerator (Fig. 18.12). This minimizes subsequent destruction of vitamin C and other changes due to oxygen.

Additional Steps

Generally, fruit juices are pasteurized to decrease microbial growth and to inactivate natural enzymes. All natural juices are low in solids, and so it is common to concentrate many of them whether they are to be frozen or not. When this is done, low-temperature vacuum evaporation generally is employed to retain maximum flavor. Nevertheless, removal of water is always accompanied by the evaporation of some of the juices' volatile essences. Therefore, the evaporated water and essence coming from the vacuum evaporator is not discarded but is passed through an essence-recovery unit. Such units distill the essence from the water and recondense it. The essence is then added back to the concentrated juice to enhance flavor. Various methods of concentrating juices, including pumping them through reverse-osmosis membranes (Fig. 18.13), are currently being studied. Reverse osmosis can be less costly than vacuum evaporation; further, special membranes can retain the essence in the juice concentrate since many essence molecules are too large to pass through them. Membrane concentration requires

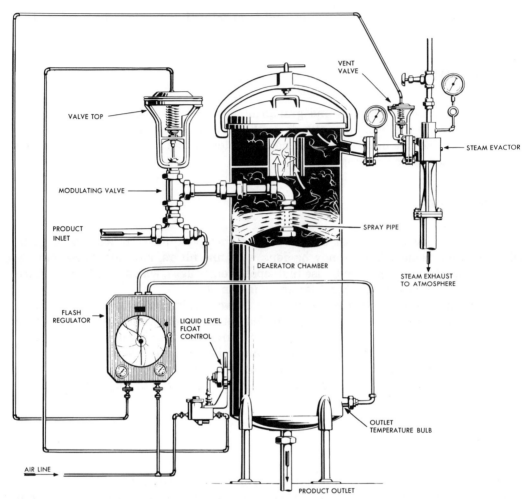

Figure 18.12. Rex microfilm deaerator for removing dissolved oxygen from juices and other liquids. *Courtesy of FMC Corp.*

that pulp first be centrifuged from the juice to prevent membrane clogging. Pulp can then be added back to the concentrated juice. The juice concentrate may then be frozen, or it may be shipped for subsequent reconstitution and packaging as single-strength juice.

In recent years it has become popular to blend different juice types together to form new beverages. Mango may be blended with apple and cranberry or citrus juices with raspberry. Pear juice is a popular base for many juices because it has a strong fruit flavor but is not highly characteristic of any one fruit.

Many juices and juice blends are rich in vitamin C. Apple juice, which normally is low in vitamin C, may be fortified with this vitamin. There is increasing demand for high quality juice as a nutritious beverage and much is now being aseptically packaged in paper cartons of individual serving size.

Whenever fruit is processed or juice produced, there remain peels, pits, and other nonjuice solids. Some of this finds its way into confectionery and jelly products, pectin manufacture, recovery of chemicals, and animal feeds.

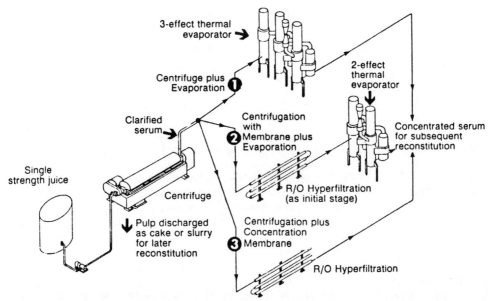

Figure 18.13. Various schemes for concentrating juices. *Courtesy of T. C. Swafford, Alfa-Laval, Inc.*

BIOTECHNOLOGY

For many decades, plants have been bred or mutated in order to produce offspring which have desirable characteristics such as disease and insect resistance or eating and processing qualities. This is done by selective breeding or direct mutation, both of which alter the genetic makeup of the plant by altering its DNA. It is this breeding that gives us so many types of apples, for example.

In recent years, scientists have learned how to more directly influence the genetics of plants by inserting specific pieces of DNA into a plant's DNA. Thus, inheritable and specific properties or traits can be incorporated into the genetics of plants. Desirable traits such as resistance to plant diseases and viruses or quality and storage attributes can be specifically "engineered" into plants.

A couple of examples illustrate the usefulness of such techniques. Squash, like other plants, are commonly infected by a pathogenic virus that causes decreased yields and quality. By incorporating a portion of the DNA from the virus into the squash's DNA, the plant acquires resistance to the virus. Another example comes from a bacterium which is pathogenic to certain insects. This bacterium is known as *Bacillus thuringiensis* and produces a protein which is quite toxic to insects but harmless to mammals. Insertion of the DNA which tells the bacterium to make this protein into a plant gives the plant the ability to form the protein as a normal part of growing. When the protein is produced by the plant, insects consume the plant, they also consume some of the toxic protein. Thus, insects can be controlled without the application of pesticides. It can be expected that the use of biotechnology to improve crops will accelerate in the next decade. This will likely result in increased crop yields and improved quality.

References

Addy, N.D. and Stuart, D.A. 1986. Impact of biotechnology on vegetable processing. Food Technol. *40*(10), 64–66.

Arthey, D. and Ashurst, P. 1995. Fruit Processing, Chapman & Hall, London, New York.

Arthey, D. and Dennis C. 1991. Vegetable Processing. Chapman & Hall, London, New York.

Ashkar, A. 1993. Quality Assurance in Tropical Fruit Processing. Springer-Verlag, New York.

Dennis, C. 1983. Post-Harvest Pathology of Fruits and Vegetables. Academic Press, London.

Downing, D.L. 1989. Processed Apple Products. Chapman & Hall, London, New York.

Gould, W.A. 1992. Tomato Production, Processing & Technology. 3rd ed. CTI Publications, Baltimore, MD.

Jagtiani, J. 1988. Tropical Fruit Processing. Academic Press, San Diego, CA.

Kader, A.A. 1992. Postharvest Technology of Horticultural Crops. 2nd ed. University of California Division of Agriculture and Natural Resources, Oakland, CA.

Kays, S.J. 1991. Postharvest Physiology of Perishable Plant Products. Chapman & Hall, London, New York.

Lecos, C.W. 1988. An order of fries hold the sulfites. FDA Consumer *22*(2), 8–11.

Luh, B.S. and Woodroff, J.G. 1988. Commercial Vegetable Processing, 2nd ed. Chapman & Hall, London, New York.

Nagy, S. and Attaway, J.A. 1980. Citrus Nutrition and Quality. Based on a symposium sponsored by the Division of Agricultural and Food Chemistry at the 179th meeting of the American Chemical Society. American Chemical Society, Houston, TX.

Nagy, S., Shaw, P.E., and Wardowski, W.F. 1990. Fruits of Tropical and Subtropical Origin. Composition, Properties, and Uses. Florida Science Source, Lake Alfred, FL.

Nelson, P.E. and Tressler, D.K. 1980. Fruit and Vegetable Juice Processing Technology. AVI Publishing Co., Westport, CT.

Pattee, H.E. 1985. Evaluation of Quality of Fruits and Vegetables. AVI Publishing Co., Westport, CT.

Peleg, K. 1985. Produce Handling, Packaging, and Distribution. AVI Publishing Co., Westport, CT.

Richardson, D.G. and Meheriuk, M. 1982. Controlled Atmospheres for Storage and Transport of Perishable Agricultural Commodities. Timber Press in Cooperation with School of Agriculture, Oregon State University, Beaverton, OR.

Salunkhe, D.K., Bolin, H.R., and Reddy, N.R. 1990. Storage, Processing, and Nutritional Quality of Fruits and Vegetables. 2nd ed. CRC Press, Boca Raton, FL.

Smith, O. 1987. Transport and storage of potatoes. *In* Potato Processing. 4th ed., W.F. Talburt and O Smith (Editors). Chapman & Hall, London, New York.

Talburt, W.F. and Smith, O. (Editors). 1987. Potato Processing. 4th ed. Chapman & Hall, London, New York.

Wardowski, W.F., Nagy, S., and Grierson, W. 1986. Fresh Citrus Fruits. Chapman & Hall, London, New York.

Woodroff, J.G. and Luh, B.S. 1986. Commercial Fruit Processing. 2nd ed. Chapman & Hall, London, New York.

19

Beverages

Some beverages are consumed for their food value (e.g., milk), yet others are consumed for their thirst-quenching properties, for their stimulating effects, or simply because consumption is pleasurable. This chapter will discuss three major groups: carbonated nonalcoholic beverages or soft drinks of which "soda pop" is characteristic; carbonated or noncarbonated mildly alcoholic beverages such as beer and wine; and nonalcoholic, noncarbonated stimulating beverages such as coffee and tea.

Each of these beverages must be considered foods in the broad sense, since all are made from food ingredients, all are subject to food laws and regulations, and all are consumed in large quantities. The annual per capita consumption of soft drinks in the United States in 1991 was 43 gal according to the U.S. Department of Agriculture, which was the highest of any category of beverage and well above the per capita consumption of milk (Table 19.1). It is likely that soft drink consumption exceeds even tap water. In other countries and areas, packaged beverages may be safer to consume than the local water supply. Further, beer, wine, and carbonated soft drinks (with the exception of dietetic formulations) furnish calories; coffee and tea, although noncaloric, frequently are consumed with cream or sugar and, thus, are vehicles of caloric intake. The technologies of each of these beverages and their ingredients are comprehensive topics in themselves.

CARBONATED NONALCOHOLIC BEVERAGES

Carbonated nonalcoholic beverages are generally sweetened, flavored, acidified, colored, artificially carbonated, and sometimes chemically preserved. Their origin goes back to Greek and Roman times when naturally occurring mineral waters were prized for "medicinal" and refreshing qualities. But it was not until about 1767, when the British chemist Joseph Priestley found that he could artificially carbonate water, that the carbonated beverage industry got its start. An early method of obtaining the carbon dioxide was by acidification of sodium bicarbonate or sodium carbonate, and from the use of these sodium salts came the name "soda" which remains today, although most carbon dioxide is no longer generated in this fashion. Gradually, fruit juices and extracts were added to carbonated water for improved flavor.

Ingredients and Manufacture

The major ingredients of carbonated soft drink beverages in addition to water and carbon dioxide are sugar, flavorings, colors, and acids. Typical levels of sugar, carbon

Table 19.1. Estimated Annual Per Capita
Beverage Consumption by U.S. Residents

	Gallons/Person/Year
Soft drinks	43
Coffee	28
Milk	26
Fruit juices	7.8
Tea	5.5

SOURCE: ERS-USDA, 1991. Food and Nutrient Consumption Food Review *14*(3) 2–18.

dioxide, and acidity for various beverages are given in Table 19.2, although the products of different manufacturers may vary somewhat from these values.

Sugar

The most common sugar used in soft drinks is high-fructose corn syrup or related corn sugars. Initially, sucrose, purchased as a pure colorless syrup from the manufacturer or made into a syrup at the beverage plant from high-purity crystalline sugar, was most commonly used and is still widely used. Increasingly, however, sucrose has been replaced with high-fructose corn sugars which are sweeter and, thus, less costly on an equal-sweetness basis. The corn sugar (or sugar syrup) is supplemented with flavoring, coloring, and acidic ingredients and may be stabilized with a preservative. Finished beverages contain about 8–14% sugar. The sugar not only contributes sweetness and calories to the drink but also adds body and mouthfeel. For this reason when dietetic beverages are made with a non-nutritive or low-calorie sweetener to replace all or much of the sugar, an agent such as carboxymethyl cellulose or a pectin is sometimes added to give the same mouthfeel as the sugar product.

Reduced Calorie and Non-Nutritive Sweeteners

Soft drinks which provide no calories are sweetened with non-nutritive sweeteners such as saccharin, Acesulfame K, or cyclamate, whereas reduced-calorie soft drinks

Table 19.2. Composition of Carbonated Beverages

Flavor	Sugar °Brix	Carbonation Gas Volume	Acid %	pH
Cola flavors	10.5	3.4	0.09	2.6
Root beer	9.9	3.3	0.04	4.0
Ginger ales	9.5	3.8	0.10	—
Cream (vanilla)	11.2	2.6	0.02	—
Lemon and lime	12.6	2.4	0.10	3.0
Orange	13.4	2.3	0.19	3.4
Cherry	12.0	2.4	0.09	3.7
Raspberry	12.3	3.0	0.13	3.0
Grape	13.2	2.2	0.10	3.0

have sweeteners that have calories but also are high-intensity sweeteners. This means that considerably less sweetener must be used to get the same degree of sweetness, so the drink ends up with fewer calories. For example the common artificial sweetener aspartame (trademark NutraSweet) is a dipeptide that yields 4 kcal/g—the same as sugar—but is about 150–200 times sweeter than sugar (sucrose) and so can sweeten in very small amounts. Thus, it is a nutritive sweetener but contributes very few calories.

Flavorings

Synthetic flavor compounds, natural flavor extracts, and fruit juice concentrates are used to flavor soft drinks. These flavors must be stable under the acidic conditions of the beverage and on exposure to light for a year or more, since bottled drinks may be held this long or longer. The flavors do not have to be stable to heat much over 38°C, since beverages are not commonly heat-sterilized or pasteurized.

An artificial fruit flavor made from synthetic flavor compounds and natural flavor extracts (Table 19.3) may contain over two dozen components contributing several hundred distinct compounds. Cola flavors may be as or more complex, and their compositions are guarded secrets, sometimes formulated to contain ingredients that will add to the difficulty of chemical analysis and duplication by competitors. Cola flavors may contain a source of caffeine, which is a mild stimulant. There also is a growing market for caffeine-free colas. When fruit derivatives that contain flavor oils are used, it is necessary to employ an emulsifying agent to keep the oils from separating out in the beverage. Water-soluble gums at low levels are the principal emulsifiers employed for this purpose.

Colors

Some important coloring agents for soft drinks are the synthetic colors, particularly U.S. certified food colors, which have been approved by the Food and Drug Administration. All certified batches of such colors must meet stringent chemical purity standards in their manufacture. Caramel from heated sugar, a nonsynthetic color, is also commonly used in dark beverages such as colas. These coloring materials are much preferred to the natural fruit colors because of their greater coloring power and color stability. Even when natural fruit extracts or juices are used, their colors are generally supplemented with the synthetic colors.

Acid

Carbon dioxide in solution contributes to acidity, but this is supplemented with additional acid in most carbonated drinks. The main reasons for acidification are to enhance beverage flavors and to act as preservatives against microbial growth. The principal acids used are phosphoric, citric, fumaric, tartaric, and malic acids. Citric, tartaric, and malic are important natural acids of fruits and so they are used, along with fumaric acid, mainly in fruit-flavored drinks, with citric being the most widely employed. Phosphoric acid is preferred for use in colas, root beer, and other nonfruit drinks.

In addition to flavor enhancement, acid acts as a preservative in non-heat-treated beverages. However, unless a very high degree of sanitation is employed in soft drink

Table 19.3. Raspberry Flavor Formulation

Ingredient	Parts
Ethyl methylphenylglycidate	400
Benzylidene isopropylidene acetone	100
Methoxyacetoxyacetophenone	60
Benzyl acetate	50
Phenethyl alcohol	50
Essence of Portugal	50
Isobutyl acetate	40
Vanillin	30
Methylionone	25
beta-Ionone	25
Coumarin substitute	10
Iris concrete essence	15
Ethyl acetate	10
Ethyl caproate	10
Isoamyl caproate	10
Hexanyl acetate	10
Hexenyl acetate	10
Methyl salicylate	10
Ethyl benzoate	10
Methyl butanol	10
Bornyl salicylate	10
Essence of clove	10
Essence of geranium	10
Hexyl alcohol	5
Hexenol	5
Anisaldehyde	5
Benzaldehyde	5
Acetylmethylcarbinol	3
Biacetyl	2

Courtesy of L. Benezet.

manufacture, the pH imparted by the acid, even in combination with acidic fruit juices, is not sufficient to ensure long-term microbial stability. For this reason, an additional preservative may be necessary; the most common is sodium benzoate at a level of about 0.03–0.05% in the final beverage. In the acid drink, sodium benzoate is converted to benzoic acid, which is more effective as a preservative.

Water

The major ingredient in carbonated soft drinks—accounting for as much as 92% by volume—is water. It is essential that the water be as nearly chemically pure as is commercially feasible, since traces of impurities react with other constituents of the drink. In this respect, municipal drinking water, although satisfactory from a bacteriological standpoint, generally is not chemically pure enough for use in soft drinks. The standards for beverage water listed in Table 19.4 would not be met by most municipal water supplies.

Table 19.4. Laboratory Standards for Water to Be
Used in Preparing Fruit Juice Beverages

	Maximum
Alkalinity	50 ppm
Total solids	500 ppm
Iron	0.1 ppm
Manganese	0.1 ppm
Turbidity	5 ppm
Color	Colorless
Residual chlorine	None
Odor	None
Taste	No off-taste
Organic matter	No objectionable content

Courtesy of G. F. Phillips.

The alkalinity of beverage water must be low to prevent neutralization of the acid used in the beverage, which would alter flavor and decrease the preservative property of the beverage. Iron and manganese must be low to prevent reaction with coloring agents and flavor components. Residual chlorine must be virtually nonexistent since it adversely affects the flavor of the drink. Turbidity and color must be low for an attractive appearance of the drink. Organic matter as well as inorganic solids must be low since colloidal particles provide nuclei for carbon dioxide accumulation and release from solution, which results in beverages boiling and gushing when containers are filled or opened.

To achieve these high water standards, bottling plants generally condition water with additional treatments such as chemical precipitation of minerals, deionization, addition of activated charcoal to remove odors, flavors, and residual chlorine, final paper filtration to remove traces that may pass the carbon filter, and deaeration to remove oxygen. Although the water supply in a bottling plant can be adequately controlled by these methods, the big problem occurs when the syrups and flavor bases are shipped to various locations to be used in fast-food restaurants and vending machines. In these locations the quality of the water will vary and frequently not meet the tight specifications of the bottling plant. The quality of the drink may suffer and vary from location to location even though the syrup formula is constant.

Carbon Dioxide

The sparkle and zest of carbonated beverages stems from the carbon dioxide gas. Carbon dioxide can be obtained from carbonates, limestone, the burning of organic fuels, and industrial fermentation processes. Soft drink bottlers buy carbon dioxide in high-pressure cylinders from manufacturers who produce the gas to comply with food purity regulations. In the cylinders, the gas under pressure exists as a liquid. The amount of CO_2 used in beverages depends on their particular flavor and brand. CO_2 improves flavor, contributes acidic preservative action, produces tingling mouthfeel, and gives the sparkling effervescent appearance to the beverage.

The amount of carbon dioxide in beverages is measured in volumes of gas per volume of liquid. A volume of gas is the volume occupied by the gas under standard temperature and pressure. Thus, a beverage containing 2 liters of CO_2 (at STP) per liter of beverage is carbonated at 2 volumes. Most beverages are carbonated in the range of about 1.5–4 volumes. This is done with a carbonator, of which there are several designs. In all, however, carbonation is speeded by providing intimate contact between the liquid and the CO_2 gas, cooling the liquid since the solubility of CO_2 in water is greater the lower the temperature, and applying pressure to force more CO_2 into solution. In practice, the entire flavored drink may be carbonated, or only the water may be carbonated for subsequent mixture with the flavored syrup.

Plant Layout

A common installation for a soft drink mixing, carbonating, and bottling operation is outlined in Fig. 19.1. Flavored syrup containing all of the drink ingredients except the remaining water and CO_2 is pumped to a metering device called a synchrometer. Treated and deaerated water also is pumped to the synchrometer. This device then meters the syrup and water in fixed proportion to the carbonator. The carbonated beverage then goes to the bottling or canning line where it is admitted to sanitized containers under a CO_2-pressurized atmosphere to prevent loss of CO_2 and beverage boiling. The containers are then capped; they are not subsequently heat-treated. In restaurants and similar establishments, concentrated syrups containing the sweetener and flavors is directly mixed with carbonated water as the drink is being drawn. Syrup from the beverage manufacturer is held in one tank and pressurized carbon dioxide in a second tank.

BEER

Brewing is a general term for the hot water extraction of plant materials. Thus, making coffee or tea is brewing. Brewing is a critical step in making beer and the entire process is termed brewing, but this does not completely describe the many steps

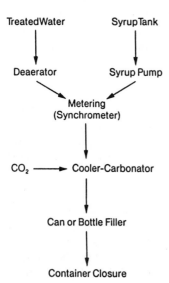

Figure 19.1. Soft drink mixing, carbonating, and bottling sequence.

involved. Brewing of beer goes back over 6000 years, and today's practices are similar to those used in earliest times. What has been gained is an understanding of the principles of biochemistry and microbiology underlying the beer-making process and a high degree of sanitation and efficiency in manufacturing practices.

Raw Materials and Manufacture

The principal raw materials of beer manufacture are water, hops, and malted cereal grains, principally barley. In many cases, rice, corn, or other unmalted grains are also added as sources of additional or "adjunct" carbohydrate for fermentation by *Saccharomyces* yeast into ethyl alcohol and carbon dioxide. Hops are used to add the characteristic flavor of most beers, and additional carbon dioxide may be added to the amount naturally produced by fermentation.

Malt

The most important ingredient in beer-making from the standpoint of quantity and function is barley malt. Malt is barley grain that has been germinated to the point were roots and stems just begin to appear. The green malt is then gently dried to stop growth yet leave the enzyme activity intact. Germination results in activation of enzymes which convert starches in the malted barley and in other cereal grains into sugars, which can be easily fermented by yeast during the fermentation step. This is necessary because yeast cannot utilize the starch in the cereal grains for conversion to ethanol and CO_2.

Hops

Hops are plants, the flowers of which contain resins and essential oils that contribute a characteristic bitter flavor and pleasant aroma to beer. Hops also contain tannins, which add to beer color. Hops are added during brewing and after the enzymes of the malt have converted the starch to the sugar maltose. Hops also have mild preservative properties and add foam-holding capacity to the beer. All of these functions, however, are secondary to the role of hops in flavor and aroma.

Cereal Adjuncts

Corn, rice, and other cereals are used in beer-making to provide supplemental carbohydrates, principally starch, for conversion to sugar for subsequent fermentation. Without these adjunct cereals, the limiting nutritional factor for yeast in fermentation would be protein. This means that carbohydrate would remain after fermentation and produce a heavier type beer. In some cases, this is desirable, but most breweries prefer lighter-type beers.

Mashing

The first step in beer-making is to combine the malted barley and cereal adjuncts with water and mildly cook the mixture, known as mash, to extract readily soluble materials and to gelatinize the starches, thus making them more susceptible to extraction and enzymatic breakdown into dextrins and maltose. The mild cooking also releases proteins from the grains; these proteins also undergo enzymatic breakdown into compounds of lower molecular weight.

These changes are brought about in specially designed vessels, and the overall operation is known as mashing. Mashing may begin at about 38°C with the temperature gradually raised to about 77°C. This heating is done in steps with rest periods of about 30 min between each temperature increment; this stepwise heating permits specific amylases and proteinases to function before they are heat inactivated. The mash vessel (called a "tun") is designed so that on completion of mashing, the liquid fraction, now high in yeast-fermentable sugars, can be separated from the spent grains. The liquid fraction is known as "wort."

Brewing

The liquid is next pumped to the brew kettle. The hops are added to the wort, and the mixture is brewed by boiling in the kettle for about 2.5 h. After brewing, the hops residue is allowed to settle and the wort is drawn from the kettle through the bed of hops, which partially filters the wort. The wort is then cooled, the solids allowed to precipitate, and it is ready for fermentation.

The boiling or brewing of the wort with the hops serves several purposes. It concentrates the wort, nearly sterilizes it, inactivates enzymes, precipitates remaining proteins that would otherwise contribute to beer turbidity, caramelizes sugars slightly, and extracts the flavor, preservative, and tanninlike substances from the hops.

Fermentation

The cooled wort is inoculated with *Saccharomyces* yeast and fermentation of the sugar formed from starch during mashing proceeds. Fermentation in tanks, under near-sterile conditions with respect to contaminating microorganisms, is carried out at temperatures from about 3 to 14°C, depending on the strain of yeast and the brewery. The fermentation is complete in about 9 days. It produces an alcohol content in the wort of about 4.6% by volume, which would be 9.2 proof. Fermentation also lowers the pH of the wort to about 4.0 and produces dissolved carbon dioxide in the wort to the extent of about 0.3% by weight.

Storage

After fermentation is complete, the beer is quickly chilled to 0°C, passed through filters to remove most of the yeast and other suspended materials, and pumped into pressure storage tanks. The young or "green" beer is stored in these tanks for several weeks to several months. This storage is known as "lagering." During this period of storage at 0°C there is further settling of finely suspended proteins, yeast cells, and other remaining materials, and development of esters and other flavor compounds, all of which contribute to improved body and a more mellow flavor.

Generally, additional carbon dioxide is added to the beer during storage to increase the level developed and absorbed during fermentation and to purge the beer of any oxygen that may be present and would adversely affect storage life. This may be done by periodically pumping the beer through a carbonator or bubbling carbon dioxide into the storage tank.

Chill haze is a condition caused by remaining traces of degraded proteins and tannins that form a colloidal haze when beer is cooled to low temperatures. To prevent this from occurring in the finished product, various chill-proofing treatments may be given to the beer during the storage period. These generally include the addition of earths or clays to adsorb the colloidal materials or use of proteolytic enzymes to further solubilize the protein fraction.

Finishing and Packaging

After storage, the beer is given a final "polishing" filtration to remove traces of suspended materials and give the beer a crystal clear appearance. Additional CO_2 may be added and the beer packaged. Analysis of beer at this point shows that it is quite complicated (Table 19.5).

Although clear, the beer is not sterile and a few viable yeast cells and low levels of fermentable sugars remain in the product. These yeasts and other microorganisms could continue to grow during storage and produce considerable pressure within the bottles when stored at room temperature. Beer is, therefore, pasteurized at a temperature of 60°C for several minutes after packaging. In the case of beer which is packaged in kegs (so-called draft beer), it is held under refrigeration and does not need pasteurization. Because it is not pasteurized, draft beer has a better flavor than pasteurized beer.

Beer may also be microbiologically stabilized by filtration processes which are fine enough to remove residual yeast and bacteria (Fig. 19.2). This is called "cold pasteurization" or filtration and achieves microbiological stability without the heat of conventional pasteurization. Similar cold pasteurization processes can be applied to other products such as fruit juices and wines.

Light Beer

Light beer contains about one-third to one-half fewer calories than regular beer and also less alcohol. It is prepared from a mash lower in solids than that used for regular beer. Its alcohol content can be further modified by changing the ratio of fermentable to nonfermentable solids in the mash. Enzymes can be added to further break down any remaining starches that would not be fermentable and thus contribute to caloric content. The various operations and quality control activities in the manufacture of beer are given in Fig. 19.3.

WINE

Like beer-making, the fermentation of grapes to make wine goes back to at least 4000 B.C. In 1991, annual worldwide industrial production of wine was 29 million metric tons, of which more than one-third was produced in France and Italy. Annual wine production in the United States is about 2.6 billion liters. The principal U.S. wine-producing regions are California, the Finger Lakes area of New York, and the Pacific Northwest, although wines are now made in many regions of the United States. Wine can be made from many fruits and berries, but the grape is by far the most popular and the most often used raw material.

Table 19.5. Analysis of a Typical Beer

Specific gravity	1.0121
Beer balling, %	3.093
Saccharometer, %	3.10
Alcohol by weight, %	3.63
Alcohol by volume, %	4.60
Real extract, %	4.73
Extract of original wort (2A + E), %	11.99
Original balling, %	11.80
Reducing sugars, %	1.160
Degree wort sugar, %	71.30
Degree attenuation, % (real degree of fermentation)	60.00
pH	4.35
Color,° L	2.94
Air, cm^3	1.20
Nitrogen, cm^3	1.01
Oxygen, cm^3	0.19
Oxygen/air, %	15.80
CO_2 % by wt	0.460
Acidity, %	0.135
Erythro-dextrins (iodine reaction)	0
Amylo-dextrins (iodine reaction)	0
Dextrins, %	2.73
Iron, ppm	0.175
Indicator-Time test, sec	290.0
Surface tension, dynes	46.00
Surface activity	0.367
$CaSO_4$, ppm	256.00
NaCl, ppm	153.00
Foam sigma	109.00
Foam density	20.40
Tannins, ppm	55.40
Viscosity, cP	1.057
SO_2, ppm	13.20
Ash, %	0.148
Diacetyl, ppm	0.210
Copper, ppm	0.245
Fractional carbohydrates, %	
glucose	0.001
fructose	Trace
sucrose	Trace
maltose	0.10
maltotriose	0.20
maltotetraose	0.45
higher saccharides	3.04
total saccharides	3.83
Fractional proteins, %	
total protein	0.299
high molecular	0.0710
medium molecular	0.100
low molecular	0.0951
nonprotein N	0.0290
high/total	23.30
medium total	33.20
low total	31.60
nonprotein/total	9.6
Calories (kcal/400 ml)	168.30

SOURCE: Ohlmeyer and Matz (1970).

Figure 19.2. Filtration assembly housing microporous membrane disks ahead of canner. *Courtesy of Millipore Corporation.*

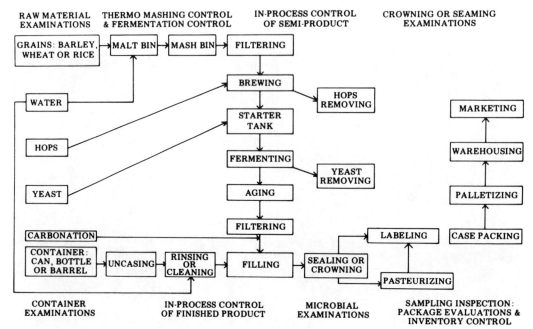

Figure 19.3. Operations and quality control activities in the manufacture of beer. Source: *Woodroof and Phillips Beverages: Carbonated and Noncarbonated, rev. ed. AVI Publishing Co., Westport, CT, 1981.*

Wine Varieties

The varieties and names given to wines are legion and reflect their region of origin, varieties of grape used in their manufacture, and certain properties such as degree of sweetness, color, alcohol content, and effervescence. In the United States, wines can be grouped into five classes—appetizer wines, red table wines, white table wines, sweet dessert wines, and sparkling wines. Although these terms are somewhat descriptive, greater insight is gained by considering some of the general characteristics of grape wines and how they are brought about.

Color

Grape varieties range in skin color from deep purple through red to pale green. Red wines result when the crushed grape skins, pulp, and seeds of purple or red varieties are allowed to remain with the juice during the fermentation period. The alcohol produced contributes to pigment extraction, and the longer the skins, pulp, and seeds are allowed to stay with the fermenting juice, the deeper the color becomes. Pink or rosé wines can be produced by removing the nonjuice "pumace" from the liquid or "must" early in the fermentation period. Thus, only a small amount of pigment is extracted. White wines can be made from pigmented grapes by removal of the skins, pulp, and seeds before juice fermentation, by ion exchange and activated charcoal treatments to remove pigment, and by the use of anthocyanase enzymes, which decolor pigments. White wines are also made from white varieties of crushed grapes with removal of nonjuice solids prior to fermentation. Pink wines can also be prepared by blending white wines with small amounts of red wines. Final wine color is determined also by pigment stability during storage, which is dependent to a considerable degree on grape variety.

Sweetness and Alcohol Content

The sweetness and alcohol content of wines are interrelated because fermentation converts the grape sugars to ethanol. As more alcohol is produced, sweetness decreases; when virtually all of the sugar is fermented, the wine is without sweetness and is said to be "dry." Dry wines contain all of the alcohol that the specific grape is capable of yielding under the conditions of fermentation. This generally is 12–14% alcohol by volume.

The relationship between disappearance of sweetness and increase in alcohol content cannot be used to characterize wines, however, because both alcohol content and sweetness of finished wines can be further and independently adjusted. Thus, a completely fermented dry wine of 14% alcohol can be made sweet after yeast removal by the addition of some unfermented juice or sugar. Similarly, a sweet wine does not necessarily mean that the alcohol content is low since additional alcohol can be added to a sweet wine in the form of distilled spirits. The wines in the U.S. designated table wines and sparkling wines generally contain 10–14% alcohol by volume; appetizer and sweet dessert wines contain 14–21% alcohol. Any of these classes may be red or white and possess varying degrees of sweetness. The terms "natural" and "fortified" also have been used in relation to alcohol content. Depending on the sugar content of the grapes, characteristics of the yeast culture, and fermentation practices employed, natural fermentation generally yields an alcohol concentration of less than 16% by volume even if more sugars are added. This is because this amount of alcohol is toxic to the yeast and it stops fermentation. The wine must often still be pasteurized after bottling to ensure against growth of unwanted microorganisms.

The term *light* wine also is used to describe a wine having an alcohol content from about 5% to 10% (the term light has nothing to do with color). Fortified wines are those that have received additional distilled spirits to bring their alcohol content up to 17–21% by volume. They are less perishable and may be stable without pasteurization.

Effervescence

Wines are termed *still* or *sparkling* depending on the amount of CO_2 they contain. The CO_2 occurs naturally as a result of fermentation (natural sparkling wines) but also can be added artificially ("carbonated wines").

During normal vat fermentation, not enough CO_2 is retained by the wine to give it effervescence when bottled. Natural sparkling wines are made by adding about 2% sugar and a special alcohol-tolerant strain of wine yeast to previously fermented wine. This causes a second fermentation under conditions that prevent CO_2 loss. This second fermentation may be in the final bottle or in closed tanks from which the wine is subsequently filtered and bottled.

The preparation of bottle-fermented wines such as champagne involves an interesting technique for removing yeast, tartrates, and other fine particles that settle in the sealed bottle and would otherwise cloud the finished product. After secondary fermentation that may last a month, the tightly stoppered bottles may be further rested for periods up to several years. Removal of sediment then involves placing the bottles neck-down in racks and periodically twirling them so the sediment moves down the neck toward the special stopper. Next the sediment-containing wine near the stopper is frozen by placing the necks of the inverted bottles in a refrigerant. The bottle is now set right side up and the stopper loosened; carbon dioxide pressure below the frozen plug forces the plug containing the sediment out of the bottle. A small amount of wine or champagne is added to make up volume and the bottle is reclosed with its permanent cork.

Fermentation and Other Operations

As grapes mature, the wine yeast *Saccharomyces ellipsoideus* naturally accumulates on the skins. When the crushed grapes or filtered juice is placed at a temperature of about 27°C, the juice proceeds to ferment, yielding essentially equal molar quantities of ethyl alcohol and CO_2 and traces of flavor compounds.

In commercial operation, special strains of *S. ellipsoideus* are used to supplement the natural inoculum and better control fermentation. Wine yeast is relatively resistant to SO_2 and so this agent commonly is added to the grapes or must to help control undesirable microorganisms, particularly bacteria. Sulfur dioxide also is effective in inhibiting browning enzymes of the grapes and providing reducing conditions by reacting with oxygen. The SO_2-treated must may next be fermented directly or after pumace removal. Fermentation causes a rise in temperature, and so cooling is required to prevent yeast inactivation. Fermentation under conditions of limited exposure to air may continue until the sugar is entirely consumed, when it stops naturally, or fermentation may be interrupted prior to this point. At around 27°C, fermentation may last for some 4–10 days depending on wine type.

After fermentation has been completed naturally or stopped by addition of distilled spirits, the next step is the first "racking," which involves allowing the wine to stand until most of the yeast cells and fine suspended materials settle out. The wine is then drawn off without disturbing the sediment or "lees." If lees are not quickly removed, yeast will autolyze and contribute off-flavors to the wine. After the first racking, the wine may be further aged in casks or tanks that prevent entrance of air for periods of several months to years, during which last traces of sugar ferment and flavor further develops. During ageing, additional rackings may be performed; these are followed by final clarification and stabilization treatments to produce brilliantly clear wines.

In addition to filtration or centrifugation of last traces of colloidal materials that impair clarity, stabilization also requires removal of the salts of tartaric acid. These tartrates, present in grape juice, tend to crystallize in wine casks, and if not completely removed from the wine before bottling, they slowly reappear as glasslike crystals in the final bottles on storage. Stabilization with respect to tartrates may involve chilling to promote crystallization for efficient removal, or removal of these salts by ion exchange treatments.

If a wine is not above 17% alcohol, it may be heat-pasteurized, or cold-pasteurized through microporous membrane filters, just before bottling. Sparkling wines, whether secondary fermentation is carried out in bottles or in bulk, are not heat-pasteurized even though they generally contain not more than 14% alcohol. In this case, depletion of nutrients from the previous double fermentation, a high concentration of carbon dioxide in solution, extreme cleanliness, and sometimes SO_2 addition before bottle closure all help to make microbial growth unlikely.

Naming of Wines

Originally, wines were named for the region where the grapes were grown and the wine was produced. Such famous names as sherry originated in Jarez, Spain; port came from Oporto, Portugal; champagne from the district of Champagne near Paris; Chablis and Burgundy from districts to the south of Champagne; sauterne from Bordeaux in western France; Rhine wines from districts along the German Rhine; Marsala from Sicily; Chianti from the Italian district of Tuscany; and so forth. Grape varieties, soil, and climate contributed to the different characteristics of these wines. Today, wines with similar characteristics are produced in many parts of the world, and to maintain identity as to type they frequently retain the original names or derivatives thereof. In many countries, by international agreement and wine laws, when the original name is used, it must be accompanied by the actual place of manufacture, for example, New York State Port, California Champagne, and Australian Sherry. The Bureau of Alcohol, Tobacco, and Firearms (BATF) of the U.S. Treasury Department issues regulations governing such labeling and taxing of wines and other alcoholic beverages in the United States. Regions of the United States can apply to BATF for designation as a viticulture area and, if granted, reserve the right to label wines as being from a specific region of the United States. Labels also must be consistent with certain labeling requirements of the Food, Drug, and Cosmetic Act.

It now is common in much of the world to name wines after the variety of grape from which the wine was made. Thus, chardonnay, zinfandel, and cabernet sauvignon are wines named after the variety of grape from which they were made.

Among the most popular distinct wine types in the United States are appetizer wines (sherry and vermouth, which is aromatically flavored with herbs or spice), red table wines (claret, Burgundy, and Chianti), white table wines (Rhine wine and sauterne), sweet dessert wines (port, white port, muscatel, and Tokay), and sparkling wines (champagne and sparkling Burgundy). The price of wines includes federal taxation based on alcohol content and for sparkling wines whether they are carbonated artificially or by natural fermentation.

COFFEE

Although there are considerable differences in their starting materials, growing, and processing, coffee and tea share several common characteristics. Both contain virtually no food value in themselves and are consumed entirely for their refreshing and stimulating beverage properties. Both contain caffeine, which provides their physiologically stimulating effect. Both are grown in regions of tropical or near-tropical climate and are important exports of these regions. Both are processed to develop flavor in the harvested beans or leaves, which then are brewed to obtain the flavored beverage.

In many parts of the world, coffee is one of the most popular beverages. In the United States, annual consumption in 1991 was approximately 28 gal per person per year, whereas annual per capita consumption of tea was 5.5 gallons. Whereas consumption of tea in the United States is much less than that of coffee, in England, China, Japan, the Soviet Union, and certain other countries, the picture is reversed.

Production Practices

Coffee trees are started in nurseries as seedlings that are later transferred to the plantation. After about 5 years, the trees bear fruit that turns red as it ripens and is referred to as cherries. When ripe, the cherries are hand-picked. One coffee tree yields about 2000–4000 cherries per year. Each cherry contains only two coffee beans; some 3000 beans yield only about 454 g (1 lb) of finished ground coffee. According to the U.S. Department of Agriculture, the 1991 worldwide production of coffee beans was 6.3 million metric tons.

The structure of the coffee cherry is shown in Fig. 19.4. The two coffee beans are covered by a thin parchmentlike hull, which is further surrounded by pulp. Both the pulp and hull are removed before the coffee beans are roasted for use. Ripe coffee cherries are first passed through pulping machines that break and separate the pulp from the rest of the bean. Separation of the pulp leaves a mucilaginous coating on the beans, which must be removed. This is done by various methods including microbial fermentation of beans heaped in large piles, use of commercial pectin-digesting enzymes, and various washing treatments. After mucilage removal, the beans still contain an outer hull.

The coffee beans are now partially dried either by being spread out in the sun or by machine driers. The object is to decrease the moisture level from about 53% down to about 12%. Drying must be uniform throughout; when sun drying is used, beans must be turned frequently. Drying by this method may take 5 days but is dependent on the weather. During drying, color and flavor attributes are modified within the beans; overdrying or wide fluctuations in temperature give variable coffee bean quality. Machine drying permits good control of temperature and has several other advantages. After the beans are dried to about 12% moisture, hulls are removed by machines that apply friction to the hulls and then remove them in a current of air.

Hulling is followed by sorting of the beans for color and defects. Hand sorting of beans moving along a belt is still practiced to some extent, but modern electronic sorting is less costly and gives better quality control. In this case, the beans are picked up individually by vacuum and sorted by an electric eye (Fig. 19.5).

The sorted beans are graded for size and color, and cup-tested to determine their potential brewing quality. Up to this point, the beans are still green; that is, they have not yet been roasted. For cup testing (Fig. 19.6), small samples are roasted, ground, and brewed. For the most part, though, graded coffee beans are shipped as green beans for further processing by coffee manufacturers.

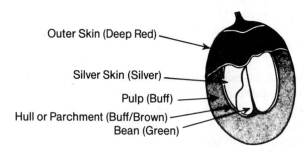

Outer Skin (Deep Red)

Silver Skin (Silver)

Pulp (Buff)

Hull or Parchment (Buff/Brown)

Bean (Green)

Figure 19.4. Structure of the coffee cherry. *Courtesy of the Squier Corporation.*

Figure 19.5. Electronic sorting of coffee beans. Other products such as peanuts can be similarly separated. *Courtesy of Elexso Corp.*

Coffee Processing

Blending

Different manufacturers favor various coffee blends and buy their beans from countries producing the required coffee types. The manufacturer then custom-blends products for special market outlets.

Roasting

During roasting, the characteristic flavor of coffee is developed. Both batch- and continuous-roasting equipment is available. Newer types of continuous roasters can automatically control temperature and humidity, recirculate roaster gases, and control residence time of beans in the roaster. Some are being fed with green bean blends that are formulated and combined under computerized control.

Figure 19.6. Organoleptic testing of coffee for brewing quality. *Courtesy of Pan American Coffee Bureau.*

Much research has been done on the roasting step since various blends require different heat treatments to develop optimum flavor. Further, a given blend roasted to various degrees will yield coffees of different color and taste qualities favored by different markets. Current roasting practices employ gas temperatures of about 260°C for about 5 min. The bean temperature rises to about 200°C during roasting. All of the free moisture is removed from beans during roasting; in addition, beans lose about 5% more of their green bean weight as volatile chemical substances. One of the newer roasting processes employs heated nitrogen under pressure; among the advantages claimed is improved flavor due in part to removal of oxygen.

Grinding

Following roasting, the beans are cooled and ground. This is not as simple a step as might appear. The size to which coffee is ground depends on its intended end use (Table 19.6): home use in a vacuum, drip, or percolator brewer; restaurant use in a larger urn; vending machine use where extremely fast brewing may be required; or use in the manufacture of instant coffee. In each case, average particle size and particle size distribution affect the brewing time, the degree of turbidity in the cup, and other properties of the brewed beverage. Since the aroma and flavor properties of ground

Table 19.6. Particle Size Versus Number of Particles Per Unit Weight

Particle Size Description	Size (mm)	No. Particles per g	Increase Particles/g	Ratio of Increase	Total Area (cm^{2}g)
Whole bean	6.0	6	—	—	8
Cracked bean	3.0	48	42	1	16
Instant R & G					
for percolation	1.5	384	336	8	32
Regular	1.0	1,296	912	22	48
Drip	0.75	3,072	1,776	42	64
Fine	0.38	24,572	21,500	512	128

Courtesy of M. Sivetz.

coffee are highly unstable to oxygen and to loss of volatiles, coffee that is to be stored for long periods generally is packed in hermetic cans and jars under vacuum or under inert gas. Coffee for restaurants, which is consumed more rapidly, may be packed in sealed bags. Storage stability in each case also is affected by grind size. Ground coffee gives off considerable CO_2 and so must be allowed to outgas before packaging or the CO_2 will accumulate and distend the package.

Brewing

As pointed out earlier, brewing is the hot water extraction of plant materials. Brewing coffee to the correct strength and flavor depends on several variables. These include the ratio of coffee to water, particle size of the ground coffee, temperature of the water, mixing action in the brewer, and time. All will affect the amount of coffee solubles that is extracted from the ground bean. There is an optimum degree of extraction for best flavor; extraction beyond this point removes bitter constituents from the bean and ruins the brew.

Optimum extraction can be measured by determining the soluble solids in the brew. This is done by measuring the brew density with a floating hydrometer. Such a hydrometer has been calibrated by the Coffee Brewing Institute and a chart has been developed relating extracted soluble solids to coffee strength (Fig. 19.7). Such measurements are very useful in developing brewing equipment, of which there are scores of designs, and in quality control measurements on brewed coffee.

Decaffeinated Coffee

Coffee is a major source of caffeine and related stimulants in the diet, although it is present in several other beverages and foods, including tea leaves, cacao beans, and kola nuts. Products made from these materials have different levels of caffeine depending on the method of processing, kind of brewing in the case of coffee and tea, and other factors. Brewed coffee generally contains about 75–150 mg caffeine per 150–ml (5-oz) cup; brewed tea about 30–45 mg per 150-ml cup; cola beverages about 30–65 mg per 360-ml (12-oz) can; and milk chocolate about 6 mg per 28 g (1 oz).

Since caffeine, in addition to its stimulating effect, may produce insomnia, nervousness, and other physiological responses in some persons, coffee and tea may be decaffeinated. Decaffeinated coffee contains about 3 mg caffeine per cup. Decaffeination

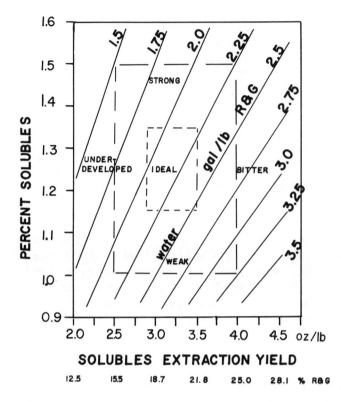

Figure 19.7. Relationships between soluble solids and coffee strength at different water to coffee brewing ratios. *Courtesy of Coffee Brewing Institute.*

involves steaming of green coffee beans followed by water extraction prior to the usual roasting step. In some cases, organic solvents are used to extract the caffeine from the bean. This leaves the problem of removing the residual solvent and of recovery of the solvent vapors. A recent advanced process utilizes high-pressure CO_2 and is known as supercritical CO_2 extraction. Under the proper conditions of high-pressure and reduced temperature, CO_2 (and other gases) has the solvent power of a liquid and the penetrating ability of a gas. This means that lower temperatures can be used and there is no worry about leaving behind traces of solvent in the coffee.

Instant Coffee

Instant coffee, or as it is technically known "solubilized" coffee, is made by dehydrating the brewed coffee; manufacture of this product is carried out in plants that incorporate the most advanced extraction, dehydration, and essence-recovery equipment to be found anywhere in the food industry.

Extraction. Extraction of roasted ground beans is accomplished in an extraction battery that may consist of as many as six to eight percolators connected to be operated as a single unit (Fig. 19.8). Percolators are run at different temperatures, and extract is pumped from one to another at various stages of the brewing operation. Conditions are set to obtain maximum extraction without heat damage or overextraction of bitter constituents. Extraction is also designed to filter the brew through the coffee grounds and thereby remove fats and waxes which otherwise would adversely affect subsequent drying and storage stability. Efficient extraction using a temperature profile decreasing

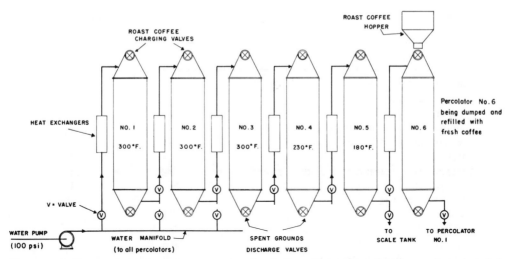

Figure 19.8. Six percolator countercurrent extraction battery. Extraction columns are steam-jacketed stainless steel, ASME coded, internal pressure at least 250 psi. All contact parts stainless steel. Source: *J. H. Nair and Sivetz, Coffee and Tea, in Food Dehydration, 2nd ed., Vol. 2., W. B. Van Arsdel, M. J. Copley, and A. I. Morgan, Jr. (Editors), AVI Publishing Co., Westport, CT, 1973.*

from about 150 to 70°C removes most of the readily soluble solids and hydrolyzes less soluble coffee bean carbohydrates (Table 19.7) resulting in a total extraction of about 40% of the weight of the roasted and ground bean. Without high-temperature hydrolysis (150°C), only about 20% of the bean weight would be extracted, which is about what is obtained in home and restaurant brewing.

The extract from the percolators is rapidly cooled and, when possible, dehydrated immediately, since coffee aroma and flavor can deteriorate in as little as 6 h even when cooled to 4°C.

Dehydration. The principal method of dehydrating the extract is spray drying, and spray driers have been designed especially for coffee. As in spray drying of other products, the size, shape, density, moisture content, solubility, and flavor properties of the dried particles depend on the droplet size sprayed into the drier, the time required for the particle to descend, the temperature exposure, the trajectory of the droplet to prevent sticking to the drier wall, and so on. A flow diagram of a typical instant coffee plant including spray drier is shown in Fig. 19.9. Spray-dried particles commonly are agglomerated to appear more like roasted and ground coffee and to improve solubility and minimize foam in the cup. Spray-dried particles also may be heated between rollers to produce fused particles, which are cooled and ground to give a crystalline flakelike appearance.

Since the late 1960s, increasing quantities of coffee extract have been dehydrated by freeze-drying to retain maximum flavor and aroma. This has included the use of freeze concentration to produce very high quality concentrated extracts for the freeze-drying process. Currently, a significant portion of all instant coffee produced in the United States is freeze-dried. Freeze-drying is a milder treatment than spray drying and produces a higher quality instant coffee but at a higher price due to the expense of the process.

Table 19.7. Percentage Composition of the Soluble and Insoluble Portions of Roast Coffee (Approximate, Dry Basis)

	Solubles	Insolubles
Carbohydrates (53%)		
reducing sugars	1–2	—
caramelized sugars	10–17	7–0
hemicellulose (hydrolyzable)	1	14
Fiber (not hydrolyzable)	—	22
Oils	—	15
Proteins (N × 6.25); amino acids are soluble	1–2	11
Ash (oxide)	3	1
Acids, nonvolatile		
chlorogenic	4.5	—
caffeic	0.5	—
quinic	0.5	—
oxalic, malic, citric, tartaric	1.0	—
Volatile acids	0.35	—
Trigonelline	1.0	—
Caffeine (Arabicas 1.0%; Robustas 2.0%)	1.2	—
Phenolics (estimated)	2.0	—
Volatiles		
carbon dioxide	Trace	2.0
essence of aroma and flavor	0.04	—
Total	27–35	73–65

Source: Sivetz and Desrosier (1979).

Aromatization. Even the best instant coffee from the drier lacks the full flavor and aroma of freshly brewed coffee. An enormous amount of work has been done to develop treatments of various kinds to improve flavor and aroma; these are referred to as aromatization. This generally involves adding back favor and aroma constituents recovered during processing to the dry state. These flavor and aroma constituents have been trapped and recovered during roasting, grinding, and extraction and have been obtained from oils pressed from the coffee bean. Hundreds of patents have been granted in this area alone. One interesting technique involves extraction of roasted and ground coffee with a coffee oil solvent such as liquid carbon dioxide. The cold CO_2 does not damage flavor and aroma compounds in the coffee oil and is easily separated from the extracted oil for recompression and reuse. The extracted oil is then sprayed onto the instant coffee. Such an aromatization scheme, currently being used commercially, is shown in Fig. 19.10. After CO_2 removal of the oil, the roasted and ground coffee is still highly suitable for extraction of water-soluble solids in the regular extraction battery operation.

TEA

Although coffee is derived from the beans (seeds) of the tree, true teas come from the young leaves of the tea plant which is a bush. The term *tea* is somewhat incorrectly also used for hot water extracts of other plant materials such as jasmine. These are

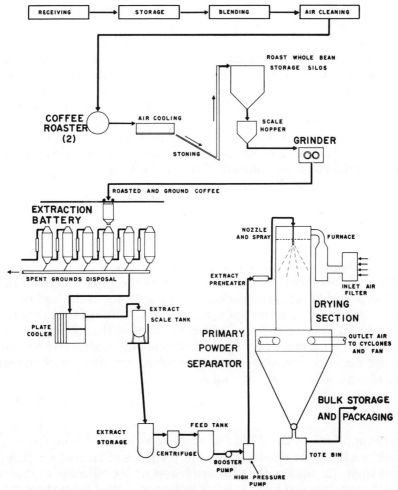

Figure 19.9. Flow diagram of an instant coffee plant. Source: *Nair and Sivitz, Food Dehydration, 2nd ed., Vol. 2, W. B. Van Arsdel, M. J. Copley, and A. I. Morgan, Jr. (Editors). AVI Publishing Co., Westport, CT, 1973.*

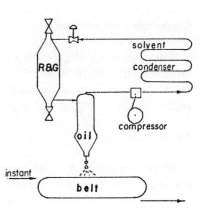

Figure 19.10. Closed-cycle system for the aromatization of instant coffee. Source: *Sivetz and Desrosier, Coffee Technology. AVI Publishing Co., Westport, CT, 1979.*

often known as "herbal teas." The tea plant is an evergreen and it is ready to yield tea leaves after about 3 years of growth. It then may yield for 25–50 years depending on growing conditions. The leaves are hand-plucked from new shoots and about 6000 leaves are needed to make 1 lb of manufactured tea. Depending on plant cultivar, climate, soil, and cultivation practices, there are about 1500 slightly different kinds of tea leaves; these can be further modified in processing and contribute to differences in the final brew and provide opportunity for custom blending to satisfy regional preferences. Tea leaves contain three important kinds of constituents that affect brew quality: caffeine, which gives tea its stimulating effect; tannins and related compounds, which contribute color and strength, often associated with the terms body and astringency; and essential oils, which provide flavor and aroma.

Leaf Processing

The three major classes of teas are known as green, black, and oolong. These three types can be made from the same tea leaves, depending on how the leaf is processed. The differences result largely from enzymatic oxidations of the tannin compounds in the leaf. If the enzymes are allowed to act, they turn the green leaf black in much the same way that a freshly cut apple blackens. If the enzymes in the leaf are inactivated by heat, as in blanching, then the leaf remains green. If a partial oxidation is allowed to occur by delayed heating, then an intermediate tea of the oolong type is obtained. The enzymatic oxidation of tea leaves is referred to as fermentation. Fermented leaves give black tea; partially fermented leaves give oolong tea. Along with the color differences there also are subtle flavor variations.

Black Tea

The processing of tea to the dried leaf stage involves relatively few steps. In the case of black tea these include (1) withering the plucked leaves to soften them and partially dry them, (2) passing the withered leaves under rollers to rupture cell walls and release the enzymes and juices, (3) fermenting the rolled leaves by exposing them to the air at about 27°C for 2–5 h (this relatively short time would be insufficient to bring about the desired color and flavor changes were the enzymes and juices not freed from the cells in the rolling step), and (4) drying the fermented leaves in ovens at about 93°C, which inactivates the enzymes and decreases leaf moisture to about 4%. This drying step in the case of tea is known as firing.

Green and Oolong Tea

The processing of green tea involves (1) steaming the plucked leaves to inactivate enzymes, (2) rolling the leaves to rupture the cell walls, which makes the leaves easier to extract during subsequent brewing, and (3) drying or firing the leaves. Oolong tea receives less steaming and a partial fermentation step before being fired.

Further Subtypes

Within each of the three broad classes of tea, there are several subtypes and styles depending on the country, region, and estate the tea came from, the season when the leaves were plucked, the size and shape of leaves (straight, curled, or twisted), whether the leaves are from the growing tip or lower on the branch, and other characteristics, all of which affect cup quality. The percentages of the dry weight of leaves extractable as caffeine, tannin, and total soluble solids on brewing some of these teas are indicated in Table 19.8. Regardless of type, the dried tea leaves are packed into chests and exported. The importer cup-tests various lots for body, flavor, color, and clarity of the brew, and then custom-blends for different markets. In the United States, about 50% of the tea imported goes into tea bags, about 10% is sold as loose tea, and the rest is used for the manufacture of instant tea, mixes, and other tea products.

Instant Tea

The manufacture of instant tea is in several ways like that of instant coffee. Instant tea processing begins with extraction of the selected tea leaf blend. Generally, a fermented black tea type is used—one chosen for reddish color, relative freedom from haze, and strong flavor when brewed. About 10 parts of water are combined with 1 part of tea leaves by weight in the extractors, and extraction is carried out at temperatures between about 60 and 100°C for 10 min. The final extract contains about 4% solids, which represents approximately 85% of the soluble solids in the leaves. Departures from these times and temperatures are common with different manufacturers. This rather dilute extract is concentrated for more efficient dehydration; just before concentration, aromatics are distilled from the extract with specially designed flavor-recovery equipment. The dearomatized extract is then concentrated in low-temperature evaporators to between 25% and 55% solids for subsequent drying.

Instant tea has grown in popularity largely because of its convenience in making iced tea. However, tea leaves and tea extract contain both caffeine and tannins. These are in solution in hot water, but in cold water some of the caffeine and tannins form a complex that imparts a slight haze or turbidity to the brew known as "cloud" or "cream." Since iced tea is generally consumed in glasses rather than cups, this turbidity is readily seen and detracts from the desirable quality attribute of clarity. This haze-forming property can be removed from the concentrate prior to drying. One method involves cooling the tea concentrate to about 10°C to encourage caffeine—tannin complex formation. The complex, which gives a fine precipitate, can be removed by filtration

Table 19.8. Caffeine, Tannin, and Total Soluble Solids Extracted by 5-Min Infusion of Various Tea Leaves

Infusion of	Caffeine (%)[a]	Tannin (%)[a]	Soluble Solids (%)[a]
Java black teas	2.7–4.4	6–20	16–26
Japan green teas	2.0–3.3	4–12	16–26
China black teas	2.0–3.7	5–10	16–22
Formosa oolong teas	3.2–3.7	12–23	23–25
India black teas	2.0–3.0	6–10	22–25

Courtesy of J. B. B. Deuss.
[a]Percentage of dry weight of leaves.

or centrifugation. This is similar in principle to the chill-proofing of beer and the winterizing of oils, although the hazes in each case differ in composition.

The concentrate with haze removed is supplemented with the tea essence distilled earlier and is ready for dehydration. Some manufacturers also supplement the tea concentrate with dextrins before drying, to provide a 50% tea solids and 50% carbohydrate solids mixture. This tends to protect the delicate tea aroma during drying and yields a dried product with quicker solubility in cold water. Bulk density of the product is controlled so that a teaspoonful gives about the same level of tea solids as does an instant tea dried without dextrins. A flow sheet of the overall instant tea manufacturing process is presented in Fig. 19.11.

Instant tea is dried primarily in spray driers and low-temperature vacuum belt driers. Tea flavor and aroma is even more sensitive than that of coffee, and so the spray driers are operated under milder heat conditions than those used for coffee, which cuts down on their capacity. Freeze-dried tea appears to offer few advantages.

In recent years, bottled and canned teas and in some parts of the world coffees have become popular. These products are often sweetened with sugars or reduced-calorie sweeteners and may have fruit flavor essences added. Their manufacture is similar to other canned and bottled beverages except that they start with a brewed product which may initially be in the form of a concentrate. If their pH is above 4.6, they would require retort processing. Many are aseptically processed and packaged.

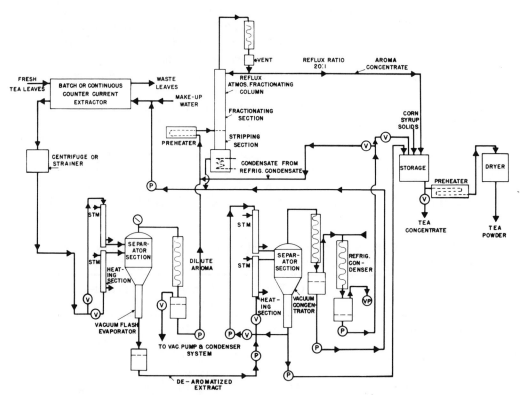

Figure 19.11. Flow sheet of instant tea process. Source: *Nair and Sivitz, Food Dehydration, 2nd ed., Vol. 2, W. B. Van Arsdel, M. J. Copley, and A. I. Morgan, Jr. (Editors), AVI Publishing Co., Westport, CT, 1979.*

References

Clifford, M.N. and Willson, K.C. (Editors). 1985. Coffee: Botany, Biochemistry, and Production of Beans and Beverages. Chapman & Hall, London, New York.

Green, L.F. 1978. Developments in Soft Drinks Technology. Applied Science Publishers, London.

Hicks, D. 1990. Production and Packaging of Non-Carbonated Fruit Juices and Fruit Beverages. Chapman & Hall, London, New York.

Houghton, H.W. 1988. Low calorie soft drinks. *In* Low-Calorie Products. G.G. Birch and M.G. Lindley (Editors). Elsevier Applied Science, London, pp. 11–21.

Lewis, M. J. 1995. Brewing. Chapman & Hall, London, New York.

Lipinski, R.A. 1992. The Complete Beverage Dictionary. Van Nostrand Reinhold, New York.

Mitchell, A.J. 1990. Formulation and Production of Carbonated Soft Drinks. Chapman & Hall, London, New York.

Ohlmeyer, D.W. and Matz, S.A. 1970. Brewing. *In* Gereal Technology. S.A. Matz (Editor). AVI Publishing Co., Westport, CT.

Priest, F.G. and I. Campbell (Editors). 1987. Brewing Microbiology. Elsevier Applied Science, London.

Sivetz, M. and Desrosier, N.W. 1979. Coffee Technology. AVI Publishing Co., Westport, CT.

Varnam, A.H. 1994. Beverages: Technology, Chemistry and Microbiology. Chapman & Hall, London, New York.

Vine, R.P. 1981. Commercial Winemaking, Processing and Controls. Chapman & Hall, London, New York.

Wells, A.G. 1989. The use of intense sweeteners in soft drinks. *In* Progress in Sweeteners. T.H. Grenby (Editor). Elsevier Science Pub., New York. pp. 169–214.

Woodroff, J.G. and Phillips, G.F. 1981. Beverages Carbonated and Noncarbonated. rev. ed. AVI Publishing Co., Westport, CT.

Wurdig, G., Woller, R., and Breitbach, K. 1989. Wine Chemistry. Ulmer, Stuttgart.

Zoecklein, B.W., Fugelsang K.C., Gump. B.H., and Nury, F.S. 1995. Wine Analysis and Production. Chapman & Hall, London, New York.

20

Confectionery and Chocolate Products

Confections (i.e., candy) can be divided into two broad categories: those in which sugar is the principal ingredient and those which are based on chocolate. Differences in sugar-based candies depend largely on manipulating the sugar to achieve special textural effects. This is accomplished primarily by controlling the state of crystallization of the sugar and the sugar–moisture ratio. Examples of sugar-type confections include nougats, fondants, caramels, taffees, and jellies. Examples of chocolate-based confections include chocolate-covered confections, chocolate-panned confections, chocolate bars, and chocolate-covered fruits, nuts, and cremes. Many ingredients, including milk products, egg white, food acids, gums, starches, fats, emulsifiers, flavors, nuts, fruits, and others are used in candy-making.

SUGAR-BASED CONFECTIONS

When the sugar in confections is crystalline, the crystals may be large or small, or the sugar may be noncrystalline, that is, amorphous and or glasslike. Whether crystalline or not, the sugary structure may be hard or soft, softness being favored by a higher level of moisture, by air whipped into the sugary mass, and by the modifying influences of other ingredients.

Table 20.1 is a simplified classification of some major candy types based on the physical state of the sugar. Candies that have sugar in the crystalline form include rock candy, in which the entire confection is a large sugar crystal, and fudges and fondants, which contain smaller sugar crystals. A fondant is a saturated sugar solution in which small sugar crystals are dispersed. Examples of fondants would be cream centers, crystallized creams, and thin mints. Candies that contain the sugar in various degrees of crystallization are also referred to as grained candies.

Candies that have sugar in noncrystalline form include sour balls, butterscotch, and nut brittles, all of which contain sugar in an amorphous glasslike state, and all of which are hard, containing 2% moisture or less. Noncrystalline candies also include chewy types, such as caramel and taffy, with about 8–15% moisture, and gummy candies, such as marshmallows, gumdrops, and jellies, with about 15–22% moisture. Marshmallows are further softened by having air whipped into them. Candies in which the sugar is noncrystalline are referred to as nongrained.

Table 20.1. Major Candy Types

Texture	Example
Crystalline sugar	
large crystals	Rock candy
small crystals	Fondant, fudge
Noncrystalline sugar	
hard candies	Sour balls, butterscotch
brittles	Peanut brittle
chewy candies	Caramel, taffy
gummy candies	Marshmallow, jellies, gumdrops

Although the candy types listed in Table 20.1 include the major varieties, there also are intermediate types; the preparation of these, though intermediate, follows the same principles that govern sugar crystallization and water removal in the major types. The wide use of garnishes (e.g., fruits, nuts, flavors, colors, and chocolate) add interest and variety to the different candy types, but the condition of the sugar and the degree of moisture are still recognizable.

The state of crystallinity and the percentage of moisture in finished confections is determined largely by their functional ingredients, the heat used in cooking and concentrating the sugar syrups, and the way in which these syrups are cooled, including whether or not they are agitated. All of these factors may be controlled by the candy-maker.

INGREDIENTS

Many ingredients are available to the confectionery manufacturer; some of these are listed in Table 20.2 along with their gross compositions. From these, the high-energy value of the concentrated foods represented by various confections also can be judged.

Sucrose

The principal ingredient in sugar-based candies and a major component of chocolate is the sweetener. The most common sweetener in candy-making is sucrose, the sugar from sugar cane or sugar beets. At room temperature, about two parts of sucrose can be dissolved in one part of water, giving a concentrated solution of approximately 67%. If the solution is cooled without agitation, it becomes supersaturated. Upon further cooling, especially with agitation, the sucrose crystallizes. Crystallization can be speeded enormously if even a single minute sucrose crystal is added to the supersaturated solution.

Greater concentrations of sucrose can be dissolved by raising the temperature of the water. The higher the sucrose concentration, the higher the boiling point of such solutions. Candy-makers take advantage of the precise relationship between boiling point and sucrose concentration to control the final degree of water in confections. This is done by heating a sugar syrup to a selected temperature corresponding to the

Table 20.2. Gross Compositions of Common Food Ingredients Used in Confectionery Manufacture

Ingredient	Caloric Value (kcal/100 g)	Protein (%)	Fat (%)	Carbohydrates (%)	Ash (%)
Almonds	640	18.6	54.1	19.6	3.0
Coconuts (dry)	579	3.6	39.1	53.2	0.8
Chocolate (bitter)	570	5.5	52.9	18.0	3.2
Chocolate (sweet)	516	2.0	29.8	60.0	1.4
Chocolate (sweet milk)	542	6.0	33.5	54.0	1.7
Cocoa (average)	329	9.0	18.8	31.0	5.2
Corn starch	365	9.1	3.7	73.9	1.3
Cream (heavy)	337	2.3	35.0	3.2	0.5
Dairy butter	733	0.6	81.0	0.4	2.5
Eggs (total edible)	158	12.8	11.5	0.7	1.0
Fruits (fresh)					
apples (edible portion)	64	0.3	0.4	14.9	0.29
lemons (edible portion)	44	0.9	0.6	8.7	0.54
peaches (edible portion)	51	0.5	0.1	12.0	0.47
pears (edible portion)	70	0.7	0.4	15.8	0.39
oranges (edible portion)	50	0.9	0.2	11.2	0.47
pineapple (edible portion)	58	0.4	0.2	13.7	0.42
Figs (dried)	300	4.0	1.2	68.4	2.4
Raisins (seedlesss and seeded)	298	2.3	0.5	71.2	2.0
Gelatin (plain, dry)	343	85.6	0.1	0.0	1.3
Milk					
whole	69	3.5	3.9	4.9	0.7
condensed	327	8.1	8.4	54.8	1.7
evaporated	139	7.0	7.9	9.9	1.5
skim	36	3.5	0.2	5.0	0.8
Milk (dried)					
whole	496	25.8	26.7	38.0	6.0
skim	359	35.6	1.0	52.0	7.9
Milk (malted)	418	14.6	8.5	70.7	3.6
Nuts					
fiberts	670	12.7	60.9	17.7	2.7
peanuts (roasted, edible portion)	600	26.9	44.2	23.6	2.7
pecans	747	9.4	73.0	13.0	1.6
walnuts (edible portion)	702	15.0	64.4	15.6	1.7
Sugars					
cane or beet (sucrose)	398	—	—	99.5	—
corn (refined dextrose, anhydrous)	398	—	—	99.5	—
maple	360	—	—	90.0	0.9
brown	382	—	—	95.5	1.2
Syrups					
cane	268	—	—	67.0	1.5
corn (commercial)	322	—	—	80.6	—
maple	256	—	—	64.0	0.7
sorghum	268	—	—	67.0	2.5
Honey (strained or extracted)	319	0.3	0.0	79.5	0.2
Molasses (light)	260	—	—	65.0	3.0

Courtesy of M. Schoen.

sugar and water concentrations desired (Table 20.3). When the boiling syrup reaches temperature, it will have the desired sugar concentration.

More concentrated solutions, on cooling, become highly supersaturated and may solidify as an amorphous glass, a totally crystalline mass, or a partially crystalline mass with the crystals suspended in a glass; or they may partially solidify as a viscous or semiplastic crystalline suspension in the remaining saturated solution. An amorphous glass might be made into a sour ball; a totally crystalline mass could be used for rock candy; a partially crystalline mass with small crystals suspended in a glass would be suitable for the manufacture of partially grained confections; and a crystalline suspension in a saturated sugar solution could become a fondant cream center or a thin mint of the kind that is usually chocolate coated.

Invert Sugar

Invert sugar is related to sucrose and common in confections. Sucrose can be hydrolyzed by acids or enzymes into two monosaccharides, glucose and fructose, according to the following equation:

$$C_{12}H_{22}O_{11} + H_2O \rightarrow C_6H_{12}O_6 + C_6H_{12}O_6$$

Sucrose	water	Glucose	Fructose
(342 g)	(18 g)	(180 g)	(180 g)

The confectionery trade refers to glucose as dextrose, and fructose as levulose. The hydrolyzed mixture of dextrose and levulose is called invert sugar. Invert sugar can prevent or help control the degree of sucrose crystallization. It can do this for at least two reasons. First, both dextrose and levulose crystallize more slowly than sucrose, and so substitution of part of the sucrose with invert sugar leaves less sucrose for rapid crystallization during cooling of syrups, when most of the crystals are formed,

Table 20.3. Boiling Points of Sucrose–Water Syrups of Different Concentrations[a]

% Sucrose	% Water	Boiling Point °C
30	70	100
40	60	101
50	50	102
60	40	103
70	30	106
80	20	112
90	10	123
95	5	140
97	3	151
98.2	1.8	160
99.5	0.5	166
99.6	0.4	171

[a]The boiling point corresponding to each sugar concentration differs for different sugars.

and during subsequent storage, when additional crystals precipitate and grow in size. Second, a mixture of sucrose and invert sugar has greater solubility in water than sucrose alone; increased solubility is equivalent to less crystallization.

Invert sugar may be obtained commercially and substituted for part of the sucrose in the candy formula, or it may be formed directly from sucrose during candy-making by including a food acid such as cream of tartar in the formula. During boiling of the sugar syrup, the acid hydrolyzes part of the sucrose; the resulting effects on crystallization and other candy properties are related to the concentration of invert sugar produced.

Invert sugar not only limits the amount of sucrose crystallization but it encourages the formation of small crystals essential to smoothness in fondant creams, soft mints, and fudges. Because it is hygroscopic, invert sugar helps prevent more chewy candies from drying out and becoming overly brittle. In terms of sweetness, the components of invert sugar differ from sucrose: Dextrose is less sweet and levulose is sweeter than sucrose. A mixture of invert sugar and sucrose is sweeter than sucrose alone.

Other sweeteners such as maple sugar generally are used for their particular flavor properties rather than for special functional attributes. Brown sugar is obtained from the cane sugar refining process and is made up of sucrose with greater amounts of ash, invert sugar, and compounds derived from the process that give the sugar its characteristic color and flavor. Brown sugar is used in several confections such as caramels, toffees, and butterscotch. Molasses is similar to brown sugar in that it is also a product of the refining process but contains less sucrose, more invert sugar, and more of the color and flavors. Honey is also used in confections as a sweetener. It contains about 31% glucose and 38% fructose.

Corn Syrups and Other Sweeteners

Corn syrups are viscous liquids containing dextrose, maltose, higher sugars, and dextrins. They are produced by the hydrolysis of corn starch using acid or acid–enzyme treatments. The extent of hydrolysis or conversion to lower-molecular-weight substances is influenced and controlled by the time, temperature, pH, and enzymes used. A wide variety of syrup compositions is commercially available.

Corn syrups retard crystallization of sucrose, and do so with less tendency toward hygroscopicity than invert sugar. Corn syrups further add viscosity to confections (largely because of their dextrin content), reduce friability of the sugar structures from temperature or mechanical shock, slow the dissolving rate of candies in the mouth, and contribute chewiness to confections.

As mentioned in Chapter 3, glucose can be enzymatically converted to its isomer fructose. This glucose or dextrose commonly has its origin in starch, which may be hydrolyzed to corn syrup or very largely to dextrose. The dextrose then may be enzymatically converted to fructose or levulose. The degree of such conversion determines the properties of these sugar syrups, which together with sucrose, invert sugar, and corn syrups, give confectionery manufacturers considerable choice with respect to sweeteners and their functional properties. Corn syrup solids are simply corn syrup that has been dried and granulated.

Sugar substitutes

The ability of sucrose to cause dental cavities and its calorie content have led to the use of sugar substitutes in some confections. These substitutes can be divided into two types: bulk sweeteners and high-intensity sweeteners. The latter can be divided into caloric and noncaloric high-intensity sweeteners.

The major bulk sweeteners are alcohol derivatives of sugars made by chemically reducing the sugar to the alcohol. Sugar alcohols are not fermentable by the bacteria in the mouth, so do not contribute to cavities. They are 50–75% less sweet than sucrose, depending on the specific sugar alcohol. Commonly used examples of sugar alcohols are sorbitol, xylitol, and mannitol. Confections using these sweeteners are often labeled "sugar-free" but this does not mean that they do not contain calories. The caloric content of sugar alcohols is the same as sucrose.

High-intensity sweeteners are used in confections to reduce the caloric content. They do this in one of two ways: They may contain calories, but because they are used in such small amounts due to their intense sweetness, they reduce the caloric content of the confection. They may also be chemicals which have sweetness but are not metabolized so they do not add caloric content. Examples of both include saccharin, sucralose, thaumatin, aspartame, glycyrrhizin, and Acesulfam K. Not all of these sweeteners are approved for use at this time. High-intensity sweeteners often do not have other functional properties such as bulking or mouthfeel of sugar, and so additional additives are often used to impart desirable characteristics.

Some Additional Ingredients

Other ingredients often used in candy-making may influence sucrose crystallization, although this effect may be secondary to the main reason for their use. Thus, besides the thickening and chewiness properties of starch, the whipping and toughening properties of egg white and gelatin, the favor and coloring properties of milk, and the flavor, tenderizing, and lubricating qualities of fats, all of these ingredients interfere with sucrose crystal formation. This is due to adsorption of these materials onto crystal surfaces during formation. A barrier between attractive forces of the crystal lattice and sucrose molecules in solution is produced, limiting the crystals from growing in size.

Some softer candies (e.g., marshmallows, gumdrops, and jellies) owe their chewiness in part to pectins, gums, and gelatin. The chewiness of caramel is due largely to prevention of the grained condition by corn syrup and invert sugar plus the chewiness of dextrins. These and other soft candies also are characterized by a moderate level of moisture as indicated earlier. When the moisture content of a candy is 20% or less, slight drying during storage will have marked effects on optimum textures. In addition to protective packaging, humectants are used to hold moisture within such confections. Common humectants, in addition to invert sugar, include glycerin (glycerol) and sorbitol. Colloidal materials such as pectins and gums, which are hydrophilic, also have humectant properties in confections.

Thus, the candy-maker can combine a wide range of functional ingredients into an almost unlimited number of formulations to affect confectionery properties. The possibilities are further enlarged by the order of ingredient addition. If crystal inhibitors are added together with sucrose to the cooking kettle, a different result will be obtained than when some of these ingredients are subsequently mixed into a smooth fondant produced by seeding, cooling, and agitation to promote rapid formation of minute crystals. The hardness–softness aspect of texture, largely controlled by the amount of water lost from the cooking kettle prior to cooling and solidifying the batch, is obviously affected also by the choice of ingredients. The incorporation of flavors, nuts, and fruits into the sugary mass further modifies the confection in a more easily predictable manner.

CHOCOLATE AND COCOA PRODUCTS

Chocolate is not only one of the principal ingredients used by the confectioner, but its widely enjoyed flavor properties make it a favorite material of bakers, ice cream producers, and other food manufacturers. In its many forms, chocolate may be consumed as a beverage, a syrup, a flavoring, a coating, or a confection in itself. It, therefore, warrants brief consideration before proceeding with some of the processing practices of the confectionery manufacturer.

Cacao Beans

Chocolate and related products begin with cacao beans, which grow in elongated melon-shaped seed pods attached to the cacao tree. The pods each contain about 25–40 cacao beans arranged in rows along the length of the pod around a central placenta. The rows of beans are surrounded by mucus and a pulpy layer beneath the pod husk.

The beans, which may be white or pale purple and are slightly larger than coffee beans, are removed from the pod and fermented microbiologically and enzymatically. This may be done by heaping the beans and covering them with leaves. Fermentation removes adhering pulp and mucus, kills the germ of the bean, and modifies the flavor and color of the bean. After fermentation, the beans, which are now cinnamon to brown in color, are sun-dried or machine-dried to about 7% moisture to give them good keeping quality. Fermentation and drying also alter the seed coat, changing it to a friable skin, which can be easily removed in a subsequent operation. The beans are now ready to be exported for further processing.

Cacao Bean Processing

At a chocolate and cocoa manufacturing plant, the beans are roasted to further develop flavor and color. They are then passed through winnowing machines to remove seed coats and separate the germ. The hulled and degermed beans are called nibs. The nibs are passed through various types of mills where they are torn apart and ground, releasing fat from the cells. The heat of grinding melts the fat, and the ground nibs acquire a liquid consistency. The liquid discharged from the mill is known as chocolate liquor. These and subsequent manufacturing operations are outlined in Fig. 20.1.

Chocolate Liquor

Chocolate liquor contains approximately 55% fat, 17% carbohydrate (most of which is digestible), 11% protein, 6% tannin compounds, 3% ash, 2.5% organic acids, 2% moisture, traces of caffeine, and about 1.5% theobromine, an alkaloid related to caffeine that is responsible for the mildly stimulating properties of cocoa and chocolate.

This chocolate liquor solidifies on cooling and is the familiar bitter chocolate used in baking and other applications. It can be further processed with sugar to yield sweet chocolate, or with sugar and milk to produce milk chocolate. The chocolate liquor also may be partially defatted in a hydraulic press.

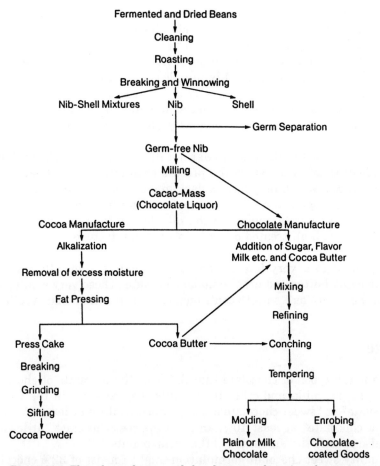

Figure 20.1. Flow sheet of cocoa and chocolate manufacturing plant operations. Source: *Chatt, Cocoa Cultivation, Processing, Analysis. Interscience Publishers, New York, 1953.*

Cocoa Butter

The fat removed from chocolate liquor is known as cocoa butter. The brittle snap of chocolate at room temperature and its quick melting properties in the mouth (releasing maximum flavor) are due to the rather narrow melting range of cocoa butter (30–36°C). This temperature range is the basis for selecting tempering conditions for molten chocolate and subsequent storage temperatures for solidified chocolate. This is to prevent uncontrolled fat crystallization which gives chocolate an impaired texture and a gray surface appearance referred to as "fat bloom." Fat bloom is not to be confused with "sugar bloom" which also occurs on chocolate surfaces from crystallization of sugar under poor temperature and humidity conditions.

Cocoa

After much of the cocoa butter is pressed from the chocolate liquor, the remaining press cake is the raw material for the manufacture of cocoa or cocoa powder. The amount of fat left in the press cake can be varied by the conditions of pressing; grinding of the press cake produces cocoa, which is classified according to its fat content. The fat content of different types of cocoa is fixed by law. In the United States, for example, "breakfast cocoa" must contain a minimum of 22% cocoa fat, medium fat "cocoa" must contain 10–22% fat, and products containing less than 10% fat must be labeled "low-fat cocoas." It is possible to remove nearly all the cocoa fat by solvent extraction to give special-use cocoas. One such use is in the manufacture of chocolate-flavored angel cakes, in which traces of fat would adversely affect the whipping properties of egg whites.

Some cocoa is treated with alkali to darken its color and modify its flavor. This is called "Dutch Process" cocoa since the process originated in Holland. The flavor of Dutch Process cocoa, which may have a dark mahogany color, generally is somewhat more bitter and astringent than the same material not treated. The "Dutching" treatment with alkali is usually applied to the nibs before they are made into chocolate liquor. One use for alkali-treated cocoa is in the manufacture of dark-colored devil's food cake.

Cocoa is ground to a very fine powder so that it has a smooth mouthfeel when suspended in cocoa butter and used to make chocolate. These very small particles are perceived in the mouth as a smooth continuous material rather than a solid suspended in a liquid.

Chocolate

There are many types of chocolate that differ in the amounts of chocolate liquor, cocoa butter, sugar, milk, and other ingredients they contain. In the United States, "sweet chocolate" or "sweet chocolate coating" must contain at least 15% of chocolate liquor, "milk chocolate" at least 10%, and "bittersweet chocolate" at least 35%. The standards also specify the amounts of other components.

A high-quality sweet chocolate formulation might consist of 32% chocolate liquor, 16% additional cocoa butter, 50% sugar, and minor quantities of vanilla bean plus other materials. After the ingredients are combined, the mixture is subjected to fine grinding (referred to as "refining") by being passed through close-clearance revolving rollers (Fig. 20.2). These reduce sugar crystals and other particulates to about 25μm or less in size to ensure smoothness, and the mixture scraped from the rolls takes on the character of a flaky powder.

Chocolate is next "conched" or kneaded in special heated mixing tanks. These tanks have pressure rollers that grind and aerate the melted mass to develop increased smoothness, viscosity, and flavor (Fig. 20.3). Conching may be done at about 60°C for 96–120 h. Conching is not essential to chocolate manufacture but is rarely omitted in producing a high quality product.

Following conching, the liquid chocolate is tempered by being stirred in a heated and then cooled kettle to promote controlled crystallization of the cocoa fat. The object here is to melt all the glycerides of the fat and then initiate uniform crystallization of the different glyceride fractions. This is in contrast to uncontrolled crystallization in which the higher-melting-point glycerides solidify within an oily mass. When the latter occurs, as mentioned earlier, uneven crystallization results in impaired chocolate texture and development of fat bloom on subsequent cool storage. Tempering conditions vary but may involve stirring at 54°C, cooling to about 32°C, and continued stirring for about 1 h more. The thickened chocolate mass is then poured into molds for subsequent hardening or into tanks maintained at about 32°C for coating of confections.

Figure 20.2. Five-roll chocolate refiner. *Courtesy of Baker Perkins Ltd.*

Imitation Chocolate

Imitation chocolate is made by replacing some or all of the cocoa fat with other vegetable fats. Imitation chocolates are formulated for special applications such as the coating of ice cream bars, crackers, or candies, where selected vegetable fats can give the chocolate product improved coating properties or resistance to melting. In the latter case, a hydrogenated vegetable fat with a higher melting point than cocoa fat

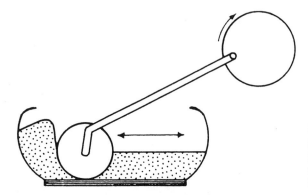

Figure 20.3. Diagram of a chocolate conche machine showing roller within curved tank. Source: *Minifie, Chocolate, Coca, and Confectionery Science and Technology, 2nd ed. AVI Publishing Co., Westport, CT, 1980.*

also will impart to the product a greater melt resistance during summer storage conditions. Imitation chocolates generally are less costly than full cocoa fat chocolate and must be appropriately labeled.

CONFECTIONERY MANUFACTURING PRACTICES

In modern confectionery manufacturing, batch or continuous processes may be used to prepare and cook the basic fondants, taffies, brittles, and hard candies. A number of specialized machines further extrude, divide, enrobe, and otherwise process these confections. For example, in the preparation of thin mints, the supersaturated, partially crystallized sugar mixture from the boiling kettle is flavored with mint and cooled to about 70°C. At this temperature it is semiliquid and can be easily deposited as small dabs onto a moving belt. The mints quickly solidify on further cooling.

Generally, firmer chewy centers are extruded by being pressed through dies. The candy pieces are then cut off by the movement of a thin wire (Fig. 20.4). Like thin mints, these may travel on a moving belt to be covered or enrobed with molten chocolate.

Candies formed from a highly liquid mixture are shaped by molding before they harden. This may be done in a starch-molding machine known as a Mogul (Fig. 20.5). In this case, trays of powdered corn starch are continuously imprinted with concave impressions. The hot liquid candy is filled into the impressions as the trays are conveyed under a hopper. Quick cooling solidifies the candies, which are then automatically dumped over a screen that separates the candies from the starch. A brush further removes the starch from the candies, and the starch is returned to the machine to be imprinted again. This is the way certain jellies, gum drops, marshmallows, and Easter egg centers are formed. Another type of forming utilizes metal, plastic, or rubber molds.

Other candies are aerated to give them softer texture. In the case of marshmallows and nougats, the formulations contain gelatin, egg white, or vegetable proteins, which impart whipping (i.e., foaming) properties; aeration is achieved in batch or continuous mixers before the confections are molded. On the other hand, taffy is aerated by pulling and folding. With each fold of the taffy, air is entrapped, and with subsequent folds, it is subdivided.

Various kinds of small and round candies are glazed by coating nuts and other centers with sugar. This is done by a process known as panning. The centers are placed

Figure 20.4. Extruding plastic candy centers which will be subsequently covered with chocolate. *Courtesy of Werner Machinery Co.*

Figure 20.5. Starch-molding machine. *Courtesy of Baker Perkins Ltd.*

in revolving heated pans and a sugar syrup is sprayed into the pan. As the centers gently tumble, they become uniformly coated with the syrup, which dries as water is evaporated from the heated pan. The thickness of the glasslike sugar coating can be easily varied by continued syrup addition. This is the way candy-coated chocolate centers that do not melt in the hand are made. Candies also are coated with chocolate by this method except that the pans are chilled with cool air to solidify the chocolate coating. Chocolate-panned items frequently are further polished or glazed by spraying a solution of gum arabic or zein into the pan after the chocolate coating is applied. Another polish, known as confectioners glaze, is an edible shellac preparation. These glazes not only improve the glossy appearance of chocolate items but protect the chocolate from the effects of humidity and air during storage.

Larger candy pieces and those that are not round are coated with molten chocolate by the method known as enrobing. In this case, the candy centers first are "bottomed" by passing on a screen over a layer of molten chocolate. They then pass through a tunnel in which they are showered by molten chocolate. Excess liquid chocolate is drained and returned to the tunnel, and the emerging pieces quickly cool, solidifying the coating. It is in enrobing that special chocolate compositions with closely specified melting, covering, and solidifying properties are important. Uniform coating at high speeds requires close control of the temperatures of the incoming candy centers as well as the molten chocolate.

A special type of confection of particular interest has a liquid center and is typified by chocolate-covered cherries and fruits in a syrup. Since the center must be firm to be enrobed, the method of getting the liquid inside the chocolate shell is a good example of food processing ingenuity. First, the fruit is covered with a sugar fondant in a form such as a starch mold and the fondant cools and solidifies. The firm fondant is then enrobed with chocolate in the usual way. However, the fondant is prepared with an invertase enzyme, which slowly hydrolyzes sucrose to invert sugar. This inversion takes place during the normal storage of the candy. Because invert sugar is more soluble than sucrose in the moisture of the fondant, it melts under the chocolate layer and converts the firm center to a creamy liquid.

References

Alikonis, J.J. 1979. Candy Technology. AVI Publishing, Westport, CT.

Beckett, S.T. (Editor). 1994. Industrial Chocolate Manufacture and Use. 2nd ed. Chapman & Hall, London, New York.

Dziezak, J.D. 1989. Ingredients for sweet success. Food Technol. *43*(10), 94–116.

Evans, E.R. 1982. Confectionery products from peanuts. Cereal Foods World *27*(12), 593–594.

Jackson, E.B. 1995. Sugar Confectionery Manufacture. 2nd ed. Chapman & Hall, London, New York.

Kruto, S.A. and Al, Z. 1992. Natural and Synthetic Sweet Substances. Ellis Horwood, New York.

Lees, R. and Jackson, E.B. 1975. Sugar Confectionery and Chocolate Manufacture. Chapman & Hall, London, New York.

Minifie, B.W. 1989. Chocolate, Cocoa, and Confectionery. Science and Technology. 3rd ed. Chapman & Hall, London, New York.

Mossu, G. 1992. Cocoa. MacMillan, London.

Pennington N. and Baker C.W. 1990. Sugar: Users Guide to Sucrose. Chapman & Hall, London, New York.

Santerre, C. (Editor). 1994. Pecan Technology. Chapman & Hall, London, New York.

Wiggall, P.H. 1982. The use of sugars in confectionery (chocolates, boiled candy, fondants, toffees, caramels, fudges). *In* Nutritive Sweeteners. G.G. Birch and K.J. Parker (Editors). Applied Science, London, pp. 37–48.

Woodroff, J.G. 1979. Coconuts: Production, Processing, Products. 2nd ed. AVI Publishing Co., Westport, CT.

Woodroff, J.G. 1979. Tree Nuts: Production, Processing, Products. 2nd ed. AVI Publishing Co., Westport, CT.

Woodroff, J.G. 1983. Peanuts: Production, Processing, Products. 3rd ed. AVI Publishing Co., Westport, CT.

Young, C.T. and Heinis, J.J. 1989. Manufactured peanut products and confections. *In* Food Uses of Whole Oil and Protein Seeds. E.W. Lusas, D.R. Erickson, W.-K. Nip (Editors). American Oil Chemists' Society, Champaign, IL, pp. 171–190.

21

Principles of Food Packaging

INTRODUCTION

Functions of Food Packaging

Packaging is an essential part of processing and distributing foods. Whereas preservation is the major role of packaging, there are several other functions for packaging, each of which must be understood by the food manufacturer. Indeed, faulty packaging will undo all that a food processor has attempted to accomplish by the most meticulous manufacturing practices. Packaging must protect against a variety of assaults including physical damage, chemical attack, and contamination from biological vectors including microorganisms, insects, and rodents. Environmental factors such as oxygen and water vapor will spoil foods if they are allowed to enter packages freely. Contamination of foods by microorganisms can spoil foods or cause life-threatening diseases. Many foods would not survive distribution without physical damage were it not for the protection afforded by packaging.

In order to be successful, packaging must also aid consumers in using products. Food packaging should have features which make the product easier to utilize and add convenience. This may be as simple as reclosure after partial use or as complicated as aiding in the microwave cooking of a product (Fig. 21.1). Many new food products are in reality standard foods packaged in a new way that aids in preparation or storage. Aseptically packaged milk is an example.

Packaging also serves to unitize or group product together in useful numbers or amounts. In some cases this might be an amount to be used at a single time like most canned foods, or in other cases, multiple servings are grouped together such as a six-pack of sodas. Products such as condiments are seldom totally consumed at one time and so reclosure for storage becomes important.

Food packaging must also be able to communicate and educate. It is the package which identifies the product for the consumer. In addition to convincing consumers to buy a product, the package must also inform consumers about how to prepare or use the product, contents or amount of product contained, ingredients, nutritional content, and other pertinent information. Much of this information is required by specific laws in many countries, including the United States.

The package is also an important part of the manufacturing process and must be efficiently filled, closed, and processed at high speeds in order to reduce costs (Fig. 21.2). It must be made of materials which are rugged enough to provide protection during distribution but be of low enough cost for use with foods. Packaging costs,

Figure 21.1. Microwavable soup and sandwich combination in which the package and product are specifically designed for thawing and heating in a home microwave oven.

which include the materials as well as the packaging machinery, are a significant part of the cost of manufacturing foods, and in many cases, these costs can be greater than the cost of the raw ingredients used to make the food. Therefore, packaging materials must be economical, given the value of the food product.

Packaging of foods has become so complex that an entire industry has developed to satisfy the need. In fact, the packaging industry as a whole is one of the largest industries in the United States. About half of the packaging used in the United States goes for foods, with about 23% being used for industrial products (Table 21.1). Today, most sizable food companies have a packaging division, and universities offer special curricula leading to a degree in package engineering. The food scientist does not have to become an expert in packaging, but increasingly he/she will be called on to assist with packaging decisions and problems. This commonly involves defining the kinds of protection essential to a specific food product and specifying in quantitative terms what the package must do. There will be considerable help available from suppliers of packaging materials and equipment, but they, in turn, will depend on the food scientist to make them aware of the peculiarities and subtleties of a particular food system.

Requirements for Effective Food Packaging

Some of the more important general requirements of food packages are that they (1) be nontoxic, (2) protect against contamination from microorganisms, (3) act as a barrier to moisture loss or gain and oxygen ingress, (4) protect against ingress of odors or environmental toxicants, (5) filter out harmful UV light, (6) provide resistance to physical damage, (7) be transparent, (8) be tamper-resistant or tamper-evident, (9) be easy to open, (10) have dispensing and resealing features, (11) be disposed of easily, (12) meet size, shape, and weight requirements, (13) have appearance, printability

Figure 21.2. A modern high-speed packaging line for filling paperboard-based cartons with liquid foods and beverages such as milk or soups. *Courtesy of Cornell University Photo Services.*

features, (14) be low cost, (15) be compatible with the food, and (16) have special features such as unitizing groups of product together.

Products must be protected against introduction of microorganisms as well as dirt. In many cases there should be resistance against boring insects and rodents although only glass and metal cans are insect and rodent proof.

All common polymers used in food packaging allow the transfer of moisture and gases such as O_2 or water vapor directly through them by a process known as permeation. However, polymers can inhibit permeation to different degrees depending on their chemical makeup and physical structure. Some polymers are high barriers, whereas others offer little resistance. The same holds true for moisture and oxygen transfer through films. Moisture protection is a two-way affair. Dry foods should not absorb moisture from the atmosphere, and moist foods should not lose moisture and dry out. There are exceptions such as permeable films that allow the escape of moisture from respiring vegetables. Barrier against fat migration is needed to keep oils and fats from passing through wrappings. A material that is a high moisture barrier is

Table 21.1. Percentage of Packaging Used for
Different Products

Consumer Products	% of Total Dollars
Foods and beverages	53
General products	16
Consumer chemicals	8
Industrial products	23

SOURCE: The Rauch Guide to the Packaging Industry
(1986). Rauch Associates, Inc., Bridgewater, N.J.

not necessarily impervious to fat. Similarly, a greaseproof material is not necessarily impervious to moisture.

Gas and odor protection also works two ways. Off-odors should be sealed out, but desirable odors such as the aroma of coffee or the essence of vanilla should be sealed in. For storage stability of many foods, oxygen must be prevented from entering the package. Yet some products generate carbon dioxide, which should escape from the package; this is the case with certain gas-evolving dough.

Physical protection prevents breakage of the package and subsequent product contamination. Resistance to product damage from impact or other physical stress (such as protection of crackers from breaking) is often a function of the secondary package.

Transparency and protection from light are contradictory objectives. A transparent package is desirable because it allows consumers to see what they are purchasing. Whereas most foods are light sensitive, at least to some degree, the choice of container must take into account the probable normal shelf life of the product and how much damage light will do in this length of time. Colored bottles for beer, wine, and juices are a common compromise.

Tamper-resistance or tamper-evident features are especially important for food packaging. Consider the practice of a shopper opening foods in screw-capped jars and tasting them with the finger for acceptability, and then closing the jar and replacing it on the shelf. This occurred enough times in the past to cause virtually a universal shift away from simple screw top covers to packaging which cannot be easily opened without breaking a seal or leaving other evidence that the jar has been opened. In recent years, many cases of malicious tampering have occurred where poisonous or harmful objects were placed inside contains. Tamper-evident packaging is now seen as a primary deterrent to such practices.

Tamper-indicating devices include plastic bands that seal the closure to the container and membranous films sealed across the mouth of a container beneath the removable lid (Fig. 21.3). These also minimize chances of product leakage, gas transfer, and aroma loss.

Ease of opening is perhaps best exemplified by the pull-tab beer and soda cans and the twist-off crown caps whose forerunners required a can or bottle opener. These technological developments had to balance minimum force for ease of opening against the potential of bursting from the internal pressure of carbon dioxide.

Dispensing features apply to containers for many granular, liquid, and particulate solids, from breakfast cereals to salt, as well as liquids. The flow properties of these materials determine the size and type of dispenser. Resealability has long been provided in screw-type bottle caps and lids. Resealability has been applied to baked goods,

Figure 21.3. Plastic tubs containing dairy products with inner membrane seals for tamper evidency.

cheese, and specialty foods in the form of plastic bags with "zipper" or press adhesive seals and clip or twist ties.

Lastly, but importantly, packaging should not have adverse environmental costs. Disposal of solid waste is an environmental problem, and packaging makes up a significant part of the total solid waste stream.

This partial list of requirements and functions is sufficient to illustrate the variability called for in packaging, especially when the tens of thousands of different food items stocked by a modern supermarket are considered. Packaging requirements are even more complex for products destined for harsher conditions of handling and storage than exist in air-conditioned supermarkets. These range from package survival during military or emergency air drops to resistance of the package to moisture and mold deterioration under tropical jungle conditions. Fortunately, very few packages must have all the properties described here.

TYPES OF CONTAINERS

Primary, Secondary, and Tertiary

Food packaging can be divided into primary, secondary, and tertiary types. A primary container is one that comes in direct contact with the food, for example, a can or a jar. Obviously, primary containers must be nontoxic and compatible with the food and cause no color, flavor, or other foreign chemical reactions. A secondary container is an outer box, case, or wrapper that holds or unitizes several cans, jars, or pouches together but does not contact the food directly. Secondary containers are a necessary part of food packaging. It would not be possible to distribute products in glass jars, for example, without the corrugated secondary carton to protect against breakage. As pointed out above, primary containers (i.e., those contacting foods) have several important functions. Secondary containers have fewer but no less important functions. Secondary containers must protect the primary containers from damage during shipment and storage. They must also prevent dirt and contaminants from soiling the primary containers and must unitize groups of primary containers. Corrugated fiberboard is most commonly used to make secondary shipping cartons. There are strict standards on the construction and use of secondary containers to ship products. The size and strength of carton used must be selected for each type of product to be shipped. Cartons come in several different designs, each of which is intended for different types or forms of products (Fig. 21.4). Damage during shipment which can be shown to result from inadequate secondary containers are the responsibility of the food manufacturer, not the shipping agent. Except in special instances, secondary containers are not designed to be highly impervious to water vapor and other gases. Dependence for this is placed on the primary container.

Tertiary containers group several secondary cartons together into pallet loads or shipping units. The objective is to aid in the automated handling of larger amounts of products. Typically, a forklift truck or similar equipment is used to move and transport these tertiary loads (Fig. 21.5).

Form-Fill-Seal Packaging

Containers may be preformed, that is, fabricated by the packaging manufacturer; or they may be formed in-line by assembly from roll stock or flat blanks just ahead of

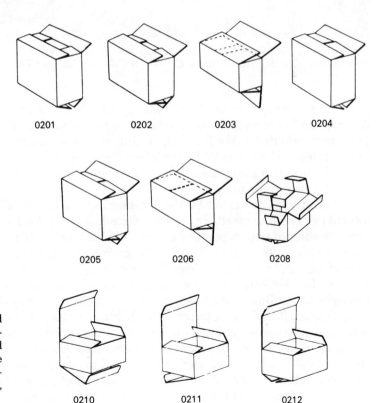

Figure 21.4. Some standard designs for secondary corrugated shipping cartons based on international case code 0200. Source: *Paine, The Packaging Media, Blackie & Sons, London, 1977.*

0201   0202   0203   0204

0205   0206   0208

0210   0211   0212

Figure 21.5. Automated palletizing and rotating stretch wrapping system for stabilizing unit (tertiary) loads prior to removal with a fork lift truck. *Courtesy of Columbia Machine, Inc.*

the filling operation in the food-handling line. This latter approach is called "form-fill-seal" and is one of the most efficient ways to package food. Today most flexible containers, whether made from paper, foil, or plastic, are in-line formed, resulting in great savings in handling labor, container transportation costs, and warehouse storage space. Figure 21.6 illustrates one common way a roll of flexible packaging material can be formed into a package.

Preforming is illustrated by the can-making machine of Fig. 21.7, which may be located in a can-manufacturing facility distant from the food plant. Formed cylindrical cans are shipped with separate can lids. The lids are seamed onto the cans by the food manufacturer after filling. The handling problems and expense are obvious in this type of operation. For this reason many large food companies set up can-making facilities in a building or area close by the can-filling line, and cans are continuously conveyed to the line. The same problems exist with preformed glass bottles.

In contrast, flexible foils, papers, plastic films, and laminates in the form of rollstock lend themselves to an endless number of in-line high-speed package forming, filling, and sealing operations. Common techniques employed with particulate foods are illustrated in Fig. 21.8. Related methods are used to package a wide variety of liquid and solid foods, from individual portions of jam and single slices of American cheese to the multipacks of sausage products made skin-tight by vacuum forming.

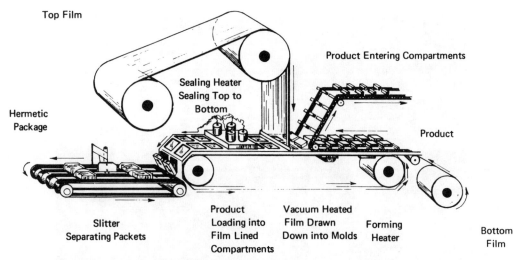

Figure 21.6. A vacuum form-fill-seal (FFS) process commonly used to package processed meats such as frankfurters, mozzarella cheese, and other products. Such packages may be gas flushed or vacuum drawn. Source: A. L. Brody, *Flexible Packaging of Foods*. CRC Press, Cleveland, OH, *1970*.

One of the earliest in-line packaging operations was the milk carton system in which cartons were assembled from coated fiber flats, filled, and sealed. Although this is still one of the most important in-line packaging processes, it is now being joined by various systems for aseptically packaging previously sterilized liquids. One such system, utilizing hydrogen peroxide and radiant heat to sterilize heat-sensitive roll-stock material is described later in this chapter. Some of the advanced methods for packaging milk and other liquids in plastic bottles call for the in-line conversion of thermoplastic resins in powder or pellet form into bottles by high-temperature blow-molding techniques, which at the same time heat sterilize the containers. The automatically filled containers are then heat sealed, again taking advantage of the thermofusing properties of the plastic.

Hermetic Closure

The term *hermetic* refers to a container that is sealed completely against the ingress of gases and vapors. Such a container, as long as it remains intact, will also be impervious to bacteria, yeasts, molds, and dirt from dust and other sources, since all of these agents are considerably larger than gas or water vapor molecules. On the other hand, a container that prevents entry of microorganisms, in many instances, will be nonhermetic; that is, it will allow some gases or vapors to enter. Hermetically sealed containers not only protect the product from moisture gain or loss and oxygen pickup from the atmosphere but are essential for vacuum and pressure packaging.

The most common hermetic containers are rigid metal cans and glass bottles, although faulty closures can make them nonhermetic. With rare exceptions flexible packages are not truly hermetic for one or more of the following reasons: (1) thin flexible films, even when they do not contain minute pinholes, generally are not completely impermeable to gases and water vapor, although the rates of transfer may be exceptionally slow; (2) the seals are sometimes imperfect; and (3) even when film materials are very high barriers to gases and water vapor (e.g., laminants containing aluminum

Figure 21.7. Metal sheets being prepared for can-making. *Courtesy of U.S. Steel Corporation.*

foil), flexing of packages and pouches can lead to minute pinholes and creases which allow gas and vapor transmission.

FOOD-PACKAGING MATERIALS AND FORMS

There are a relatively few materials used in food packaging: metal, glass, paper and paperboard, plastics, and minor amounts of wood and cotton fiber (Table 21.2). However, within each of these categories many types of packaging materials or combinations of materials are available. In the case of polypropylene film alone, there are dozens of types of films and laminates varying in moisture permeability, gas permeability, flexibility, stretch, burst strength, and so on. Often, a new food product requires its own special package since optimum protection, economic considerations, and merchandising requirements change rapidly with variations in product composition, weight and form, and performance demands.

Packaging materials are found in a wide variety of forms including the following:

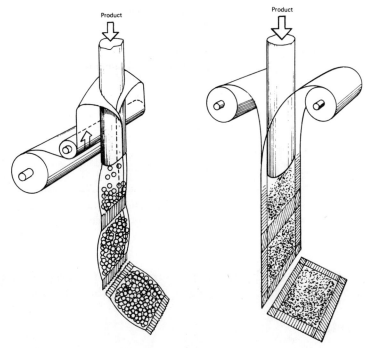

Figure 21.8. One method of forming pouches using vertical form-fill-seal machines. Pouches can also be formed on horizontal machines. *Courtesy of T. E. Wolmsley and M. Bakker (editor).* Encyclopedia of Packaging Technology. John Wiley & Sons, N.Y. 1986.

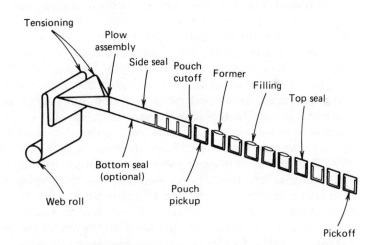

Figure 21.8. *Continued*

rigid metal cans and drums; flexible aluminum foils; glass jars and bottles; rigid and semirigid plastic cans and bottles; flexible plastics made from many different films used for bags, pouches, and wraps; paper, paperboard, and wood products in boxes, pouches, and bags; and laminates or multilayers in which paper, plastic, and foil are combined to achieve properties unattainable with any single component.

In addition to the many forms of packaging, food packaging also encompasses the equipment and machinery for producing or modifying certain packaging materials, for forming them into the final containers, for weighing and dispensing of food materials, for vacuumizing or gas flushing the containers, and for sealing the final packages. Food packages in many instances must withstand additional processing operations

Table 21.2. Projected Dollar Value of
Packaging in the United States in 1990 in
Billions of Dollars

Paperboard and molded pulp	24.6
Metal	18.6
Plastics	16.0
Paper	5.6
Glass	5.2
Wood	2.0
Textile	0.6
Total	72.6

SOURCE: The Rauch Guide to the Packaging
Industry (1986). Rauch Associates Inc.,
Bridgewater, N.J.

such as heat sterilization in pressure retorts, freezing and thawing in the case of frozen
foods, and even final cooking or baking in the package.

Metal

Two basic types of alloyed metals are used in food packaging: steel and aluminum.
Steel is used primarily to make rigid cans, whereas aluminum is used to make cans
as well as thin aluminum foils and coatings. Until a few years ago, nearly all steel
used for cans was coated with a thin layer of tin to inhibit corrosion; hence, the name
"tin can." The tin was applied electrolytically at a rate of as little as 0.25 lb per 440
ft^2. The reason for using tin was to protect the metal can from corrosion by the food.
Tin is not completely resistant to corrosion, but its rate of reaction with many food
materials is considerably slower than that of steel.

Because of the expense, tin has been replaced in the United States and elsewhere
by corrosion-resistant steel alloys, steel alloys with other thinner metallic coatings,
as well as improved interior polymeric coatings which help the steel resist corrosion.
This material is termed tin-free steel. For example, baseplate steel is coated with very
thin layers of chromium followed by chromium oxide that are much thinner than a
layer of tin but equally protective. The chromium oxide layer is further covered with
an organic coating that is compatible with the food. This means that the nature of the
baseplate steel is of major importance; various steels are used depending on the product
to be canned. The specifications for five types of base steels used in food canning are
listed in Table 21.3. The classes of foods requiring these different steels are also
described in Table 21.3. Thus, Type L plate is used with the most highly corrosive
foods, which are generally acidic. At the other extreme are mildly corrosive or noncorro-
sive low-acid foods and dry products that may require cans with Type MR or MC steel.

The strength of the steel plate is another important consideration especially in larger
cans that must withstand the pressure stresses of retorting, vacuum canning, and
other processes. Can strength is determined by the temper given the steel, the thickness
of the plate, the size and geometry of the can, and certain construction features such
as horizontal ribbing to increase rigidity. This ribbing is known as beading. The user
of cans will find it necessary to consult frequently with the manufacturer on specific

Table 21.3. Types of Steel Base Required for General Classes of Food Products

Class of Foods	Characteristics	Typical Examples	Steel Base Required
Mostly strongly corrosive	Highly or moderately acid products, including dark-colored fruits and pickles	Apple juice Berries Prunes Cherries Pickles	Type L
	Acidified vegetables	Sauerkraut	Type MS
Moderately corrosive	Mildly acid fruit products	Apricots Figs Grapefruit Peaches	Type MR
Mildly corrosive	Low-acid products	Peas Corn Meats Fish	Type MR or MC
Noncorrosive	Mostly dry and nonprocessed products	Dehydrated soups Frozen foods Shortening Nuts	Type MR or MC

SOURCE: R. F. Ellis

applications, since metal containers like all other materials of packaging are undergoing constant change.

Aluminum is lightweight, resistant to atmospheric corrosion, and can be shaped or formed easily. However, aluminum has considerably less structural strength than steel at the same gauge thickness. This means that aluminum has limited use in cans such as those used with retorted foods. Aluminum works well in very thin beverage cans that contain internal pressure such as soda or beer. This internal pressure from the CO_2 gives rigidity to the can. Common types of beverage cans use a ring riveted to the lid which is scored to facilitate easy opening. Scored aluminum pulls apart with less force than comparably scored steel. Scored aluminum lids also are being sealed to steel can bodies in great numbers. When this is done, special care must be taken that there is an unbroken enamel coating between the two metals in contact, otherwise bimetallic reactions can occur that can be harmful to the contained food. Aluminum in contact with air forms an aluminum oxide film which is resistant to atmospheric corrosion. However, if the oxygen concentration is low, as it is within most food-containing cans, this aluminum oxide film gradually becomes depleted and the underlying aluminum metal is then no longer highly resistant to corrosion. This can be overcome with enamel coatings similar to those used to protect steel and tin.

Aluminum is used in very thin gauges (approximately 35 ten-thousandths of an inch or 9 μm) as a foil in many packaging applications. When rolled this thin, aluminum acts as a very good barrier to O_2 and water vapor transmission but is very fragile. Strength is added by laminating the foil to a stronger material such as paper or plastic films. In this way, the strength of the film or paper can be combined with the barrier of the aluminum foil to produce a high quality package.

The principle disadvantage to aluminum is its requirement for large amounts of electricity for isolation from aluminum-containing ores. For this reason, recycling of aluminum containers has been successful.

As mentioned above, the inside and outside of metal cans is coated with organic coatings to further inhibit corrosion. Common coating materials approved by the FDA and their uses are indicated in Table 21.4. The coatings not only protect the metals from corrosion by food constituents but also protect the foods from metal contamination, which can produce a host of color and flavor reactions depending on the specific food. Particularly common are dark-colored sulfides of iron and tin produced in low-acid foods that liberate sulfur compounds when heat-processed. Bleaching of red plant pigments can occur on contact with unprotected steel, tin, and aluminum.

Metal Cans

The hermetic property of the metal can is a remarkable engineering achievement when one considers that cans are manufactured and later sealed at speeds exceeding 1000 units per minute and defective cans are fewer than one in many tens of thousands. The hermetic property of steel is extended to the seals by five folds of metal at the double-seam can ends. Between these folds is an organic sealing compound to ensure gas-tightness.

Can Construction.　Metal cans for food and beverage packaging can be divided into two basic types based on method of construction; three-piece and two-piece. Three-piece cans are comprised of a cylindrical body and two end pieces. Two-piece cans are made from one single body and end unit and one can end piece which is applied after

Table 21.4 General Types of Can Coatings

Coating	Typical Uses	Type
Fruit enamel	Dark-colored berries, cherries and other fruits requiring protection from metallic salts	Oleoresinous
C-enamel	Corn, peas, and other sulfur-bearing products, including some sea foods	Oleoresinous with suspended zinc oxide pigment
Citrus enamel	Citrus products and concentrates	Modified oleoresinous
Seafood enamel	Fish products and meat spreads	Phenolic
Meat enamels	Meat and various specialty products	Modified epons with aluminum pigment
Milk enamel	Milk, eggs, and other dairy products	Epons
Beverage can enamel (non-carbonated beverages)	Vegetable juices; red fruit juices; highly corrosive fruits; noncarbonated beverages	Two-coat system with oleoresinous type base coat and vinyl top coat
Beer can enamel	Beer and carbonated beverages	Two-coat system with oleoresinous or polybutadiene type base coat and vinyl top coat

SOURCE: R. F. Ellis

the can is filled with product (Fig. 21.9). Two-piece cans do not have side seams. The side seam of cans made from three pieces is most commonly welded in the United States. Side seams which are soldered with a tin–lead alloy are still in common use in parts of the world. Two-piece cans do away with the need for solder, which can contribute undesirable traces of lead to the food. Rigid aluminum containers can also be readily formed without side seams or bottom end seams by draw and ironing techniques. This is the type of can which is commonly used for carbonated beverages. The absence of one of the end seals and the side seal reduces the risk of can failure.

Can Corrosion. In years past, the steel used to make cans was protected from corrosion such as rust pitting by a thin electrolytically deposited coating of tin. The tin reduced the chance of corrosion of the steel (i.e., iron) by acting as an anode in a galvanic-type cell with the food serving as the electrolyte (Fig. 21.10). As the tin dissolves, electrons are transferred to the iron which prevents dissolution,

Tin has largely been replaced by other nonrusting metals such as chromium or the base steel may be given special rust-inhibiting treatment called "passivation." These types of metals are termed *tin-free* and used because of their lower cost. The inside and often outside of the can is further protected against rusting, pitting, or reaction with the food by a thin layer of nonrusting metal and a baked-on resin. There are several resins that are selected based on the type of food to be canned (Table 21.4).

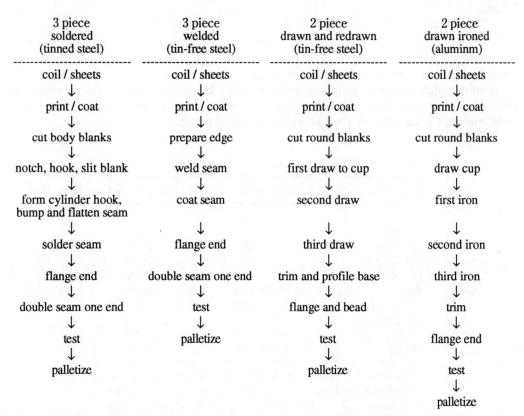

3 piece soldered (tinned steel)	3 piece welded (tin-free steel)	2 piece drawn and redrawn (tin-free steel)	2 piece drawn ironed (aluminm)
coil / sheets	coil / sheets	coil / sheets	coil / sheets
↓	↓	↓	↓
print / coat	print / coat	print / coat	print / coat
↓	↓	↓	↓
cut body blanks	prepare edge	cut round blanks	cut round blanks
↓	↓	↓	↓
notch, hook, slit blank	weld seam	first draw to cup	draw cup
↓	↓	↓	↓
form cylinder hook, bump and flatten seam	coat seam	second draw	first iron
↓	↓	↓	↓
solder seam	flange end	third draw	second iron
↓	↓	↓	↓
flange end	double seam one end	trim and profile base	third iron
↓	↓	↓	↓
double seam one end	test	flange and bead	trim
↓	↓	↓	↓
test	palletize	test	flange end
↓		↓	↓
palletize		palletize	test
			↓
			palletize

Figure 21.9. Comparison of the steps in manufacturing three-piece soldered side seam, three-piece welded side, drawn and redrawn two-piece, drawn and ironed two-piece metal cans.

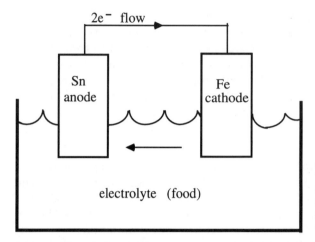

2e⁻ flow

Sn anode

Fe cathode

electrolyte (food)

Figure 21.10. A tin-coated can acts as a galvanic cell where tin (Sn) is a "sacrificial anode" and protects the iron (Fe) from corrosion. The food is the electrolyte. Rapid corrosion of the steel occurs if the polarity is reversed as in the presence of oxygen.

Can Sizing. Cans are given standard size designations based on their diameter and height in whole inches plus sixteenths of an inch. Thus, a 303 × 404 can has a diameter of 3³⁄₁₆ in. and a height of 4⁴⁄₁₆ in. Table 21.5 lists several standard can sizes, their volume, and their standard name.

Glass

As a food-packaging material, glass is chemically inert and an absolute barrier to the permeation of O_2 or water vapor. The principal limitations of glass are its susceptibility to breakage, which may be from internal pressure, impact, or thermal shock, its weight which increases shipping costs, and the large amounts of energy required for forming into containers. Glass is primarily formed from oxides of metals, with the most common being silicon dioxide which is common sand.

Table 21.5. Selected Standard Can Sizes and Volumes

Can Name	Dimension[a]	Fill Volume[b]
62	203 × 308	9.42
82 short	211 × 300	12.34
No. 1 picnic	211 × 400	17.05
211 cylinder	211 × 414	21.28
No. 300	300 × 407	23.71
303	303 × 406	26.31
303 cylinder	303 × 509	34.11
No. 2 vacuum	307 × 306	22.90
No. 2	307 × 409	32.00
No. 10	603 × 700	170.71

[a] The first digit gives the dimension in whole inches. The second and third digits give the fraction in sixteenths of an inch. Thus, a 303 × 406 can is 3³⁄₁₆ in. in diameter and 4⁶⁄₁₆ in. high.
[b] In cubic inches. 1 cubic inch = 0.554 fluid ounces.

Forming glass containers from a carefully controlled mixture of sand, soda ash, limestone, and other materials made molten by heating to about 1500°C is seen in Fig. 21.11. After forming, the containers are sent through curing (annealing) ovens to impart toughness or temper to the glass. Apart from influences of chemical composition, optimum shaping of the container, times and temperatures of forming, annealing, and cooling of jars and bottles, and other production practices, the breakage properties of glass containers can be minimized by proper choice of container thickness and coating treatments.

The heavier a jar or bottle of a given volume is, the less likely it is to break from internal pressure. A heavier jar, however, is more susceptible to both thermal shock and impact breakage. The greater sensitivity to thermal shock of heavier jars is due to wider temperature differences which cause uneven stresses between the outer and inner surfaces of the thicker glass. The greater susceptibility to impact breakage of heavier jars is due to less resiliency in thicker walls.

Coatings of various types can markedly reduce breakage by protecting the surface from scratches and nicks. Scratches and nicks substantially weaken glass. These coat-

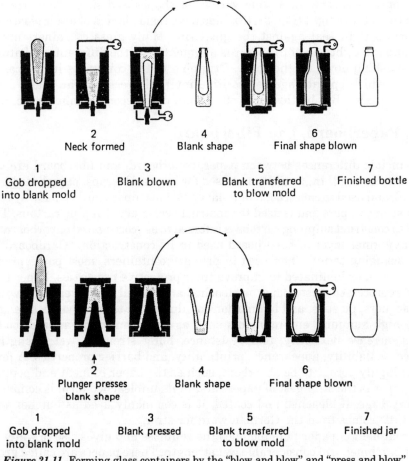

Figure 21.11. Forming glass containers by the "blow and blow" and "press and blow" techniques. Source: G. L. Robertson, *Food Packaging: Principles and Practice.* Marcel Dekker, NY. *1993.*

ings, commonly of special waxes and silicones, impart lubricity to the outside of glass containers. As a result, impact breakage is lessened because bottles and jars glance off one another rather than sustain direct hits when they are in contact in high-speed filling lines. Further, after coming from the annealing ovens, the glass surfaces, virtually free of abrasions, quickly acquire minute scratches in normal handling. These scratches are weak points where many of the subsequent internal pressure and thermal shock breaks originate. Surface coating also improves the high-gloss appearance of glass containers and is said to decrease the noise from glass-to-glass contact at filling lines, probably due to the increased rate of glancing blows rather than direct impacts.

To help prevent thermal shock, it is good practice to minimize temperature differences between the inside and outside of glass containers wherever possible. Some manufacturers recommend that the temperature difference between the inside and outside should not exceed 44°C. This requires slow warming of bottles before they are used for a hot fill, and partial cooling before such containers are placed under refrigeration.

Glass Containers

Glass containers come in a wide variety of shapes and sizes. They are hermetic, provided the lids are tight (Fig. 21.12). Lids have inside layers of a soft plastic material which form a tight seal against the glass rim. Many glass containers are vacuum packed, and the tightness of the cover is augmented by the differential of atmospheric pressure pushing down on the cover. Crimping of the covers, as in the case of soda bottle caps which operate against positive internal pressure, can make a gas-tight hermetic seal also. But bottles more often than cans become nonhermetic.

Paper, Paperboard, and Fiberboard

The principle differences between paper, paperboard, and fiberboard are thickness and use. Papers are thin, flexible, and used for bags and wraps; paperboard is thicker, more rigid, and used to construct single-layer cartons; fiberboard is made by combining layers of strong papers and is used to construct secondary shipping cartons. The material used to construct shipping cartons is referred to as "corrugated paperboard" because of the wavy inner layer of paperboard used in its construction. "Cardboard" is not a correct packaging term. When used in primary containers, most paper products are treated, coated, or laminated to improve their protective properties. Paper from wood pulp and reprocessed waste paper is bleached and coated or impregnated with waxes, resins, lacquers, plastics, and laminations of aluminum to improve its strength, especially in high-humidity environments such as are often found around foods. Other additives increase flexibility, tear resistance, burst strength, wet strength, grease resistance, sealability, appearance, printability, and barrier properties. A few papers are made highly porous to be absorbent, such as the paper in meat and poultry trays.

Kraft paper is the strongest of papers and in its unbleached form is commonly used for grocery bags. If bleached and coated, it is commonly used as butcher wrap. The word "Kraft" comes from the German word for strong.

Acid treatment of paper pulp modifies the cellulose and gives rise to water- and oil-resistant parchments of considerable wet strength. These papers are called greaseproof or glassine papers and are characterized by long wood pulp fibers which impart increased physical strength.

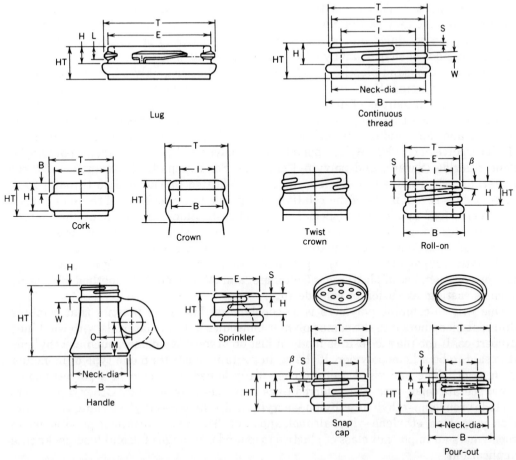

Figure 21.12. Typical glass container closures commonly used for foods and beverages. Source: M. Bakker (Editor) *Encyclopedia of Packaging Technology*, John Wiley & Sons, NY *1986*.

Paper that comes in contact with food must meet FDA standards for chemical purity and its coatings must be nontoxic. Additionally, the microbiological condition of paper products is rigidly specified by food manufacturers and in certain food ordinances. Thus, the Grade A Pasteurized Milk Ordinance of the U.S. Public Health Service– FDA states that paper for milk cartons and caps be made from sanitary virgin pulp and contain no more than 250 colonies per gram of disintegrated stock by a standard bacteriological test.

Plastics

The term *plastics* refers to a broad group of materials that have the common property of being composed of very large long-chain molecules. These molecules may have molecular weights of 100,000 or more and are made by connecting small repeating molecules called "monomers" together in a head-to-tail fashion. Polymer chemists have copied much from nature in which many polymers occur naturally, such as starches, proteins, and natural rubber. This molecular arrangement gives plastics some unusual physical properties. Thermoplastic polymers can be melted repeatedly, for example. This means that they can be melted, formed into a desired shape by one of several processes, and solidified on cooling. This allows plastics to be formed in an almost infinite number of shapes, many of which are useful as packages.

Of the few thousand plastics which have be synthesized, only 20 or so are used to make food packaging. However, these 20 polymers are combined in a variety of ways so that several hundreds of different plastic-containing structures are commercially available for food-packaging applications. Among the more important plastics used for films and semirigid containers for food packaging are cellulose acetate, polyamide (Nylon), polyesters (PET, Mylar), polyethylene, polypropylene, polystyrene, polyvinylidene chloride (Saran), and polyvinyl chloride. Some important properties of these materials, when made into flexible films are indicated in Tables 21.6 through 21.8. These tables reveal many of the relative strengths and weaknesses of these materials for specific food applications (special uses and restrictions will be mentioned later). These tables do not begin to convey the variety of products that can be made from these materials depending on many variables of their manufacture [e.g., the identity and mixture of polymers, degree of polymerization and molecular weight, spatial polymer orientation, use of plasticizers (softeners) and other chemicals, method of forming such as casting, extrusion or calendering].

One way to combine polymers is as copolymers. These are plastics that combine different monomers into the same polymer molecules to form materials with combined properties. If the plastic resin is made of just one type of monomer, such as ethylene, it is said to be a homopolymer. If the resin contains more than one type of monomer such as ethylene and vinyl acetate, chemically joined, it is termed a polyethylene–vinyl acetate copolymer, also referred to as ethylene–vinyl acetate. Other copolymers include propylene–ethylene, ethylene–acrylic acid, ethylene–ethyl acrylate, vinyl chloride–propylene, ethylene–vinyl alcohol, and so on. The many variations possible make copolymers an important class of plastics to extend the range of useful food-packaging applications.

Another new class of plastic materials, the ionomers, further illustrates how the properties of plastics can be modified. Carboxylic acid groups can be added to the polymer chains of polyethylene. These acid groups form strong interactions between polymer chain molecules and affect the physical properties of the resulting plastic. The interactions can impart such improved functional properties as greater oil, grease, and solvent resistance, and higher melt strength. The range of applications for ionomers in food packaging is expanding.

Laminates

As noted already, packages made of polymer films are not absolute barriers against the transfer of water and O_2 through the package, although they may be excellent barriers against microorganisms and dirt. Fortunately, not all foods need absolute hermetic protection. Various flexible materials (papers, plastic films, thin metal foils) differ with respect to water vapor transmission, oxygen permeability, light transmission, burst strength, pinhole and crease hole sensitivity, and so on. Multilayers or laminates of these materials that combine the best features of each can be used to produce packaging materials with combined properties such as the strength of paper, heat seal ability of plastics, and barrier properties of aluminum foils (Fig. 21.13; Table 21.9).

Table 21.6. General Characteristics of Packaging Films[a]

Film Material	Thickness (in.)	Clarity[b]	Specific Gravity	Use Limits					Heat Sealing Range (°F)
				Max Temp		Min Temp		Resistance to Sunlight	
				(°C)	(°F)	(°C)	(°F)		
Cellophane, lacquered[b]	0.0009–0.0017	Transparent	1.40–1.55	About 149	About 300	−4	24	Good	200–350
Cellophane, polymer coated[b]	0.001–0.002	Transparent	1.44	About 149	About 300	About −18	About 0	—	200–350
Cellophane, polyethylene coated[b]	0.002 and up	Transparent to translucent	1.2	82	180	—	—	Good	230–300
Cellulose acetate	0.00088–0.250	Transparent	1.28–1.31	66–93	150–200	−26	−15	Good	350–450
Nylon-6	0.0005–0.030	Transparent to translucent	1.13	93–204	200–400	−73	−100	Fair to good	380–450
Polyester	0.00012–0.014	Transparent	1.38–1.41	149	300	−73	−100	Fair	425–450
Polyethylene, low density	0.0003 and up	Transparent to opaque	0.910–0.925	82–93	180–200	−57	−70	Fair to good	250–400
Polyethylene, high density	0.0004 and up	opaque	0.941–0.965	121	250	−46	−50	good	275–400
Polypropylene, unoriented	0.00087–0.010	Transparent	0.885–0.905	132–149	270–300	−18	0	Fair	285–400
Polypropylene, oriented	0.0005–0.00125	Transparent	0.902–0.907	140–146	285–295	−51	−60	Fair	300–320
Polystyrene, oriented	0.00025–0.020	Transparent	1.05–1.06	79–96	175–205	−57 to −70	−70 to −94	Fair	250–350
Polyvinyl chloride	0.0005–0.100	Transparent to opaque	1.20–1.80	66–93	150–200	−29 to −46	−20 to −50	Good	300–420
Vinylidene chloride-vinyl chloride copolymer	0.0004–0.006	Transparent	1.59–1.71	143	290	—	—	Fair	240–300

[a]Representative values from Modern Plastics Encyclopedia, McGraw-Hill Co., 1983.
[b]Data from miscellaneous sources.

Table 21.7. Permeability and Chemical Properties of Packaging Films[a]

Film Material	Gas Transmission (cm³/100 in.²/24 hr/mil) at 25°C			Water Vapor Transfer (g/100 in.²/24 hr/mil) at 37.8°C 90% RH	% Water Absorption in 24 Hr	Resistance to				
	O_2	N_2	CO_2			Strong Acids	Strong Alkalies	Greases and Oils	Organic Solvents	Water
Cellophane, lacquered[b]	1	1	13	0.2–1.0	High	Poor	Poor	Good	Poor	Fair
Cellophane, polymer coated[b]	0.5	0.5	0.5	0.4–0.9	High	Variable	Good	Good	Good	Good
Cellophane, polyethylene coated[b]	—	—	—	1.2 and up	—	Good	Good	Variable	Good	Good
Cellulose acetate	117–150	30–40	860–1000	About 150	3–9	Poor	Poor	Good	Poor	Good
Nylon-6	2.6	0.9	10–12	16–22	9.5	Poor	Good	Good	Good	Variable
Polyester	3.0–4.0	0.7–1.0	15–25	1.0–1.3	<0.8	Good	Poor	Good	Good	Good
Polyethylene, low density	500	180	2700	1.0–1.5	<0.01	Good	Good	Poor	Good	Good
Polyethylene, high density	185	42	580	0.3	Nil	Good	Good	Good	Good	Good
Polypropylene, unoriented	150–240	40–48	500–800	0.7	<0.005	Good	Good	Good	Good	Good
Polypropylene, oriented	160	20	540	0.25	<0.005	Good	Good	Good	Good	Good
Polystyrene, oriented	250–350	—	900	7.0–10.0	0.04–0.10	Good	Good	Variable	Variable	Good
Polyvinyl chloride	8–160	1–70	20–1900	4–10	Nil	Good	Good	Good	Variable	Good
Vinylidene chloride–vinyl-chloride copolymer	0.8–6.9	0.12–1.5	3.8–44	0.2–0.6	—	Good	Good	Good	Variable	Good

[a]Representative values from Modern Plastics Encyclopedia, McGraw-Hill, Co., 1983.
[b]Data from miscellaneous sources.

Table 21.8. Mechanical Properties of Packaging Films[a]

Film Material	Tensile Strength (100 psi)	Elongation (%)	Tearing Strength (g/mil)	Bursting Strength, 1 Mil Thick (psi)	Folding Endurance
Cellophane, lacquered[b]	70–180	15–25	2–10	55–65	Good
Cellophane, polymer coated[b]	70–180	25–50	7–15	—	—
Cellophane, polyethylene coated[b]	70–180	15–25	2–10	40–50	Good
Cellulose acetate	70–164	15–70	4–10	30–60	Fair
Nylon-6	90–180	250–550	50–90	Elongates	Very high
Polyester	200–350	60–165	12–27	55–80	Very high
Polyethylene, low density	15–30	100–700	50–300	10–12	Very high
Polyethylene, high density	24–61	10–650	15–300	—	Good
Polypropylene, unoriented	45–70	550–1000	—	—	Very high
Polypropylene, oriented	75–400	35–475	3–10	—	Good
Polystyrene, oriented	80–120	3–40	5	16–35	—
Polyvinyl chloride	14–160	3–500	10–1400	20–40	—
Vinylidene chloride– vinyl chloride copolymer	80–160	30–80	10–<100	25–35	Very high

[a]Representative values from Modern Plastics Encyclopedia, McGraw-Hill Co., 1983.
[b]Data from miscellaneous sources.

Commercial laminates containing up to as many as eight different layers are commonly custom-designed for a particular product. In the case of a quality instant tea mix, for example, the laminate (progressing from the exterior of the package inward) may have a high quality paper exterior that is printable, a layer of polyethylene to serve as an adhesive to the next layer, and, in the middle, a layer of aluminum foil that serves as the gas barrier, and an innermost layer of polyethylene to provide the thermoplastic material for heat-sealing the package's inner surfaces. Laminations of different materials may be formed by various processes including bonding with a wet adhesive, dry bonding of layers with a thermoplastic adhesive, hot melt laminating where one or both layers exhibit thermoplastic properties, and special extrusion techniques.

Another new technique for combining different plastics is coextrusion. Coextrusion simultaneously forces two or more molten plastics through adjacent flat dies in a manner that ensures laminar flow and produces a multilayer film on cooling (Fig. 21.14). Such structured plastic films may be complete in themselves or be further bonded to papers or metal foils to produce more complex laminates.

Retortable Pouches and Trays.

Flexible materials can be combined to withstand even the adverse conditions of retorting encountered with low-acid foods. Such "flexible cans" have become standard

Figure 21.13. Flexible laminants are used to package a complete meal including retorted entree for military use by soldiers in the field. Source: S. Sacharow and R. C. Griffin *Principles of Food Packaging, 2nd ed., AVI Publishing Co., Westport, CT, 1980.*

containers for some applications such as providing foods to soldiers in the field. The advantages of pouches and trays over cans and jars of equivalent volume include shorter retort times, which can produce higher quality products and save on energy, lighter weight, increased compactness, easier opening, and easier disposability. Retortable pouches are constructed of a three-ply laminate consisting of (1) an outer layer of polyester film for high-temperature resistance, strength, and printability, (2) a middle layer of aluminum foil for barrier properties, and (3) an inner layer of polypropylene film that provides heat-seal integrity. Retortable trays are constructed from multilayers of polymers, one of which is ethylene–vinyl alcohol to provide an oxygen barrier. These trays are often sealed with a polymer–foil laminant film.

Edible Films

Edible films have been used for centuries. Sausage casings are one example. More recently there has been renewed interest in such films. For example, food materials can be protected from loss of volatiles or reaction with other food ingredients by being encapsulated in protective edible materials. This can be done by spray drying various flavoring materials emulsified with gelatin, gum arabic, or other edible materials to form a thin protective coating around each food particle (microencapsulation). The coating of raisins with starches to prevent them from moistening a packaged breakfast cereal and the coating of nuts with monoglyceride derivatives to protect them from oxidative rancidity are additional examples of edible coatings.

Table 21.9. Water Vapor Transmission Rate (WVTR) of Aluminum Foil Laminates

Material	Thickness (in.) Foil	Thickness (in.) Laminant	WVTR[a] (g/100 in.^{2}/24 h) Flat	WVTR[a] (g/100 in.^{2}/24 h) After Creasing[b]
Al. foil laminated to moisture-proof cellophane	0.00035	0.0009	0.00	—
			0.01	0.03
			0.01	0.01
Al. foil laminated to cellulose acetate	0.00035	0.0012	0.02	0.07
Al. foil laminated to rubber hydrochloride	0.00035	0.0008	0.01	0.01
			0.01	—
Al. foil laminated to vinyl polymer	0.00035	0.0012	0.01	0.02
Al. foil laminated with wax to 30-lb glassine	0.00035	—	0.00	0.04
Al. foil BEIS-O	0.00035	—	0.07	0.42
Al. foil laminated to moistureproof cellophane	0.001	0.0009	0.00	0.00
Al. foil laminated to vinyl polymer	0.001	0.0012	0.00	0.00
Al. foil laminated with wax to 35-lb glassine	0.001	—	0.00	0.02
Al. foil BEIS-O	0.001	—	0.00	0.40

SOURCE: Aluminum Co. of America.
[a]At 100°F, 100% relative humidity.
[b]"Creasing" means creased with four equidistant parallel folds and then with four more folds at right angles to the first.

Food materials such as amylose starch and the proteins zein and casein when solubilized can be cast to give sheets of edible films on drying. These films may then be used to fabricate small packets to hold other food ingredients. One application of such films has been to package baking ingredients which can then be added directly to the mixing bowl as an intact packet; on addition of water, the edible film dissolves and releases the packaged ingredients.

Edible films are also used to coat fresh fruits and vegetables to reduce moisture loss and to provide increased resistance to growth of surface molds. The most common and oldest edible film is wax. A wide range of products such as apples are waxed for appearance and improved keeping quality. Newer edible films are being developed which can keep produce longer. All edible films must be approved by the FDA for human consumption.

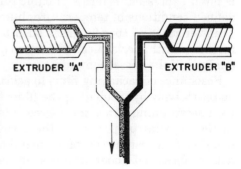

EXTRUDER "A"   EXTRUDER "B"

Figure 21.14. Two or more different polymers can be combined into a multiple-layer film or sheet by forcing different melted plastics through a single slit die. This process is known as coextrusion. Source: J. F. Hanlon *Handbook of Package Engineering, 2nd ed.,* Technomic Publishing, Co. Lancaster, PA. *1992.*

Wood and Cloth Materials

Woven cloth such as jute bags (burlap) and cotton bags are used to a limited extent, mostly for bulk shipment of grains and flours. Wire-wound wood strips have been used to make crates for fresh fruits and vegetables. Solid wooden crates are also used for transporting iced fish.

PACKAGE TESTING

Many test procedures exist to measure quantitatively the protective properties of packaging materials and entire containers. These can be divided into chemical and mechanical parameters. Examples of chemical tests are those used to identify plastics, determine if portions migrate to foods, and measure resistance to greases. Mechanical tests include such things as barrier properties, strength, heat-seal ability, and clarity. The tables in this chapter contain data from several of these tests. Mechanical properties of packaging films (e.g., tensile strength, elongation, tearing strength, bursting strength) are determined on specially designed instruments that precisely measure the forces required to produce these effects. Unfortunately, reporting of test data has lacked standardization and various forms of English and metric units continue to be used. Test data in the tables of this chapter have been kept in their original units, since these remain meaningful to suppliers and users of packaging materials. The packaging industry, however, can be expected to gradually replace many of its present designations with standardized metric units. One of the best sources of methods for testing packaging materials are the publications of the American Society for Testing and Materials (ASTM).

Water vapor transmission rates (WVTR) can be measured by sealing sheets and films across the opening of a vessel that contains a weighed quantity of a desiccant material. The vessel is then placed in an atmosphere of controlled temperature and humidity. Periodic weighing of the container of desiccant to determine water pickup gives a measure of water vapor transfer. This measure is commonly expressed in terms of grams per 100 in.^{2} of film, of 1 mil (0.001 in.) thickness, per 24 h under the defined conditions of temperature, humidity, and atmospheric pressure.

Gas transmission rates can be measured by an instrument which uses the test film to separate an inert gas from the test gas(es). The instrument then continuously measures increase in concentration of oxygen in the inert gas (Fig. 21.15). This increase in concentration with time can be used to calculate gas transmission rates. Gas transfer is often expressed in terms of cubic centimeters per 100 in.^{2} of film per 24 h under defined conditions of temperature, humidity, and pressure on both sides of the film. Transfer rates of specific gases such as oxygen, carbon dioxide, or nitrogen can be measured with special electrodes fitted into the sealed vessel or by gas chromatographic analysis of the vessel contents.

Resistance of packaging films to acids, alkalies, and other solvents can be measured quantitatively by incubating the films in the solvent under controlled conditions and then determining either the degree of leaching of the film into the solvent or changes in the physical properties of the recovered films by some of the methods already mentioned. Resistance of coated metal cans to acid can be estimated by a colorimetric test for dissolved underlying iron in the acid test solution. Resistance of metal cans

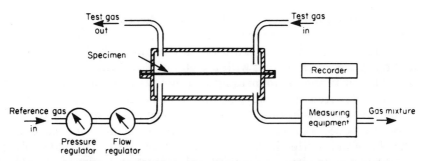

Figure 21.15. Method for determining the gas transmission rate (i.e., permeability) of a plastic film (specimen) by measuring the concentration of test gas in a reference gas which is separated from the test gas by the plastic film. Source: F. A. Paine and H. Y. Paine. *Handbook of Food Packaging. Blackie & Sons, London, 1983.*

to acid also can be established in terms of the rate at which hydrogen is given off by the corroding metal.

These are just a few of the approaches used to test package materials. But in the final analysis, although such data permit intelligent initial screening of suitable packaging materials for a particular food application, the final package and product are best evaluated in actual or simulated use tests. This is especially so when the food will receive additional processing (retorting, freezing, etc.) in the final package.

Actual use tests consist of sending limited numbers of food-filled packages through the processing, shipping, warehousing, and merchandising chain where they will be exposed to naturally occurring vibrations, humidities, temperatures, and handling abuses. Such packages are then recovered for analysis. Simulated use tests involve machines and devices for producing physical stresses, and incubation cabinets where packages can be subjected to various temperature and humidity cycles comparable to what the packaged food will subsequently experience in trade channels. Simulated use test conditions can often be intensified to arrive more quickly at a judgment of package performance.

PACKAGES WITH SPECIAL FEATURES

As pointed out, one of the newer requirements of food packaging is that it help with the product's use. This usually means that the package will have some type of added convenience feature. The "boil-in-bag" package is one of many examples. In addition to protecting the food against microorganisms and dirt, and to a certain degree against moisture and gas transfer, it also is impermeable to grease, nontoxic, compatible with the food, transparent, capable of being evacuated and heat-sealed under vacuum, attractive, tamper-evident, easy to open and dispose of, light in weight, requires little storage space, and is low in cost. But this is not all. Its material and seals withstand freezing temperatures and the expansion of foods frozen within it. It then survives frozen storage and the extreme shock of being taken from the home or restaurant freezer and plunged into boiling water for cooking. During boiling, the bag does not burst from steam or allow boiling water in to dilute the food. All of this is made possible by the exceptional properties of polyester and Nylon films including high tensile strength and stability over a range of temperatures from −73 to 150°C (Tables 21.6 through 21.8).

The plastic shrink package shown in Fig. 21.16 protects food against contamination and yet lets the customer see the meat. In addition, it keeps the meat from drying out. It is made to fit skin-tight by first drawing a vacuum on the bagged item, twisting the bag and tying a knot or sealing with a metal clip, and then passing the package through a mild heat tunnel or immersing it in hot water to shrink the plastic. The package may be made from polypropylene film if the product can tolerate or is favored by a moderate oxygen transfer rate. The polypropylene film is specially treated during manufacture to produce a biaxial orientation of its molecules. This contributes to a uniform shrinkage in all directions on heating to about 82°C. Oriented polyethylene and several other plastic films also have shrink properties. Cryovac "Type L" film is a shrinkable polyester. The shrink property is particularly useful in packaging poultry to be frozen because the skin-tight fit excludes pockets of air around the irregularly shaped bird and minimizes voids where water vapor can migrate to the package surface and result in desiccation (i.e., freezer burn) to the skin below. The shrink property also is exploited in the packaging of fragile vegetables and fruits to keep these items from becoming damaged. In this case, several individual items may be packed in a paper tray overwrapped with the shrinkable plastic; when shrunk, this type of package firmly holds the items in place, preventing bruising from loose movement. This differs somewhat from the skin-tight packages used for meat since it does not usually employ vacuum prior to the shrinking step.

Microwave Oven Packaging

One of the most rapidly developing areas of added convenience is packaging designed for the microwave oven (Fig. 21.1). Such packaging must meet all other standard

Figure 21.16. Cooked and processed meats packaged in shrink films. Many times meats are cooked in the bags in order to retain moisture. *Courtesy of Cryovac Division of W. R. Grace.*

requirements for food packaging, and also must be transparent to microwaves and able to withstand the temperatures encountered in heating foods in the microwave oven. The most commonly used materials for this application are made of plastics. Several plastics such as polyester and Nylon which are capable of withstanding higher temperatures have been used to package microwave foods. These plastics do not deform or char when exposed to temperatures in excess of 100°C.

One disadvantage of microwave heating is that the heating surface, in this case the package, does not get hot itself. This means that heat is not transferred to the food by conduction and the food does not brown or otherwise behave like conventionally cooked foods. In order to solve this problem, packaging engineers have constructed high-temperature polymeric packaging materials that contain very small aluminum particles (Fig. 21.17). These particles get hot during microwaving, which, in turn, heats the polymer which further heats the food by conduction. These materials are called "susceptors." Susceptors cause foods to brown in a microwave oven. This technology improves the quality of popcorn which is popped in a microwave oven, for example. There are many other examples of innovations resulting from the use of packaging in microwave ovens.

High Barrier Plastic Bottles.

An important development has been the recent introduction of squeezable plastic bottles that have very high barrier properties and are less than one-fourth the weight of glass, do not break when dropped, and can be incinerated without the production of toxic, corrosive, or noxious compounds beyond those found in burning household or municipal trash. This means that products which required the barrier properties of glass can be packaged in plastic. The reduced breakage of plastic bottles not only benefits consumers but also reduces costs throughout production and shipping channels. Bottle breakage is a frequent cause of complete disruption of filling-line operations. Resistance to breakage also permits use of lighter, less expensive corrugated shipping

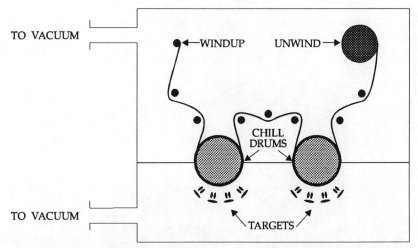

Figure 21.17. Typical manufacturing technology for depositing very small amounts of aluminum on high-temperature plastic films used to heat foods by conduction in a microwave oven. The technique is commonly referred to as "sputtering." Source: G. L. Robertson *Food Packaging: Principles and Practice,* Marcel Dekker, NY. *1993.*

containers. These containers also allow for easier dispensing of viscous products such as ketchup.

Aseptic Packaging in Composite Cartons.

Another development of worldwide significance has been the composite paper carton which is capable of being sterilized and then aseptically filled with sterile liquid products. This process is called aseptic packaging even though it is both a packaging and processing technology. This technology allows foods such as milk to be packaged in relatively inexpensive flexible containers which do not require refrigeration. This means that milk and juices can be distributed in parts of the world where refrigeration is not common. The packaging material is made from laminated roll stock consisting (from the outside inward) of polyethylene, paper, polyethylene, aluminum foil, polyethylene, and a coating of ionomer resin. With equipment shown in Fig. 21.18, the roll stock enclosed in a cabinet at floor level is drawn upward as a continuous sheet through a hydrogen peroxide bath near the top of the machine. The sheet is passed through squeeze-rollers to remove excess peroxide, and the descending sheet is formed into a tube that is exposed to radiant heat to complete the sterilization and remove traces of peroxide. Next, the tube is further formed into a rectangular shape, end-sealed at package-size intervals, filled with presterilized liquid food, top-sealed, and separated into individual package units in a continuous operation. Commercially sterile liquids have a shelf life of several months at room temperature in the exceptionally lightweight

Figure 21.18. Aseptic packaging system using form-fill-seal technique. *Courtesy of Tetra Pak, Inc.*

form-fill-seal package. Several form-fill-seal systems have been developed to take advantage of the rapidly growing aseptic package market.

Military Food Packaging

Special packaging problems have always confronted the military. In addition to providing protection, packages that simplify preparation and consumption of the food under adverse circumstances often are required. Thus, one type of military food container is designed with a chemical system separate from the food, which can be made to undergo a rapid exothermic reaction, thus heating the can on opening. It is also possible to design a self-cooling container that might be based on the rapid expansion and release of a compressed refrigerant gas. Such containers presently would be too expensive for general commercial use.

Newer Methods of Cooking and Foodservice

These create special package needs. An example is ovenable paperboard, which is impervious to moisture and fat staining and is heat resistant to 218°C for short periods. These properties, as well as a chinalike gloss, have been obtained by coating paperboard with PET polyester. Such paperboard containers, in the form of serving dishes, are now used for frozen foods to be reconstituted in conventional and microwave ovens.

Packaging and Communication.

The package designer knows that the message conveyed by the package is often the most important single factor that determines a product's sale or rejection. Among details to be considered are the package's color and symbolism. In one's own country errors are less likely to be made, but when packages are designed for distribution in a less familiar region, special care must be taken and the services of a consultant often employed. Pitfalls are many. Purple is an unlucky color to the Chinese and white connotes mourning. The cherry blossom is a favored symbol of the Japanese, but the chrysanthemum, which connotes royalty in Japan, is to be avoided on package labels. Further examples applying to Asian markets are listed in Table 21.10. One of the more subtle examples of regional psychology was experienced in Japan where a U.S.-designed tuna fish can pictured a tuna with nose turned down toward the water. When the product did not sell, it was learned that to Japanese a tuna with nose turned down meant a fish that was dead. When the picture of the tuna was modified, sales increased.

Distribution Packaging

Tertiary packaging combines several secondary units or cartons into a single unit for greater efficiency in distribution. The most common unit is usually a full pallet load. Related commercial handling methods involving ship-to-rail, truck-to-air, and other combinations are becoming more sophisticated. The newer methods are not obvious extensions of previous trucking and railroad practices but represent a systems approach to integrating and optimizing packaging, loading, transporting, and un-

Table 21.10. Color and Symbolism in Packaging for Asian Markets

Country	Color	Connotation	Symbol	Connotation
China	White	Mourning (avoid)	Tigers, lions, & dragons	Strength (use)
Hong Kong	Blue	Unpopular (avoid)	Tigers, lions, & dragons	Strength (use)
India	Green & orange	Good (use)	Cows	Sacred to Hindus (avoid)
Japan	Gold, silver, white, & purple	Luxury & high quality (use);	Cherry blossom	Beauty (use)
	black	use for print only, prefer gay, bright colors.	Chrysanthemum	Royalty (avoid)
Malaysia (population is mixed Malay, Indian, Chinese)	Yellow	Royalty (avoid)	Cows	Sacred to Hindus (avoid)
	Gold	Longevity (use)	Pigs	Unclean to Moslems (avoid)
	Green	Islamic religion (avoid)		
Pakistan	Green & orange	Good (use)	Pigs	Unclean to Moslems (avoid)
Singapore	Red, red & gold, red & white, red & yellow, yellow	Prosperity and happiness (use); Communist (avoid)	Tortoises Snakes Pigs & cows	Dirt, evil (avoid) Poison (avoid) Same as for India & Pakistan (avoid)
Taiwan	Black	(Avoid)	Elephants	Strength (use)
Thailand			Elephants	National emblem (avoid)
Tahiti	Red, green, gold, silver, and other bright colors	(Use)		
Arab & Moslem States	White	(Avoid)	Animals	(Avoid)
			Pigs	Religious pollution (avoid)
			Star of David	Political (avoid)

SOURCE: Hygrade Products Co., New Zealand.

loading practices for greatest efficiency. This involves improved palletizing and stacking techniques which often do away with intermediate-size cartons, drums, and sacks, and put new demands on unit package sizes and shapes for maximum utilization of cargo container space. In many instances the newer packing methods also shift unit package requirements from abrasion, tearing, or puncture resistance to greater emphasis on compression and stacking strength. Major food companies have become increasingly involved in how these advanced modes of containerization and distribution affect package needs and their contents. With some products, pallet loads are delivered directly to supermarkets and customer purchases are made directly from the pallet, further cutting down on handling costs.

SAFETY OF FOOD PACKAGING

Migration from Plastics

It is important to know that plastics are not completely inert to foods. Aside from the permeation of gases and vapors, it is also possible that components of the plastic can migrate to the food and would then be consumed with the food. This raises concern for the safety of some plastics. For this reason, all plastics used in food contact must have specific approval from regulatory agencies for the intended use. Food manufacturers must get written assurance from the plastic manufacturer that their container wrap meets all requirements for use in food contact.

Contamination

It is primarily packaging which acts as a barrier to contamination of foods. Preventing recontamination of thermally processed low-acid foods which are stored at room temperature is especially serious. Recontamination with pathogenic bacteria such as *Clostridium botulinum* can lead to outbreaks of food-borne disease. One example occurred when fish had been processed in defective metal cans which contained small holes. Several people ended up with botulism, which is often fatal.

ENVIRONMENTAL CONSIDERATIONS

Packaging of all forms makes up approximately 33% of the disposable solid waste in the United States. Foods use about one-half of all packaging (Fig. 21.19). The most common way of disposing of solid waste is landfilling. The need to find better ways to dispose of solid municipal waste has prompted interest in both increased recycling of food packaging as well as using less packaging in the first place. Neither of these objectives are as simple as they may seem. It is likely that recycled materials will have more chemical and microbiological contaminants than virgin materials. For example, some consumers might use empty plastic bottles to dilute and mix pesticides before recycling the bottle. Traces of the pesticide could remain in the plastic and later enter the food during storage. Recycled paperboard could contain more microbial spores due to contamination.

Another strategy to reduce packaging waste is to use less material when packaging

COMPOSITION OF MUNICIPAL WASTE BY MATERIAL, 1986

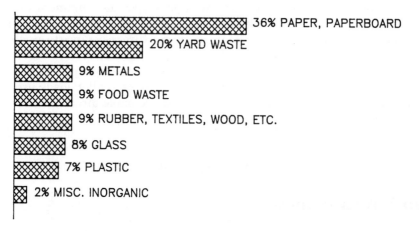

Franklin Assoc. (EPA), 1988

COMPOSTION OF MUNICIPAL WASTE BY PRODUCT, 1986

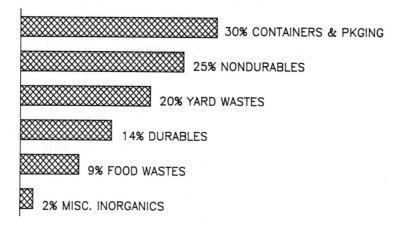

Franklin Assoc. (EPA) 1988

Figure 21.19. Composition of solid municipal waste by (A) type of material and (B) by type of product. Source: *Based on data from the Environmental Protection Agency, 1988.*

foods. This is termed *source reduction.* However, to do so takes considerable care. Making packages thinner and/or smaller is one approach that has been taken for many products. This reduces the amount of packaging used, but care must be taken so that the package will not break during shipment and allow the food to become contaminated. Making products more concentrated so that they can be packaged into smaller containers is another approach to reducing the amount of packaging used (Fig. 21.20). In some cases, extra packaging is used not to protect the food but for marketing reasons. In these cases, packaging can be reduced, but food manufactures are concerned that

Figure 21.20. Reduction in the amount of packaging used (i.e., source reduction) by making products more concentrated. Both detergent bottles contain the same amount of cleaning power but the smaller bottle is twice as concentrated and hence requires a smaller bottle.

they will be at a competitive disadvantage if they eliminate packaging which helps sell their products.

Metal and glass containers are recyclable without worry about recontamination, but they are heavy and require considerable energy to transport and melt and, therefore, have their own negative environmental costs. For this reason, lighter-weight plastics or combined plastic, paper, foil containers are sometimes preferred. Another alternative is to incinerate the trash. This too has negative environmental costs. Highly efficient and clean incinerators can be built, but they are expensive. Burning must not produce toxic or otherwise noxious fumes to pollute the air. Innovation in plastics technology holds promise of yielding more plastics that incinerate cleanly. Plastics that can be recycled are another possibility, but the economics of gathering and sorting plastic containers from other trash may limit the feasibility of this approach. Admirable progress continues to be made in the recycling of steel, aluminum, and polyester containers. Numerous steel can collection stations have been set up by major steel companies and more are planned. Some municipalities mine steel cans from dumps using magnetic separating equipment before or after garbage incineration and then sell the recovered steel scrap. Major aluminum companies continue to invest heavily in reclamation and recycling. Aluminum has greater scrap value than other components of solid waste, about 10 times that of steel and 15 times that of glass. Collectors bringing aluminum cans to Alcoa reclamation centers receive about $0.50 cents per kilogram for the cans. The Reynolds Metals Co. has been operating a fleet of mobile recycling units in the form of large truck trailers that carry a magnetic separator, electronic scale, and can shredder. The mobile units buy aluminum cans from the public and prepare them for smelting plants. Melting cans to recycle aluminum requires only about 5% of the energy needed to make aluminum from bauxite ore, and the aluminum industry is approaching 70% recycled aluminum in new can production.

Because of its lower economic value and certain other considerations, recycling of glass is not always feasible. One of the largest markets for salvaged glass is in conventional glass-making where crushed waste glass can supply about 30% of the raw materials needed to make new bottles. But such waste glass must be free of food, metal, and other forms of contamination and needs to be color sorted. Where costs for these handling operations are prohibitive, crushed glass of lesser purity can be used by the construction industries in making glass bricks and for admixture with asphalt to pave roads. After incineration to form a partially melted mass, waste glass also is suitable as an undersurface for road construction. With or without crushing and incineration, waste glass can provide useful landfill.

Paper and paperboard make up nearly half of all packaging materials used. Paper is easily recycled to produce fiber or is burned as fuel. Recycled fibers are used in the manufacture of cartons, boxes, newspaper, and other paper goods. However, more must be learned about recycling of paper and elimination of toxic contaminants that can find their way into paper goods not made from virgin pulp. Recent findings of polychlorinated biphenyls in paperboard food containers emphasize this point. Presently, about 20% of the U.S. production of paper is recycled. More economic means of separating paper from other waste materials are being researched to increase this figure.

Innovations that make packaging more "environmentally friendly" are to be encouraged. Development and use of such materials require input by the food scientist.

References

American Society for Testing and Materials. 1991. Selected ASTM Standards on Packaging. 3rd ed. American Society for Testing and Materials, Philadelphia, PA.

Bakker, M. (Editor-in-Chief). 1986. The Wiley Encyclopedia of Packaging Technology. John Wiley & Sons, New York.

Brody, A.L. (Editor). 1989. Controlled/Modified Atmosphere/Vacuum Packaging of Food. Food & Nutrition Press, Trumbull, CT.

Brown, W.E. 1992. Plastics in Food Packaging: Properties, Design, and Fabrication. With Contributions by C.F. Finch, A. Speigel, and J.H. Heckman. Marcel Dekker, New York.

Fellows, P. 1993. Appropriate Food Packaging. Tool Publications, Amsterdam.

Hanlon, J.E. 1992. Handbook of Package Engineering. 2nd. ed. Technomic Publishing Co., Lancaster, PA.

Hirsch, A. 1991. Flexible Food Packaging: Questions and Answers. Chapman & Hall, London, New York.

Holdsworth, S.D. 1992. Aseptic Processing and Packaging of Food Products. Chapman & Hall, London.

Jenkins, W.A. and Harrington, J.P. 1991. Packaging Foods with Plastics. Technomic Publishing Co., Lancaster, PA.

Koros, W.J. (Editor). 1990. Barrier Polymers and Structures. American Chemical Society, Washington, DC.

Labuza, T.P. 1982. Shelf-life Dating of Foods. Food & Nutrition Press, Inc., Trumbull, CT.

Leonard, E.A. 1987. Packaging: Specifications, Purchasing, and Quality Control. 3rd ed. Marcel Dekker, New York.

Manypenny, G.O. (Editor). 1988. Glossary of Packaging Terms: Standard Definitions of Trade Terms Commonly Used in Packaging. 6th ed. The Packaging Institute International, Stamford, CT.

Ooraikul, B. and Stiles, M.E. (Editors). 1991. Modified Atmosphere Packaging of Food. E. Horwood, New York.

Paine, F.A. 1987. 1987. Modern Processing, Packaging and Distribution Systems for Food. Chapman & Hall, London, New York.

Paine, F.A. and Paine, H.Y. 1992. A Handbook of Food Packaging. 2nd ed. Chapman & Hall, London, New York.

Parry, R.T. (Editor). 1993. Principles and Applications of Modified Atmosphere Packaging of Foods. Chapman & Hall, London, New York.

Robertson, G.L. 1993. Food Packaging: Principles and Practice. Marcel Dekker, New York.

Singh, R.K. and Nelson, P.E. (Editors). 1992. Advances in Aseptic Processing Technologies. Chapman & Hall, London, New York.

Stilwel, E.J., Canty, R.C., Kopf, P.W., and Montrone, A.M. 1991. Packaging for the Environment, A Partnership for Progress. Arthur D. Little, Inc., Boston.

22

Food Processing
and the
Environment

The conversion of raw agricultural commodities into finished foods, like other manufacturing processes, can be detrimental to the environment if precautions are not taken. There are several special environmental problems associated with food processing that must be considered. Food processing can have adverse environmental effects on both air and water quality as well as producing toxic side products. In general, the major environmental problem associated with food processing is the large amounts of solid and liquid waste product produced. These wastes are nearly always of biological origin and can pollute waterways and soil if not properly treated.

Water quality is primary to the manufacture of foods. Many food manufacturing operations use large amounts of water which can become heavily contaminated with processing by-products. Vegetable processing and brewing are but two of many examples. Cheese manufacturing produces vast amounts of liquid whey which must be further utilized or processed so as not to pollute large amounts of water. The ability to raise foods depends on the continuous circulation of the earth's water. This movement of water from the oceans, lakes, and streams as water vapor to the atmosphere, its condensation and precipitation back to the earth's surface, its flow over the land and return to the ocean, its penetration into the ground and use by plants, which return part of the water to the atmosphere through transpiration, and its subsurface return to the sea is known as the hydrologic cycle or water cycle (Fig. 22.1). Food scientists must be aware of the importance of water and understand the ways in which contamination can be eliminated or reduced.

The present industrial rate of water usage is about double the rate of water usage for irrigation in the United States and represents about 10 times the amount used for domestic purposes. Nearly all of this water becomes contaminated in one way or another, requires purification, and is then reused. The water we drink has a history of reuse; if it could be traced, a given water molecule may well have experienced passage through a domestic animal, partial purification in route to the sea, further purification in a potable water treatment plant, reuse as cooling water in a canning plant or a paper mill, repurification on passage through a municipal sewage system, and so on repeatedly. Municipal sewage is over 99% water. Industrial wastes, including wastes from food production and processing, are highly variable in composition but commonly are composed largely of water.

The food production and processing industries are concerned particularly with three

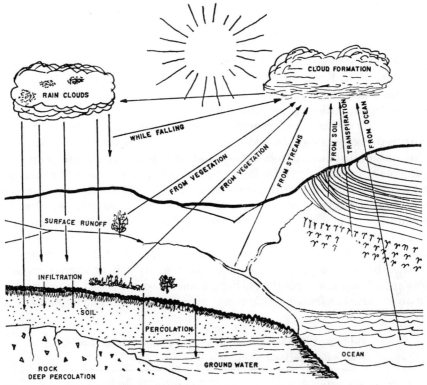

Figure 22.1. The hydrologic cycle describing the movement of water in the biosphere. *Courtesy of U.S. Department of Agriculture.*

broad aspects of water technology: microbiological and chemical purity and safety; impurities that affect suitability for processing use; and decontamination after use. Contamination affects the difficulty and cost of disposing of waste water and ultimately affects the cost of manufacturing food. There are strict environmental regulations regarding the discharge of polluted water from processing plants in most of the world. Plants which contaminate water with food processing wastes must treat the water to return it to an uncontaminated state before discharging it into a lake, river, or the ocean. In smaller plants the waste water is sometimes discharged to a municipal sewage system which repurifies the water. The plant will likely be assessed a fee for this treatment.

The availability of sufficient and suitable water and of means for disposal of plant wastes always have been prime factors in determining food plant locations. Enforcement of antipollution laws is now challenging the economic feasibility of many existing food production and processing operations. Today, materials that were formerly considered wastes are converted to useful by-products in order to dispose of them more economically. In the past, the food scientist primarily was concerned with the purity and chemical composition of water as it affected processing and food properties; problems of waste disposal were left mostly to the sanitary engineer. Now the food scientist and the sanitary engineer commonly plan and work together, since the increasing problems in handling food wastes are having direct effects on acceptable methods of food processing and disposition of the less desirable fractions of food raw materials.

PROPERTIES AND REQUIREMENTS OF
PROCESSING WATERS

Water entering a food processing plant must meet health standards for potable (drinking) water. The Environmental Protection Agency has issued National Primary Drinking Water Regulations (Table 22.1). Regulations covering radioactive contaminants and certain volatile synthetic organic chemicals have been added. These regulations are primarily concerned with health. Secondary regulations deal with color, taste, and other aesthetic qualities. In addition to the chemical limits for safety of potable water, this water must be free from contamination with sewage, pathogenic organisms, and organisms of intestinal origin. Regulations call for such water to contain no more than one coliform organism (statistical value) per 100 ml. Coliform organisms of the type assayed are not pathogenic in themselves but serve as a sensitive index of possible sewage contamination, which if present could harbor many kinds of human pathogens. Such water from municipal supplies or from private wells meeting these EPA recommendations for drinking purposes may not be suitable for certain food processing uses. On the other hand, this same water may be used as a heat exchange medium to condense vapors from an evaporator, to heat canned food in a retort, or to prechill orange concentrate enroute to a freezer. It then may still be quite suitable for subsequent plant

Table 22.1. EPA National Interim Primary Drinking Water Regulations

Characteristic	Limit Not to Be Exceeded
Inorganic chemicals	
arsenic	0.05 mg/liter
barium	1.0
cadmium	0.010
chromium	0.05
lead	0.05
mercury	0.002
nitrate (as N)	10.0
selenium	0.01
silver	0.05
fluoride	1.4–2.4[a]
Organic chemicals	
Endrin	0.0002 mg/liter
Lindane	0.004
Methoxychlor	0.1
Toxaphene	0.005
2,4-D	0.1
2,4,5-TP Silvex	0.01
Turbidity	1 unit (statistical value)
Coliform bacteria	1/100 ml (statistical value)

SOURCE: Adapted from Environmental Protection Agency (1975). Applies to community water systems. Regulations for radioactive contaminants were added July 9, 1976 (Environmental Protection Agency 1976).
[a]Depends upon air temperature.

reuse without further purification, for cleaning or conveying fruits and vegetables, or for plant cleanup purposes. Such reuse of water within the plant cuts down on water costs, minimizes the volume of plant waste water that is discharged, and represents efficient operation. Reuse is not acceptable, however, if it contributes a threat to the production of clean, wholesome food.

Water Hardness

Among the soluble materials that potable water may possess, calcium and magnesium ions are of major importance. These ions form precipitates with bicarbonates in the water on heating and with sulfates and chlorides when water is evaporated from solution. This is known as hardness and causes scale on equipment, which acts as an insulating layer against efficient heat transfer and may eventually clog pipes and foul valves. These deposits can harbor bacteria and add to the difficulty of equipment cleanup and may affect food products directly. The hardness of waters may be expressed quantitatively in terms of parts per million (ppm) of calcium, ppm of calcium carbonate, grains per gallon of calcium carbonate, or equivalent values for magnesium (Table 22.2).

The firming effect of calcium ions on certain fruits and vegetables was discussed previously. This may be used to advantage under controlled conditions, but in excessive amounts, the calcium from hard water can cause various textural defects. Thus, 200 ppm of hardness as calcium carbonate in the brine of canned peas may lead to excessive toughening. Certain cultivars of beans are more sensitive in this regard. Another type of problem occurs when such water is used as boiler-feed water. Here the continued evaporation of water as steam leaves behind a growing layer of scale, which not only reduces boiler efficiency but tends to contaminate steam generated in the boiler. Such steam can become alkaline and corrosive to aluminum and tin cans.

There are various ways to soften water, depending primarily on the nature of the hardness. Water hardness may be looked on as one kind of contamination and the softening treatment as a form of purification. When calcium or magnesium exists in water as bicarbonates, temporary hardness results since these salts can be easily precipitated from the water by heating prior to use. Chemical treatments involving the addition of hydrated lime are also effective in softening such waters. Hardness caused by sulfates and chlorides of magnesium or calcium, referred to as permanent

Table 22.2. Range of the Hardness of Water

Degree of Hardness	Expressed as Calcium Carbonate[a]		Expressed as Calcium (ppm)
	(ppm)	(gr per Gal.)	
Soft	Less than 50	Less than 2.9	Less than 20
Slightly Hard	50–100	2.9–5.9	20–40
Hard	100–200	5.9–11.8	40–80
Very Hard	Above 200	Above 11.8	Above 80

SOURCE: National Canners Assoc.
[a]One grain of calcium carbonate per U.S. gal. is equivalent to 17.1 ppm; 100 ppm of calcium carbonate is equivalent to 40 ppm of calcium. In reference to magnesium carbonate, 100 ppm of carbonate is equivalent to 24 ppm of magnesium.

hardness, is not removed by these treatments but can be removed by ion exchange techniques. Permanently hard water is passed over special ion exchange resins or similar materials that contain sodium or hydrogen as loosely bound cations. The resin, having a greater affinity for calcium and magnesium than for its bound sodium or hydrogen, exchanges these ions by taking calcium and magnesium from the water and giving up its sodium or hydrogen to the water. The water is thus softened.

Other Impurities

Water meeting health standards for drinking may be unsatisfactory for plant use for several other reasons. A detectable taste or odor of chlorine from tap water is not uncommon. This may be due to overchlorination (satisfactory chlorination will leave residual chlorine of about 1 ppm or less), but more commonly it is due to the presence of traces of phenol in the water, which reacts with chlorine and gives a strong medicinal odor. Frequently, this can be removed by filtration of water through a bed of carbon or adsorbent clay. The amount of chlorine to produce a sterile or nearly sterile water is quite variable, depending on other substances present in the water. This will be discussed further later in this chapter.

Off-flavors and off-odors may also be due to decomposition of organic matter in water by nonpathogenic bacteria. (Drinking water and the pipes through which it passes are seldom sterile.) When water contains sulfates and reducing types of bacteria, production of sulfide odors may occur. Although off-odors may be removed by filtering water through carbon, the underlying cause of the problem can be more difficult to eliminate and may require pipeline sanitation or repair to eliminate pockets of microbial concentration.

Absolutely pure water is colorless, but drinking water may have a slight color. For example, iron salts in water can be oxidized to red-brown ferric hydroxides. Manganese hydroxides are gray-black in color. Suspended colloidal matter of either organic or inorganic origin can produce excessive turbidity. Dissolved mineral impurities can be effectively removed by in-plant ion exchange treatments, and colored particulates can be removed by carbon and clay adsorbents and filtration. Colloidal impurities may be flocculated with alum (aluminum and potassium sulfates) and the precipitate removed by filtration or centrifugation.

Municipal and well waters from different locations vary in acidity and pH. It is not uncommon for hard waters to have a pH of about 8.5 or for acidic waters to have a pH as low as 5. This is generally easily corrected by in-plant water softening or direct neutralization.

Occasionally, well waters and municipal waters entering a plant will be contaminated with moderate numbers of proteolytic or lipolytic food spoilage organisms. Such waters have caused problems when used in direct contact with foods, as in the washing of cheese curds and butter granules. Waters with microbial contaminants often must be chlorinated or treated with ultraviolet light, and many food processors have installed automatic chlorinators and ultraviolet systems for routine continuous treatment of water for such uses.

Chlorination

In addition to chlorination of water for the purpose of rendering it safe from a health standpoint, the food processor often further chlorinates water in the plant for use in disinfecting products and machinery. Whether water is rendered bacteriologically pure at the municipal treatment plant or subsequently made to have disinfectant properties for special uses, effective chlorination must take into account the different chlorine demands of various waters before a germicidal effect can be achieved.

Materials may be present in water that react with chlorine and inactivate it before the chlorine can exert its germicidal effect. Hydrogen sulfide and organic impurities are particularly objectionable in this regard. Only after these interfering substances are satisfied in terms of their chlorine demands can residual free chlorine have a significant killing or inhibitory effect on microorganisms. This gives rise to the concept of breakpoint chlorination illustrated in Fig. 22.2. If water contains no interfering substances, then the level of residual chlorine in the water is directly proportional to the amount added; such water is said to have zero chlorine demand. Waters higher in organic materials and other interfering substances will begin to show a residual chlorine level only after all of their chlorine demand is satisfied. The point where this occurs is known as the breakpoint for that particular water. Additional chlorine can then be added beyond the breakpoint to achieve any desired level of residual chlorine.

For drinking purposes, a residual chlorine level in excess of about 0.4 ppm is seldom required. Tables and conveyors for food products may need frequent or continuous rinsing with water containing about 5 ppm of residual chlorine to maintain sanitary operation. Water used for general food plant cleanup may require a residual chlorine level of 25 ppm since much of this chlorine will be used up in satisfying the chlorine demand of soil before it can have disinfecting properties. Chlorine for chlorination may be discharged into the water from a cylinder of the gas, or the chlorine may be derived from hypochlorite preparations. Although chlorine as a disinfectant has been in common use for many years, the safety of chlorine in direct contact with foods continues to be studied by the FDA and the U.S. Deptartment of Agriculture. Toxicological evaluations include possible formation of mutagenic and carcinogenic organochlorine compounds.

PROPERTIES OF WASTEWATERS

Since food plant wastewaters may contain a wide variety of materials (e.g., meat and bone scrap, animal or fish entrails and excreta, blood and dairy wastes, pulp and

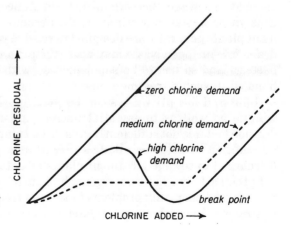

Figure 22.2. Chlorine demand characteristics of water. *Courtesy of M. A. Joslyn.*

peels of vegetable origin, spent coffee grounds and distillery wastes, soils and detergents from washing), their compositions and contamination loads will vary greatly. Nevertheless, food plant wastewaters can be broadly classified according to the nature of their impurities and pollution potentials, which determine what treatment methods will be suitable. It is convenient to consider wastewaters according to the physical, chemical, and biological natures of their impurities.

Physical Nature of Impurities

Materials in food wastewaters vary in size from coarse floating or sinking solids down to colloidally suspended matter, which will neither settle nor rise on standing undisturbed. Beyond this size limit are substances in true solution. Water-insoluble liquids such as oils and certain solvents may be present also.

Generally, gross particulates must be removed before plant wastewaters are sent to treatment plants or dumped. Many treatment plants will not accept sizable solids since they contribute substantially to pollution load and may easily be the cause of exceeding the capacity of the plant. Floating solids also are prohibited from discharge into streams and lakes because of their high pollution potential and unaesthetic appearance. Grinding to smaller size is not a solution. Such materials are generally easily removed at the food plant and treated separately. After removal of gross particulates, colloidal and dissolved impurities may still constitute a greater pollution load in wastewaters than will be acceptable to many municipal sewage treatment plants, or be permissible for discharge into streams, and so further treatment by the food plant is often required.

Chemical Nature of Impurities

Colloidal and dissolved impurities in wastewaters may be divided into organic and inorganic materials. Organic impurities are further differentiated according to their ratio of nitrogenous constituents to carbohydrate materials. Meat, poultry, and seafood wastes have the highest nitrogen-to-carbohydrate ratios. Many vegetable wastes are intermediate. Fruit wastes generally are higher in carbohydrate materials and lower in nitrogenous constituents. This becomes significant in terms of the end products of microbial degradation of these wastes both in treatment plants and when discharged onto land or into bodies of water. Wastes rich in nitrogen contribute this important element to sewage microorganisms, which need it for growth and continued activity; thus, these wastes may stimulate the decomposition process. Municipal sewage treatment plants generally are designed to receive wastes rich in nitrogen. A high-carbohydrate–low-nitrogen waste may upset the pH and metabolic activities of decomposition bacteria and so has to be supplemented with nitrogenous material before it can be handled in such a treatment plant.

High- and low-pH wastes can be particularly damaging to fish and other aquatic wildlife as well as to essential microorganisms in sewage treatment plants if the dilution factor is not sufficiently great. Food plant wastes usually require neutralization by addition of acids or alkalies to within a pH range of 6–9 before they may be discharged to sewage treatment plants or natural waters.

In the past, synthetic detergents and surface active chemicals, which tend to foam, have caused operating problems in sewage treatment plants and unsightly conditions in streams. The problem was related to the persistence of these foams and froths which were only slowly broken down by the action of microorganisms. This problem has now been largely overcome by a major shift on the part of detergent manufacturers to the production of readily biodegradable detergent types. Froths from such materials are quickly dissipated in sewage plants and streams.

Food plant wastes generally are not as corrosive as the wastes from many chemical and mining operations but may be more odorous. Offensive odors frequently necessitate additional treatment of these wastes; less odorous wastes of otherwise acceptable pollution load may be dumped.

Biological Nature of Impurities

Food plant wastes are largely organic and are decomposed in treatment plants and in nature by biological degradation. This degradation is carried out mostly by aerobic microorganisms that require large quantities of oxygen to completely oxidize carbohydrates and other organic materials to carbon dioxide and water and to convert nitrogenous residues to their highest state of oxidation (nitrate). To the extent that these oxidations are incomplete and leave intermediate products such as alcohols, acids, amines, and ammonia, waste decomposition may be said to be incomplete. These intermediates generally are odorous, may be toxic in themselves to plant and fish life, and, in any event, will undergo further degradation in nature. When this occurs in streams and lakes, it often does so at the expense of fish life due to consumption of oxygen.

Biological Oxygen Demand

Perhaps the most significant property of water containing organic waste is the capacity to consume oxygen in the course of microbial decomposition. The higher the organic contaminant level, the higher the demand for oxygen to consume the organics. The amount of oxygen required is termed the biological oxygen demand (BOD) and is an important measure of the dissolved organic waste in water. When the BOD of waste discharged into a stream is excessive, depletion of the stream's oxygen to satisfy this BOD causes fish suffocation, death of the fish's natural food supplies, and a general upset of the stream's ecology. BOD of a waste can be measured quantitatively, as can the BOD of stream or lake water at any given time.

The BOD test measures the quantity of oxygen in ppm (or milligrams per liter) required by aerobic microorganisms to stabilize waste or polluted water under specific conditions (generally incubation at 20°C for 5 days). Another useful test is the chemical oxygen demand (COD) test, which indicates the oxygen demand of organic materials after treatment with a strong chemical oxidizing agent such as potassium dichromate. The BOD and COD tests both measure oxygen demand; the COD is broader in that it determines total demand without regard to whether the organics can serve as substrates for microorganisms. For this reason, results of the two tests do not often correlate. Both tests are fully described in *Standard Methods for the Examination of Water and Wastewater,* published by the American Public Health Association.

The BOD and other values of various food processing wastes are given in Table 22.3. Generally, the higher the BOD values, the more difficult and costly are waste treatment and disposal practices. Sewage and waste treatment plants can be rated in terms of their BOD-removing capacity. Antipollution regulations are virtually always written

Table 22.3. Volume, Biological Oxygen Demand (BOD), and Suspended Solids of Some Food Processing Wastes

Commodity	Volume (gal/case)	5-Day BOD (ppm)	Suspended Solids (ppm)
Apples		1,700–5,500	300–600
Apricots	57–80	200–1,000	260
Beans, green or wax	26–44	160–600	60–150
Beans, limas, dried	17–22	1,740–2,880	160–600
Carrots	23	520–3,030	1,830
Citrus	1,000[a]	1,000–5,000	1,200
Corn, whole kernel	25–70	1,120–6,300	300–4,000
Cherries, sour	12–40	700–2,100	20–600
Grapefruit	5–56	310–2,000	170–280
Meat packing house	2,000–8,000	600–1,600	400–720
Milk processing industry	3–5[b]	20–650	30–363
Peaches	1,300–2,600[a]	1,350–2,240	600
Peas	14–75	380–4,700	270–400
Potato chips	4,000[a]	730–1,800	800–2,000
Potatoes, white		200–2,900	990–1,180
Poultry packing industry	1½[c]	725–1,148	769–1,752
Sauerkraut	3–18	1,400–6,300	60–630
Spinach	160	280–730	90–580
Tomatoes	3–100	180–4,000	140–2,000

SOURCE: J. W. Casten.
[a]Per ton.
[b]Per gallon (3.785 liters) of milk.
[c]Per chicken.

to include maximum permissible BOD loadings into natural waters. These maximum loadings differ widely in accordance with the volume, flow, and other characteristics of the body of water.

On rare occasions, food plant wastes have become contaminated with highly toxic materials such as pesticides and disinfectants. The need to keep these out of streams is obvious. Such wastes and wastewaters, however, also should not be sent on to sewage treatment plants without prior warning and permission, since the toxic substances may be sufficiently concentrated to kill the plant's normal microbial flora essential to sewage and waste treatment. In such situations, the waste may have to be diluted substantially, sent on to the treatment plant a little at a time, or possibly tanked and trucked to a suitable waste storage facility. The latter becomes essential when toxic substances are not biodegradable or otherwise subject to detoxification.

WASTEWATER TREATMENT

The final treatments applied to potable water for special food processing uses—such as water softening, ion exchange, and carbon filtration, which are referred to as advanced treatment, tertiary treatment, or "polishing treatments"—are generally not needed or used to treat food plant wastewaters. Rather, primary and secondary treatments are chosen for removal of gross particulates and coagulable colloidal matter and reduction of BOD sufficient for ultimate discharge onto land or into streams.

The extent to which pollution load must be decreased before wastewater leaves a food plant is highly variable, especially from one location to another. It depends on many factors including the following: (1) Will the wastewater be discharged to a municipal sewage or commercial waste treatment plant, and if so what is the maximum pollution load this plant can treat? (2) What will be the cost for such treatment and can it be done more economically by the food plant itself? (3) What dumping privileges does the food plant have and what pollution laws apply? Because the answers to these questions are so varied, all degrees of wastewater purification are currently being performed by food plant and outside facilities.

Primary Treatment

Gross particulates generally are removed at the food processing plant by screening through vibrating sieves. Smaller particles may be removed by filtering or centrifuging. Minute particles may be allowed to settle or rise in large tanks. Scum and oil are readily skimmed from such tanks, and settled solids are concentrated for removal and subsequent treatment by pumping the supernatant liquid away. Colloidal materials commonly are coagulated or flocculated with the aid of alum, which promotes settling. These primary treatments may remove some 40% of the wastewater's BOD and perhaps 75% of total solids, depending on the nature of the waste.

Secondary Treatment

Secondary treatment often is performed by large food plants in plant site facilities similar to those of municipal sewage installations. In other cases, the partially treated wastewaters are discharged to the municipal stations. Secondary treatment commonly involves the use of trickling filters, activated sludge tanks, and ponds of various types. Sometimes, these are preceded by use of anaerobic digesters.

Trickling Filters

The several types of trickling filters available bring together wastewater and waste-digesting bacteria under highly aerobic conditions. Commonly, a trickling filter consists of a bed of crushed rock or other material of large surface area in contact with air or through which air is blown. Wastewater trickling through the crushed rock soon develops a highly aerobic microbiological growth around the rocks, or this may be initially established by seeding with an appropriate sewage culture. One type of trickling filter is illustrated in Fig. 22.3. Oxidation of organic materials passing through several trickling filters in series can often reduce the BOD of incoming waters by 90–95%.

Activated Sludge Tanks

Activated sludge tanks are essentially large aeration tanks in which air is bubbled through wastewater. The tanks soon develop a highly aerobic flocculent microbial growth, which continues to be nourished by the incoming waste (Fig. 22.4). If wastes

Figure 22.3. Tricking filter system used for purifying waste water from food plants as well as municipal waste systems. *Courtesy of Dorr-Oliver, Inc.*

Figure 22.4. Activated sludge process for purifying waste water from food plants as well as municipal waste systems. *Courtesy of Dorr-Oliver, Inc.*

are low in nitrogen or phosphorus needed by the microorganisms, these may be added in the form of sewage or other supplements. Wastewater may be continuously added and removed from such tanks, providing a residence time of several hours as required to achieve a substantial decrease in BOD.

Anaerobic Digesters

Anaerobic digesters differ from trickling filters and activated sludge tanks in that waste materials are largely converted to carbon dioxide, methane, and other organic compounds in the absence of air by an anaerobic microbial flora. The methane may be recovered for its fuel value and the wastewater with decreased BOD then further conveyed to the trickling filters or activated sludge tanks.

Ponds and Lagoons

Waters from trickling filters and activated sludge tanks frequently are pumped to concrete tanks or man-made ponds and lagoons. These are shallow (about 1–2 m) to maintain an aerobic condition (Fig. 22.5). In these ponds, further microbiological decrease of remaining BOD occurs as well as additional settling of traces of solids. Water

Figure 22.5. Lagoons containing treated wastewater from a New York winery. *Courtesy of Y. D. Hang.*

from ponds and lagoons, when clarified of solids, usually is sufficiently low in BOD for approved discharge into lakes and rivers. Many municipalities require that such water be lightly chlorinated to ensure its freedom from pathogens and to reduce coliform counts before such discharge. Frequently, such water without chlorination is used to advantage, as are some waters from trickling filters and activated sludge tanks, for land irrigation.

Where a food plant has no convenient means for disposing of treated wastewaters, it may choose to discharge these waters into sewer lines for further handling by a municipal plant. Municipal plants are always situated for easy disposal of treated wastewaters, which are chlorinated for disposal into waterways or used for irrigation. Except under conditions of emergency, municipal plants virtually never route treated wastewaters to be purified directly into potable water supplies. Plants that produce drinking water draw their supplies from protected reservoirs or unpolluted streams and purify it in accordance with rigid standards for potable water.

WASTE SOLIDS UPGRADING
AND TREATMENT

Under favorable economic conditions most wastes from a food processing plant can be processed or altered to a more useful and valuable material. Fruit and vegetable skins, pulp, and pits can be pressed to further remove water and be converted into compost for improving soils or animal feeds. Cacao bean hulls are sold as a high-priced ornamental mulch. Dried spent coffee grounds have been used to dress tennis courts. Packing house scraps are ground, rendered, and dried for animal feeds. Entrails are similarly treated or shipped "wet" to mink farms. Spray- and drum-dried blood meal is an important component of feeds. Inedible portions of fish are steamed, ground, and dehydrated to yield fish meal for animal consumption. The liquid pressed from such fish during processing (fish "stick" water) is concentrated and may be desalted and dried to yield a high quality protein for human consumption. Poultry feathers may be collected, cleaned, and further processed for pillow stuffing. Animal hides yield edible gelatin. Hulls from almond processing plants are burned to provide energy to run the plant.

Dairy "wastes" such as cheese whey and buttermilk (from the churn) should be more correctly called by-products. Both are especially rich in nutrients and are produced under highly sanitary conditions. For many years most buttermilk and whey were treated as waste. Double losses result from the throw-away of valuable nutrients and the subsequent cost of waste treatment. In some cases this has been economically justifiable, especially when whey had to be hauled great distances from a cheese plant to a dehydration or concentration plant and the market for the processed product was limited. Much research has been directed at modifying whey and finding new food uses to remedy this situation; however, many problems have been encountered.

Dried whey from acid-coagulated cheese (such as cottage) are normally highly acidic and can be consumed by livestock in only limited quantities. If whey is neutralized with alkali, a high level of salt is produced. Even when not neutralized, whey is too high in salts for direct consumption in large amounts. Acid and salts can be removed from whey by various ion exchange and desalting methods, but costs of these treatments have not always been justified. Acid whey and sweet whey from rennet-coagulated cheese (such as Cheddar) are high in lactose. Lactose may act as a laxative when consumed in large amounts, and some people suffer from lactose intolerance. Many of these problems can be minimized by blending whey with other food materials as a supplement, but the amount so consumed has always been relatively small. Similar problems are encountered and have been overcome with many food wastes and by-products. The problems associated with whey disposal are large because of the large quantities of whey produced, its high BOD value, its relatively low level of solids, and the marginal economic value of these solids without further upgrading.

In recent years several factors have had an influence on the disposal of whey. Membrane technology, including reverse osmosis and ultrafiltration, have made it economically feasible to remove salt from whey and separate whey into its lactose and protein components. Tighter antipollution laws threaten the existence of numerous cheese-making plants if alternatives to whey dumping are not adopted. Modified and blended whey products have unique properties and so are finding new markets. Today, considerable whey (and buttermilk) is modified, concentrated, and drum or spray dried for use in animal feed, pet food, and numerous other items including baked goods, frozen desserts, confectionery, and sausage products. One of the newest uses is the fermentation of the lactose of whey to produce alcohol for mixture with gasoline to extend this fuel (Fig. 22.6).

Some materials cannot be economically upgraded to fit into an existing market structure. These, too, may still have some value. Corn cobs, nut shells, coffee grounds, and contaminated fats and oils commonly are burned as fuel to generate steam in food plants. Fruit and vegetable wastes have been used as a source of fermentable carbohydrate. Some wastes are plowed into the ground as fill. When none of these uses

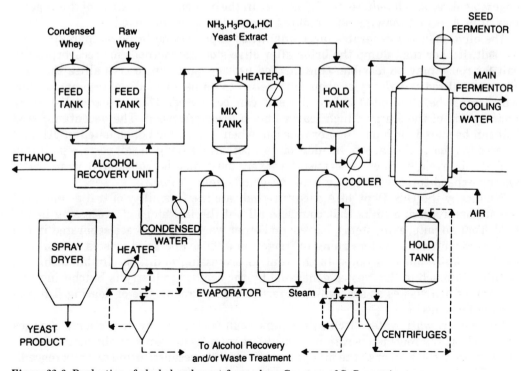

Figure 22.6. Production of alcohol and yeast from whey. *Courtesy of S. Bernstein.*

are feasible, then waste may be burned as garbage where ordinances permit. Sludge and residues remaining after wastewater and sewage treatment have been dried and sold as fertilizer or used wet for this purpose. Sometimes they are incinerated, leaving only a small amount of ash to dispose of. Still other possibilities are indicated in the recycling scheme for municipal refuse in Fig. 22.7.

LOWERING DISCHARGE VOLUMES

There are many alternatives to conventional processing operations that can substantially lower the volumes of wastewaters currently being discharged from food plants. Reuse of water for less demanding operations within the plant has already been mentioned.

It is possible to peel certain fruits and vegetables by a "dry caustic" method in which concentrated sodium hydroxide combined with vigorous mechanical action replaces large volumes of dilute caustic solution to loosen the skins. In the case of potatoes, about one-fourth as much water, and in the case of peaches about one-fifteenth as much water is required with the dry caustic method. Further, BOD of the water per weight of potato or peach peeled by the dry caustic method is only about one-third that generated by the more conventional method.

Another operation that contaminates large quantities of water is conventional hot water blanching of vegetables. A newer method exposes diced vegetables in a monolayer to live steam for about 30 sec followed by a 60-sec or longer holding time for heat equilibration and enzyme inactivation. The blanched dice may next be cooled with chilled air. In addition to the use of less water, fewer solids are dissolved from the vegetable dice, which reduces the BOD level in the effluent, and texture of the vegetables after freezing, thawing, and cooking is reported to be improved.

The olive industry generates large quantities of processing brine of approximately 8% salt. Rather than dump the brine after olive storage, methods for recovering the salt for reuse may be feasible. One method uses evaporation to crystallize the salt. The crystallized salt slurry of about 58% solids contains about 6% of organic matter, which must be eliminated before the salt can be reused. This is accomplished by incineration of the slurry which leaves only a trace of carbon. The decontaminated salt can be stored for reuse the next season, when it is dissolved, adjusted in pH, and filtered to remove the carbon. In addition to a potential saving in salt costs, a pollution problem is avoided. Such processes, however, must also be considered in terms of energy expenditures, which may make trade-offs unfavorable.

Studies by the U.S. Dept. of Agriculture indicate the feasibility of trailer-mounted tomato preprocessing units that complete all but the last steps of processing in the field. Tomato pulp, skin, stems, leaves, and other waste material are separated in the field area with wash waters going to irrigation ditches and solid wastes spread over the land. Juice alone is taken to the central processing plant. In similar fashion a solution to much of the wastes generated by the seafood industry is butchering and disposal of fish wastes at sea, which is facilitated by greater use of floating fish processing factories.

Frequently, wastewater discharge volumes can be reduced by very simple changes in a food plant's routine operations. An ice cream manufacturer might better return rinsings from the freezer to the mix preparation vat for reconstituting dry mix ingredients. This not only decreases waste load but recovers food solids of economic value.

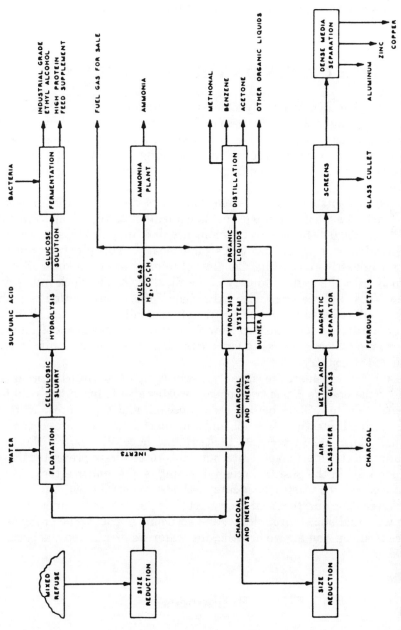

Figure 22.7. Flow chart for potential recycling systems for municipal refuse. *Courtesy of R. F. Testin.*

The segregation of waste streams is often sound; frequently, it is better to recover a concentrated high BOD waste such as blood or offal and handle it separately than to allow it to contaminate large volumes of more dilute wastewater. The purchasing of certain preprocessed ingredients can often reduce waste loads needing treatment. These and related practices are not new but are becoming increasingly important to pollution control.

A CONTINUING RESPONSIBILITY

Waste disposal has always meant problems and expense to food producers and processors. With the passing of the years, these problems generally have grown more severe. In the United States, the population is increasing and people continue to move to suburbs close by areas of food production and processing. The ability of any geographical area to absorb wastes is limited and frequently exceeded when controls are not intelligently administered.

Tables of equivalents have been developed to show the contribution of wastes from various sources. The waste from one cow is equivalent to that from 16.4 humans, a poultry farm of 200,000 birds, not uncommon today, may produce a waste disposal problem equivalent to a city of 20,000. A sizable meat packing plant or cannery during peak season can easily exceed this. Conflicts of interest are common. Polluted waters have seriously hurt the oyster industry of the East Coast. The duck industry of Long Island, New York is threatened with extinction if their contribution to pollution in this highly populated area cannot be adequately controlled. A multimillion dollar per year poultry industry in southeastern New York State is in the heart of a multimillion dollar resort area which is averse to foul odors. This type of situation is increasing in many areas of the country.

In the past, it was considered reasonably safe to dump wastes into the ocean. Growing evidence that the ocean's capacity to absorb wastes also is finite has resulted in increased legislation to protect both the deep oceans and the shores. In the United States, the Marine Protection Research and Sanctuaries Act regulates ocean dumping. International efforts are in progress to study and hopefully control the problem of ocean pollution worldwide. Forceful international protests have prevented the dumping of lethal chemicals in the past. All ocean dumping is now under attack.

For most food-related wastes, adequate methods of control exist, but they often are costly. Yet there is no question that they must be legislated and employed stringently, in keeping with realistic environmental and economic impacts. The problem is everybody's since it affects the air we breathe, the water we drink, and the price we must pay for food.

References

American Public Health Association. 1981. Standard Methods for the Examination of Water and Wastewater. 15th ed. American Public Health Association, Washington, DC.

Biswas, A.K. and Arar, A. 1988. Treatment and Reuse of Wastewater. Food and Agriculture Organization of the United Nations. Butterworths, London.

Bradshaw, A.D., Southwood, Sir R., and Warner, Sir F. 1992. The Treatment and Handling of Wastes. Chapman & Hall, for the Royal Society, London, New York.

Calabrese, E.J., Gilbert, C.E., and Pastides, H. 1989. Safe Drinking Water Act. Amendments, Regulations, and Standards. Lewis Publishers, Chelsea, MI.

Cheremisinoff, N.P. 1993. Water Treatment and Waste Recovery: Advanced Technology and Applications. Prentice-Hall, Englewood Cliffs, NJ.

Environmental Protection Agency. 1975. National interim primary drinking water regulations. Fed. Reg. *41*(298), 59566–59574.

Environmental Protection Agency, Office of Technology Transfer. 1975. Pollution Abatement in the Fruit and Vegetable Industry, Vol. 2. In-Plant Control of Food Processing Wastewaters. Prepared by A.M. Katsuyama, N.A. Olson, and W.W. Rose. U.S. Environmental Protection Agency, Office of Technology Transfer, Washington, DC.

Environmental Protection Agency. 1989. Bibliography of Municipal Solid Waste Management Alternatives. U.S. Environmental Protection Agency, Solid Waste Emergency Response, Washington, DC.

Environmental Protection Agency. 1989. The Solid Waste Dilemma. An Agency for Action: Final Report of the Municipal Solid Waste Task Force, Office of Solid Waste, United States Environmental Protection Agency. U.S. Environmental Protection Agency, Washington, DC.

Environmental Protection Agency. 1991. Ensuring Safe Drinking Water. Environmental Protection Agency, Office of Research and Development, Office of International Activities, Office of Water, Washington, DC.

Erwin, L. and Healey, Jr., L. H. 1990. Packaging and Solid Waste: Management Strategies. American Management Association, New York.

Jones, J.G. 1993. Agriculture and the Environment. Ellis Horwood, New York.

Novotny, V. 1994. Water Quality: Prevention, Identification, and Management of Diffuse Pollution. Van Nostrand Reinhold, New York.

Rockey, J. and Zaror, C. 1988. Alternative strategies for the treatment of food processing waste waters. Dev. Food Microbiol. 4, 187–221.

Ruttan, V.W. 1994. Agriculture, Environment, and Health: Sustainable Development in the 21st Century. University of Minnesota Press, Minneapolis.

Selke, S.E.M. 1994. Packaging and the environment: alternatives, trends, and solutions. Technomic Publishing Co. Lancaster, PA.

Strauss, S.D. 1986. Wastewater management. Power Power 130(6), S1–S16.

Vernick, A.S. and Walker, E.C. 1981. Handbook of Wastewater Treatment Processes. Marcel Dekker, New York.

World Health Organization. 1993. Guidelines for Drinking-Water Quality. 2nd. ed. World Health Organization, Geneva.

Young, G.J. 1994. Global Water Resource Issues. Cambridge University Press, Cambridge.

23

Food Safety, Risks and Hazards

INTRODUCTION

As consumers, we have several expectations of the food supply, including that it be nutritious, wholesome, pure, and "safe." We also expect that it be plentiful, offer wide choices, and be a reasonable value. In recent years consumers have placed increased emphasis on food safety and expect that foods not contribute to chronic diseases such as cancer or heart disease. In fact, recent scientific evidence suggests that some types of diet can help prevent chronic diseases. The potential adverse environmental impact of agriculture and, in some cases, concerns for animal welfare are also current consumer issues.

The general safety of the food supply has been debated for many years although the issues change with time. In the 1970s the safety of food additives was hotly debated. In the 1980s pesticide residues and irradiation were major issues, and in the 1990s biotechnology is a major issue.

SAFETY, HAZARDS, AND RISKS

In order to understand what "food safety" means, we must first define the terms *safe, hazard,* and *risk*. "Safe" means that nothing harmful happens when we do something such as driving a car or consuming a food. But this is not a very satisfactory viewpoint when considering foods. We have learned that exposure to certain toxicants can harm us years after the exposure; cancers induced by tobacco use are examples. Safety also means an absence of effect; that is, nothing harmful happens. Science does not deal well with the absence of effect or negative outcomes because they are hard to interpret and cannot be quantified. In the absence of harmful effects, scientists can only conclude there was no harm under certain conditions.

Scientists think of food safety in terms of hazards and risks. A hazard is the capacity of a thing to cause harm. This is not to say something *will* cause harm, only that under some conditions it *could* cause harm and what the harm is. The probability that a defined harm will occur is the risk associated with the hazard. Falling off a ladder is a good analogy. The hazard in falling off a ladder is that one will be injured, perhaps by breaking a leg. However, the probability of being injured depends on the height of the fall. If one falls from the bottom rung, the risk or probability is low. Falling from the top rung results in a greater risk. In both cases, the hazards are similar, but the risks differ.

Many times in life we recognize hazards and take actions which will reduce risk to an acceptable or improved level. For example, we realize that riding in a car is hazardous in

532

that we could be injured, but we might wear a seat belt to reduce the likelihood (i.e., risk) that we will be injured.

Scientists use this same thinking when evaluating food safety. They first identify hazards related to foods or food components and then estimate the size of the risk that the hazard will occur. They go through a multistep process which starts with hazard identification. For example, a pesticide is required to undergo a battery of tests to first determine what hazards it might present. It will be tested for its ability to cause cancer in laboratory animals, for example. If it turns out to be positive, it can be said that one of its hazards is the ability to cause cancer. Note that this says nothing about the size of the risk however. Another hazard might be the ability to induce nerve damage.

The next step is to evaluate the size of the risk associated with the pesticide or food chemical. Scientists ask what is the statistical probability that this hazard will occur. This is done by using statistical estimates of how large the risk to humans could actually be under the worst-case scenario. The process of identifying hazards and estimating their size is termed *risk assessment* and attempts to quantify the size of food-related risks. Risk assessment is a scientific process that depends heavily on toxicologists, microbiologists, and statisticians.

The next step in the process is to decide what to do about the risk; that is to decide whether or not the risk is acceptable. This is a judgment and termed *risk management*. This process is less scientific and more social or political. In most cases, government representatives make such decisions on the acceptability of risks. Risk managers have the option of banning things in which the risk is considered too high, or restricting the use of something in order to reduce the risk. In some cases, benefits are considered when determining the acceptability of a risk. Often, consumer activists pressure government to make certain decisions.

It is important to note that the process outlined above considers all foods to have some degree of risk and that no food is absolutely "safe." The important consideration becomes the size of the risk and how the size of the risk can be reduced without eliminating the food source. The goal of food safety is to reduce the size of risks to the lowest reasonable level without severe disruption of the food supply.

FOOD-RELATED HAZARDS

There are five broad general categories and several subcategories of hazards associated with foods. These are presented in the following subsections.

Biological Hazards

Biological hazards include bacterial, fungal, viral, and parasitic (protozoa and worms) organisms and/or their toxins. There are many microorganisms which are pathogenic in humans but relatively few are associated with foods (Table 23.1). Those that are, are termed food-borne pathogens. Diseases caused by these organisms are sometimes incorrectly called food poisonings. There are two types of food-borne disease from microbial pathogens: infections and intoxications. Infections result from ingestion of live pathogenic organisms which multiply within the body and produce disease (Fig. 23.1). Intoxications occur when toxins produced by pathogens are consumed. Intoxications can occur even if no viable microorganisms are ingested. This often occurs when foods are stored under conditions which allow the pathogens to grow and produce toxin. Subsequent processing of the food may destroy the microorganisms but not the toxin.

Table 23.1. Hazardous Microorganisms and Parasites
Grouped on the Basis of Risk Severity

I. Severe Hazards
   Clostridium botulinum types A, B, E, and F
   Shigella dysenteriae
   Salmonella typhi; paratyphi A, B
   hepatitis A and E
   Brucella abortis; B. suis
   Vibrio cholerae O1
   Vibrio vulnificus
   Taenia solium
   Trichinella spiralis
II. Moderate Hazards: Potentially Extensive Spread[a]
   Listeria monocytogenes
   Salmonella spp.
   Shigella spp.
   Enterovirulent *Escherichia coli* (EEC)
   Streptococcus pyrogenes
   rotavirus
   Norwalk virus group
   Entamoeba histolytica
   Diphyllobothrium latum
   Ascaris lumbricoides
   Cryptosporidium parvum
III. Moderate Hazards: Limited Spread
   Bacillus cereus
   Campylobacter jejuni
   Clostridium perfringens
   Staphylococcus aureus
   Vibrio cholerae, non-O1
   Vibrio parahaemolyticus
   Yersinia enterocolitica
   Giardia lamblia
   Taenia saginata

SOURCE: Pierson and Corlett (1992).
[a]Although classified as moderate hazards, complications and sequelae may be severe in certain susceptible populations.

As pointed out in Chapter 1, the frequency of food-borne disease in the United States is surprisingly high. It is estimated that several million cases occur each year. Healthy individuals nearly always recover from these bouts, but an estimated 1000 or more elderly, very young, or people with medical problems die each year from food-borne diseases. Most food scientists think these numbers could be reduced if more care was taken in the handling, storage, and preparation of foods.

Nutrition-Related Diseases

There is a relationship between health and diet and the major objective of eating is to maintain or even improve health. Foods, components of foods, or diets which detract

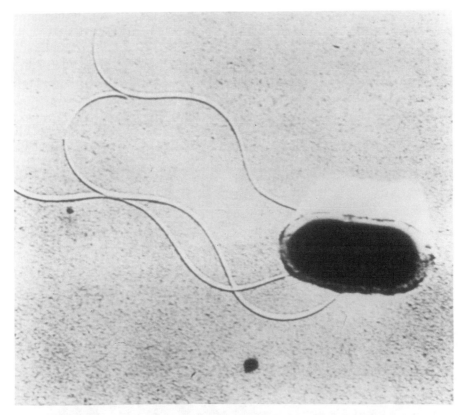

Figure 23.1. Photomicrograph of *Listeria monocytogenes*, a food-borne pathogenic bacterium which has caused several deaths. *Courtesy of R. B. Gravani.*

from health represent diet hazard. In recent years, knowledge about the connections between diet and health has increased and it has become apparent to many food scientists and nutritionists that food and diet play a role in chronic diseases such as heart disease and cancer. When the incidence of diseases such as cancer and heart disease is compared among human populations with different dietary habits, it becomes clear that certain diets can increase the risk of heart disease and cancer. Other diets appear protective. In addition, many people overconsume foods, and diseases related to obesity are common. This amounts to a misuse of foods and is considered a major food-related hazard because it affects large numbers of people. Many health organizations both within and outside of governments have advocated changes in the diet to promote better health in order to reduce these risks (Fig. 23.2). The relationships between diet and diease are discussed in more detail in a subsequent chapter.

Trace Chemicals

Chemical hazards associated with foods can be subdivided into naturally occurring, indirectly added, and directly added. Some type of chemical hazards are listed in Table 23.2. Toxic chemicals occur in foods at levels which cause acute (i.e., immediate) food poisoning symptoms or intoxications. They also occur at lower levels which may

Figure 23.2. Diet and Health, published by the U.S. National Research Council, is one of several publications that reviews the evidence showing that changing dietary habits would promote better health.

represent chronic or long-term risks. In most but not all cases, humans have learned to avoid foods which contain toxicants which can cause illness shortly after being eaten. Occasionally these lessons are forgotten, as in the case of poisonous mushrooms.

Foods also contain low levels of toxicants that when consumed over long periods of time could potentially promote chronic diseases. Heavy metals such as lead and mercury are prime examples. Low levels of lead or mercury in foods do not represent an immediate threat to health, but when consumed over longer periods, they are a real and serious risk.

Trace toxicants can be further divided into several types based on their source and route of entry into foods. For example, it is clear that the single largest source of toxicants in the diet are the naturally occurring toxic substances in foods themselves. Many fast-acting toxicants have been identified in foods and recant evidence indicates that natural foods also contain slower-acting toxicants, including carcinogens. Com-

Table 23.2. Types of Chemical Hazards

I. Naturally Occuring Chemicals
    mycotoxins (e.g., aflatoxin)
    scombrotoxin (histamine)
    ciguatoxin
    mushroom toxins
    shellfish toxins
        paralytic shellfish poisoning (PSP)
        diarrheic shellfish poisoning (DSP)
    neurotoxic shellfish poisoning (NSP)
        amnesic shellfish poisoning (ASP)
        Pyrrolizidine alkaloids
    Phytohemagglutinin
II. Added Chemicals
    agricultural chemical
        pesticides, fungicides, fertilizers, insecticides, antibiotics
          and growth hormones
    prohibited substances (21 CFR, Section 189)
        direct
        indirect
    toxin elements and compounds
        lead, zinc, arsenic, mercury, and cyanide
    food additives
        direct—allowable limits under GMPs
          preservatives (nitrite and sulfiting agents)
          flavor enhancers (monosodium glutamate)
          nutritional additives (niacin)
          color additives
        secondary direct and indirect
          plant chemicals (e.g., lubricants, cleaners, sanitizers,
            cleaning compounds, coating and paint)
    chemicals intentionally added (sabotage)

SOURCE: Pierson and Corlett (1992).

pounds found in citrus oils or many spices, for example, will produce tumors in lab animals if given in high enough amounts for a long period. This does not necessarily indicate that they are a significant risk to humans, however. Foods also contain many antitoxicants which may decrease the risks associated with toxicants.

Because most foods are grown in the open environment, they can become contaminated with natural and human-derived environmental toxicants. Lead, in most cases, is one such toxicant. PCBs, dioxin, and other pollutants resulting from human activity are further examples.

In addition to environmental pollutants, foods become contaminated with trace toxicants which are unintentionally or intentionally added to foods. The use of pesticides to control insects, unwanted plants, or fungi can result in trace residues of the pesticide in the food. In some cases, components of packaging materials migrate from the package to foods. Oils from processing machinery or other processing aids can leave trace residuals in foods. These substances are sometimes called processing aids and are regulated for safety.

Traces of drugs which are given to food-producing animals to treat diseases in these animals or make them grow more quickly could, under some circumstances, remain in the food. Traces of antibiotics in milk are one example. These are also considered trace toxicants.

Direct Food Additives and Macrocomponents of Foods

The intentional addition of chemicals (e.g., preservatives) has been practiced since before recorded history. Prehistoric man added chemicals to meats for preservation. The later practices of salting fish and flesh, of fermenting plant and animal substances, and of improving the palatability of insipid diets with spices are other examples of introducing chemicals. Food additives must be thoroughly tested before use, and use of those with significant risk is not permitted so the risk from this category is considered extremely small.

Physical Hazards

Foods may contain physical hazards such a stones, seeds, glass fragments, or small bits of metal (Table 23.3). These materials can become part of foods from the natural environment in which they are grown or they may be contaminated during processing and packging. Small pieces of metal can come loose from processing machinery, for example. For this reason, many food processing operations have an electronic metal detector which screens each package for metals. Foreign objects represent one of the largest categories of complaints by consumers.

All substances within each of the above five categories can be thought of as carrying some degree of health risk because no food is completely risk-free. The acceptability of each risk will depend on a number of factors, including the alternatives available, cost, benefit, and size of the actual risk. It is the job of government and industry to

Table 23.3. Main Materials of Concern as Physical Hazards and Common Sources

Material	Injury Potential	Sources
Glass	Cuts, bleeding; may require surgery to find or remove	Bottles, jars, light fixtures, utensils, gauge covers
Wood	Cuts, infections, choking; may require surgery to remove	Fields, pallets, boxes, buildings
Stones	Choking, broken teeth	Fields, buildings
Metal	Cuts, infection; may require surgery to remove	Machinery, fields, wire, employees
Insects and other filth	Illness, trauma, choking	Fields, plant postprocess entry
Insulation	Choking; long term if asbestos	Building materials
Bone	Choking, trauma	Fields, improper plant processing
Plastic	Choking, cuts, infection; may require surgery to remove	Fields, plant packaging materials, pallets, employees
Personal effects	Choking, cuts, broken teeth; may require surgery to remove	Employees

SOURCE: Pierson and Corlett (1992).

ensure that these risks are minimal and acceptable. Very often it is possible to reduce the risk by eliminating the substance or choosing less risky alternatives. In other cases, the only way to eliminate risks is to ban whole foods.

Risks from processed foods can be reduced through use of good manufacturing practices and careful analysis of the most important steps in processing. This latter approach is often termed "hazard analysis and critical control point" or HACCP and is discussed in detail later in this chapter.

MICROBIOLOGICAL CONSIDERATIONS IN FOOD SAFETY

As indicated above, one of the most serious and widespread risks from foods is the occurrence of pathogenic microorganisms. In the United States, the FDA and U.S. Department of Agriculture, as well as state and local governments, have strict requirements for the handling of foods to reduce microbial hazards. Unfortunately, this does not completely eliminate these hazards as was demonstrated in a food-borne disease outbreak caused by the occurrence of a deadly type of *E. coli* bacteria in hamburger that was undercooked by a fast-food restaurant. Several people became seriously ill and four died. Outbreaks such as these have resulted in an increased interest in ways to prevent microbial related food-borne disease.

The Centers for Disease Control (CDC) investigates each documented outbreak of food-borne disease and attempts to determine not only the specific microorganism and food(s) involved but also the events which led to the outbreak. In most cases there are a series of events which contribute to the outbreak. These events include the following:

1. The occurrence of the pathogen in the product either because it was in the raw product or because the food became contaminated during processing or preparation. Some pathogens such as *Salmonella* spp. are commonly associated with certain foods in their raw form. These pathogens can be carried throughout the processing of the food into the final product. In other cases, cross-contamination occurs when a raw food contaminates a food which has been cooked or will be consumed raw. This might occur when contaminated raw chicken is cut up on a surface on which salads will be prepared. If the surface is not sanitized, cross-contamination is likely to occur.

2. Mishandling of precooked frozen, refrigerated, catered, or vended foods during storage or preparation. Foods may be contaminated with numbers of pathogens which are too low to cause disease; however, when not properly refrigerated or held too long at room temperature, the microorganisms may grow to sufficient numbers to cause disease. Contamination may come from unsanitary foodservice worker practices such as failure to wash hands or having open wounds. Mishandling is probably the most common problem in many food-borne outbreaks.

3. Inadequate cooking, processing, or preparation can also result in disease. In the case related to pathogenic *E. coli* mentioned above, the hamburger was not heated sufficiently to kill all the pathogenic microorganisms. The surviving organisms infected people eating the sandwiches. In this case, the outbreak could have been prevented if either the hamburger did not contain the pathogens or if it had been cooked sufficiently to kill the organisms.

4. Some people suggest that modern animal agriculture, where large numbers of animals are raised in relatively confined areas, intensifies food-borne disease because the opportunity for one animal to infect the next with a human pathogen is greater. Considerable research is underway to determine how animal husbandry and other agricultural practices could be improved to reduce the occurrence of pathogens.

5. The centralization of plant operations and large-scale production of individual food products. This has the potential for greater microbiological control but also for exposure of a large segment of the population in the event of a single mishap.

EFFECTS OF PROCESSING AND STORAGE
ON MICROBIAL SAFETY

As pointed out, mishandling by both the consumer and the food processor is a key factor in many food-borne disease outbreaks. It is important to understand the effects of different processes on the microbiology of foods.

Freezing and Refrigeration

Freezing does not inactivate microorganisms in foods. Although total bacterial counts can be lowered during freezing of food, substantial numbers of pathogens may survive freezing. The degree of survival will be greater, the quicker the freezing process. Faster freezing also improves food quality, so newer liquid-nitrogen and carbon dioxide freezing processes offer even less in the way of microbial destruction than the slower freezing methods, which are poorly bactericidal. Thus, frozen foods, whether precooked or not, must be safe before freezing if they are to be safe after freezing. In the case of precooked foods, this obviously requires prevention of contamination after cooking and minimum time of holding between cooking and freezing. Frozen foods also must not be allowed to remain at elevated temperatures during the thawing process.

After processing, frozen and refrigerated foods, with few exceptions, must remain cold during warehousing, transportation, and marketing. This is particularly true of low-acid foods. Unfortunately, many trucks and frozen food cabinets in current commercial use are not capable of maintaining proper temperatures. Generally, the pathogens common to food cannot grow below 3.3°C, and grow slowly between 7 and 10°C. In the case of refrigerated, catered, and vended foods, this range must be considered with caution, particularly with precooked foods that will receive no subsequent cooking or only reconstitution with minimal heat to bring them to serving temperature.

Staphylococcus aureus and certain Salmonella species can grow down to 7°C. *Clostridium botulinum* type E may grow at 3.3°C under appropriate conditions of anaerobiosis and freedom from chemical or other microbial antagonists. Recently, microbiologists have found that *Listeria monocytogenes*, which occurs with high frequency in some foods and is capable of causing death in certain groups of people, can grow at temperatures as low as 4°C. At the same time, commercial and household refrigerators are often at temperatures somewhat above 7°C. This means that what has been traditionally considered adequate refrigeration temperatures can still allow the growth of deadly microorganisms in foods. Where precooked foods are marginally heated or reconstituted, it is not uncommon for times and temperatures to be insufficient to guarantee destruction of pathogenic organisms or their toxins. It is not uncommon in fast-food establishments to encounter deep-fat fried items that retain cold spots in their centers. Electronic ovens of the microwave type are not inherently unsafe, but different constituents of food absorb microwave energy at different rates; precooked foods can emerge from these ovens unequally heated.

Minimally Processed and New Foods

Minimally processed foods often receive less harsh processing in order to make them more "freshlike," milder tasting, or increase their shelf life. Any processing condition that lowers total bacterial counts but does not achieve commercial sterilization can be expected to cause a shift in the types of microorganisms present in a food from what it otherwise would be. Processes that cause desirable shifts include fermentation and acidification, nitrate and nitrite additions to cured meat, salting and smoking, and pasteurization of milk. These treatments coupled with proven methods of subsequent handling of the products constitute little hazard.

However, whenever variations of these proven techniques are introduced and considered to replace them, caution must be exercised. In recent years there has been a move to change foods to meet changing preferences for lower salt, smoke, and acid content of many foods. Mildly smoked fish, less highly cured meats, less acidic cheeses, and milder natural ripened cheeses are examples. To some extent these changes are offset by greater sanitary control during manufacture. However, lower salt levels, for example, generally makes a food more favorable for growth of staphylococci, and lower acidity increases the likelihood of survival and growth of many pathogens, including clostridia. Mild heat or other treatments sufficient to substantially reduce total counts often are not fully effective in destroying bacterial spores, among which some of the most resistant belong to the genus *Clostridium*. When such treatments substantially decrease the numbers of less-resistant organisms (the acid-forming streptococci and lactobacilli, and the lipolytic and proteolytic members of such genera as *Pseudomonas, Achromobacter, Proteus,* and the molds), *Clostridia* find many foods more favorable substrates for growth. This is also true for many non-spore-forming pathogens, including members of the genera *Staphylococcus* and *Salmonella,* that may survive the mild treatment or find their way into foods as contaminants following treatment. Not only may these pathogens grow in the subsequently less competitive environment, but with a gross reduction in the normal spoilage flora, mishandling of the food frequently will not result in the familiar spoilage patterns that warn consumers of potential danger.

Many kinds of mild processing conditions favor microbial selection. Of particular concern is the increased use of milder heat treatments to better retain food texture, color, flavor, and nutrients. Numerous high-temperature–short-time and ultra-high-temperature–short-time regimens have been introduced. Irradiation pasteurization has recently received considerable attention as a potential preservation process for seafood, fruits and vegetables, cereal grains, and other commodities. These practices are subjected to intensive investigation before they are approved for commercial use. In the case of irradiation, no previous preservation method, including canning, had been so thoroughly studied prior to its proposed introduction. Yet there remains much that is not known about the microbiology of irradiated foods.

There is a concern among microbiologists that the introduction of minimally processed foods increase microbial risks. This also is true of the microbiology of other cold pasteurization techniques, including the use of membrane filters. These are being successfully used for the cold pasteurization of beer, but beer contains bacteriostatic substances and has a pH that makes this medium quite unfavorable to the growth of pathogens. On the other hand, similar membranes have been suggested as a possible means of pasteurizing milk and other low-acid foods, and this is a different matter.

It also should be reemphasized that such preservation methods as conventional canning have substantial safety factors built into their standard procedures. In the case of low-acid foods, for example, the heat treatment is selected to give a 12 log cycle decrease in the heat-resistant spores of *Clostridium botulinum* or kill more resistant spore-formers. Generally, little is known about the quantitative relationships in food systems between intensity of treatment and death rates of bacteria when a new destructive method is proposed. Certain ultraviolet, X-ray, dielectric, ultra-high-pressure, and microwave processes fall into this category. To this must be added the engineering problems associated with today's requirements for automatic high-speed processing. It is one thing to achieve an average effect, such as an average heat exposure of 150°C for 1 sec; it is quite another to assure that every particle of food experiences this exposure. The engineering of high-temperature–short-time systems for liquid foods has advanced to a high degree, but the problems associated with laminar flow, product burn-on, changes in viscosity, and uniform heat transfer continue to challenge the development engineer. The situation becomes more complex where suspensions of particulate foods and solid foods are involved.

The microbial safety of freeze-dried foods is unique. This is because freeze-drying is the most gentle commercial dehydration procedure with respect to the food as well as the microbial population. No drying method known today, capable of producing a food of high organoleptic and functional quality, produces a sterile product. In the case of freeze-drying particularly, a high bacterial count in the frozen food prior to drying will be reflected in the dried food. In the dry state no microbial growth will occur, but on reconstitution such food is perishable and must be held under appropriate conditions of refrigeration if not soon consumed. This also applies to foods dehydrated by other methods. In the case of freeze-drying, however, it is easy to visualize a sequence of events that could produce a potential danger. Such a sequence could involve contamination of precooked food with pathogenic organisms before or after freezing, freeze-drying, reconstitution with cool water, and holding without adequate refrigeration prior to consumption. Abuses in the preparation and use of precooked freeze-dried shrimp and similar products could thus lead to danger.

Changes in the composition of foods can create additional uncertainties. In the development of new products there has been much recent interest in eliminating food additives and ingredients that may be harmful or may diminish a food's perception as natural. Examples, which include salt, sugar, nitrites, and sulfur dioxide, have antimicrobial properties. New products and food analogs also are being made with ingredients and physicochemical properties that can influence microbial growth in ways not yet well understood.

Potential dangers from imperfect closure or mechanical failure of plastic, foil, and laminated packages are always present, but the safety record of these newer packages has been generally excellent. A related problem is the increased use of easy-open reclosable packages. These packages could be more vulnerable to failure. However, greater use of tamper-resistant and tamper-evident closures has reduced the possibility that products will be intentionally or unintentionally contaminated before purchase.

The potential for changing microbial population balances by vacuum and modified atmosphere packaging (MAP) are even greater. The objective of MAP is to produce a minimally processed food that is freshlike yet has sufficient shelf life for distribution with a short storage time. This is achieved by surrounding the packaged food with a gas atmosphere that has been changed from the normal 20/80 O_2/N_2 composition of air to some other mixture. In nearly all cases these products are distributed and held under refrigeration, but it is now known that some pathogenic bacteria can survive and even grow, albeit slowly, at temperatures as low as 3.3°C, well below the normal refrigeration temperature of 7.2°C. One of the major determinants of microorganism growth is the composition of the atmosphere. MAP extends shelf life by reducing the O_2 content and increasing the CO_2 content of the atmosphere. This can greatly inhibit the growth rate of some microorganisms while having little or even a stimulatory effect on others. In general, spoilage-type gram-negative organisms are inhibited, whereas gram-positive organisms are unaffected or stimulated. Thus, a large change in the normal microorganisms present in these products is seen.

The interest in the safety of these and related minimally processed products is because inhibition of spoilage organisms will prevent the normal bad odors and appearance that accompanies food spoilage. Many pathogenic microorganisms do not change the odor, taste, or appearance of foods. Given the ability of these technologies to extend the time that a product remains in edible condition, it may be possible for slowly growing pathogens to develop.

The answer to the uncertainty about some of the newer foods is to produce them under very strict conditions in which the likelihood of being contaminated with pathogenic microorganisms is very small. Control of temperature and open shelf life dating of products are likewise important safety measures.

MICROBIOLOGICAL METHODOLOGY

One of the ways to ensure food safety is to improve the methods used to test foods for the presence of pathogenic microorganisms. There are several drawbacks to standard plating methods for determining the number and type of microorganisms in a food product. The principal drawback is the time required for preenrichment and incubation of the product and plates. Often, identification can be confirmed only after further growth on selective media and specific biochemical tests. The time to complete these tests may be several days or longer. Foods cannot be held during these tests and so the food is often distributed before any results are obtained. Such tests often only reflect what the microbial condition of the product was before distribution.

Not surprisingly, there has been considerable work on making microbial analyses faster. Tests are being developed for the detection of specific types of microorganisms in a matter of hours rather than days. Most of these tests are based on recent discoveries from genetics and molecular biology. Systems are being developed which isolate, separate, and detect a specific type of DNA from the microorganisms of interest. These DNA-based tests are considerably faster and more specific than traditional plating and enumeration methods (Fig. 23.3).

HACCP AS A METHOD TO PREVENT FOOD-BORNE ILLNESS

Although microbiological testing of foods is an important tool to ensure safety, such testing has the disadvantages that it normally requires time and it often detects

1. SAMPLE LYSIS

rRNA TARGET

2. HYBRIDIZATION

CAPTURE PROBE   AAAAAA

DETECTOR PROBE

FL

FL

FL

AAAAAAA

FL

FL

rRNA TARGET

3. CAPTURE

rRNA TARGET

FL

FL

4. DETECTION

HRP
ANTI-FL
FL

HRP
ANTI-FL
FL

rRNA TARGET COMPLEX

POSITIVE

NEGATIVE

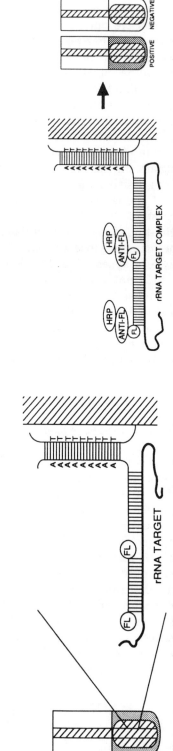

Figure 23.3. A rapid microbial test system in which the RNA or DNA of specific microorganisms is rapidly detected. RNA from the bacteria is linked to detector molecules. *Courtesy of Gene-Trak Systems Inc.*

544

problems only after they occur. Preventing food-borne disease is more a matter of understanding where food-borne diseases originate and how food manufacturing and storage can increase the risk of disease. Once these factors are understood, steps can be taken to ensure that such risks are minimized. Prevention is the preferred way to reduce microbial risks in foods.

The best and most effective method of assuring food safety is to establish a systematic approach to raw materials screening and to identifying food manufacturing and handling procedures which result in the lowest possible risk. One of the major tools for achieving a high degree of reliability and safety is called the Hazard Analysis and Critical Control Point (HACCP) system. HACCP is an approach to food manufacture and storage in which raw materials and each individual step in a process is considered in detail and evaluated for its potential to contribute to the development of pathogenic microorganisms or other food hazards. HACCP principles apply to microbiological, chemical, and physical hazards associated with foods but are most widely applied to microbiological hazards because they are the leading cause of food-borne disease. There is not unanimous agreement on all details of HACCP, but, in general, a series of seven principles or steps have been developed:

1. Hazards associated with growing, harvesting, raw materials, ingredients, processing, manufacturing, distribution, marketing, preparation, and consumption of a given food are each assessed in detail. Areas of potential microbiological, chemical, or physical contamination are determined. This includes both incoming ingredients as well as finished product.
2. The critical control points for controlling each hazard identified above are identified. Those steps in the process in which loss of control could result in an unacceptable health risk are considered to be critical control points. Control of these points must be maintained in order to ensure the safety of the product. Careful control of temperature after processing a product might be one such critical control point.
3. After each critical control point is identified, the limits on that point must be defined. This may be a minimum or maximum temperature or the addition of a minimum amount of acid or salt, for example.
4. Specific procedures for monitoring the control point(s) must be established next. It is of little value to have a maximum temperature for a control point unless there is specific procedure for collecting data on the critical control point. If a product depends on the addition of acid for safety, then the limits on the pH or acid content must be defined and monitored.
5. The next step is establishing an action plan for taking corrective action when monitoring indicates that the critical control point's limits have been exceeded. This corrective action will likely involve adjusting the process, dealing with non-complying product, correcting the cause(s) for noncompliance, and maintaining records of the incidence.
6. It is important that the entire HACCP plan be thoroughly documented. This includes not only the plan itself but also the keeping of records of measurements for all critical control points.
7. Finally, it is important to have procedures for verifying that the HACCP plan is being followed and that it is working according to plan.

A better understanding of HACCP can be gained by considering an example product such as refrigerated chicken salad. Chicken salad is a very complex product which has been implicated as a vehicle for several outbreaks of microbial food-borne disease. The

HACCP process should start with a description of the product and its ingredients. Refrigerated chicken salad, intended for the retail and foodservice industries, is made from approximately two dozen processed and raw ingredients (Fig. 23.4). Each of these ingredients brings a different type and level of risk to the final product and many require different processing/handling conditions to remain safe and of high quality. Chicken salad is packaged in tubs or polymer bags and stored at or below 7.2°C. It is code dated and marked "Keep Refrigerated." Chicken salad has several microbiological, chemical, and physical risks associated with the ingredients and completed product.

Next, the process for preparing the salad is described in detail (Fig. 23.5). Not surprisingly given the complexity of this product, the number of critical control points is large. There are 15 critical control points which cover microbiological, chemical, and physical hazards (Fig. 23.6). Each hazard has a limit as well as a control strategy. For example, control point 5 relates to the preparation of ingredients and has several points of temperature control. In each case, product temperature must be maintained at 7.2°C or less in order to maintain microbiological safety. Control point 11 relates to the potential for foreign metal contamination (physical hazard) and uses a metal detector to monitor the product. The last control point (15) relates to the use of labels indicating to consumers how the product should be stored to maintain microbiological safety.

As the chicken salad example shows, HACCP is a detailed study of individual products and the risks associated with the product as well as a study of how to minimize those risks. HACCP and related systems may become mandatory for some products such as raw poultry, raw fish, shellfish, hamburger, and similar foods.

Raw Material or Ingredient	How Received or Prepared	Raw Material or Ingredient	How Received or Prepared
Cooked chicken (from USDA Establishment)	•Frozen, or •Canned	Starch	•Flour
Dressing	•Ready-made, or, •Prepared in manufacturing facility	Gums and stabilizers	•Dehydrated
Celery (diced)	•Fresh stalk celery in crates, or •Frozen, or •Canned	Lemon juice	•Concentrated
Bread Crumbs/Cracker Meal	•Ready-to-use	Horseradish	•Prepared/acidified
Diced sweet pickles and pickle relish	•Ready-to-use	High fructose corn syrup	•Liquid
Red or green bell peppers	•Fresh, or •Frozen, or •Dehydrated, or •Canned	MSG/HVP	•Dry powder
Hard boiled eggs (diced)	•Purchased hard boiled and peeled	Salt	•Crystalline
Diced onion	•Fresh, or •Frozen, or •Dehydrated •Concentrated	Sugar	•Crystalline
Chicken broth	•Concentrated	Citric acid	•Crystalline
Onion powder	•Dehydrated	Titanium dioxide	•Powder
Garlic powder	•Dehydrated	Textured vegetable protein	•Frozen, or •Dehydrated
Spices	•Dehydrated	Natural and artificial flavors	•Powder

Figure 23.4. Ingredients of refrigerated chicken salad. Source: M.D. Pierson and D.A. Corlett Jr. *HACCP Principles and Applications, Van Nostrand Reinhold, New York, 1992.*

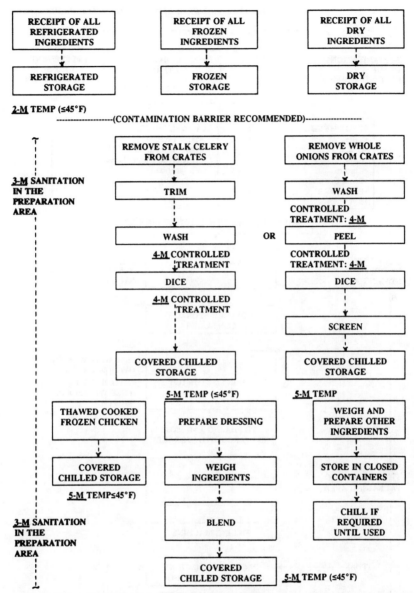

Figure 23.5. Production of refrigerated chicken salad showing many critical control points. (M=microbiological, C=chemical, P=physical, S=sanitation). Source: M.D. Pierson and D.A. Corlett Jr. *HACCP Principles and Applications, Van Nostrand Reinhold, New York, 1992.*

CHEMICAL HAZARDS ASSOCIATED WITH FOODS

Chemicals which cause a harmful response when consumed by animals or humans are said to be toxic. It turns out that almost everything is a toxicant or "poison" if consumed at a high enough level. Even table salt and vitamins are toxic in large amounts. Thus, nearly everything can be considered a "toxicant" without regard for the origin of the substance. The factors which determin toxicity are the dose or amount

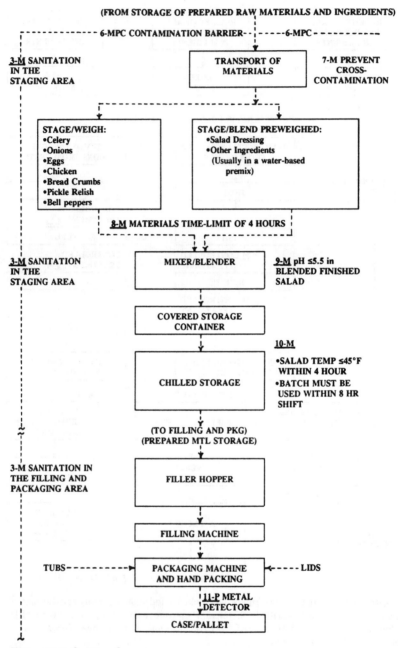

Figure 23.5. Continued

of exposure and the potency of the chemical. Acute toxicants act within short periods after exposure (minutes, hours, days), whereas chronic toxicants produce an adverse effect after longer periods, often years. Food-borne toxicants can be divided into three categories:

1. Those coming from natural sources, including the food itself. Naturally occurring toxicants are found in plants, microorganisms, and animals. We have learned to

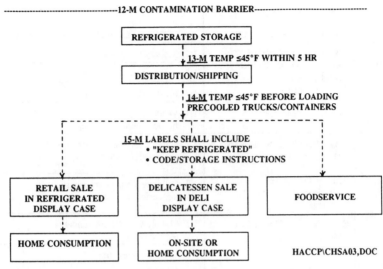

Figure 23.5. Continued

avoid foods which contain naturally occurring acute toxicants. In other cases we process foods in order to remove the toxicant. However, it is clear that many foods also contain low levels of naturally occurring chronic toxicants. It is unknown if these naturally occurring chronic toxicants pose any risk for human health.

2. Those toxicants which become food contaminants because of the way in which foods are grown, produced, processed, stored, or prepared. Toxic environmental contaminants like lead, polychlorinated biphenyls (PCBs), and pyrolysis products from cooking are examples of unintended toxicants in foods. Most of these toxicants are in trace amounts and do not pose an acute risk. Often there is little we can do to control these toxicants except to limit the amount of some foods eaten or limit the amount of toxicant permitted in a food. As with naturally occurring toxicants, the size of the risk involved from these contaminants is often unknown.

3. Those that are intentionally added to foods for some desirable function. This may be a food additive which acts as a preservative or it may be a pesticide which is used to reduce insect or mold damage or it may be a drug which is used to treat disease in food-producing animals. These intentional additives are often the most controversial. Some feel that they should not be used no matter how small the risk, whereas others feel the risks are very small and the benefits large. Because we have control over intentional additives, they are closely regulated by governments around the world.

Natural as well as highly processed foods contain chemicals which can be toxic at some dose. Many types of toxicity can occur from food substances, including nerve damage, organ toxicity, antinutritional effects, birth defects, and cancer. Our bodies cannot tell if a toxicant comes from natural sources or if it is synthetic. Both are treated similarly by the body.

As pointed out, often there is little that can be done about the naturally occurring toxicants in foods. We have more control over toxicants that enter foods either intentionally or unintentionally. Food chemicals that are intentionally added to foods as direct additives or become part of foods as unintentional additives are tested and regulated

CCP Number	CCP DESCRIPTION	CRITICAL LIMIT(S) DESCRIPTION
1-MPC	**HAZARD CONTROLLED;** Microbiological, Physical and Chemical Point or Procedure: Incoming Inspection	1.1 Sanitary Condition
		1.2 Refrig. Material ≤45°F
		1.3 Frozen Material ≤32°F
		1.4 Vendor met all safety specifications before shipping
2-T	**HAZARD CONTROLLED:** Microbiological Point or Procedure Refrigerated Ingredient Storage	2.1 Material internal temperature not to exceed 45°F
		2.2 Calibrate temperature-measuring devices before shift
3-M	**HAZARD CONTROLLED:** Microbiological Point or Procedure: Sanaitation Requirements in - Preparation area - Staging area - Filling/Packaging area	3.1 Comply with USDA sanitation requirements
		3.2 Sanitation crew trained
		3.3 Each area must pass inspection
		before shift start-up
	HAZARD CONTROLLED: Point or Procedure: *Listeria*	3.4 Food contact surface Microbiological test
		3.5 Environmental area Microbiological tests
		(USDA Methodology for 3.4 & 3.5)
4-M	**HAZARD CONTROLLED:** Microbiological Point or Procedure: Controlled treatment to reduce microbiological contamination on raw celery and onions	Application of alternative approved treatments 4.1 Wash product with water containing - Chlorine, or - Iodine, or - Surfactants, or - No process additives
		4.2 Hot water or steam blanch followed by chilling
		4.3 Substitute processed celery or onions: - Blanched frozen - Blanched dehydrated - Blanched canned

Figure 23.6. The limits for many critical control points for refrigerated chicken salad manufacture. Source: *HACCP Principles and Applications, Van Nostrand Reinhold, New York, 1992.*

by authorities in order to ensure that the risks are minimal and acceptable in the context of the benefits (Fig 23.7).

The regulation of pesticides in the United States is a good example. "Pesticides" are chemicals that are applied to food crops and animals to control insects, molds (fungi), or unwanted plants. As such, they are of benefit because they reduce the competition for the food. They also have risks because they, by definition, are toxic. In the United States, the Environmental Protection Agency (EPA) is responsible for requiring the

5-M	**HAZARD CONTROLLED:** Microbiological Point or Procedure: Chilled storage temperature of prepared celery, onions and chicken	5.1 Not to exceed 45°F
		5.2 Refrigerator not to exceep 45°F
		5.3 Daily calibration of temperature measuring devices
6-MPC	**HAZARD CONTROLLED** Microbiological, Physical and Chemical Point or Procedure: Physical barrier to prevent cross-contamination from raw material preparation area	6.1 Physical barrier in-place
		6.2 Doors kept closed when not in use
		6.3 Color-coded uniforms
		6.4 Supervision in-place
7-M	**HAZARD CONTROLLED:** Microbiological Point or Procedure: Cross-contamination prevention from transfer equipment from raw material area	7.1 Comply with USDA sanitation requirements
		7.2 Prevent entry of soiled pallets cart wheels, totes, and other equipment
8-M	**HAZARD CONTROLLED:** Microbiological Point or Procedure: Time limit for in-process food materials	8.1 Time limit not to exceed four hours for any materials in staging area
9-M	**HAZARD CONTROLLED:** Microbiological Point or Procedure: Maximum pH limit on finished salad before packaging	9.1 Product pH must not exceed a pH of 5.5
		9.2 pH meter must be calibrated with approved standards before each shift
10-M	**HAZARD CONTROLLED;** Microbiological Point or Procedure: Chilled product storage temperature and time before packaging	10.1 Internal temperature not to exceed 45°F
		10.2 Product must not be held more than one shift before filling/packaging
11-P	**HAZARD CONTROLLED:** Physical Point or Procedure: Metal detector for packages	11.1 Ferrous metal detection device for individual packages
		11.2 Calibration or inspection not to exceed every four hours

Figure 23.6. *Continued*

pesticide industry to test extensively each pesticide for it toxic potential and to submit the results to EPA for evaluation. In the United States as well as elsewhere, specific pesticides can be used only if permission is granted for a specific use. This is called "registration" and a given pesticide must be registered for each use or crop. The process of registration starts with a company who wants to gain approval for a specific pesticide. The company must submit data to the EPA which indicates, first, how much residue will remain on the food so an estimate of the human exposure can be made and, second, what the potential of the pesticide is for causing disease. This latter toxicological data is based on animal studies and is used to determine what the maximum size the risk would be to humans. The EPA will allow the pesticide to be used if it determines the size of the risk to be very small. If the risk is too large, then the chemical will not be

12-M	**HAZARD CONTROLLED:** Microbiological	12.1 Physical barrier in place
	Point or Procedure	12.2 Doors kept closed when not in use
	Physical barrier to prevent cross-contamination from warehouse area	12.3 Color coded uniforms
		12.4 Supervision in-place
13-M	**HAZARD CONTROLLED:** Microbiological Point or Procedure: Refrigerated storage of cased/palleted finished product	13.1 Product internal temperature not to exceed 45°F in four hours
		13.2 Temperature measuring devices calibrated before shift
14-M	**HAZARD CONTROLLED:** Microbiological Point or Procedure: Truck and shipping containers for distribution of finished product	14.1 Shipping compartments must be pre-cooled to 45°F or less before loading product
15-M	**HAZARD CONTROLLED:** Microbiological Point or Procedure: Label Instructions	15.1 Each package or bulk case shall have label instructions
		15.2 Each laabel shall include: - Keep Refrigerated - Code - Storage Instructions

Figure 23.6. *Continued*

approved. If the chemical is approved, it is given a "tolerance" which is the maximum amount of the pesticide that is allowed on the food product. The FDA is then given the job of enforcing the tolerance to make sure that the chemical does not occur in foods for which it is not registered or in amounts larger than approved (Fig. 23.8).

One major function, but not the sole function, of food additives is to aid in preserving foods. Other chemicals added to foods are not preservatives, but are added for functional properties associated with food color, flavor, or texture. Still other additives are incorporated as nutritional supplements and as processing aids in manufacturing. Food additives impart several important funtions to foods, several of which are listed in Table 23.4.

Most of additives are added to foods in very small amounts, often in amounts equaling 1 to 100 ppm. One part per million means that for every million pounds of the food, a pound of the food additive is used.

Broad Classes of Intentional Food Additives

There are approximately 3000 intentional food additives in 12 major groups. A few representative types from each group are listed below.

Preservatives

These are substances added to foods to inhibit the growth of bacteria, yeasts, and/or molds. Examples include sodium benzoate used in soft drinks and acidic foods,

Steps in the Risk Assessment of Additives in Food at the U.S. Food and Drug Administration

Toxicological Evaluation

```
• Input Obtained from Internal Experts
    - Toxicologists     - Pathologists
    - Chemists          - Biostatisticians
    - Other Experts     - Epidemiologists
• Input Obtained from External Experts (where need is
  indicated)
```

↓

Cancer Assessment Committee (CAC) Evaluation

```
• CAC Reviews input from Internal and External Experts
    - Is the substance a likely carcinogen?
    If Yes: CAC recommends the studies, tissue sites,
    species, and sex suitable for quantitative risk evaluation if
    risk assessment is allowed under the statute.
    If No: No further consideration by CAC or QRAC is
    needed
```

↓

Quantitative Risk Assessment Committee (QRAC) Evaluation

```
• QRAC Reviews Data and Exposure Potential
• QRAC Chooses Risk Assessment Model and Procedure
• QRAC Estimates Magnitude of Potential Human Risk
    - Calculate the upper bound lifetime risk
```

↓

Action Taken by Director of CFSAN, FDA

```
• Makes Risk Management and Policy Recommendations to the
  Commissioner
```

Figure 23.7. Four risk assessment steps undertaken by the Food and Drug Administration when considering the safety of a food additive. Source: *Toxicological Principles for the Assessment of Direct Food Additives and Color Additives Used in Food, Food and Drug Administration, Washington, DC, 1993 [Draft].*

sodium and calcium propionate used in breads and cakes as a mold inhibitor, and sorbic acid used on cheese and in moist dog foods to control mold. Substances such as sulfur dioxide, that control browning of fruits and vegetables caused by enzymes, also are considered preservatives.

Antioxidants

These substances are used to prevent oxidation of fats by molecular oxygen. Without them, potato chips, breakfast cereals, salted nuts, fat-containing dehydrated foods, crackers, and many other fat-containing foods could not be stored very long without developing rancidity. Principal among these antioxidants are butylated hydroxyanisole

Figure 23.8. Chemist testing for pesticide residues in foods by liquid chromatography. *Courtesy of Hasleton Laboratories America, Inc.*

(BHA), butylated hydroxytoluene (BHT), tertiary butylated hydroquinone (TBQH), and propyl gallate. The antioxidants also include such diverse materials as ascorbic acid, stannous chloride, and tocopherols (vitamin E). Sulfur dioxide, listed as a preservative, is further listed as an antioxidant. Many other food chemicals also exhibit dual roles.

Sequestrants

These are the chelating agents or sequestering compounds which serve to scavenge metal ions. They do this by combining with trace metals such as iron and copper and remove them from solution. The trace metals are active catalysts of oxidation and also contribute to off-color reactions in foods. Their removal by chelating additives such as ethylenediamine tetraacetic acid (EDTA), polyphosphates, and citric acid prevents these defects.

Surface Active Agents

These include the emulsifiers used to stabilize oil-in-water and water-in-oil mixtures, gas-in-liquid mixtures, and gas-in-solid mixtures. In addition to emulsifiers of natural origin such as lecithin and to emulsifiers that can be prepared synthetically such as monoglycerides and digylcerides and their derivatives, other emulsifying agents include certain fatty acids and their derivatives and bile acids, which are important in digestion. Also included among the surface active agents are numerous defoaming compounds and detergent chemicals.

Table 23.4. Some Functional Properties of Common Food Additives

I. *Purpose*: Maintain/improve nutritional quality
   1. *Class*: Nutrients
     Function: Fortify: add nutrients; enrich; replace nutrients

Examples	*Where found*
Vitamin C	Beverages
Iron	Baked goods

II. *Purpose*: Facilitate production, processing, or preparation
   1. *Class*: Acidulants, alaklies, buffers
     Function: Multi, flavor, pH control, processing, antimicrobial

Examples	*Where Found*
Acetic acid	Dressings
Sodium bicarbonate	Chocolate
Phosphoric acid	Soft drinks
Phosphates	Meats

   2. *Class*: Anticaking agents
     Function: Keep powders free-flowing

Examples	*Where Found*
Silicates	Salt, baking powder
Talc	Rice

   3. *Class*: Bleaching, maturing agents, dough conditioners
     Function: Accelerate dough aging, machinability

Examples	*Where Found*
Peroxides	Flour
Gums	Bread dough

   4. *Class*: Bulking, bodying, thickening, firming
     Function: Change consistency, stabilize, mouthfeel

Examples	*Where Found*
Pectin	Jams, jellies
Alginates	Dairy desserts
Carboxymethylcellulose	Ice cream, bread
Modified food starch	Sauces, soups
Calcium lactate	Apple slices

   5. *Class*: Chelating agents
     Function: Sequester metal ions

Example	*Where Found*
EDTA	Beer, canned fish

Stabilizers and Thickeners

These substances include gums, starches, dextrins, protein derivatives, and other additives that stabilize and thicken foods by combining with water to increase viscosity and to form gels. Gravies, pie fillings, cake toppings, chocolate milk drinks, jellies, puddings, and salad dressings are among the many foods that contain such stabilizers and thickeners as gum arabic, carboxymethyl cellulose (CMC), carrageenan, pectin, amylose, gelatin, and others.

Bleaching and Maturing Agents, Starch Modifiers

Freshly milled flour has a yellowish tint and suboptimum functional baking qualities. Both the color and baking properties improve slowly in normal storage. These improvements can be obtained more rapidly and with better control through the use of certain oxidizing agents such as benzoyl peroxide which bleaches the yellow color. Oxides of nitrogen, chlorine dioxide, and other chlorine compounds both bleach color and mature the flour. Starch modifiers include compounds such as sodium hypochlorite, which oxidizes starches to a higher degree of water solubility.

Buffers, Acids, Alkalies

As pH-adjusting and pH-controlling chemicals, buffers, acids, and alkalies affect an endless number of food properties, many of which have been previously described. The acids in particular may be derived from natural sources, such as fruits, and from fermentation, or they may be chemically synthesized.

Food Colors

Colors are added to thousands of food items to produce appetizing and attractive qualities in foods. Both synthetic and naturally occurring colors are added to foods. Artificial colors, each batch of which must be tested and certified by the U.S. FDA, must be labeled as artificial. Many natural coloring agents such as extract of annatto, caramel, carotene, and saffron are widely used in foods. Synthetic colors generally excel in coloring power, color uniformity, color stability, and lower cost. Further, in many cases, natural coloring materials do not exist for a desired hue. Carbonated beverages, candies, and gelatin desserts are among items colored with certified synthetic dyes. In addition to the organic synthetic dyes and the natural organic coloring agents, food colors also include such inorganic materials as iron oxide to give redness and titanium dioxide to intensify whiteness. Organic coloring materials also can be coated onto metallic salts to produce suspensions of colored particles, known as lakes, as opposed to dissolved colored solutions.

Interest in the use of natural reds from grapes, beets, and cranberries has increased. Another natural red food color, cochineal, used to produce carmine, is extracted from females of the insect *Coccus cacti.*

Artificial Sweeteners

Chemicals which taste sweet but which contain few or no calories at the levels required for sweetness are considered non-nutritive or reduced-calorie sweeteners. Some compounds such as aspartame contain the same number of calories as sucrose (sugar) but are hundreds of times sweeter; hence, only a fraction of the amount is used compared to normal sugar. Thus, they are "reduced"-calorie sweeteners. Other compounds, such as saccharin, contain no calories and are truely non-nutritive. The largest group of users of non-nutritive and reduced-calorie sweeteners have been the low-calorie soft drink manufacturers. But such sweeteners also have been important in the manufacture of other low-calorie foods such as candies, frozen desserts, salad dressings, gelatin desserts, and some baked goods. Other non-nutritive sweeteners, which have sweetness ranges up to 3000 times that of sucrose, are under study. Among them are neohesperidine dihydrochalcone from citrus rinds and L-isomers of common sugars, which have comparable sweetness to the naturally occurring D-isomers but are not metabolized and therefore do not contribute calories. Aspartame is currently the most widely used artificial sweetener in the United States. Aspartame is a synthetic aspartic acid–phenylalanine dipeptide.

Nutritional Additives

Vitamins, minerals, and other nutrients are added as supplements and enrichment mixtures to a number of products. Major examples are the following: vitamin D added to milk; B vitamins and iron added to cereal products; iodine added to salt; vitamin A added to margarine; and vitamin C added to fruit juices and fruit-flavored desserts. Several amino acids also are listed in this group. The amino acid lysine has received considerable study since it is the only essential amino acid absent from the protein of wheat flour in sufficient amount to prevent wheat flour from being a nutritionally complete protein source.

Flavoring Agents

Natural flavoring substances include spices, herbs, essential oils, and plant extracts. Typical of the synthetic flavor additives are benzaldehyde (wild cherry/almond), ethyl butyrate (pineapple), methyl anthranilate (grape), and methyl salicylate (wintergreen). Currently, there are over 1200 different flavoring materials used in foods, making this the largest single group of food additives. These are typically used in trace amounts and are similar to the chemicals found in natural sources. Also included among the flavor additives are flavor enhancers or potentiators, which do not have flavor in themselves in the low levels used but intensify the flavor of other compounds present in foods. Monosodium glutamate (MSG) and the 5'-nucleotides (related to components found in nucleic acids) are examples of this kind of material.

Miscellaneous Additives

Many additional food additives provide functions other than those already discussed. Included are substances to promote growth of bakers' yeast such as ammonium sulfate, firming agents (for fruits and vegetables) such as calcium chloride, anticaking agents for salt and granular foods such as calcium phosphate, antisticking agents such as hydrogenated sperm oil, clarifying agents for wine such as bentonite, solvents such as ethanol, acetone, and hexane, machinery lubricants such as mineral oil, meat curing agents such as sodium nitrite and sodium nitrate, crystallization inhibitors such as oxystearin, plant growth stimulants used in malting barley such as gibberellic acid, enzymes for a wide variety of uses, and many more.

Macrocomponents and Foods Substitutes

In recent years, a special type of additive known as a macrocomponent is being used with increasing frequency. Macrocomponents substitute a large portion of the food with another component. For example, there is interest in reducing the fat content of some foods such as ice cream without changing texture and flavor. This means substituting a less caloric fat replacer for all or most of the fat in ice cream. This is a macrosubstitution.

Food additives are regulated by governments in a manner similar to pesticides. In the United States, the FDA has responsibility for food additives. Like pesticides, they must be approved for specific uses in foods with considerable scientific data showing that the risks are very low and that the additive imparts some desired function to foods. Additives can only be used for specific functions in approved foods, and levels permitted for a given additive differ for different foods.

References

Chaisson, C.F. 1991. Pesticides in Food: A Guide for Professionals. American Dietetic Association, Chicago, IL.

Clayson, D.B., Krewski, D., and Munro, I. (Editors). 1985. Toxicological Risk Assessment. CRC Press, Boca Raton, FL.

Concon, J.M. 1988. Food Toxicology. Marcel Dekker, New York.

CRC Handbook of Naturally Occurring Food Toxicants. 1983. CRC Press, Boca Raton, FL.

Farrer, K.T.H. 1987. A Guide to Food Additives and Contaminants. Parthenon Publishing Group, Park Ridge, NJ.

Food and Agriculture Organization of the United Nations and World Health Organization. 1992. Codex Alimentarius. Food and Agriculture Organization of the United Nations and World Health Organization, Rome.

Food and Drug Administration. 1990. Food Risk: Perception vs. Reality. Food and Drug Administration, U.S. Public Health Service, Dept. of Health and Human Services, Rockville, MD.

Food Protection Committee. 1981. Food Chemicals Codex. 3rd ed. National Academy of Science–National Research Council, Washington, DC.

Francis, F.J. 1992. Food Safety: The Interpretation of Risk. Council for Agricultural Science and Technology, Ames, IA.

Harlander, S.K. and Labuza, T.P. 1986. Biotechnology in Food Processing. Noyes Publications, Park Ridge, NJ.

Janssen, W.F. 1985. The U.S. Food and Drug Law: How it Came, How it Works. U.S.Dept. of health and Human Services, Food and Drug Administration, Rockville, MD.

Lewis, R.J. 1989. Food Additives Handbook. Van Nostrand Reinhold, New York.

Miller, K. (Editor). 1987. Toxicological Aspects of Food. Chapman & Hall, London, New York.

Persley, G.J. 1992. Biosafety: The Safe Application of Biotechnology in Agriculture and the Environment. World Bank, Washington, DC.

Pierson, M.D. and Corlett, D.A. Jr. 1992. HACCP: Principles and Applications. Van Nostrand Reinhold, NY. 212 pp.

Schultz, H.W. 1981. Food Law Handbook. Chapman & Hall, London, New York.

Smith, J. (Editor). 1991. Food Additive User's Handbook. Chapman & Hall, London, New York.

Taylor, S.L. and Scanlan, R.A. 1989. Food Toxicology: A Perspective on the Relative Risks. Marcel Dekker, New York.

Thonney, P.F. and Bisogni, C.A. 1992. Government regulation of food safety: Interaction of scientific and societal forces. Food Technol. 46(1), 73–80.

Watson, D.H. 1993. Safety of Chemicals in Food: Chemical Contaminants. Ellis Horwood, New York.

Winter, C.K. 1992. Pesticide tolerances and their relevance as safety standards. Regul. Toxicol. Pharmacol. 15(2)pt.1, 137–150.

24

Governmental Regulation of Food and Nutrition Labeling

INTRODUCTION

Governments worldwide regulate foods with two general objectives: The first is to ensure the safety and wholesomeness of the food supply. The second is to prevent economic fraud or deception. These objectives encompass such concepts as safety, purity, wholesomeness, and value. Recently, a third objective, to inform consumers about the nutritional content of foods, has been added. Because in a highly complex society individual consumers are not in a position and usually do not have the specialized knowledge to protect themselves, the responsibility rests on the food industry and on government. Industry and government must cooperate in the role of providing protection. Furthermore, the food industry looks to government to set high standards and to enforce these standards in order to protect itself against unethical competition.

Economic regulations are intended to prevent consumers from being defrauded. The major intent here is to allow consumers to judge the value of products before purchase. This means that food must be labeled honestly and that its packaging must not be deceptive.

Recent years have seen an increased concern by the public over the safety of foods, especially with respect to intentional and unintentional chemical additives and the incidence of microbial food-borne diseases. Arguments have both defended and attacked food production practices including the use of food additives, pesticides, biotechnology, and irradiation. No less controversial have been debates over proper labeling and honest representation.

In the United States the primary responsibility for ensuring the safety and labeling of foods lies with the Food and Drug Administration (FDA) for most foods and with the U.S. Department of Agriculture (USDA) for meat and poultry products. Other agencies also play a role. The Bureau of Alcohol, Tobacco, and Firearms in the Department of Commerce regulates alcoholic beverages and the Environmental Protection Agency has responsibility for ensuring that the pesticides used on foods are safe. Many countries have similar agencies that regulate food safety and prevent economic fraud.

FEDERAL FOOD, DRUG, AND COSMETIC ACT

The main law giving the U.S. FDA authority for food regulation is the Food, Drug, and Cosmetic Act of 1938. Several amendments have been added to the law over the years. Among them, the Miller Pesticide Amendment of 1954, the Food Additives Amendment of 1958, and the Color Additive Amendments of 1960 have been enacted to keep the law up to date. Additional amendments can be expected in the future. The FDA currently employs over 7000 scientists, legal advisers, and inspectors with offices throughout the country. That part of the Federal Food, Drug, and Cosmetic Act which deals with foods has as its purpose to ensure that foods entering interstate commerce are safe, pure, wholesome, sanitary, honestly packaged, and honestly labeled. Like many laws, the Food, Drug, and Cosmetic Act is short and simple and lays out general principles but does not deal with specific foods or issues. That portion of the law dealing directly with foods is only 25 pages long. It is FDA's responsibility to write regulations for each specific area requiring regulation based on the law. These regulations which are contained in a set of books called the Federal Code of Regulations, Title 21 give precise regulations and rules. For example, the Food, Drug, and Cosmetic Act does not specify what pH defines a low-acid food and the way in which it must be processed to be safe. The FDA's responsibility is to interpret the law and make regulations which cover specific products and processes such as low-acid food canning regulations. These regulations tell what pH constitutes a low-acid food and details how such foods must be processed.

To carry out the provisions of the Federal Food, Drug, and Cosmetic Act, the FDA conducts the following activities: (1) works with industry to help interpret the regulations; (2) helps industry in establishing control measures for product protection; (3) makes inspections of food plants; (4) examines food samples from interstate shipments; (5) issues and enforces regulations on food additives; (6) approves and certifies acceptable food colors; (7) tests for pesticide residue to ensure tolerances are met; (8) examines imported foods for acceptability; (9) works with state and local food inspection agencies in an advisory capacity; (10) works with state and local agencies in times of disaster to detect and dispose of contaminated foods; and (11) sets up "standards of identity" for manufactured foods to promote honesty and value of products. For example, standards of identity define the allowable ingredients and compositions of macaroni products. If a product is sold as macaroni and does not meet its standard, it is considered adulterated and misrepresented and the manufacturer is prosecuted.

In order for the Federal Food, Drug, and Cosmetic Act to protect the public and ensure that foods are not adulterated, mislabeled, deceptive, falsely packaged, or falsely guaranteed, the law scrupulously defines these and other terms. For example, food is considered adulterated if it (1) contains poisons or harmful substances at detrimental concentrations, (2) contains filth, is decomposed, or is otherwise unfit, (3) was prepared or handled under unsanitary conditions such that it may have become contaminated, (4) is derived from a diseased animal, (5) was subjected to radiation, other than where permitted, (6) has any valuable constituent omitted, (7) has a specified ingredient substituted by a nonspecified ingredient, (8) has a concealed defect, (9) is increased in bulk weight or reduced in its strength making it appear better than it is, or (10) contains a coloring agent that is not approved or certified.

Similarly, the law says that a food is misbranded if it (1) is falsely or misleadingly labeled, (2) goes by the name of another food, (3) is an imitation of another food, unless the label indicates that it as an imitation, (4) is packaged so as to be misleading, (5) is packaged and the label fails to list the name and address of the manufacturer, packer, or distributor, and if the label fails to give a statement of net contents, (6) fails by labeling to declare the common name of the product, and the names of each ingredient, (7) has information which is not legible and easily understood, (8) if the food is represented as a food for which there is a standard of identity but the food does not conform, (9) is represented to conform to a quality standard or to a fill of container and does not conform, and (10) is represented for special dietary use but the label fails to give information concerning its dietary properties as required by law.

ADDITIONAL FOOD LAWS

The FDA has also published regulations that have become known as "Good Manufacturing Practices" or code of GMPs. This code defines requirements for acceptable sanitary operation in food plants. Appendices are issued periodically to cover specific GMPs expected of different segments of the food industry. In the case of GMPs for manufacturers of smoked fish, for example, originally specified along with sanitation requirements were temperatures to be used before, during, and after processing as well as salt content. These GMPs were intended to minimize the hazard from botulism. These very specific GMPs were deleted in 1984. GMPs that have become regulations cover frozen raw breaded shrimp, low-acid canned foods and acidified foods, cocoa and confectionery products, bottled drinking water, and others. Guidelines also have been issued defining maximum allowances for natural filth in foods since complete elimination is impossible; pesticide, heavy metal, and other environmental contaminant limits; aflatoxin limits in various products; and recall procedures for the retrieval of food from distribution channels that is in violation of the law.

FDA inspectors are authorized to inspect any and all aspects of food production, manufacture, warehousing, transportation, and sale, so long as the product enters interstate commerce. Malpractice can result in confiscation of the food, closing down of the plant, and fines or imprisonment. With regard to imported foods, all items must meet the Federal Food, Drug, and Cosmetic Act requirements. Further, exports must not violate the laws of the country for which the food is intended. In 1966, the Fair Packaging and Labeling Act was enacted to help strengthen the laws governing misbranding. This law defines in precise terms the kinds of information that must appear on the labels of food packages, the size of type and the positions on the labels and packages that may be used to convey information, and the kinds of designations and terms that would be misleading to the consumer and are therefore forbidden.

Besides the Food, Drug, and Cosmetic Act and its amendments, there are additional federal laws covering certain specific foods and food-related areas.

Federal Meat Inspection Act of 1906

This provides for mandatory inspection of animals, slaughtering conditions, and meat processing facilities. It helps ensure that meat and products containing meat are clean, wholesome, unadulterated, free from disease, and properly represented. Such products are stamped "U.S. Inspected and Passed by Department of Agriculture." All meat and meat products must bear this stamp to be permitted shipment in interstate commerce. The law also applies to imported meats. The Food Safety and Inspection Service (FSIS) of the Department of Agriculture enforces the act. Federal standards for meat under this law have been mandatory only for products crossing state lines. State and city meat regulations have covered products not shipped interstate. The Wholesome Meat Act requires that all state and city meat inspection regulations must meet federal standards and meat plants that do not come up to these standards are closed down.

Federal Poultry Products Inspection Act of 1957

This is essentially the same as the Meat Inspection Act but applies to poultry and poultry products. The new Wholesome Meat Act has been essentially paralleled to cover poultry by the Wholesome Poultry Products Act of 1968.

Federal Trade Commission Act (Amended for Food in 1938)

This protects the public as well as the food industry against false advertising related to foods or their alleged properties. This extends to newspapers, radio, television, or any other advertising media. The Fair Packaging and Labeling Act, mentioned earlier, deals more specifically with the package itself.

Infant Formula Act of 1980

This provides that manufactured formulas contain the known essential nutrients at appropriate levels for infant health. It was passed following adverse effects due to omission of the chloride ion from certain commercial products. It has been followed by detailed quality control procedures to help assure compliance with the law.

Nutrition Labeling and Education Act of 1990

This provides protection against partial truths, mixed messages, and fraud with regard to nutrition information. More will be said about it later in this chapter.

Federal Grade Standards

These are primarily standards of quality to help producers, dealers, wholesalers, retailers, and consumers in marketing and purchasing of food products (Fig. 24.1). Establishment of grade standards and inspection programs come under the authority of the Agricultural Marketing Service of the Department of Agriculture. The grade standards, which were discussed in previous chapters, are not aimed at health protection but rather at ensuring value received by labeling products according to uniform quality standards. Federal law does not require this type of grading. Inspection and grading are on a voluntary basis and may be performed by federal inspectors for a fee on request of food producers or marketers.

Figure 24.1. Many fruits and vegetables are hand sorted into different grades as set by the U.S. Department of Agriculture. Source: K. Peleg *Produce Handling Packaging, and Distribution,* AVI Publishing Co., *Westport, CT, 1985.*

State and Municipal Laws

Every state and many cities in the United States have their own food laws in addition to the federal regulations. This is important since food produced and consumed within a state and not entering interstate commerce generally is not subject to federal laws. In practice, these state and city laws commonly are adopted or patterned after the federal laws. In addition, where needed, special regulations have been created to protect health, ensure sanitary practices, and prevent economic deception at specific operating levels, including retail food outlets and eating establishments.

LEGAL CATEGORIES OF FOOD SUBSTANCES

In the United States, substances which become part of foods, either intentionally or unintentionally, can be legally divided into several categories. The division of these categories is not always logical but is derived from laws which have been amended several times over a nearly 90-year period. This means, for example, that artificial colors added to foods are not food additives because color additives are a separate category.

GRAS Substances

GRAS stands for "generally recognized as safe." These are substances added to foods that have been shown to be safe based on a long history of common usage in food. Typical GRAS substances include the common spices, natural seasonings, and numerous flavoring materials; baking powder chemicals such as sodium bicarbonate and monocalcium phosphate; fruit and beverage acids such as citric acid, malic acid, and phosphoric acid; gums such as agar-agar and gum karaya; emulsifiers such as fatty acid monoglycerides and diglycerides; and many, many more. Many GRAS substances are actually foods in their own right. The purpose of establishing a GRAS list is to recognize the safety of basic substances without the requirement for rigorous safety testing. To require all such primary substances to be "proved safe" would bring the food production process to a halt.

Substances are sometimes added to the GRAS list and several substances have been removed when new scientific evidence raises questions about their safety. A substance removed from the GRAS list may be placed under more rigid control of its use, or prohibited from further use if warranted. Thus, the GRAS status, as well as the safety of any food additive, is subject to continued review in the light of new scientific knowledge and other factors, including possible changes in consumption patterns with time.

Prior Sanctioned Substances

Both the FDA and the USDA had granted specific approval to use several food substances as additives before the 1958 Food Additives Amendment to the law was enacted. In order not to burden and disrupt the food industry unfairly, the law allowed these substances to be used without meeting the very stringent requirement for direct food additives, as long as there was not any direct toxicological information which seriously questioned the safety of these substances. This is termed "grandfathering" and is a common legal practice. It has led to some confusion, however. For example, the substance nitrite is prior sanctioned when added to red meats and regulated by the U.S. Department of Agriculture. The same substance in fish is a legal food additive and regulated by FDA because it is not prior sanctioned in fish.

Food Additives

The term *food additive* legally refers to a very specific group of food substances. Included are substances which are added intentionally and directly to foods in regulated amounts, to impart some function such as texture, flavor, or microbial preservation. Also included are indirectly added substances which unintentionally become part of foods. These might be migrants from packaging materials, lubricants from processing equipment, or processing aids such as enzymes.

Approval by FDA to use food additives, whether direct or indirect, is granted on submission of scientific data clearly showing that the intended chemical is harmless in the intended food application at the intended use level. This is done by petition to the FDA. The FDA then sets limits with respect to the kinds of foods in which the additive may be used and the maximum concentrations that may be employed. Commonly, a food additive may be permitted at a level of, say, 100 µg/kg in one food, only 25 µg/kg in another food, and be prohibited from use in a third food. At present, over 3000 food additives may be used in foods within the limits that have been deemed safe by the FDA. Changes in the status of food additives are published in the *Federal Register*. The burden of proving the safety of a new chemical substance is on the company that wishes to market or use the chemical. This can require a period of testing of several years, since the FDA usually requires that a new additive undergo at least a 2-year feeding test in two species of animals to reveal long-term as well as short-term effects. In some cases, feeding tests of over 7 years have been required, as well as limited tests in human subjects. Other chemicals may be cleared rapidly if data on related substances are applicable. In many cases, proving the safety of an additive may cost several million dollars. This work generally is not done by food companies, or even by manufacturers of food chemicals or other materials that may impart incidental additives to foods, but is contracted out to specialized toxicology laboratories.

Most substances considered to be food additives are chemicals or can be defined chemically. However, ionizing radiation is considered to be a food additive since such treatment can induce chemical changes in foods. In this case, the FDA approves the foods that may be irradiated, and the irradiation sources and sets limits on the maximum permitted irradiation dosage.

Color Additives

Substances added to foods to impart color have a special category. Color additives may either be naturally occurring chemicals or synthetic. If synthetic, then a sample of each batch must be sent to the FDA for analysis and certification. These colors are known as FD&C certified colors and are then permitted for use in foods and cosmetics.

Pesticide Residues

The occurrence of pesticide residues in foods have been highly controversial. Some feel that these residues are responsible for a significant portion of human disease, whereas others have dismissed the risks. The majority of toxicologists do not believe that the small residues found in foods as consumed represent a significant risk.

Pesticide residues are regulated as pesticides under the law. In some cases they become food additives and must meet the stringent requirements for this class of food substances. When a pesticide residue occurs in a raw product such as a carrot, it is considered a pesticide residue. However, if it occurs and is concentrated (i.e., is in higher concentration in the processed product than the raw), it becomes a food additive. For example, a pesticide on raw carrots might increase in concentration when the raw carrots are dehydrated. This is significant because food additives which can be shown to cause cancer in man or animal without regard to mechanism or amount are not allowed to be added to foods. This is a result of a controversial clause in the law known as the "Delaney Clause." This clause in the Food, Drug, and Cosmetic Act states that no additive may be permitted in food if the additive at any level can produce cancer when fed to man or animals or can be shown to be carcinogenic by any other appropriate test. One problem with this clause is that some rather harmless and common substances may possibly induce cancer in one animal or another under special conditions that may have little to do with normal food consumption. Some desirable and harmless substances such as saccharin are animal carcinogens under special situations that are not applicable to humans. Other examples include materials from charred meat, burned fat, and pepper which have been implicated as potential carcinogens. In recent years, considerable attention has been focused on possible modification of the Delaney Clause, the definition of safe in regard to foods, and the use of informed scientific judgments in regulatory decisions.

Food additives must meet the following additional requirements:

1. Intentional additives must perform an intended and useful function.
2. Additives must not deceive consumers or conceal faulty ingredients or defects in manufacturing practices.
3. An additive must not substantially reduce the food's nutritional value.
4. An additive cannot be used to obtain an effect that could be obtained by otherwise good manufacturing practices.
5. A method of analysis must exist with which to monitor the use of the additive in foods or its incidental occurrence in foods (e.g., trace quantities of permitted migrating chemicals from packaging materials).

The FDA may reject a food additive if any one of these requirements is not met.

TESTING FOR SAFETY

Most commonly, safety testing for substances added to foods begins with the exposing of two or more species of animals, such as the rat and dog, to the substance using high levels of the test material administered as a single dose orally or by injection into the bloodstream. This is known as an acute toxicity test and is intended to reveal indications of harmful effects that might occur within hours or days of exposure. Death or abnormalities seen on autopsy give preliminary information as to harmful dose level and mode of action.

This usually is followed by subacute toxicity testing, again with two or more animal species. Here large numbers of animals are fed daily for 90 days amounts of the test substance just below the lethal dose. This is to find the maximum level that will not produce an adverse effect and to further observe the nature of the adverse effects at higher levels. Male and female animals are observed for changes in behavior, appearance, weight, blood and urine chemistry, and organ and tissue pathology. The highest level of the test material that produces no adverse effects (called the No Observable Effect Level or NOEL) may then be divided by a safety factor of 100 or more to set the maximum daily intake level (in milligrams per kilogram of body weight) allowed for humans. This safety factor takes into account possible differences in sensitivity between humans and the test animals as well as differences in size. In most cases, additional chronic testing is required.

A chronic toxicity test involves feeding the test material to experimental animals throughout their lifetime (2–3 years for rats) and 2 years or more for dogs. Daily feeding of many animals commonly includes amounts that may be 100 to more than 1000 times the human allowable levels calculated from subacute tests. Animals are killed at intervals and examined physiologically and biochemically in great detail. This kind of testing has been used to reveal tumors and other chronic disorders. Results of such tests provide information on which to base conservative judgments of safety or to prohibit cancer-causing materials in accordance with the Delaney Clause.

Additional tests include pharmacokinetic studies to determine the absorption, distribution among tissues and organs, metabolism, and elimination of test substances. Reproductive tests determine the effects on male or female fertility or on litter size, litter weight, number of surviving young, and possible birth defects (teratogenicity). Adverse effects may be initially absent but show up in subsequent generations. Teratogenic and mutagenic effects may be missed because they may occur at a very low-frequency rate. Newer testing procedures involve testing with microorganisms that grow through many generations rapidly, in vitro mammalian cell cultures, isolated organ culture, mathematical modeling from chemical structure analysis, and computer simulations of biological systems. Animal behavioral testing is aimed at revealing wide-ranging subtle effects such as changes in response to stimuli, learning ability, and dexterity (Fig. 24.2). By no means is there general agreement with respect to test methods or safety criteria.

The use of animals to test the safety of foods, drugs, and other products has been controversial. It is not practical to test products on humans before release, so animals must act as surrogates for humans. New testing methods which are done on cultured cells or bacteria are reducing the numbers of animals required for testing.

Increasingly, genetic engineering or recombinant DNA (rDNA) technology is being explored to produce food ingredients and additives. Enzymes derived from rDNA are currently used in cheese-making and some animal drugs such as recombinant bovine growth hormone (rBGH) are approved and being used. Disease and insect resistant food plants are being developed using rDNA technologies. Like many previous new ingredients used in foods, genetically engineered materials are not without controversy with regard to safety and testing.

FOOD LABELING

One of the main goals of governmental regulation of foods is ensuring that consumers are given complete and useful information about the food products they purchase. This information is important for both economic and health reasons. Truthful and complete information allows value comparisons to be made among competing products. For example, requiring all packages to state the net quantity of contents allows a calcula-

Figure 24.2. Ingredient testing for behavioral effects. *Courtesy of Takeda Chemical Industries, Ltd.*

tion of the price per unit weight. Ingredient labeling also helps people avoid foods to which they may be allergic. For example, some people can have life-threatening responses to some food colors. These colors must be clearly labeled to prevent such problems.

The information on a food package can be divided into three types: the first is the mandatory information required by the Fair Packaging and Labeling Act and the Nutrition Labeling and Education Act, as well as the Food, Drug, and Cosmetic Act and others. The second type of information is optional or voluntary information but which is often regulated if present. Last is information that is provided by the manufacturer to help the consumer use or understand the product. This last type contains such information as instructions for preparation and additional recipes.

Most food packages have at least two panels which are required to carry the mandatory information (Fig 24.3). The principle display panel (PDP) is the panel most likely to be displayed or presented to the consumer and is considered the "front" side. This panel must have the common or usual name of the product and the net quantity of contents. To the right of the PDP is the information panel (IP). Whenever possible, the IP contains most of the required additional information. Some of the information may be elsewhere on the package if the size or shape of the package do not allow it to be on the IP. The remainder of the package can be thought of as the consumer panel (CP) and contains much of the optional information.

The information appearing on a label is as follows:

1. **Food Name.** All foods must be labeled on the PDP with the common or usual name without adjectives that might be confusing. Thus, "whole corn" and not "super whole corn" is required. "Honey," "sliced peaches," and "whole peaches" are required. Note that these are not brand names which are sometimes confused with product names. "Tang" is a brand name for "orange-flavored instant breakfast drink," the latter being the common or usual name.

Figure 24.3. Typical food label showing general information, principle display, and nutrition information panels as well as the mandatory information.

2. **Net Quantity of Contents.** This tells consumers how much food is contained in the package and allows price comparisons to be made. It is required to be on the PDP and should account for only that portion of the food commonly consumed. The units used to indicate net quantity are specified in the regulations. Pounds, ounces, quarts, pints, and fluid ounces are examples.
3. **Ingredients.** A listing of all ingredients in descending order by weight is required. This applies to nearly all products, including those whose ingredients are dictated by a standard of identity. Ingredient listings usually appear on the IP.
4. **Company Name.** The company making a product or the distributor and their address must be given so that consumers have a contact should there be a problem.
5. **Product Dates.** It is common for food products to be dated in some form. "Open date" refers to a date which can be read by consumers. "Code dates" are dates that can only be read by a manufacturer because they are coded. Several types of open dates are used including "pull date" (last date product should remain for sale), "best if used by date" (shelf life for optimal quality), "pack date" (date food was packaged), and "expiration date" (last date on which product should be eaten).
6. **Nutrition Information.** Three types of nutrition-related information are now closely regulated on package labels: "Nutrition Facts," Nutrient Content Claims, and Health Claims. Nearly all food labels require "Nutrition Facts" information to be given (see below).
7. **Other Information.** In addition to the above, several types of voluntary information may appear on a food product label. However, if the information appears, it is often regulated. U.S. Department of Agriculture grades or standards for products such as butter, eggs, orange juice, and meat may be displayed as long as the standards are official and the product meets the standards. Packages often carry trademark or copyright symbols. Religious symbols may be used to indicate that a product meets certain religious qualifications. Nearly all products now contain a universal product code (UPC), which is the bar code that allows for automated scanning of products. Perishable products may be required to contain safe handling or preparation information. In other cases, such information is voluntary. Products containing alcohol, sulfites, yellow number 5, or aspartame carry special warning labels to alert consumers to specific hazards that may occur in selected groups of consumers.

NUTRITION LABELING

It has been known since the early 1900s that humans require certain nutrients to prevent diseases such as scurvy. More recently, however, there has been growing scientific evidence that the kinds and amounts of foods eaten affect the risk of chronic diseases such as heart disease and cancer. Specifically, diets low in fat and high in fiber and fresh fruits and vegetables are believed to reduce these risks. In order for this scientific knowledge to be used, consumers must understand both what type of diet reduces risk and the nutritional content of foods in order to choose healthier diets. Several government agencies with health and nutrition responsibility have published guidelines for improved diets. One such publication is the nutrition pyramid of the U.S. Department of Agriculture (Fig 24.4). This pyramid is a way of depicting an optimum diet based on grains, fruits, and vegetables with smaller amounts of meats and dairy products, and even smaller amounts of high-fat foods.

Consumers must know what the nutritional contents of foods are so they can purchase

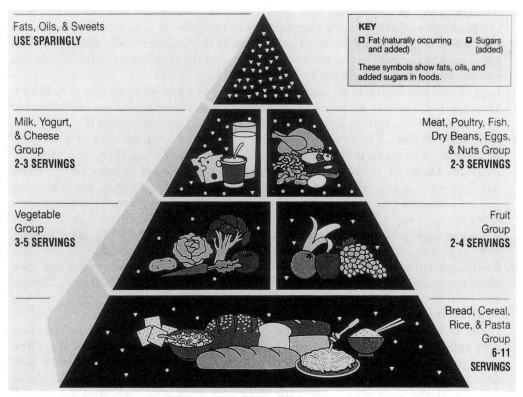

Figure 24.4. USDA Food Guide Pyramid: A Guide to Daily Food Choices. Source: *U.S. Department of Agriculture.* Home and Garden Bulletin No. 252. 1992.

foods which are of better nutritional quality. With this in mind, in 1990 the U.S. Congress passed the Nutrition Labeling and Education Act (NLEA). This law required the FDA to develop requirements for the nutritional labeling of nearly all foods sold in grocery stores in the United States. Foods sold in restaurants and fresh meats and poultry where excluded, but the U.S. Department of Agriculture has since proposed regulations to cover meats. Most of the regulations went into effect in 1994.

As pointed out above, "Nutrition Facts," Health Claims, and Nutrient Content Claims are now closely regulated on nearly all foods in the United States. "Nutrition Facts" labeling is required on nearly all foods, whereas the other two are voluntary. "Nutrition Facts" contains required and voluntary information on dietary components and nutrients (Fig 24.5). If a nutrient content or other type of claim is made about any voluntary component, then nutrition information for that component becomes mandatory. Only those components listed in Fig 24.5 are permitted in the "Nutrition Facts" section. Information on other nutrients can be added in other parts of the label.

All nutrient amounts in the "Nutrition Facts" section are based on a standard serving size which has been determined by the FDA. Most nutrients are labeled based on "percent (%) of daily value." This means the percent of a recommended amount of the nutrient in a total daily food intake of 2000 cal that the standard serving of the product contributes. For example, the total daily value for fat based on 30% of calories coming from fat and a 2000 cal/day food intake is 600 calories (0.30 × 2000 cal = 600 cal). One standard serving of a food containing 13 g of fat would thus represent 117 cal

New heading sig-
nals a new label.

More consistent
serving sizes, in
both household and
metric measures,
replace those that
used to be set by
manufacturers.

Nutrients required
on nutrition panel
are those most im-
portant to the health
of today's consum-
ers, most of whom
need to worry about
getting too much of
certain items (fat,
for example), rather
than too few vita-
mins or minerals, as
in the past.

Conversion guide
helps consumers
learn caloric value
of the energy-pro-
ducing nutrients.

New mandatory
component helps
consumers meet
dietary guidelines
recommending no
more than 30 per-
cent of calories
from fat.

% Daily Value shows
how a food fits into
the overall daily
diet.

Reference values
help consumers
learn good diet
basics. They can be
adjusted, depending
on a person's calo-
rie needs.

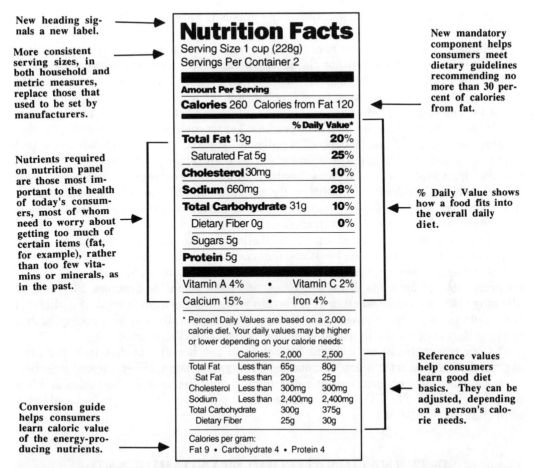

Figure 24.5. The "Nutrition Facts" label for foods. Source: *Focus on Food Labeling, FDA Consumer, 1993.*

from fat or 20% of the daily value of fat [(117 cal/600 cal) × 100 = 20%]. A standard serving of a food containing 140 mg of sodium would be labeled as 6% of the daily value which is 2400 mg for sodium in a 2000 cal/day diet.

Serving size is no longer at the discretion of the product manufacturer. The FDA has set some 139 reference serving sizes based on what is believed to be an amount commonly consumed at a single time. For products where a single unit is more than 50% but less than 200% of the reference, the serving size is one unit (e.g., one can of cola) of the product. Thus, a 12 fl oz serving of soda is one serving because it is less than 200% of the 8 fl oz reference value for soda.

Desirable nutrient intake is now based on "daily value" rather than the U.S. recom-mended daily allowance (U.S. RDA) which was previously required on nutritionally labeled products. A daily value takes into consideration not only the minimum recom-mended amount of required minerals and vitamins but also the recommended amount of macronutrients such as fat, carbohydrate, and protein, as well as cholesterol, sodium, and potassium. Daily values for components which contain calories (e.g., fat) are based on the recommended amount that should be contained in a 2000 cal/day intake. As pointed out above, the recommendation for fat based on 30% of calories per day from fat and a 2000 cal/day diet is 600 cal/day from fat. At 9 cal/g, this equals approximately

67 g of fat per day. Other calorie-containing components are similarly calculated (Table 24.1).

In addition to the "Nutrition Facts" section, regulations now dictate what nutrition-related terms or nutrient content claims can be used on labels and what those terms mean. For example, terms such as "free," "light," "low," and "high" have specific meanings (Table 24.2). In order to be labeled "low fat" a product must have less than 3 g of fat per standard reference serving, "light" means that a product contains one-third fewer calories than the reference product. Terms such as "good source of," "more," "contains or provides," and "modified" are all now defined. Implied claims which suggest that a nutrient or ingredient are important in maintaining a healthy diet and are made with an explicit claim such as "healthy, contains 3 grams of fat" are not allowed unless specifically permitted. Additionally, the term "fresh" can only be used on products that have not been frozen, heat processed, or otherwise preserved. "Fresh frozen" is a "fresh" food that was frozen quickly.

The last area of regulation of nutrition labeling relates to direct and implied health claims for foods. Traditionally, health benefit claims such as "reduces the risk of cancer or heart disease" have not been allowed for foods. It is common to make such claims for drugs but not for foods. However, based on recent scientific findings, FDA is now allowing six health-related claims for certain foods and has indicated that additional health claims may be allowed in the future. The major criterion for making such a claim is that it be based on sound scientific evidence that it is true.

The six currently allowed claims all relate to the benefits of diet in preventing cancer, coronary heart disease, osteoporosis, or hypertension. When viewed together, these health claims suggest eating a diet high in fiber, fruits and vegetables and low in sodium and fat. The FDA has suggested the type of wording that is acceptable for each claim (Table 24.3).

INTERNATIONAL FOOD STANDARDS AND CODEX ALIMENTARIUS

Two changes related to the international trade in food and food products have recently occurred. First, there has been a significant increase in the international trade of both raw agricultural commodities and processed foods over the last decade. Thus, it is not uncommon to find foods such as grapes or tomatoes year-round in parts of the world where they only grow seasonally. Second, there has been a decrease in the traditional economic trade restrictions and tariffs that have been imposed on foods and agricultural products (as well as other goods). This has increased the possibility that countries will

Table 24.1. Calculation of Amounts of Caloric Nutrients in Foods Based on a 2000 cal/day Intake

Nutrient	Calories	Amount (g)
Fat[a]	2,000 cal × .30 = 600 cal	600 cal / 9 cal/g = 67 g
Total carbohydrate	300 g × 4 cal/g = 1200 cal	—
Protein	25 g × 4 cal/g = 100 cal	—

[a]Assumes a maximum fat intake of 30% of calories.

Table 24.2. Some Examples of Nutrient Definitions Allowed on Food Labels

Sugar

Sugar free: less than 0.5 grams (g) per serving

No added sugar, without added sugar, no sugar added:

- No sugars added during processing or packing, including ingredients that contain sugars (for example, fruit juices, applesauce, or dried fruit)
- Processing does not increase the sugar content above the amount naturally present in the ingredients
- The food that it resembles and for which it substitutes normally contains added sugars
- If the food doesn't meet the requirements for a low- or reduced-calorie food, the product bears a statement that the food is not low-calorie or calorie-reduced

Reduced sugar: at least 25% less sugar per serving than reference food

Cholesterol

Cholesterol free: less than 2 milligrams (mg) of cholesterol and 2 g or less of saturated fat per serving

Low cholesterol: 20 mg or less and 2 g or less of saturated fat per serving and, if the serving is 30 g or less or 2 tablespoons or less, per 50 g of the food

Reduced or Less cholesterol; at least 25 percent less and 2 g or less of saturated fat per serving than reference food

Fiber

High fiber: 5 g or more per serving. (Foods making high-fiber claims must meet the definition for low fat, or the level of total fat must appear next to the high-fiber claim.)

Good source of fiber: 2.5 g to 4.9 g per serving

More or Added fiber: at least 2.5 g more per serving than reference food

Calories

Calorie free: Fewer than 5 calories per serving

Low calorie: 40 calories or less per serving and if the serving is 30 g or less or 2 tablespoons or less, per 50 g of the food

Reduced or Fewer calories: at least 25 percent fewer calories per serving than reference food

Fat

Fat free: less than 0.5 g of fat per serving

Saturated fat free: less than 0.5 g per serving and the level of trans fatty acids does not exceed 1 percent of total fat

Low fat: 3 g or less per serving, and if the serving is 30 g or less or 2 tablespoons or less, per 50 g of the food

Low saturated fat: 1 g or less per serving and not more than 15 percent of calories from saturated fatty acids

Reduced or Less fat: at least 25 percent less per serving than reference food

Reduced or Less saturated fat: at least 25 percent less per serving than reference food

Sodium

Sodium free: less than 5 mg per serving

Low sodium: 140 mg or less per serving and, if the serving is 30 g or less or 2 tablespoons or less, per 50 g of the food

Very low sodium: 35 mg or less per serving and, if the serving is 30 g or less or 2 tablespoons or less, per 50 g of the food

Reduced or Less sodium: at least 25 percent less per serving than reference food

SOURCE: FDA Consumer (1993).

Table 24.3. Examples of Model Health Claims

Food and Disease	Model Health Claim
Fiber grains, fruits, vegetables & cancer	Low-fat diets rich in fiber-containing grain products, fruits, and vegetables may reduce the risk of some types of cancer, a disease associated with many factors
Sodium & hypertension	Diets low in sodium may reduce the risk of high blood pressure, a disease associated with many factors
Fruits, vegetables, grains & coronary heart disease	Diets low in saturated fat and cholesterol and rich in fruits, vegetables, and grain products that contain some types of dietary fiber may reduce the risk of heart disease, a disease associated with many factors
Dietary fat & cancer	Development of cancer depends on many factors. A diet low in total fat may reduce the risk of some cancers
Calcium & osteoporosis	Regular exercise and a healthy diet with enough calcium helps teens and young adult white and Asian women maintain good bone health and may reduce their high risk of osteoporosis later in life
Dietary saturated fat & cholesterol & risk of coronary heart disease	While many factors affect heart disease, diets low in saturated fat and cholesterol may reduce the risk of this disease
Fruits & vegetables & cancer	Low fat diets rich in fruits and vegetables (foods that are low in fat and contain dietary fiber, vitamin A, or vitamin C) may reduce the risk of some types of cancer, a disease associated with many factors. Broccoli is high in vitamins A and C, and is a good source of dietary fiber

SOURCE: Nutrition Reviews *51*(3), 90–93 (1993).

impose "nontariff" trade barriers. Nontariff barriers are actions which restrict trading because products may not meet one country's safety, quality, or product standards. For example, a country may restrict the importation of rice because it does not meet certain grain size or shape standards. For these reasons, the setting of international standards has become a high priority in world trade.

In matters of international scope two important agencies are the World Health Organization of the United Nations (WHO) and the Food and Agriculture Organization of the United Nations (FAO). These are organized to increase and improve food resources, nutrition, and health throughout the world.

The need for coordination in setting standards has long been recognized, and in 1962 an international body operating under the auspices of the United Nations through FAO/WHO was established and designated as the Codex Alimentarius Commission. The object of this commission has been to develop international and regional food standards and publish them in a Codex Alimentarius (which is latin for "code concerned with nourishment"). The Codex Alimentarius Commission was established to also develop agreements on international standards and safety practices for foods and agricultural products. The standards are designed to protect consumers from health hazards and economic fraud, promote international trade, and ensure fair business practices in foods. The codex sets minimum quality, safety and hygienic standards that countries voluntarily adhere to in importing and exporting food products. The commission has 27 committees chaired by different countries which, in addition to drafting standards on specific foods, also draft standards on such topics as permitted and prohibited additives, food hygiene, pesticide residues, analytical methods and sampling, labeling, and general principles. There are 17 commodity committees, each dealing with a separate food group. Besides safety and wholesomeness aspects, Codex Standards are aimed at overcoming misuse of alleged food standards set up by some countries in the past, more to protect their products from the competition of imports than to protect the health of their citizens.

By 1990, Codex Alimentarius Commission membership had increased to include 137 countries. Several U.S. food standards of identity are being revised to bring them into agreement with proposals of the Codex Alimentarius and, thus, support the movement to international standards. The continuing work of the commission, highly valuable to developing nations with as yet few food standards of their own, as well as to other countries whose standards may need updating, is another step toward protecting consumer health and facilitating world trade.

There are other important food or ingredient standard setting groups. The Joint FAO/WHO Expert Committee on Food Additives (JECFA) sets standards for purity of food additives. These standards do not have the force of law but are used by many countries in developing their own standards. The European Commission sets standards for foods and food additives for members of the European Economic Community.

Standards for the identification and purity of food additives and chemicals are also developed in the United States by the Food Chemicals Codex Committee. This committee operates as part of the National Academy of Sciences Food and Nutrition Board. Many of the standards set by this committee are used by several governmental agencies around the world in regulating food additives. The standards also become important in commerce. For example, a company that wants to use a particular antioxidant such as BHT in a food product will often specify that it be "FCC grade."

It can be expected that world trade in foods and food products will continue to increase. Trading agreements such as the North American Free Trade Agreement (NAFTA) among Canada, the United States, and Mexico, and the trade agreements among European countries ensure this. This will necessitate the need for more and better standards as well as cooperation between countries and governments.

References

Anon. 1981. Food Chemical Codex. 3rd ed. National Academy Press, Washington, DC.

Anon. 1993. Mandatory nutrition labeling–FDA's final rule. Nutr. Rev. *51*(4), 101–105.

Anon. 1993. The FDA's final regulations on health claims for foods. 1993. Nutr. Rev. *51*(3), 90–93.

Burns, J. 1983 Overview of safety regulations governing food, drugs and cosmetics in the United States. *In* Safety Evaluation and Regulation of Chemicals. F. Homburger. (Editor). Karger, Basel pp. 9–15.

FDA Consumer. 1993. Focus on Food Labeling. May, 1993. Food and Drug Administration, Rockville, MD.

Foulke, J.E. 1992. Food labeling. FDA Consumer Jan.–Feb. 26(1): 9–13.

Hodgson, E. and Levi, P.E. 1987. A Textbook of Modern Toxicology. Elsevier, New York.

Kimball, E.F. 1982. Codex alimentarius food standards and their relevance to U.S. standards. Food Technol. *36*(6), 93–95.

McKenna and Cuneo, Technology Services Group, Inc. 1993. Pesticide Regulation Handbook. McGraw-Hill, New York.

Middlekauff, R.D. and Shubik, P. 1989. International Food Regulation Handbook: Policy, Science, Law. Marcel Dekker, New York.

Morrison, R.M. 1983. Codex alimentarius commission. Natl. Food Rev. *21,* 14–16.

Pimentel, D. and Lehman, H. 1993. The Pesticide Question: Environment, Economics, and Ethics. Chapman & Hall, New York.

Segal, M. 1993. What's in a food? FDA Consumer March–April, *27*(2) 14–18.

Smith, B. 1985. Food standards and controls: Necessity for export. Food Technol. *39*(7), 72–75.

Tardiff, R.G. and Rodricks, J.V. (Editors). 1987. Toxic Substances and Human Risk: Principles of Data Interpretation. Plenum Press, New York.

BNA International. 1991. World Food Regulation Review. BNA International, London.

25

Hunger, Technology, and World Food Needs

BACKGROUND

Nature, history, and economics have been generous to a portion of the world's population by providing food security and freeing societies from the burden of the constant search for food. A combination of productive land, favorable climate, natural resources, stable politics, and the application of science and technology under circumstances of advantageous marketing opportunity have produced a seemingly boundless and abundant food supply at a remarkably low cost. In the United States, approximately 1.5% of the population works in farming yet produces sufficient food to feed more than 270 million people all at an average cost of less than 11.5% of disposable income.

Unfortunately, this is not the situation in most of the world. The contrasts between the richest and poorest countries are strikingly cruel and almost beyond the comprehension of societies with such wealth. Tables 25.1, 25.2, and 25.3 compare indicators of population growth, general health, nutrition, and welfare for a few selected regions and countries. A look at the Gross National Product (GNP, a measure of the overall wealth of a country) illustrates the disparity. The per capita GNP for sub-Saharan Africa in 1991 was $350, whereas the average GNP for the industrialized nations was $21,530. The difference is even greater when individual countries are compared. Mozambique had a per capita GNP of $80, whereas Switzerland's was $33,610. This difference in wealth is reflected in measures of food intake and health such as the caloric intake as a percent of the minimum requirement for health. On the average, people in sub-Saharan Africa consumed 93% of necessary calories, whereas consumption in the industrialized countries was 113%. The U.S. food system supplies 138% of the minimum caloric intake. These differences in nutritional intake translate into differences in health, particularly among children. In sub-Saharan Africa, 16% of children are born with low-birth-weight versus 6% in the industrialized world. The infant mortality rate in sub-Saharan Africa is 102 per 1000 births, whereas in the industrialized world it averages 12. Mali in Africa has the world's highest infant mortality rate of 159, whereas Japan has the lowest at 5. Without question, there is a strong link between health and the ability to provide sufficient food.

Table 25.1. Indicators of Population Growth for Selected Countries and Regions

	Population (Millions)		Population Annual Growth Rate 1980–1991	Fertility (Children per Woman) 1991	Life Expectancy 1990–1995	Infant Mortality per 1000 Live Births 1990–1995	Under 5 Mortality per 1000 Live Births		Maternal Mortality per 100,000 Live Births 1980–1990
	1992	2000					1960	1991	
Africa (sub-Saharan)	561	713	3.0	6.5	51	102	262	180	600
Cameroon	12.2	15.3	2.9	5.8	56	63	270	126	430
Ethiopia	53.0	67.2	2.6	7.0	47	122	294	212	—
Nigeria	115.7	147.7	3.3	6.6	53	96	212	188	800
Uganda	18.7	23.4	2.9	7.3	42	104	223	190	300
South Asia	1183	1392	2.2	4.3	59	97	238	131	490
India	879.5	1018.7	2.1	4.0	60	88	236	126	460
East Asia & the Pacific	1725	1930	1.7	2.5	69	35	198	42	160
China	1180.0	1309.7	1.5	2.3	71	27	205	27	95
Korea, S.	44.2	46.9	1.2	1.7	71	21	126	10	26
Laos	4.5	5.6	2.8	6.7	52	97	233	148	—
Latin America & the Caribbean	458	520	2.1	3.2	68	47	161	57	180
Brazil	154.1	172.8	2.0	2.9	66	57	179	67	200
Cuba	10.8	11.5	0.9	1.9	76	14	91	14	39
Haiti	6.8	8.0	1.9	4.9	57	86	270	137	340
Mexico	88.2	102.6	2.3	3.3	70	35	138	37	110
Middle East & N. Africa	317	384	3.0	4.9	65	55	242	83	181
Egypt	54.8	64.8	2.5	4.2	62	57	260	85	320
Saudi Arabia	15.9	20.7	4.5	6.5	69	31	292	43	90
Yemen	12.5	16.4	—	7.3	53	106	378	182	—
Industrial countries	1203	1290	0.6	1.9	75	12	45	17	15
Australia	17.6	19.6	1.5	1.9	77	7	24	10	3
Canada	27.4	30.4	1.1	1.8	77	7	33	9	5
Germany	80.3	82.6	0.2	1.5	76	7	40	9	5
Sweden	8.7	9.0	0.3	2.0	78	6	20	5	5
United States	255.2	275.3	0.9	2.0	76	8	30	11	8
World	5479	6220	1.8	3.4	65	62	178	82	266

SOURCE: Bread for the World Institute on Hunger & Development (1994).

Table 25.2. Indicators of General Health for Selected Countries and Regions

	Mean Years of Schooling (1991)			% Infants with Low Birth Weight 1990	Food Production per Capita (1979–81=100) 1988–1990	Daily Calorie Supply (as % of Requirements) 1988–1990
	Total	Male	Female			
Africa (sub-Saharan)	1.6	2.2	1.0	16	95	93
Cameroon	1.6	2.5	0.8	13	79	95
Ethiopia	1.1	1.5	0.7	16	73	73
Nigeria	1.2	1.8	0.5	16	124	93
Uganda	1.1	1.6	0.6	—	97	93
South Asia	2.3	3.4	1.2	34	113	99
India	2.4	3.5	1.2	33	116	101
East Asia & the Pacific	4.8	5.9	3.7	11	127	112
China	4.8	6.0	3.6	9	138	112
Korea, S.	8.8	11.0	6.7	9	100	120
Laos	2.9	3.6	2.1	18	110	111
Latin America & the Caribbean	5.2	5.3	5.1	11	105	114
Brazil	3.9	4.0	3.8	11	115	114
Cuba	7.6	7.5	7.7	8	96	135
Haiti	1.7	2.0	1.3	15	83	89
Mexico	4.7	4.8	4.6	12	97	131
Middle East & N. Africa	3.3	4.4	2.2	10	110	126
Egypt	2.8	3.9	1.6	10	106	132
Saudi Arabia	3.7	5.9	1.5	7	223	121
Yemen	0.8	1.3	0.2	19	79	—
Industrial countries	10.0	10.4	9.6	6	99	133
Australia	11.5	11.6	11.4	6	92	124
Canada	12.1	12.3	11.9	6	117	122
Germany	11.1	11.7	10.6	—	114	130
Sweden	11.1	11.1	11.1	5	96	111
United States	12.3	12.2	12.4	7	94	138

SOURCE: Bread for the World Institute on Hunger & Development (1994).

But even within wealthy societies there are pockets of poverty and undernourishment. In this chapter we will explore hunger, food security, and the role that food science and technology plays in improving the availability of food for the world's population.

Like other areas related to foods, it is important to understand the specialized terms used to describe the problem. Table 25.4 is a glossary of several of these terms. Whenever statistics are presented related to food shortages it is necessary to know exactly what type of population or country is being discussed.

The magnitude and seriousness of the world's food needs is staggering. Although experts do not agree on exactly how many people are malnourished or have insufficient food and inadequate diets, there is agreement that the problem is massive. Acute episodes of famine have appeared almost daily in the world's press. These famines and

Table 25.3. Indicators of Economic Status for Selected Countries and Regions

	Per capita GNP 1991	Annual per Capita GNP Growth (%) 1965–1991	Food as % of Exports 1991	Food as % of Imports 1991	Per Capita Energy Consumption (kg of Oil Equiv.) 1991	Food as % of Household Consumption 1980–1985
Africa (sub-Saharan)	350	0.2	—	—	135	—
Cameroon	850	2.8	18.7	19.5	147	24
Ethiopia	120	-0.3	8.3	36.0	20	49
Nigeria	340	0.3	1.3	6.9	154	48
Uganda	170	—	12.4	3.2	25	—
South Asia	320	2.0	—	—	289	—
India	330	2.0	6.6	2.8	337	52
East Asia & the Pacific	715	5.2	—	—	571	—
China	370	5.9	—	—	602	61[b,c]
Korea, S.	6,330	7.1	0.8	4.3	102	35
Laos	220	0.6	18.4	6.3	42	—
Latin America & the Caribbean	2,410	1.6	—	—	1,051	—
Brazil	2,940	3.1	11.1	9.7	908	35
Cuba	—[a]	—	57.1	8.8	—	—
Haiti	370	0.1	8.8	40.6	49	—
Mexico	3,030	2.7	7.0	9.1	1,383	35[b,c]
Middle East & N. Africa	1,910	1.1	—	—	—	—
Egypt	610	4.0	9.7	20.1	594	49
Saudi Arabia	7,820	2.2	0.9	11.5	4,866	—
Yemen	540	—	2.6	29.7	96	—
Industrial countries	21,530	2.4	—	—	5,106	—
Australia	17,050	1.8	15.4	2.7	5,211	13
Canada	20,440	2.9	6.1	4.6	9,390	11
Germany	23,650	2.7	3.9	7.4	3,463	12
Sweden	25,110	1.9	1.5	4.3	5,901	13
United States	22,240	1.9	6.8	3.2	7,681	10

SOURCE: Bread for the World Institute on Hunger & Development (1994).
[a]GNP per capita estimated in $500–1499 range.
[b]Years outside range specified.
[c]Includes beverages and tobacco.

Table 25.4. Glossary of Terms Related to World Hunger

Absolute poverty—The income level below which a minimally nutritionally adequate diet plus essential non-food requirements are not affordable.

Anemia—A condition in which the hemoglobin concentration (the number of red blood cells) is lower than normal as a result of a deficiency of one or more essential nutrients, such as iron, or due to disease.

Daily calorie requirement—The average number of calories needed to sustain normal levels of activity and health, taking into account age, sex, body weight, and climate.

Famine—A situation of extreme scarcity of food, potentially leading to widespread starvation.

Food security—Assured access for every person, primarily by production or purchase, to enough nutritious food to sustain productive human life.

Foreign exchange—Currency acceptable for use in international trade, such as U.S. dollars. The value of one currency in terms of another is the **exchange rate.**

Gross domestic product (GDP)—The value of all goods and services produced within a nation during a specified period, usually a year.

Gross national product (GNP)—The value of all goods and services produced by a country's citizens, wherever they are located.

Hunger—A condition in which people lack the basic food intake to provide them with the energy and nutrients for fully productive, active, and healthy lives.

Infant mortality rate (IMR)—The annual number of deaths of infants under one year of age per one thousand live births.

Inflation—An increase in overall prices, which leads to a decrease in purchasing power.

International Monetary Fund (IMF)—An intergovernmental agency which makes loans to countries that have foreign exchange and monetary problems. These loans are conditioned upon the willingness of the borrowing country to adopt economic policies designed by IMF.

Low birthweight infants—Babies born weighing 2,500 grams (5 pounds, 8 ounces) or less who are especially vulnerable to illness and death during the first month of life.

Malnutrition—Failure to achieve nutrient requirements, which can impair physical and/or mental health. Malnutrition may result from consuming too little food or a shortage or imbalance of key nutrients, e.g. refined sugar and fat.

Poverty Line—An official measure of poverty defined by national governments. In the United States, for example, it is based on ability to afford USDA's Thrifty Food Plan.

Recession—A period in which a country's GDP declines in two or more consecutive three-month periods.

Structural adjustment—Economic policy changes, often imposed upon an indebted country by its lenders as a condition for future loans, intended to stimulate economic growth. These generally involve reducing the role of government in the economy and increasing exports.

Stunting—Failure to grow to normal height caused by chronic undernutrition during the formative years of childhood.

Sustainability—Society's capacity to shape its economic and social systems so as to maintain both natural resources and human life.

World Bank—An intergovernmental agency which makes long-term loans to the governments of developing nations.

Under five mortality rate—The annual number of deaths of children under five years of age per one thousand live births.

Underemployment—A situation where a large portion of the population is not fully employed year round.

Undernutrition—A form of mild, chronic, or acute malnutrition which is characterized by inadequate intake of food energy (measured by calories), usually due to eating too little. Stunting, wasting, and being underweight are common forms of undernutrition.

Underweight—A condition to which a person is seriously below normal weight for her/his age. The term can apply to any age group, but is most often used as a measurement of undernutrition in children under five years of age.

Vulnerability to hunger—Individuals, households, communities, or nations who have enough to eat most of the time, but whose poverty makes them especially susceptible to changes in the economy, climate, or political conditions.

Wasting—A condition in which a person is seriously below the normal weight for her/his height due to acute undernutrition.

SOURCE: Bread for the World Institute on Hunger & Development (1994).

mass starvation are more often a result of politics than the inability to produce food. War- and political-related famines are only the most visible part of the world food problem. Warring factions have used food as a political weapon. However, when viewed against the total size of food shortages around the world, these famines are a small fraction of the problem. The World Bank has estimated that more than 1 billion people are too poor to afford the basic needs of food, water, and shelter. The World Hunger Education Service estimates that approximately 10% of the earth's population (about one-half of a billion people) in 1990 lacked enough to eat. The World Health Organization (WHO) estimates that as many as 40,000 people, mostly children, die as a result of common malnutrition and diet-related diseases *each day*. The Food and Agriculture Organization (FAO) estimates that 400 million people suffer from chronic malnutrition and that another 350 million cannot afford the bare minimum diet required for health. They suggest that as many as one in eight people are thus affected. Never in history have so many lacked proper nourishment.

Whether the problem is improving or growing worse depends on the way one views the changes. The number of people without adequate diet is growing larger in absolute terms but as a percentage of the world's population it is decreasing. The situation becomes more disturbing when one looks at the populations which are most vulnerable to malnutrition, namely, women and children under 5 years. The United Nations Children's Fund (UNICEF) estimates that

• As many as 40% of all children under 5 years of age have stunted growth due to protein-energy malnutrition.
• 40 million of these children are wasting away due to malnutrition.
• approximately 20% (17 million per year) of children born in developing countries are underweight at birth, mostly due to poor maternal nutrition.
• half of all women of childbearing age in developing countries have nutritional anemia.

Meanwhile, the gap between rich and poor nations has become greater and projections for the future, if corrective actions are not taken on an international level, are less than bright.

NATURE OF NUTRITIONAL PROBLEMS

Many diseases have a nutritional component and the lack of an adequate diet directly causes disease or contributes to an individual's susceptibility to disease. Nutrition-related health problems can be divided into several categories.

Protein-calorie malnutrition in the young. Adequate nutrition is especially important in the first months and years of life. Mothers who do not receive sufficient protein or calories during pregnancy or during breast-feeding have children with increased susceptibility to disease. If protein-calorie malnutrition continues throughout the first years of life, permanent physical and mental impairment of development occurs.

Chronic energy deficiency. Individuals who are chronically deprived of sufficient caloric intake must reduce their activities. This leads to reduced economic activity and family support. This in turn leads to less economic output which further compounds the problem.

Iron deficiency. Iron deficiency is among the most widespread of the trace nutrient deficiencies. This results in both social and medical problems and less productivity.

Iodine deficiency diseases. This deficiency causes goiterism or thyroid enlargement and is still widespread. Several other related diseases are also related to iodine insufficiency. This deficiency has been all but eliminated in the developed world by the addition of iodine to table salt.

Vitamin deficiencies. The major vitamin deficiency is for vitamin A. In parts of the world this single deficiency has led to widespread blindness and related diseases. Other severe vitamin and mineral deficiencies occur in isolated populations and areas.

In many cases these deficiencies do not directly cause disease but may lower a population's abilities to contribute to a productive society. Put simply, malnourished are not able to work toward the common good. Such deficiencies also lead to increased susceptibility to other diseases.

SOME DIMENSIONS OF THE PROBLEM

Famine

Destruction of a region's or country's food supply due to natural or man-made disaster often results in outright famine. This dramatic loss of food can result in loss of hundreds of thousands of lives in a short period. Such disasters are seen around the world and often result in mobilization of massive relief efforts by individual countries as well as international organizations (Fig 25.1). At least in theory, such famines could be

Figure 25.1. Food being unloaded from an airplane at a famine relief center in west Africa. Source: 1993 *Food Aid Review, World Food Program, United Nations,* New York.

completely eliminated if relief measures were always taken quickly. Preventing such famines is not a matter of finding enough food, the amount of food consumed in famines on a worldwide basis is extremely small, but rather getting the resources mobilized in time and overcoming political and logistical problems become the major obstacles.

Short-Term Food Shortages

Temporary shortages in food can result from both shortfalls in production as well as marketplace forces. Weather is often a major factor. Shortages in rainfall, for example, affect crop yields. Price increases usually follow and availability is decreased. The poorer a country is, the more disruptive is a temporary short fall. Economic changes in prices and world markets can also cause temporary shortfalls. The more dependent a country is on food imports, the more vulnerable it is to world price fluctuations.

Long-Term Malnutrition

Chronic food shortages that plague some regions of the world is a more difficult problem than are short-term shortages. These chronic shortages are a result of a combination of political, geographic, economic, and population factors. It is this condition for which the problems are greatest and the solutions most difficult.

Existing Food Supplies

The world's existing food supply presently is adequate for the world's total population. The amount of food required to feed the world's hungry is a relatively small portion of the total food produced. It has been estimated that in 1987–1988 it would have required only 2–3% of the world's grain production to eliminate the hunger of 750 million underfed people. The world food problem in the short term is largely one of unequal distribution, not inability to produce sufficient food. This may not be the case in the longer term.

Population Pressures

In 1987 the world's population was 5 billion. By the year 2000 another billion people will have been added. The annual population growth *rate* has been declining for a couple of decades and is now at 1.6%. However, it is estimated that the world's population will not level off until there are 10–14 billion people. The question is, "Will food production be able to keep pace?" A combination of three factors will be needed for food production to keep pace with the increased population: increased commitment of land to agriculture; increased yields; and increased efficiency of utilization of current and new food resources. In many regions of the world, however, the most agriculturally productive land is already under cultivation; the remaining land is of poorer productivity, would require great expenditures of capital to improve it, and its use would require population dislocations with accompanying political and social problems. The "green revolution" of the 1960s which brought high-yielding varieties along with improved farming technologies still may contribute to increased production but certainly to less of a degree than in the past. On the other hand, biotechnology holds particular promise for increased yields and disease resistance; many marine food resources and the use of aquaculture have not yet been fully exploited; and increasing the nutrient content of conventional foods and development of palatable forms of new foods are achievable.

Population–Land Ratios

Although the world's population is increasing, the amount of agriculturally productive land remains relatively constant, and bringing substantial new land into food production may not be feasible in much of the world. This, of course, means that the arable land per person will decrease proportionately. An interesting comparison can be made by considering current and projected U.S. and world populations in terms of this food-bearing land. The present U.S. population of about 270 million divided into this country's arable land gives a figure of approximately 0.6 hectare (1.5 acres) per person. The present world population divided into current world arable land gives 0.3 hectare per person. With land remaining constant and U.S. and world populations expected to grow, there will be approximately 0.5 arable hectare per person in the United States and only slightly over 0.2 hectare per person for the world a generation from now. Since relatively little additional land will be made arable during this period, the land must be made more productive if food requirements are to be met.

Crop Yields

Based on the best current estimates, it has been calculated by the FAO that world agricultural productivity must more than double during the next 40 years to prevent mass famine in distressed areas. With regard to important crop yields, there is cause for guarded optimism if advances in several countries over the past two decades can be sustained and extended to yet other regions. Thus, in the early 1960s, wheat yields per hectare in Germany and France ran some five to seven times what they were in Tunisia and Brazil. Rice yields per hectare in Japan and the United States averaged about three times greater than those in India, Pakistan, and the Philippines. Yields of corn per hectare in the United States were more than three times those of Portugal or Brazil and five times those of Mexico and the Philippines. Equally discouraging were comparisons for other important food crops. Since about 1968, however, with the introduction of high-yielding seed, the situation has improved appreciably. Wheat yields per hectare have increased threefold in India, Bangladesh, and Mexico, and rice yields have doubled in Indonesia, the Philippines, and countries of South America. But some of these increases have been made from extremely low bases, and coupled with increased populations, have thus far contributed only slightly to the overall food picture. Further, yield gains in Africa have been far less encouraging. During this same period, yields have continued to increase in the United States and other developed countries.

It is clear that major increases in crop yields are directly related to the level of economic development and the possession of capital with which to purchase equipment, fertilizer, pesticides, and other technological devices of modern food production. Once again, the poverty cycle appears and no one believes it can be broken in many of the less developed regions without major outside help. Use of biotechnology to further improve crop yields holds promise for the future.

Yields from Livestock

Animal products represent about 7% of worldwide food production, with the remainder coming directly from plants. The quantity and quality of crop yields influence yields of food that can be derived from livestock. Crop yields and animal feeding practices are not the only determinants of animal productivity but are major factors. Inadequate feeding of animals not only reduces weight gains of animals and their further production of meat, milk, or eggs but also lowers resistance to disease. The efficiency of conversion of feedstuffs into flesh, milk, and eggs is a biological process and largely genetically determined. In developed countries, food-producing animal and poultry strains are the products of decades of breeding. This is not the case in many of the less developed countries. Striking differences between livestock productivity in various countries exist for all animal products. In the United States, for example, the average annual yield of milk per milking cow in 1992 was 7000 kg which given the large number of cows equaled 270 kg of milk per person. In Pakistan, milk yield is only about 450 kg and in parts of India a mere 250 kg. In the case of eggs, U.S. production level exceeds 240 eggs per hen per year, but in some of the less developed countries, it is believed to be no more than about 50. The contribution of animal products to the food supply of a country depends not only on the total number of livestock but also on the average productivity of various species.

Nutrient Content of Foods and Development of New Foods

One way to improve nutritional status without increasing agricultural output would be to use available nutrients more efficiently by having them consumed directly by humans rather than first by animals. For example, 1 bushel (20 kg) of grain consumed as such may supply a human with energy and protein adequate for approximately 23 days of sustenance. The same bushel of grain fed to cows and converted to milk would provide a human with energy for only about 5 days and protein for about 12 days. Similarly, the grain converted to pork or eggs would provide energy and protein in amounts far less than the grain consumed as such. This is because the animal uses a portion of the nutrients in the grain to maintain its own body needs before it returns the excess in the form of useful food. It must be remembered, however, that animals are able to convert plant materials that cannot serve as food for humans into useful foods. Cows, for example, convert cellulose, which is not digestible by humans into milk.

Another way to improve nutritional status is to raise the nutrient content of foods as consumed. Increasing the protein content or protein quality of a cereal grain will improve nutritional status. Development of new or altered foods which have an improved nutritional balance is another way of improving nutritional status. This may be accomplished by developing new foods which are made from a combination of ingredients which when consumed together have improved nutrient adequacy. Existing foods can also be fortified to provide missing essential nutrients. For example, fish "flours" can be made from fish of species not commonly eaten as a source of protein. Incaparina, a cereal formulation containing about 28% protein, is prepared from a mixture of maize, sorghum, and cottonseed flour. Miltone was developed from ingredients—peanut protein, hydrolyzed starch syrup, and cow or buffalo milk—that are readily available in India. Other examples include the development of breadlike products made from mixed grains which have an improved amino acid balance.

However, when introducing new or altered foods, it must be remembered that food habits and preferences are deeply rooted. They are passed from generation to generation and often are difficult to change, even in the face of grave food shortages. There have been instances where unfamiliar foods have been rejected by malnourished peoples.

Exports of wheat, for example, have not always been accepted by rice-eating populations. From a practical standpoint, improvement of regional food yields from plants and animals traditionally accepted appears to offer the most promising results within a reasonable time.

Fertilizer

Unquestionably, fertilizers improve crop yield. Unfortunately, fertilizer, like farm machinery and pesticides, is a worldwide commodity, and often because of shipping costs it is more expensive in less developed areas than in the country of manufacture. A further frustration is that in some parts of the world, indigenous cultivars of crops, which have not been improved by intensive breeding, are less responsive to fertilizer than they are in technologically developed areas. This is generally true at any level of fertilizer application. Thus, subsistence farmers, who can least afford it, must pay more for fertilizers yet receive fewer benefits. Improved responsiveness to fertilizer requires the introduction of improved plant cultivars and often changes in cultivation practices. But this too requires considerable reeducation and time. About 30 years ago all of Latin America used little more fertilizer than Holland, and all of Africa employed about the same quantity as Italy. Over the past two decades many developing countries have substantially increased their fertilizer usage, India by a factor of about 20-fold. But increases in the cost of energy, particularly oil, have increased the costs of producing and shipping fertilizer. There also is a growing world demand for fertilizer. These factors are making it much more difficult for developing countries to overcome food deficits.

Pesticides

As with fertilizer, pesticides increase yields but are costly. In most parts of the developing world, pesticides must be imported. Imported pesticides generally have not been developed for the specific pests and climatic conditions of unfamiliar regions. This has been especially true of tropical regions. Further, the effective application of pesticides often requires high-pressure spraying equipment not generally found in the less developed areas.

Farm Machinery

Cost also is a major deterrent to the increased use of farm machinery in many countries. Moreover, the suitability of such machinery is presently limited by the characteristically small size of farms in most less developed areas. Such "farms" in many instances may constitute a very few hectares or even less. Appropriate machines and tractors have been developed for these areas, but their use cannot be as efficient as that of machines on larger size farms. It also must not be overlooked that in addition to the costs of fuel, spare parts and machine maintenance can become limiting factors in a remote region, as has happened many times in the past.

Transportation

The lack of adequate transportation to move both agricultural inputs and finished products can hamper efforts to improve the food situation. Thus, 1000-ton quantities of imported fertilizer cannot be efficiently moved inland on primitive roads, nor are such roads suitable for moving produce to highly populated areas of consumption. The importance of adequate transportation has not always been appreciated in the past.

Storage and Packaging

In some parts of the world it has been estimated that food losses resulting from inadequate storage and packaging are as much as 50% of the food produced. In many regions even crude silos are lacking and grains are heaped uncovered in the open prior to distribution. Burlap bags and earthen jars stored in sheds offer only a little more protection against insects, rodents, birds, and other animals. One direct way of increasing food availability is to prevent these losses. This requires capital to construct modern bulk storage facilities for agricultural products. Although some governments are funding construction of bulk-storage facilities in rural areas, the need remains great.

Packaging, whether in bulk systems or consumer-sized packages, is an important component of the food supply. Packaging allows for the efficient transportation, handling, and preservation of many foods. The losses suffered by the transport of fresh fruits and vegetables are much greater if simple corrugated cartons are not available. The same is true of other foods such as eggs and fish, each of which require some form of protection. In addition, the combination of packaging and processing allows food consumption to be spread out over a longer period of time. Orange juice consumption in the United States in 1948 was approximately two glasses per person per year. The invention of the paperboard "can" and frozen concentrates made orange juice available year-round at an affordable price. Now the per capita consumption is one glass per person every other day. The paper-based aseptic carton used for milk (Fig 25.2) has

Figure 25.2. Paper/foil/plastic laminants used to aseptically packaging milk and juices at a lower price than metal cans or glass bottles have had a large impact on the availability of such products in the third world.

had a major impact on milk consumption worldwide and has benefited millions of children in the developing world by making milk available in places which lack refrigeration for storage of conventional milk.

Water

The world food problem is aggravated by an equally complex world water problem. Apart from the water requirements of crops, world health depends on potable water. It has been estimated that more than a billion people in developing countries cannot conveniently and safely drink, cook, or wash with water. Countless millions carry water from distant springs or take it from lakes or rivers at great risk. No program to improve the food production capability of a region can be meaningful without attention to the water supply. Simple irrigation systems can substantially improve productivity. However, irrigation with polluted waters can result in the transfer of pathogenic bacteria or toxic chemicals to field workers and food.

Erosion and Pollution

Erosion and pollution problems are at least as severe in the developing countries as they are in the industrialized world. Many third world regions have undergone environmental degradation including deforestation, loss of topsoil, increased soil salinity, waterlogging, and pollution of rivers and lakes from raw sewage and industrial wastes. Many of these destructive influences are far beyond the economic means of poor countries to control.

Sustainable Agriculture

In recent years, concern for the environmental effects of agriculture has increased. The green revolution required considerable inputs of energy and chemicals. Some are now questioning if these inputs can be continued indefinitely. The focus of some research is shifting to agricultural methods which do not appear to have limitations. Such systems are being termed "sustainable agriculture," which can be defined as an integrated system of plant and animal production practices that satisfy human needs while enhancing natural resources and making efficient use of nonrenewable resources. Advocates of this approach suggest that farming practices which produce less impact are desirable for developing as well as developed parts of the world. Using sustainable practices may result in lower production costs, less water contamination, and less erosion. Over the long term this may be of benefit to agriculture in developing countries because they will have to spend less on food production, and land is likely to remain productive for a longer time.

APPROACHES TO COMBAT WORLD HUNGER

From the above it is abundantly clear that the world food situation is enormously complex and that no simple answers exist. There have been considerable differences of opinion among experts with regard to the most productive approaches to the world food problem, and what priorities should be established in specific areas. Unfortunately, failure like success feeds on itself and tends to become more extreme. The causes and factors contributing to starvation generally are interrelated so as to form a descending cycle in which poverty perpetuates illiteracy and poor health, which leads to low productivity, which in turn supports only more poverty.

Internationally recognized authorities have suggested that a combination of several approaches will be required to combat world hunger. These measures can be divided into categories:

- Technological measures in food and agriculture. As pointed out above, improving the world hunger situation will require improvements in efficiency of production and utilization of agricultural products. Systems which bring appropriate technologies to farmers are necessary. Improvements in food processing, storage, and preservation are likewise required. Food scientists can play a key role in developing foods from new sources and more efficiently utilizing current resources.
- Social reforms. Several changes in public policy and increased social equity will be required in many parts of the world. Chief among the reforms is access to land. In some parts of the world as few as 2% of the population control 80% of the land. This results in an underclass of laborers who do not have sufficient income for food. Incentives to produce more food are needed. Increased economic activity and income would provide more people with the money to purchase food.
- Population control. Without some form of reduction in the population growth rate in many parts of the world, social and technical improvements will be doomed to failure. Many feel the key to population control will be improved educational opportunities, particularly for women.
- Improved public health measures. Relatively low-cost but effective improvements in health care are essential. Immunization and the use of simple rehydrating solutions for treating infant diarrhea, and improved maternal nutrition before birth and during breast-feeding would greatly improve the current situation and long-range prospects.

ROLES OF TECHNOLOGY

Technology is only a part of the answer to an improved food status, but it is an essential part. Governmental policies which encourage agricultural production, reduced interest rates, and freer trade to avoid future crises are critical.

However, history has shown that science and technology have major roles in alleviating hunger. Technologies to improve the food situation are of two types. The first are those directly related to agricultural production. New varieties of grains, improved fertilizers and pesticides, better irrigation systems, and higher nutrient content cereals are examples. However, agricultural improvements cannot be fully utilized without improved postharvest technologies. This second type of technology involves all steps occurring after crops or animal products are harvested. Postharvest technologies include transportation, storage, processing, packaging, and utilization. Food science and technology plays a critical role in helping societies preserve and store the foods they have produced. The ability to consume products long after harvest is a key development that can free a society from the ups and downs of climate and nature. Food scientists and technologists can develop improved foods manufactured with higher efficiency and systems to fortify foods with specific nutrients for areas where there are inadequacies.

CONCLUSIONS

Food shortages and hunger have been largely overcome in many parts of the world over the last century as a result of a combination of factors including remarkable advances in agriculture. Malnutrition was common in significant proportions of the population of western Europe as recently as the 1930s. Advances not only in agriculture but also in refrigeration, food processing, government policies, marketing practices, and food standards, together with increased productivity and income, contributed to the abundance that Europe now enjoys.

In other parts of the world, the increases in food availability are even more dramatic. The self-sufficiencies of China and India are truly remarkable. Since the 1950s there has been an increase in arable land, large increases in the productivity of that land, land reform policies which have encouraged increased incentive, and the application of appropriate technologies. The predictions of large-scale disasters, with the exceptions of isolated famines, made in the 1940s and 1950s have not come true.

Unfortunately, these successes have not occurred around the globe. The per capita productivity of agriculture in Africa may have actually declined over the past few decades. Parts of Latin America remain without sufficient food. The successes of the last 50 years indicate that undernutrition and malnutrition can be eliminated, but the environmental, agricultural, population control, and political challenges are enormous. It remains to be seen if the world as a whole has the wisdom and the will to ensure that everyone has at least a sufficient diet to lead a productive life.

References

Berck, P. and Bigman, D. 1993. Food Security and Food Inventories in Developing Countries. CAB International, Wallingford, Oxon U.K.

Borlaug, N.E. 1992. Lighting fires at the grass roots. Food Technol. 46(7), 84–85.

Bread for the World Institute on Hunger & Development. 1993. Hunger 1993: Uprooted People: Third Annual Report on the State of World Hunger, 1st ed. Bread for the World Institute on Hunger & Development, Washington, DC.

Bread for the World Institute on Hunger & Development. 1994. Hunger 1994: Transforming the Politics of Hunger: Fourth Annual Report on the State of World Hunger. Bread for the World Institute on Hunger & Development, Washington, DC.

Cohen, B.E. 1990. Food security and hunger policy for the 1990s. Nutr. Today 25(4), 23–27.

Conway, G. and Barbier, E.B. 1990. After the Green Revolution. Sustainable Agriculture for Development. Earthscan, London.

Department of Agriculture, Economic Research Service. 1990. Global Food Assessment: Situation and Outlook Report. Department of Agriculture, Economic Research Service, Washington, DC.

Edwards, C. 1984. Meeting world food needs. National Food Review NFR–26, 2–4.

Gajewski, G., Calvin, L., Vandeman, A., and Vasavada, U. 1992. Sustainable agriculture: What's it all about? Agricultural Outlook 185, 30–33.

Grigg, D.B. 1993. The World Food Problem. 2nd ed. Blackwell, Oxford.

Kutzner, P.L. 1991. World Hunger. A Reference Handbook. ABC-Clio, Santa Barbara, CA.

Swaminathan, M.S. 1992. Environment and global food security. Food Technol. 46(7), 89–90, 95.

Scrimshaw, N.S. 1987. The phenomenon of famine. Annu. Rev. Nutr. 7, 1–21.

Spencer, W. 1991. The Challenge of World Hunger. Enslow Publishers, Hillside, NJ.

Timmer, C.P., Joseph, S.C., and Scrimshaw, N.S. 1985. Realistic approaches to world hunger. Food Nutr. Bull. 7(7), 1–14.

Index